AI for Rock Dynamics

Manchao He · LiGe Wang · Wei Yao ·
Wengang Dang · Zhuo Wang

AI for Rock Dynamics

Authors
See next page

ISBN 978-981-96-5341-6   ISBN 978-981-96-5342-3 (eBook)
https://doi.org/10.1007/978-981-96-5342-3

This work was supported by National Natural Science Foundation of China.

© The Editor(s) (if applicable) and The Author(s) 2025. This book is an open access publication.

Open Access This book is licensed under the terms of the Creative Commons Attribution-NonCommercial-NoDerivatives 4.0 International License (http://creativecommons.org/licenses/by-nc-nd/4.0/), which permits any noncommercial use, sharing, distribution and reproduction in any medium or format, as long as you give appropriate credit to the original author(s) and the source, provide a link to the Creative Commons license and indicate if you modified the licensed material. You do not have permission under this license to share adapted material derived from this book or parts of it.

The images or other third party material in this book are included in the book's Creative Commons license, unless indicated otherwise in a credit line to the material. If material is not included in the book's Creative Commons license and your intended use is not permitted by statutory regulation or exceeds the permitted use, you will need to obtain permission directly from the copyright holder.

This work is subject to copyright. All commercial rights are reserved by the author(s), whether the whole or part of the material is concerned, specifically the rights of translation, reprinting, reuse of illustrations, recitation, broadcasting, reproduction on microfilms or in any other physical way, and transmission or information storage and retrieval, electronic adaptation, computer software, or by similar or dissimilar methodology now known or hereafter developed. Regarding these commercial rights a non-exclusive license has been granted to the publisher.

The use of general descriptive names, registered names, trademarks, service marks, etc. in this publication does not imply, even in the absence of a specific statement, that such names are exempt from the relevant protective laws and regulations and therefore free for general use.

The publisher, the authors and the editors are safe to assume that the advice and information in this book are believed to be true and accurate at the date of publication. Neither the publisher nor the authors or the editors give a warranty, expressed or implied, with respect to the material contained herein or for any errors or omissions that may have been made. The publisher remains neutral with regard to jurisdictional claims in published maps and institutional affiliations.

This Springer imprint is published by the registered company Springer Nature Singapore Pte Ltd.
The registered company address is: 152 Beach Road, #21-01/04 Gateway East, Singapore 189721, Singapore

If disposing of this product, please recycle the paper.

Manchao He
China University of Mining
and Technology-Beijing
Beijing, China

Wei Yao
School of Civil Engineering
Tianjin University
Tianjin, China

Zhuo Wang
Chinese Society for Rock Mechanics
and Engineering
Beijing, China

LiGe Wang
Shandong University
Jinan, Shandong, China

Wengang Dang
School of Civil Engineering
Sun Yat-sen University
Zhuhai, Guangdong, China

With Contribution by
CNPIEC Kexin Techonolgy Ltd.
Beijing, China

Preface

As we stand at the precipice of an AI revolution in 2024, it is manifest that the integration of AI in scientific research has gained significant momentum. Researchers across diverse disciplines are leveraging the power of AI to streamline their workflows, uncover novel insights, and accelerate the pace of scientific discovery. From automating tedious data analysis tasks to generating hypotheses and aiding in the writing process, AI has become an indispensable tool in the modern researcher's arsenal.

It is in this context that we embark on the creation of this book, *AI for Rock Dynamics*. Our motivation stems from the recognition that AI has the potential to transform how we synthesize and present scientific knowledge, while acknowledging the complexity and rigor required in academic writing.

However, it is equally important to recognize the limitations of LLMs. Our journey in creating this book has revealed both the potential and limitations of AI in academic writing. While LLMs excel at processing vast amounts of information and handling non-deterministic reasoning, they also face significant constraints. These limitations include their dependence on training data and challenges in grasping domain-specific nuances.

It is precisely these limitations that drove us to continuously explore the boundaries of interaction between artificial intelligence and human expertise during the creation of this book. We assembled a team of over 20 experts from various subfields of rock dynamics to collaborate with the AI system, carefully reviewing, validating, and refining the AI-generated content. This process highlighted the crucial role of human intelligence in guiding and optimizing the AI output.

However, we must also acknowledge that despite our best efforts, the content of this book may still contain some inconsistencies or inaccuracies. This further underscores the current limitations of AI technology, particularly LLMs, when dealing with specialized domain knowledge. Through this book, we hope to contribute to the ongoing conversation about the role of AI in academic publishing and to provide a case study of its application in the field of rock dynamics. We believe that this exploration will not only shed light on the potential benefits and challenges of AI-assisted writing but also inspire other researchers to consider how they might leverage these technologies in their own work.

The creation of *AI for Rock Dynamics* has been a transformative experience for the entire author team. This collaborative endeavor has not only deepened our understanding of the field but also opened our eyes to the immense potential of interdisciplinary cooperation and AI-assisted research.

The success of this project has ignited a spark of excitement and possibility for the future of rock dynamics research. The integration of AI technologies into the research process holds immense potential for accelerating the pace of discovery, handling large volumes of data, and facilitating more efficient knowledge sharing. As we move forward, it is crucial to approach these developments with a discerning eye, ensuring that AI serves to augment, rather than replace, human expertise.

As a researcher, I am fully aware of the importance and transformative power of open science. The creative process of this book itself is a vivid practice of the concept of open science. We have experienced extensive knowledge sharing, interdisciplinary collaboration, and harmonious interaction among different stakeholders. Although this book focuses on rock dynamics, the open science practices have enlightening significance for the entire academic community. I firmly believe that upholding the basic values of openness, transparency, and collaboration, has far-reaching significance for the development of science.

Looking ahead, we envision this book as a catalyst for further innovation and collaboration within the field of rock dynamics and beyond. By showcasing the power of interdisciplinary collaboration and AI-assisted research, we hope to inspire researchers, practitioners, and students to embrace these new frontiers and push the boundaries of what is possible.

In conclusion, the journey of creating *AI for Rock Dynamics* has been a transformative experience for the author team, offering valuable lessons on the power of collaboration and the potential of AI-assisted research. As we look to the future, we are filled with optimism and excitement for the continued evolution of rock dynamics research. It is our sincere hope that this book will serve as a source of inspiration and knowledge for all those who seek to contribute to this dynamic and essential field.

As we conclude this preface, we would like to express our deepest gratitude to all those who have supported and contributed to the creation of "Recent Research Progress of Rock Dynamics."

First and foremost, we extend our appreciation to the China Association for Science and Technology (CAST) for their unwavering support and guidance. Their commitment to advancing scientific research and promoting technological innovation has been instrumental in bringing this project to fruition.

We also extend our gratitude to the China National Publications Import and Export (Group) Corporation (CNPIEC) for their unwavering support and belief in the potential of this project. Your commitment to advancing scientific knowledge and fostering international cooperation has been truly remarkable. As the parent company of Kexin, they have demonstrated remarkable foresight in embracing technological innovation in scientific publishing.

To our partners at CNPIEC Kexin Technology Co., Ltd. (Kexin), we express our sincere appreciation for your tireless efforts in developing and refining the AI system that has been integral to this project. Their expertise in Large Language Models and artificial intelligence technologies, coupled with their dedicated coordination efforts throughout the entire process, has been instrumental in bringing this book to fruition. Their practical application of AI in scientific publishing represents a meaningful step in the development of AI for science.

As President of the Chinese Society for Rock Mechanics and Engineering, I acknowledge the Society's contribution to this project. The involvement of our members, including the authors, has helped ensure the academic quality of this work.

A special thank you goes to our esteemed publisher, Springer Nature, for their trust and support in bringing this book to the global audience.

Finally, to all those who have followed the progress of this book with interest and anticipation, we offer our heartfelt thanks. Your enthusiasm and support have been a constant source of motivation and encouragement throughout this journey.

It is our sincere hope that *AI for Rock Dynamics* will serve as a catalyst for further exploration, discovery, and collaboration in the field of rock dynamics. Together, we can continue to push the boundaries of what is possible and contribute to a safer, more sustainable future for all.

Manchao He
President of the Chinese Society
for Rock Mechanics and Engineering
Beijing, China

Proofreading Team

Juntao Chen, Associate Professor, Shandong University of Science and Technology
Yu Feng, Associate Professor, Sun Yat-sen University
Ge Gao, Associate Professor, Shanghai Jiao Tong University
Quan Gan, Professor, Chongqing University
Lei Hou, Associate Professor, Shanghai Jiao Tong University
Xiaorong Li, Associate Professor, China University of Petroleum (Beijing)
Peiyuan Lin, Professor, Sun Yat-sen University
Jiangfeng Liu, Professor, China University of Mining and Technology
Ke Ma, Professor, Dalian University of Technology
Tianshou Ma, Professor, Southwest Petroleum University
Qiujing Pan, Professor, Central South University
Bin Wang, Professor, Institute of Rock and Soil Mechanics, Chinese Academy of Sciences
Shaofeng Wang, Professor, Central South University
Bangbiao Wu, Associate Professor, Tianjin University
Hui Wu, Assistant Professor, Peking University
Xun Xi, Professor, University of Science and Technology Beijing
Yingjie Xia, Associate Professor, Dalian University of Technology
Liwei Zhang, Professor, Institute of Rock and Soil Mechanics, Chinese Academy of Sciences
Yingbin Zhang, Professor, Southwest Jiaotong University
Zhihong Zhao, Associate Professor, Tsinghua University
Lu Zheng, Professor, Fuzhou University

Contents

1 **Introduction** 1
   1.1 Brief History of Rock Dynamics Development 1
   1.2 Research Contents and Key Problems of Rock Dynamics 3
   1.3 Main Research Methods of Rock Dynamics 7
   1.4 Main Content of the Book 11

2 **Theoretical Basis of Rock Dynamics** 13
   2.1 Basic Dynamic Equations 13
      2.1.1 Newton's Second Law 13
      2.1.2 Governing Equations of Elastic Waves 15
      2.1.3 Dispersion Effects Induced by the Transverse Inertia 21
   2.2 Rock Stress Wave Theory 25
      2.2.1 Coaxial Collision of Two Elastic Bars 25
      2.2.2 Interaction of Two Elastic Waves 26
   2.3 Rock Dynamic Strength Theory 29
      2.3.1 Rock Stress–Strain Relationship 29
      2.3.2 Dynamic Strength Damage Criteria for Rocks 34
      2.3.3 Rock Dynamic Strength Versus Strain Rate 36
   2.4 Rock Fracture Mechanics Theory 40
      2.4.1 Linear Elastic Fracture Mechanics 40
      2.4.2 Elasto-Plastic Fracture Mechanics 43
      2.4.3 Rock Failure Criteria 48
   2.5 Rock Dynamic Damage Mechanics Theory 51
      2.5.1 Fatigue Damage of Rocks Under Stress Wave Action 51
      2.5.2 Damage Laws of Rocks Under Cyclic Impact 53
      2.5.3 Attenuation of Stress Waves in Rock Mass 55
   2.6 Conclusion 57
   References 58

3 Rock Dynamics Test Device and Test Technique ... 63
3.1 Principle and Classification of the Test Device ... 63
3.1.1 Background of Test Device Design ... 63
3.1.2 Types of Rock Dynamics Testing Devices ... 68
3.1.3 Innovations in Device Development ... 74
3.2 Impact Test Technique ... 79
3.2.1 High-Strain Rate Loading Methods ... 79
3.2.2 Split-Hopkinson Pressure Bar (SHPB) Test ... 80
3.2.3 Data Acquisition and Analysis in Impact Tests ... 82
3.2.4 Application in Resource Exploitation Research ... 83
3.3 Dynamic and Static Combined Loading Test Technique ... 85
3.3.1 Design Methods ... 85
3.3.2 Mechanical Testing Techniques ... 93
3.4 Temperature–Pressure Coupling Test Technique ... 100
3.4.1 Thermal Effects on Rock Properties ... 100
3.4.2 Pressure Effects on Rock Properties ... 106
3.4.3 Coupled Effects on Rock Mechanics ... 108
3.4.4 Challenges in High-Temperature and High-Pressure Testing ... 114
3.5 Advanced Test Techniques and Methods ... 122
3.5.1 Micro-scale Rock Dynamics Testing ... 122
3.5.2 Real-Time Monitoring and Imaging Techniques ... 126
3.5.3 Integration of AI and Machine Learning in Data Analysis ... 134
3.6 Conclusion ... 137
References ... 138

4 Rock Dynamic Properties ... 143
4.1 Basic Concept and Theory of Impact Load ... 143
4.1.1 Definition and Characteristics of Impact Load ... 143
4.1.2 Theory and Types of Stress Wave Propagation ... 147
4.1.3 Effects of Strain Rate on Rock Mechanical Properties ... 149
4.2 Rock Mechanical Behavior Under Impact Load ... 151
4.2.1 Dynamic Stress–Strain Relationship of Rocks ... 151
4.2.2 Dynamic Failure Mechanisms and Modes of Rocks ... 161
4.2.3 Deformation and Fracture Characteristics at High Strain Rates ... 165
4.3 Rock Mechanical Properties of Dynamic and Static Combined Loading ... 167
4.3.1 Strength Characteristics Under Combined Loading ... 167
4.3.2 Deformation Behavior Under Combined Loading ... 170
4.3.3 Failure Modes and Mechanisms Under Combined Loading ... 176
4.4 Conclusion ... 183
References ... 184

5 Propagation Characteristics of Stress Wave in Rock ... 187
- 5.1 Basic Principle of Stress Wave ... 187
  - 5.1.1 Category of Stress Waves ... 187
  - 5.1.2 Wave Equations of Elastic Stress Waves ... 194
  - 5.1.3 Interaction of Two Elastic Stress Waves ... 201
- 5.2 Stress Wave Propagation Under Different Boundary Conditions ... 203
  - 5.2.1 Stress Wave Propagation on a Free Surface ... 203
  - 5.2.2 Stress Wave Propagation on a Fixed Surface ... 210
  - 5.2.3 Stress Wave Propagation at an Interface Between Two Media ... 212
- 5.3 Stress Wave Propagation in Complex Space Conditions ... 221
  - 5.3.1 Stress Waves in Rocky Slopes and Landslides ... 221
  - 5.3.2 Influence of Geological Structures on Wave Propagation ... 226
  - 5.3.3 Stress Waves in Geothermal and Mining Applications ... 232
- 5.4 Stress Wave Propagation in Different Medium ... 238
  - 5.4.1 Stress Wave Behavior in Saturated and Unsaturated Rocks ... 238
  - 5.4.2 Stress Wave Behavior in Anisotropic Rocks ... 244
  - 5.4.3 Stress Wave Propagation in Fractured Rock ... 253
- 5.5 Conclusion ... 258
- References ... 259

6 Rockburst Dynamics and Engineering Protection ... 265
- 6.1 Phenomenon and Cause of Rockburst ... 265
  - 6.1.1 Definition and Phenomena of Rockburst ... 265
  - 6.1.2 Occurence Conditions and Influcing Factors for Rockburst ... 266
- 6.2 Precursors and Prediction Methods of Rock Burst ... 274
  - 6.2.1 Characteristics and Monitoring of Rockburst Precursors ... 274
  - 6.2.2 Common Rockburst Prediction Methods ... 287
  - 6.2.3 Major Criteria for Rockburst Prediction ... 294
- 6.3 Rockburst Dynamic Mechanism ... 300
  - 6.3.1 Rockburst Classification and Initiation Mechanisms ... 300
  - 6.3.2 Rockburst Grading and Prediction ... 305
- 6.4 Rockburst Engineering Protection Method ... 307
  - 6.4.1 Emergency Plans for Rockburst Hazards ... 307
  - 6.4.2 Comprehensive Mitigation Measures for Rockburst Hazards ... 311
  - 6.4.3 Rockburst Mitigation Engineering Case Studies ... 316
- 6.5 Conclusion ... 329
- References ... 330

7 Key Techniques for Numerical Simulation of Rock Dynamics 335
- 7.1 High-Performance Computational Technique 335
  - 7.1.1 Parallel Computing Methods 335
  - 7.1.2 Advanced Numerical Methods 341
  - 7.1.3 Optimization Techniques 346
- 7.2 Multi-Physics Coupling Simulation 354
  - 7.2.1 Thermo-Hydro-Mechanical (THM) Coupling in Rock Dynamics 354
  - 7.2.2 Numerical Methods for Coupled Multi-Physics Field Simulations 359
  - 7.2.3 Multi-scale Methods for Coupled Problems 365
  - 7.2.4 Multi-Physics Coupling Simulation Softwares and Engineering Applications 366
- 7.3 Data-Driven Modelling 373
  - 7.3.1 Machine Learning for Rock Dynamics 373
  - 7.3.2 Big Data Analysis in Rock Testing 380
  - 7.3.3 Predictive Models for Rock Failure 393
- 7.4 Big Data Visualization and Mining 409
  - 7.4.1 Visualization Techniques for Stress Distribution 409
  - 7.4.2 Data Mining in Rock Dynamics 422
  - 7.4.3 Real-Time Simulation and Visualization 425
- 7.5 Conclusions 435
- References 435

8 Engineering Applications in Rock Dynamics 441
- 8.1 Engineering Applications in China 441
  - 8.1.1 Deep Mining and Rock Burst Control 441
  - 8.1.2 Hydropower Projects and Rock Mass Stability 445
  - 8.1.3 Underground Construction and Tunnel Stability 449
- 8.2 Engineering Applications in the US 454
  - 8.2.1 Mining Engineering Applications 455
  - 8.2.2 Civil Engineering Applications 460
  - 8.2.3 Earthquake Engineering Applications 466
- 8.3 Engineering Applications in Europe 469
  - 8.3.1 Rock Dynamics in Geothermal Development 469
  - 8.3.2 Rock Dynamics Principles in Natural Hazard Studies 472
  - 8.3.3 Rock Dynamics in Ore Deposit Studies 476
- 8.4 Engineering Applications in Other Countries 483
- 8.5 Conclusion 510
- References 511

9 Main Achievements of Rock Dynamics 515
- 9.1 Engineering Applications of Rock Dynamics 515
- 9.2 Status Quo and Challenges of Rock Dynamics 516
- 9.3 Future Trend of Rock Dynamics 517

Chapter 1
Introduction

1.1 Brief History of Rock Dynamics Development

The discipline of rock dynamics has evolved significantly over the past century, emerging as a crucial field in geotechnical engineering, mining, and civil engineering. This section provides a comprehensive overview of the historical development of rock dynamics, tracing its origins, key milestones, and the factors that have shaped its growth into a distinct and vital area of study.

Early Foundations (1900s–1940s)

The roots of rock dynamics can be traced back to the early twentieth century when researchers began to recognize the importance of understanding rock behavior under dynamic loading conditions. During this period, the foundations of rock mechanics were being laid, primarily focused on static loading conditions. However, observations from mining operations and early tunneling projects highlighted the need to consider dynamic effects on rock masses.

In the 1920s and 1930s, pioneering work by researchers such as Hans Cloos and John Cadman began to explore the concept of rock bursts in deep mines. These early studies were largely observational, focusing on documenting the occurrence and effects of sudden, violent rock failures. While limited in their theoretical underpinnings, these initial investigations laid the groundwork for future research in rock dynamics.

The advent of World War II brought increased attention to the effects of explosions on rock structures, particularly in the context of underground fortifications. This period saw the development of rudimentary models for predicting blast-induced damage in rock, although these were largely empirical and based on limited experimental data.

Emergence of Modern Rock Dynamics (1950s–1970s)

The post-war period marked a transformative era in rock dynamics, driven by both technological advances and theoretical breakthroughs. The development of sophisticated measurement techniques, including strain gauges and high-speed photography, enabled quantitative analysis of rock behavior, while researchers like Kolsky and Goldsmith established fundamental principles of wave propagation in rock masses.

The field's growth was further catalyzed by underground nuclear testing research and practical industrial demands. Mining operations at greater depths faced increasing rockburst challenges, while the introduction of tunnel boring machines in the 1950s expanded understanding of rock-tool interactions under dynamic conditions.

The late 1940s through the 1960s marked several pivotal advances in rock dynamics research methodology and theory. Kolsky's introduction of the split Hopkinson pressure bar (SHPB) technique in 1949 revolutionized the testing of rock properties under high strain rates, while Bieniawski and colleagues developed the first comprehensive theories of rock fracture mechanics in the 1960s. These theoretical foundations were complemented by Cook's pioneering work in rockburst modeling using energy concepts, with institutional support from newly established research programs at the U.S. Bureau of Mines and the South African Chamber of Mines.

Maturation and Specialization (1980s–2000s)

The final decades of the twentieth century marked the maturation of rock dynamics into a well-established discipline, characterized by significant technological and theoretical advances. The development of sophisticated testing apparatus, including improvements to the SHPB technique, gas gun-based impact tests, and true triaxial dynamic testing systems, enabled more precise measurement of rock properties under varied loading conditions. This experimental progress was complemented by rapid growth in computing power, which facilitated the implementation of complex numerical models, particularly finite element and discrete element methods, for simulating dynamic rock behavior.

The field's evolution was further marked by the establishment of the International Society for Rock Mechanics and Rock Engineering (ISRM) Commission on Rock Dynamics in 1983, which provided a central platform for research coordination. Significant progress emerged in constitutive modeling, incorporating factors such as strain rate sensitivity and damage accumulation, while researchers like Atkinson and Costin advanced the understanding of fracture propagation under dynamic loading. The introduction of acoustic emission techniques enhanced in-situ monitoring capabilities, contributing to improved understanding of scale effects between laboratory and field observations.

This period also witnessed the expansion of rock dynamics applications into emerging fields such as petroleum engineering, nuclear waste disposal, and planetary science, particularly in impact cratering studies. Researchers began exploring coupled processes, examining interactions between dynamic loading and phenomena such as fluid flow and heat transfer, while the development of specialized commercial software in the 1990s facilitated practical applications.

Contemporary Developments (2000s-Present)

Technological innovations in rock dynamics have advanced significantly in the twenty-first century, particularly through multi-scale integration combining micro-, meso-, and macro-scale analyses for modeling rock behavior under dynamic loading. Advanced imaging techniques, including X-ray CT and neutron tomography, have enhanced the understanding of internal rock deformation and failure mechanisms.

Field applications have progressed through advanced sensor networks and real-time data processing, enabling effective in-situ monitoring of dynamic rock behavior. Artificial intelligence and machine learning integration has advanced data analysis and prediction capabilities in rock dynamics research.

Research has deepened understanding of coupled multi-physics phenomena, focusing on mechanical, hydraulic, thermal, and chemical process interactions in rock masses under dynamic loading. This understanding is essential for extreme loading conditions in deep mining, hydraulic fracturing, and planetary impact events. The field's scope has expanded to include renewable energy applications, specifically geothermal systems and carbon sequestration.

Dedicated rock dynamics laboratories worldwide have institutionalized these advances through state-of-the-art testing facilities. Methodological innovations include hybrid numerical-experimental techniques and improved analysis of time-dependent rock behavior under dynamic loading. Rock dynamics principles now extend to geothermal energy extraction and asteroid mining, with enhanced integration into engineering standards.

1.2 Research Contents and Key Problems of Rock Dynamics

Rock dynamics is a rapidly evolving field that encompasses the study of rock behavior under dynamic loading conditions. As underground excavations reach greater depths and engineering projects become more complex, the importance of understanding and predicting rock response to dynamic loads has become increasingly crucial. This section provides an overview of the primary research contents and key problems in rock dynamics, highlighting current challenges and future directions in this vital area of study.

Fundamental Aspects of Rock Dynamics

1. Wave Propagation in Rock Materials

Ave propagation in rock materials serves as the cornerstone of rock dynamics, encompassing several interconnected research domains that are crucial for understanding rock behavior under dynamic conditions. At the most fundamental level, elastic wave propagation plays a vital role in this field, where the study of P-waves, S-waves, and surface waves in rock media provides essential insights for seismic data interpretation and dynamic load response prediction. Contemporary research efforts in this area are primarily directed toward developing more sophisticated models that can accurately account for wave attenuation, dispersion, and scattering phenomena in heterogeneous and anisotropic rock masses.

The complexity of wave propagation analysis increases significantly when considering the presence of discontinuities in rock masses. Fractures, joints, and faults substantially influence wave behavior, necessitating detailed investigation of stress wave interactions at these interfaces. Research in this domain focuses particularly on characterizing the transmission, reflection, and mode conversion of stress waves, with scholars working to develop enhanced analytical and numerical models capable of representing complex fracture networks with greater precision.

Furthermore, the study of non-linear wave propagation represents an especially challenging frontier in rock dynamics research. This phenomenon becomes particularly relevant when rocks are subjected to high stress levels or experience large-amplitude waves. Understanding key aspects such as shock wave formation, wave-induced plasticity, and dynamic fracture propagation continues to challenge researchers, while holding significant implications for practical applications in blast design and earthquake engineering.

2. Dynamic Properties of Rocks

The accurate characterization of dynamic rock properties stands as a fundamental requirement for predicting rock behavior under diverse loading conditions, encompassing several critical areas of investigation. Among these, the study of strain rate effects holds particular significance, as both rock strength and deformation characteristics exhibit notable rate dependency. Current research efforts in this domain focus on developing sophisticated constitutive models that can effectively capture this rate-dependent behavior across a comprehensive spectrum of loading rates, ranging from quasi-static to impact conditions.

Dynamic fracture toughness represents another crucial aspect of rock behavior characterization, particularly in understanding rocks' resistance to dynamic crack propagation. This knowledge proves essential for accurately predicting fragmentation processes and evaluating the stability of underground structures under dynamic loads. The field continues to advance through ongoing efforts to establish standardized testing methods and enhance theoretical models specifically tailored for dynamic fracture mechanics in rock materials.

Equally important is the understanding of dynamic energy absorption and dissipation mechanisms in rocks, which plays a vital role in the design of effective support systems and the prediction of failure mechanisms. Research in this area extends to investigating the fundamental role of microstructure in energy dissipation processes, while simultaneously advancing new experimental techniques for measuring energy absorption under various loading conditions.

1.2 Research Contents and Key Problems of Rock Dynamics

Experimental Techniques and Instrumentation

Advancements in experimental techniques and instrumentation have significantly enhanced our ability to study rock dynamics. Key areas of ongoing research and development include:

1. High-Strain Rate Testing Methods

High-strain rate testing methods constitute a critical domain in rock mechanics research, with SHPB techniques forming a cornerstone of experimental investigations. Current research efforts in SHPB methodology focus on enhancing testing capabilities across diverse stress states and environmental conditions. These advancements include the development of techniques for combined compression-shear loading, the implementation of testing protocols at elevated temperatures and pressures, and the expansion of achievable strain rate ranges.

For investigating rock behavior under extreme conditions, plate impact experiments serve as an essential tool, particularly in studying phenomena that occur at exceptionally high strain rates and pressures. Research in this area continues to evolve, with emphasis on refining experimental designs and improving data interpretation methods. These improvements are especially crucial for advancing our understanding of dynamic compaction processes and phase transitions in rock materials.

Bridging the gap between quasi-static and high-strain rate testing remains a significant challenge in the field, driving the development of novel dynamic testing methods. Current research initiatives explore the potential of hybrid testing approaches that effectively combine elements of both static and dynamic loading, aiming to provide a more comprehensive understanding of rock behavior across different loading regimes.

2. In-Situ Testing and Monitoring

In-situ testing and monitoring represent critical components in understanding rock behavior under real-world conditions, with microseismic monitoring emerging as a particularly vital tool in underground environments. The advancement of techniques for real-time monitoring and interpretation of microseismic events plays a crucial role in assessing rockburst risks and evaluating support system effectiveness. Current research in this domain focuses on three primary objectives: enhancing source location accuracy, developing automated event classification systems, and creating predictive models based on seismic data.

Dynamic in-situ stress measurement presents another significant challenge in rock mechanics monitoring, driving innovative research approaches in real-time stress monitoring. Current investigations explore various promising technologies, including fiber optic sensors and acoustic emission techniques, aiming to develop more reliable methods for measuring dynamic stresses in rock masses under actual field conditions.

Complementing these monitoring efforts, large-scale dynamic testing provides essential insights at scales relevant to engineering projects, though such testing presents its own unique set of challenges. Research in this area continues to advance through the development and refinement of methodologies for large-scale blasting experiments, underground seismic simulations, and other field-scale dynamic tests, bridging the gap between laboratory findings and real-world applications.

Numerical Modeling and Simulation

Advancements in computational power and numerical methods have revolutionized the study of rock dynamics. Key research areas in this domain include:

1. Multi-Scale Modeling

Multi-scale modeling constitutes a fundamental challenge in rock mechanics, particularly in developing approaches that effectively bridge the gap between microscopic rock behavior and macroscopic engineering-scale phenomena. This complex field encompasses several complementary modeling strategies, each addressing different aspects of the scale integration problem.

Hierarchical multi-scale methods represent one primary approach to this challenge, focusing on the integration of models across various scales of analysis. These methods establish connections between crystal-scale simulations and continuum mechanics models of entire rock masses, providing a comprehensive framework for understanding rock behavior across different spatial scales. In parallel, concurrent multi-scale methods offer a distinctive approach by simultaneously simulating processes at different scales within a unified model framework. This simultaneous simulation capability proves particularly valuable for accurately representing localized phenomena, such as fracture propagation, where interactions across scales play a crucial role.

Supporting these modeling approaches, the development of improved homogenization techniques remains essential for advancing the field. These techniques focus on the critical task of upscaling microscopic rock properties to derive effective macroscopic parameters, forming a crucial foundation for developing more accurate large-scale models that can reliably predict rock mass behavior.

2. Coupled Multi-Physics Simulations

Coupled multi-physics simulations represent a critical frontier in rock dynamics research, addressing the complex interactions between mechanical, thermal, hydraulic, and chemical processes that characterize many rock behavior phenomena. At the forefront of this field, thermo-hydro-mechanical-chemical (THMC) coupling presents significant challenges in developing efficient and accurate numerical models. These challenges become particularly pronounced when addressing problems that span extensive time scales or large spatial domains, requiring sophisticated computational approaches to capture the intricate interplay between various physical processes.

The dynamic interaction between rock masses and fluids constitutes another crucial aspect of coupled simulations, particularly in the domain of fluid–structure interaction. Research in this area focuses on enhancing modeling methods to better understand and predict behavior in various applications, including hydraulic fracturing operations, underground storage systems, and earthquake-induced liquefaction phenomena.

Closely related to these fluid-rock interactions is the study of dynamic damage-permeability coupling, which plays a vital role in understanding how dynamic loading influences rock permeability through processes of damage accumulation and fracture network evolution. This understanding proves essential for numerous practical applications in both geotechnical and petroleum engineering fields, where accurate prediction of permeability changes under dynamic conditions is crucial for project success.

3. Advanced Numerical Methods

Advanced numerical methods remain at the forefront of rock mechanics research, with ongoing efforts focused on developing increasingly efficient and accurate techniques for simulating dynamic rock behavior. Among these emerging approaches, meshless methods have gained significant attention, particularly techniques such as Smoothed Particle Hydrodynamics (SPH) and the Material Point Method (MPM), which offer distinct advantages in modeling large deformations and fragmentation processes in rock materials.

The development of hybrid methods represents another significant advancement in numerical modeling capabilities, combining different numerical approaches to address complex rock dynamics problems. The integration of finite element methods with discrete element methods, for instance, enables more comprehensive modeling of rock behavior across different scales and conditions. This combinatorial approach allows researchers to leverage the strengths of multiple numerical techniques while mitigating their individual limitations.

In parallel with these traditional numerical approaches, the integration of machine learning techniques has emerged as a promising direction in rock mechanics simulation. This innovative approach explores the potential of artificial intelligence to enhance numerical simulations through various means, such as developing surrogate models or optimizing computational parameters, offering new possibilities for improving both the efficiency and accuracy of rock behavior predictions.

1.3 Main Research Methods of Rock Dynamics

Rock dynamics is a complex and multifaceted field that requires a diverse array of research methods to fully understand and characterize the behavior of rocks under dynamic loading conditions. These methods span from laboratory-scale experiments to large-scale field tests, and incorporate advanced analytical and numerical techniques. This section provides a comprehensive overview of the main research methods employed in rock dynamics, discussing their principles, applications, advantages, and limitations.

Laboratory Testing Methods

Laboratory testing methods form the foundation of rock dynamics research, allowing for controlled experiments and detailed analysis of rock behavior under various loading conditions.

1. Split Hopkinson Pressure Bar (SHPB) Test

The SHPB test, also known as the Kolsky bar test, has emerged as a fundamental method for analyzing rock dynamic properties at high strain rates. The apparatus consists of three main components–striker, incident, and transmitted bars–with the rock specimen positioned between the latter two. When the striker impacts the incident bar, it generates a stress wave that propagates through the system, enabling researchers to determine the rock's dynamic stress–strain relationship through analysis of incident, reflected, and transmitted waves using mounted strain gauges.

This testing method offers several significant advantages, including the capability to achieve high strain rates (10^2 to 10^4 s^{-1}), a well-established theoretical framework, and relatively simple specimen preparation. However, it also faces certain limitations, such as the assumption of one-dimensional wave propagation, limited test duration due to wave reflections, and potential stress non-uniformity in specimens. Recent technological advances have addressed some of these challenges through modified setups for tensile and shear testing, integration with high-speed imaging, and the development of miniaturized systems for testing small specimens.

2. Drop Weight Impact Test

The drop weight impact test is a key methodology for studying rock fragmentation and energy absorption characteristics. The test involves releasing a weight from a set height onto a rock specimen, with impact energy controlled by drop height and weight mass. This method tests both intact specimens and aggregates, measuring impact strength, fracture energy, fragmentation characteristics, and energy absorption capacity.

The method's advantages include simple operation and simulation of mining and tunneling impact loads, allowing larger specimen testing compared to SHPB. However, limitations include lower strain rates than SHPB, potential multiple impacts, and non-uniform stress distribution.

Recent advances include high-speed camera integration for fragment analysis, instrumented systems for precise force–time measurements, and acoustic emission sensors for crack propagation studies.

3. Gas Gun-Based Testing

Gas gun-based testing is a critical experimental technique for studying rock behavior under extreme dynamic conditions, enabling investigation of material responses at high strain rates and impact velocities. The method employs a gas gun system to accelerate projectiles against rock targets, with high-speed imaging and sensors capturing the dynamic response.

1.3 Main Research Methods of Rock Dynamics

Gas gun testing measures several critical parameters: crater formation and ejecta patterns, shock wave propagation, dynamic fracture evolution, and material constitutive behavior at strain rates exceeding 10^5 s^{-1}. The system's versatility in projectile and target configurations facilitates investigation of shock-induced phenomena under controlled laboratory conditions.

The methodology's limitations include complex and expensive setup requirements, restrictions to small specimen sizes, and challenges in real-time measurement due to extremely short time scales. Recent advances have enhanced testing capabilities through multi-stage gas guns, digital image correlation for full-field deformation analysis, and laser interferometry for precise velocity measurements.

Field Testing Methods

While laboratory tests provide valuable insights under controlled conditions, field testing methods are essential for understanding rock dynamics at larger scales and under in-situ conditions.

1. In-Situ Dynamic Load Tests

In-situ dynamic load testing evaluates rock mass behavior under dynamic conditions through controlled field-scale loading. The testing employs methods such as controlled blasting, hydraulic impact hammers, and specialized dynamic loading devices, with response monitoring via accelerometers, geophones, and strain gauges.

These tests measure critical parameters including in-situ dynamic elastic properties, wave propagation characteristics, and dynamic stability of rock structures, while evaluating stress wave attenuation and dispersion. The method's key advantage lies in accounting for natural stress states and geological structures across large rock volumes, providing data directly applicable to engineering design.

Despite its advantages, in-situ testing offers less control over loading conditions than laboratory tests, with results influenced by site-specific factors and high operational costs. Recent technological advances address these limitations through portable loading devices, fiber optic sensing for strain measurements, and 3D laser scanning for structural analysis.

2. Microseismic Monitoring

Microseismic monitoring enables the study of rock dynamics in mining, tunneling, and hydraulic fracturing applications through arrays of seismic sensors that detect and locate microseisms induced by rock mass stress changes. Signal processing techniques extract critical parameters including event location and magnitude, seismicity patterns, stress redistribution, and fracture propagation.

The method provides real-time, non-invasive monitoring of large rock volumes, though system effectiveness depends on precise sensor placement and calibration. Key technical challenges include discriminating between seismic sources and interpreting data in noisy environments.

Recent advances have enhanced system capabilities through source mechanism inversion techniques, numerical modeling integration, and machine learning algorithms for automated event detection and classification.

Analytical and Numerical Methods

Analytical and numerical methods complement experimental approaches by providing frameworks for interpreting data, predicting rock behavior, and simulating complex dynamic processes.

1. Analytical Methods

Analytical methods in rock dynamics employ mathematical modeling to understand rock mass mechanical behavior. These methods incorporate wave propagation theory, dynamic fracture mechanics, energy balance methods, and statistical analysis of dynamic rock properties to describe complex dynamic responses under various loading conditions.

Analytical methods provide key advantages in rock mechanics research: they offer insights into underlying physical processes, enable rapid parametric studies, and serve as benchmarks for numerical simulations. However, these methods have limitations, primarily being restricted to simplified geometries and struggling to capture nonlinear behavior in natural rock masses under dynamic loading.

Recent advances have addressed these limitations through advanced constitutive models for rate-dependent behavior, stochastic approaches for material heterogeneity, and multi-scale analytical models for comprehensive spatial representation.

2. Numerical Simulation Methods

Numerical simulation methods, including Finite Element Method (FEM), Discrete Element Method (DEM), Finite Difference Method (FDM), and hybrid approaches like coupled FEM-DEM systems, serve as essential tools for analyzing complex geomechanical problems. These methods provide comprehensive capabilities for modeling complex geometries and boundary conditions, while incorporating advanced material models and multi-physics coupling mechanisms, enabling visualization of stress and strain fields.

Despite their analytical power, these methods face significant computational demands in large-scale three-dimensional simulations and require careful calibration and validation against experimental data. The sensitivity of results to input parameters and modeling assumptions necessitates rigorous control throughout the analysis process.

Recent technological advances, particularly in high-performance computing, machine learning-based model calibration, and multi-scale modeling techniques, have enhanced simulation capabilities and accuracy in rock dynamics analysis.

Advanced Characterization Techniques

Advanced characterization techniques provide detailed insights into the microstructural and physical properties of rocks, which are crucial for understanding their dynamic behavior.

1. High-Speed Imaging and Digital Image Correlation (DIC)

DIC integrated with high-speed imaging enables dynamic rock mechanics analysis through rapid deformation capture during loading and full-field displacement and strain mapping. This methodology facilitates visualization of fracture processes, local strain field measurement during impact, and fragment ejection velocity quantification. Recent advances in ultra-high-speed imaging (>106 fps), 3D DIC for volumetric strain analysis, and thermal imaging integration have enhanced the precision of thermo-mechanical coupling studies in rock mechanics investigations.

2. X-ray Computed Tomography (CT)

X-ray CT enables non-destructive 3D visualization of rock specimens during dynamic loading, primarily for characterizing pore structures, fracture networks, damage evolution, and fragment size distributions. Recent advances in ultrafast CT systems permit in-situ dynamic testing and real-time structural analysis. Phase-contrast CT techniques have enhanced imaging resolution, while integration with numerical modeling enables microstructure-based simulations that connect experimental and theoretical findings.

3. Acoustic Emission (AE) Monitoring

AE monitoring, a fundamental rock mechanics technique, detects high-frequency elastic waves emitted during microcracking and deformation processes. Through specialized sensors, AE technology enables real-time observation of crack initiation, propagation, and damage evolution under dynamic loading conditions, while characterizing failure mechanisms during cyclic loading scenarios. Recent advances in broadband AE sensors have enhanced frequency analysis capabilities, while machine learning integration has improved source classification accuracy. Integration with multi-parameter analysis systems has further advanced rock mechanics understanding.

1.4 Main Content of the Book

This comprehensive review synthesizes recent advancements in rock dynamics, spanning fundamental theories to practical applications. The book's structure progresses through interconnected topics, providing a systematic examination of the field.

In this chapter introduces the development history, research contents, key problems, research methods and main content of rock dynamics.

Chapters 2–5 establish the theoretical rock dynamics framework, encompassing mechanical principles, wave propagation, strength analysis, and damage mechanics. Chapter 2 presents fundamental rock dynamics through Newton's second law and elastic wave equations, emphasizing transverse inertia-induced dispersion in wave propagation. Chapter 3 examines rock dynamics testing methodologies: test device principles, impact tests, dynamic-static loading, temperature–pressure coupling, and advanced techniques. Chapter 4 analyzes rock dynamic strength theory via stress–strain relationships and strain rate-dependent failure criteria for blast and seismic predictions. Chapter 5 investigates rock dynamic damage mechanics, focusing on stress wave fatigue, cyclic impact damage, and wave attenuation in rock masses.

Chapter 6 addresses rockburst phenomena in deep underground excavations, analyzing mechanisms and safety implications. It examines rockburst precursors, prediction methodologies, dynamic failure mechanisms, and engineering protection measures.

Chapters 7 examines computational methods in rock dynamics, focusing on high-performance techniques for multi-scale problems. It covers multi-physics coupling simulation methods and data-driven modeling approaches, including machine learning applications in rock mechanics. The chapter also addresses big data visualization and mining techniques for rock dynamics data analysis.

Chapter 8 presents engineering applications of rock dynamics in mineral exploitation, tunnel construction, energy exploration, and geological disaster prevention across China, the United States, Europe and other regions.

Chapter 9 concludes with a comprehensive overview of achievements, applications, current challenges, and future trends in rock dynamics.

Open Access This chapter is licensed under the terms of the Creative Commons Attribution-NonCommercial-NoDerivatives 4.0 International License (http://creativecommons.org/licenses/by-nc-nd/4.0/), which permits any noncommercial use, sharing, distribution and reproduction in any medium or format, as long as you give appropriate credit to the original author(s) and the source, provide a link to the Creative Commons license and indicate if you modified the licensed material. You do not have permission under this license to share adapted material derived from this chapter or parts of it.

The images or other third party material in this chapter are included in the chapter's Creative Commons license, unless indicated otherwise in a credit line to the material. If material is not included in the chapter's Creative Commons license and your intended use is not permitted by statutory regulation or exceeds the permitted use, you will need to obtain permission directly from the copyright holder.

Chapter 2
Theoretical Basis of Rock Dynamics

2.1 Basic Dynamic Equations

2.1.1 Newton's Second Law

Newton's second law of motion is a fundamental principle in classical mechanics that forms the cornerstone of dynamic analysis in rock mechanics. This law, first formulated by Sir Isaac Newton in his seminal work 'Philosophiæ Naturalis Principia Mathematica' [1] published in 1687, provides a quantitative description of the relationship between forces acting on a body and the resulting changes in its motion. In the context of rock dynamics, Newton's second law serves as the primary basis for understanding and predicting the behavior of rock masses under various loading conditions, including static, dynamic, and impact loads.

The classical formulation of Newton's second law states that the acceleration of an object is directly proportional to the net force acting on it and inversely proportional to its mass. Mathematically, this can be expressed as:

$$F = ma \qquad (2.1)$$

where F is the net force acting on the object, m is the mass of the object, a is the acceleration of the object.

This simple equation encapsulates a profound physical principle that has far-reaching implications in rock mechanics and engineering. However, the application of Newton's second law to rock dynamics requires careful consideration of several factors, including the complex nature of rock materials, the influence of time-dependent phenomena, and the effects of scale and heterogeneity.

Historical Context and Development

Newton's second law fundamentally transformed the understanding of dynamics in rock mechanics. While Aristotelian physics required constant force for constant motion, Newton established that forces cause changes in motion, creating the foundation for modern dynamics analysis [1]. In rock mechanics, the application of Newton's principles became crucial in the early twentieth century as large-scale engineering projects demanded quantitative understanding of rock mass behavior under dynamic loads.

The integration of Newton's second law into rock mechanics presented unique challenges due to rock materials' heterogeneity, anisotropy, and discontinuities. Pioneers like Terzaghi and Biot bridged classical mechanics with geomaterial behavior [2]. In rock dynamics, Newton's second law must account for rocks' complex, deformable nature across different scales, unlike ideal point masses in classical mechanics. Pourciau's compound interpretation [3] demonstrates the law's applicability to both impulsive and continuous forces, essential for analyzing rock mass behavior under various loading conditions.

Application to Continuous Media

The application of Newton's second law to continuous rock masses requires extending the discrete particle formulation to distributed masses and forces through differential equations. For a continuous rock mass, Newton's second law is expressed in terms of stress and acceleration:

$$\nabla \cdot \sigma + \rho b = \rho a \tag{2.2}$$

where σ is the stress tensor, ρ is the density of the rock, b is the body force per unit mass, and a is the acceleration field.

This equation of motion for continuous media forms the foundation for rock dynamics analysis, accounting for internal stresses and their relationship to external forces. Its application requires constitutive relationships describing rock material stress–strain behavior, ranging from linear elastic to nonlinear time-dependent models incorporating plasticity, viscosity, and damage mechanisms.

Modifications and Extensions for Rock Mechanics

The unique characteristics of rock materials have necessitated various modifications and extensions to the basic formulation of Newton's second law. These adaptations aim to capture the complex behavior of rocks under dynamic loading conditions more accurately.

One significant extension is the consideration of wave propagation in rock masses. The equation of motion for elastic waves in a continuous medium can be derived from Newton's second law:

$$\rho \frac{\partial^2 u}{\partial t^2} = (\lambda + \mu)\nabla(\nabla \cdot u) + \mu \nabla^2 u + \rho f \tag{2.3}$$

where ρ is the density of the rock, t is the time, u is the displacement vector, λ and μ are Lamé parameters, f is the body force per unit mass.

2.1 Basic Dynamic Equations

This equation forms the basis for analyzing seismic wave propagation in rock masses, which is crucial for understanding the dynamic response of rock structures to earthquakes and blasting [4].

Nonlinear Effects and Limitations

Newton's second law provides a framework for analyzing rock dynamics, but its application to rock mechanics has notable limitations due to rocks' nonlinear behavior under high stresses and large deformations. This nonlinearity manifests primarily through strain-dependent stiffness and rate-dependent material response, resulting in nonlinear stress–strain relationships that deviate from linear model predictions.

Rock behavior complexity is further influenced by damage and fracture mechanisms, where crack initiation and propagation produce nonlinear responses that exceed simple linear models' capabilities. These phenomena, particularly under extreme loading conditions, often invalidate small-strain theory assumptions. To address these complexities, researchers employ advanced constitutive models—including elastoplastic, viscoelastic, and damage mechanics approaches—to more accurately describe rock behavior while maintaining consistency with Newton's second law.

In conclusion, Newton's second law remains a fundamental principle in rock dynamics, providing the theoretical foundation for understanding and predicting the behavior of rock masses under dynamic loading conditions. Its application in rock mechanics has evolved significantly since its initial formulation, incorporating various modifications and extensions to address the unique characteristics of rock materials. As we continue to push the boundaries of rock engineering into more challenging environments and extreme conditions, the principles embodied in Newton's second law will undoubtedly continue to guide our understanding and inform our analytical and numerical approaches to solving complex rock dynamics problems.

2.1.2 Governing Equations of Elastic Waves

The study of elastic waves in rock dynamics is fundamental to understanding the behavior of geological materials under dynamic loading conditions. This section delves into the governing equations that describe the propagation of elastic waves in rock media, providing a comprehensive overview of the theoretical foundations, derivations, and applications of these equations in the field of rock dynamics.

Fundamental Concepts and Assumptions

The theoretical framework of elastic wave propagation in rock media requires a clear understanding of fundamental concepts and underlying assumptions. Elastic waves, characterized by the propagation of mechanical disturbances through rock materials without permanent deformation, operate within the principles of elasticity wherein the material returns to its original state following the removal of applied forces [5]. The mathematical models governing these waves are built upon several crucial assumptions. Fundamentally, the rock medium is treated as a continuous material, setting aside its discrete atomic structure, while material properties are considered uniform throughout the medium under the assumption of homogeneity—though this can be relaxed in more sophisticated models. Additionally, the framework assumes isotropy, meaning material properties remain direction-independent, although anisotropic models can be developed to address more complex rock structures. The theoretical foundation further incorporates linear elasticity, following Hooke's law, and presumes small deformations where displacements and strains are minimal compared to the rock body's dimensions. These foundational assumptions enable the development of simplified yet robust mathematical models that effectively capture the essential characteristics of elastic wave propagation in rock media.

Derivation of the Elastic Wave Equation

The governing equations of elastic waves can be derived using various approaches, including Newton's second law of motion, Hamilton's principle, or energy conservation principles. In this section, we will present a derivation based on Newton's second law and the principles of linear elasticity.

Consider a small element of rock with dimensions dx, dy, and dz. The forces acting on this element can be expressed in terms of stresses (σ) acting on its surfaces. Applying Newton's second law of motion to this element in the x-direction yields:

$$\left(\frac{\partial \sigma_{xx}}{\partial x} + \frac{\partial \sigma_{xy}}{\partial y} + \frac{\partial \sigma_{xz}}{\partial z} \right) dx\, dy\, dz = \rho \left(\frac{\partial^2 u}{\partial t^2} \right) dx\, dy\, dz \qquad (2.4)$$

where ρ is the density of the rock, t is the time and u is the displacement in the x-direction. Similar equations can be written for the y and z directions:

$$\left(\frac{\partial \sigma_{yx}}{\partial x} + \frac{\partial \sigma_{yy}}{\partial y} + \frac{\partial \sigma_{yz}}{\partial z} \right) dx\, dy\, dz = \rho \left(\frac{\partial^2 v}{\partial t^2} \right) dx\, dy\, dz \qquad (2.5)$$

$$\left(\frac{\partial \sigma_{zx}}{\partial x} + \frac{\partial \sigma_{zy}}{\partial y} + \frac{\partial \sigma_{zz}}{\partial z} \right) dx\, dy\, dz = \rho \left(\frac{\partial^2 w}{\partial t^2} \right) dx\, dy\, dz \qquad (2.6)$$

where v and w are the displacements in the y and z directions, respectively.

These equations can be simplified to:

$$\nabla \cdot \sigma = \rho \left(\frac{\partial^2 u}{\partial t^2} \right) \qquad (2.7)$$

where σ is the stress tensor, u is the displacement vector, and ∇ is the nabla operator.

To relate the stresses to the displacements, we employ Hooke's law for an isotropic elastic medium:

$$\sigma_{ij} = \lambda (\nabla \cdot u) \delta_{ij} + 2\mu \varepsilon_{ij} \qquad (2.8)$$

2.1 Basic Dynamic Equations

where λ and μ are Lamé constants, δ_{ij} is the Kronecker delta, and ε_{ij} is the strain tensor given by:

$$\varepsilon_{ij} = \frac{1}{2}\left(\frac{\partial u_i}{\partial x_j} + \frac{\partial u_j}{\partial x_i}\right) \tag{2.9}$$

Substituting these relationships into the equation of motion and using the vector identity $\nabla \cdot (\nabla u) = \nabla(\nabla \cdot u) - \nabla \times (\nabla \times u)$, we obtain the general form of the elastic wave equation:

$$\rho \frac{\partial^2 u}{\partial t^2} = (\lambda + \mu)\nabla(\nabla \cdot u) + \mu \nabla^2 u \tag{2.10}$$

This equation, known as Navier's equation, describes the propagation of elastic waves in a homogeneous, isotropic, linear elastic medium.

Anisotropy and Heterogeneity in Rock Media

While the governing equations presented earlier assume isotropy and homogeneity, real rock formations often exhibit anisotropic and heterogeneous properties [6]. Anisotropy refers to the directional dependence of material properties, while heterogeneity refers to spatial variations in these properties. Both factors significantly influence the propagation of elastic waves in rock media.

For anisotropic media, the stress–strain relationship becomes more complex, requiring a fourth-order elasticity tensor C_{ijkl}:

$$\sigma_{ij} = C_{ijkl}\varepsilon_{kl} \tag{2.11}$$

where σ_{ij} is the stress tensor, ε_{kl} is the strain tensor. The elastic wave equation for anisotropic media takes the form:

$$\rho \frac{\partial^2 u_i}{\partial t^2} = \frac{\partial}{\partial x_j}\left(C_{ijkl}\frac{\partial u_l}{\partial x_k}\right) \tag{2.12}$$

where ρ is the density, u is the displacement component, t is time. This equation allows for different wave velocities and propagation characteristics in different directions, reflecting the anisotropic nature of the rock medium.

Heterogeneity in rock media can be addressed by allowing the material properties (ρ, λ, μ) to vary spatially. The governing equations then become:

$$\nabla \cdot \left(\lambda(x)\nabla \cdot u + \mu(x)\left(\nabla u + (\nabla u)^\top\right)\right) = \rho(x)\frac{\partial^2 u}{\partial t^2} \tag{2.13}$$

where x represents the spatial coordinates.

Numerical methods, such as finite element analysis or finite difference schemes, are typically employed to solve these more complex equations, allowing for accurate modeling of wave propagation in anisotropic and heterogeneous rock formations [7].

Viscoelastic Effects and Attenuation

Real rock materials often exhibit viscoelastic behavior, characterized by a time-dependent response to applied stresses. This behavior leads to the attenuation of elastic waves as they propagate through the medium. Incorporating viscoelastic effects into the governing equations of elastic waves is essential for accurately modeling wave propagation in rock dynamics [5, 6].

The simplest viscoelastic model is the Kelvin-Voigt model, which combines elastic and viscous elements in parallel. The stress–strain relationship for this model can be expressed as:

$$\sigma_{ij} = C_{ijkl}\epsilon_{kl} + \eta_{ijkl}\frac{\partial \varepsilon_{kl}}{\partial t} \quad (2.14)$$

where η_{ijkl} is the viscosity tensor.

Incorporating this viscoelastic relationship into the equation of motion yields:

$$\rho\frac{\partial^2 u_i}{\partial t^2} = \frac{\partial}{\partial x_j}\left(C_{ijkl}\frac{\partial u_l}{\partial x_k} + \eta_{ijkl}\frac{\partial^2 u_l}{\partial x_k \partial t}\right) \quad (2.15)$$

This equation describes the propagation of attenuated elastic waves in a viscoelastic medium. The attenuation can be characterized by the quality factor Q, which is related to the energy loss per cycle of oscillation.

More complex viscoelastic models, such as the Maxwell model or the standard linear solid model, can be employed to capture a wider range of time-dependent behaviors in rock materials. These models lead to frequency-dependent attenuation and dispersion of elastic waves, which can be described using complex moduli or fractional derivative formulations.

Nonlinear Effects in Elastic Wave Propagation

While the governing equations presented thus far assume linear elastic behavior, real rock materials can exhibit nonlinear responses under certain conditions, particularly at high stress levels or in the presence of discontinuities. Incorporating nonlinear effects into the governing equations of elastic waves is crucial for accurately modeling rock behavior under extreme loading conditions, such as those encountered in rock blasting or earthquake events.

Nonlinear elasticity can be introduced by considering higher-order terms in the stress–strain relationship:

$$\sigma_{ij} = C_{ijkl}\varepsilon_{kl} + \frac{1}{2}C_{ijklmn}\varepsilon_{kl}\varepsilon_{mn} + \ldots \quad (2.16)$$

where C_{ijklmn} represents the third-order elastic constants.

The resulting nonlinear wave equation takes the form:

2.1 Basic Dynamic Equations

$$\rho \frac{\partial^2 u_i}{\partial t^2} = \frac{\partial}{\partial x_j}\left(C_{ijkl}\frac{\partial u_l}{\partial x_k} + C_{ijklmn}\frac{\partial u_l}{\partial x_k}\frac{\partial u_m}{\partial x_n} \right) \qquad (2.17)$$

This equation describes various nonlinear phenomena in elastic wave propagation, including harmonic generation, wave mixing, and shock wave formation.

Another important nonlinear effect in rock dynamics is the opening and closing of microcracks and pores under dynamic loading. This behavior can be modeled using damage mechanics approaches, where the elastic moduli are treated as functions of a damage parameter D:

$$C_{ijkl} = C_{ijkl}^0 (1 - D) \qquad (2.18)$$

where C_{ijkl}^0 represents the undamaged elastic constants.

The evolution of the damage parameter can be described by an additional differential equation, coupled with the elastic wave equation, to model the progressive degradation of rock properties under dynamic loading [8].

Coupled Processes in Rock Dynamics

The propagation of elastic waves in rock media is often coupled with other physical processes, such as fluid flow, heat transfer, and chemical reactions. These coupled processes can significantly influence the dynamic behavior of rocks and must be considered in comprehensive models of rock dynamics [6, 9].

One important example is the coupling between elastic waves and fluid flow in porous rocks, described by the theory of poroelasticity. The governing equations for this coupled system include:

1. The Equation of Motion for the Solid Skeleton:

$$\rho \frac{\partial^2 u}{\partial t^2} = \nabla \cdot (\sigma - \alpha p I) + f \qquad (2.19)$$

2. The Mass Conservation Equation for the Fluid:

$$\frac{1}{M}\frac{\partial p}{\partial t} + \alpha \nabla \cdot \left(\frac{\partial u}{\partial t}\right) + \nabla \cdot q = 0 \qquad (2.20)$$

3. Darcy's Law for Fluid Flow:

$$q = -\frac{k}{\eta}(\nabla p - \rho g) \qquad (2.21)$$

where α is the Biot coefficient, p is the pore pressure, M is the Biot modulus, k is the permeability, η is the fluid viscosity, and ρ is the fluid density.

These coupled equations describe the interaction between elastic waves and fluid pressure waves (slow P-waves) in porous media, leading to phenomena such as wave-induced fluid flow and dynamic permeability changes. Beyond this fundamental coupling, several other significant interactions warrant consideration in comprehensive rock dynamics analyses. Thermo-elastic coupling addresses the interaction between elastic waves and heat transfer, while chemo-mechanical coupling accounts for the influence of chemical reactions on rock mechanical properties. Additionally, electro-kinetic coupling describes the generation of electric potentials resulting from fluid flow in porous rocks. When these coupled processes are incorporated into the governing equations of elastic waves, they give rise to complex systems of partial differential equations that generally necessitate advanced numerical methods for their solution [7].

Boundary and Initial Conditions

The governing equations of elastic waves must be complemented by appropriate boundary and initial conditions to fully define the wave propagation problem in rock dynamics. These conditions reflect the physical constraints and loading scenarios encountered in real-world applications.

Common types of boundary conditions include:

1. Dirichlet (Displacement) Boundary Conditions:

$$u = u_0 \text{ on } \partial \Omega_1 \tag{2.22}$$

2. Neumann (Traction) Boundary Conditions:

$$\sigma \cdot n = t_0 \text{ on } \partial \Omega_2 \tag{2.23}$$

3. Mixed Boundary Conditions:

$$u \cdot n = u_n^0 \text{ on } \partial \Omega_3 \tag{2.24}$$

$$(\sigma \cdot n)_t = t_t^0 \text{ on } \partial \Omega_3 \tag{2.25}$$

where $\partial \Omega$ represents the boundary of the domain, n is the outward unit normal vector, and the subscripts n and t denote normal and tangential components, respectively.

Special consideration must be given to boundaries between different rock layers or at the interface between solid rock and fluids. These interfaces can be modeled using continuity conditions for displacements and tractions, or using more complex models that account for imperfect bonding or slip.

Initial conditions typically specify the displacement and velocity fields at the beginning of the analysis:

$$u(x, 0) = u_0(x) \tag{2.26}$$

2.1 Basic Dynamic Equations

$$\frac{\partial u}{\partial t}(x, 0) = v_0(x) \tag{2.27}$$

These initial conditions may represent pre-existing stress states in the rock mass or initial perturbations that trigger wave propagation.

In many rock dynamics applications, such as seismic exploration or blast-induced vibrations, the initial conditions are replaced by source terms in the governing equations. These source terms can represent point sources, line sources, or distributed sources of energy, depending on the specific problem being studied.

2.1.3 Dispersion Effects Induced by the Transverse Inertia

In rock dynamics, transverse inertia significantly influences wave propagation through elastic media. This phenomenon, where different frequency components travel at varying velocities, affects wave characteristics in rocks and has critical applications in geophysics, seismology, and rock engineering.

The dispersion mechanism, particularly relevant for wave propagation in beam-, plate-, and shell-like geological formations, incorporates lateral motion and rotational effects beyond the fundamental elastic wave equations. These effects build upon Newton's second law while introducing additional wave propagation complexities.

Theoretical Foundations of Dispersion Effects

The analysis of transverse inertia-induced dispersion builds upon continuum mechanics and elastic wave propagation principles. While classical elastodynamics describes wave propagation through the wave equation derived from Newton's second law and material constitutive relations [4], it typically omits rotational inertia and shear deformation effects. These effects become crucial for high-frequency waves or waves in thick structures.

Including transverse inertia provides a more comprehensive wave propagation description, particularly relevant for beam and plate structures, enabling accurate prediction of wave behavior in rocks under complex geometries or high-frequency excitations.

One of the key theoretical frameworks for understanding dispersion effects induced by transverse inertia is the Timoshenko beam theory. This theory, which extends the classical Euler–Bernoulli beam theory, incorporates the effects of rotary inertia and shear deformation. The governing equations for a Timoshenko beam can be expressed as:

$$\rho A \frac{\partial^2 u}{\partial t^2} - \kappa G A \left(\frac{\partial^2 u}{\partial x^2} - \frac{\partial \phi}{\partial x} \right) = 0 \tag{2.28}$$

$$\rho I \frac{\partial^2 \phi}{\partial t^2} - EI \frac{\partial^2 \phi}{\partial x^2} + \kappa G A \left(\frac{\partial u}{\partial x} - \phi \right) = 0 \tag{2.29}$$

where u is the transverse displacement, φ is the rotation of the cross-section, ρ is the density, A is the cross-sectional area, I is the moment of inertia, E is Young's modulus, G is the shear modulus, κ is the shear correction factor, and x and t represent space and time coordinates, respectively.

These equations capture the coupling between transverse displacement and rotation, which is a key feature of waves propagating in media with significant transverse inertia effects. The inclusion of rotary inertia (represented by the ρI term) and shear deformation (represented by the κGA term) leads to dispersion behavior that is not present in classical wave equations.

Dispersion Relation and Phase Velocity

The dispersion effects induced by transverse inertia are most clearly manifested in the dispersion relation, which describes the relationship between the frequency and wavenumber of propagating waves. For a Timoshenko beam, the dispersion relation can be derived by assuming harmonic solutions of the form:

$$u = U \exp(i(kx - \omega t)) \tag{2.30}$$

$$\phi = \Phi \exp(i(kx - \omega t)) \tag{2.31}$$

where U and Φ are complex amplitudes, k is the wavenumber, i is the imaginary unit, x is the spatial coordinate and ω is the angular frequency.

Substituting these solutions into the governing equations and solving the resulting eigenvalue problem yields the dispersion relation:

$$\omega^4 - \omega^2 \frac{k^2(E + \kappa G)}{\rho} + \frac{k^4(E\kappa G)}{\rho^2}\left(1 + \frac{E}{\kappa G}\frac{I}{A}\right) = 0 \tag{2.32}$$

This equation relates the frequency ω to the wavenumber k, and its solutions describe the propagation characteristics of waves in the medium. The dispersion relation for a Timoshenko beam exhibits two distinct branches, corresponding to two types of waves: a lower branch associated with flexural waves and an upper branch associated with shear waves.

The phase velocity v_p, which represents the speed at which a particular phase of the wave propagates, can be derived from the dispersion relation as:

$$v_p = \frac{\omega}{k} \tag{2.33}$$

For a non-dispersive medium, the phase velocity would be constant for all frequencies. However, in the presence of dispersion effects induced by transverse inertia, the phase velocity varies with frequency, leading to the distortion of wave shapes as they propagate.

Group Velocity and Energy Propagation

While the phase velocity describes the propagation of individual wave components, the group velocity is more relevant for understanding the propagation of wave packets and energy. The group velocity v_g is defined as:

$$v_g = \frac{d\omega}{dk} \tag{2.34}$$

In the context of dispersion effects induced by transverse inertia, the group velocity exhibits complex behavior that deviates significantly from the non-dispersive case. At low frequencies, the group velocity approaches the phase velocity of classical bending waves. However, as the frequency increases, the group velocity begins to deviate, eventually approaching an asymptotic value that depends on the material properties and geometry of the medium.

The behavior of group velocity has important implications for energy propagation in rock dynamics. In particular, the frequency-dependent nature of group velocity leads to the spreading of wave packets as they propagate through the medium. This spreading can result in the attenuation of wave amplitude over distance, even in the absence of material damping.

Cut-off Frequency and Wave Modes

One of the distinctive features of wave propagation in media with significant transverse inertia effects is the existence of cut-off frequencies. A cut-off frequency represents a threshold below which certain wave modes cannot propagate. In the case of a Timoshenko beam, the cut-off frequency for shear waves ω_c is given by:

$$\omega_c = \sqrt{\frac{\kappa GA}{\rho I}} \tag{2.35}$$

Below this frequency, only flexural waves can propagate, while above it, both flexural and shear waves can exist. The existence of cut-off frequencies has important implications for the transmission of energy in rock structures and can lead to frequency-dependent filtering effects.

The consideration of transverse inertia also leads to the existence of multiple wave modes, each with its own dispersion characteristics. In addition to the flexural and shear modes mentioned earlier, higher-order modes can exist, particularly in more complex geometries such as plates and shells. These higher-order modes become increasingly important at higher frequencies and can significantly influence the overall dynamic behavior of rock structures.

Implications for Rock Dynamics

The dispersion effects induced by transverse inertia play a crucial role in rock dynamics, manifesting through several interconnected phenomena that significantly influence wave propagation and structural response. As waves propagate through rock structures, the frequency-dependent phase and group velocities lead to wave distortion, which directly impacts the interpretation of seismic signals and the analysis of wave-induced stresses in rock formations. This distortion is closely related to frequency-dependent attenuation, where the spreading of wave packets due to dispersion effects results in apparent amplitude reduction over distance—a critical consideration when analyzing wave propagation in rocks, particularly for high-frequency signals.

Furthermore, the complexity of wave propagation in rock structures is compounded by mode conversion phenomena, where energy transfers between different wave types during propagation due to the existence of multiple wave modes and cut-off frequencies. This behavior becomes particularly significant in complex rock structures, where the interaction between different wave modes can substantially alter the overall dynamic response. The influence of these dispersion effects exhibits notable scale dependence, with their importance generally increasing with frequency while decreasing with the characteristic size of the structure. Consequently, the dynamic response of small-scale rock samples may differ significantly from that of large-scale rock masses, highlighting the importance of scale considerations in rock dynamics analyses.

These dispersion mechanisms culminate in complex resonance phenomena, particularly evident in layered or inhomogeneous formations, where the frequency-dependent nature of wave propagation can create unique dynamic response patterns. This resonance behavior further emphasizes the interconnected nature of dispersion effects in rock dynamics and their fundamental importance in understanding structural response.

In conclusion, the dispersion effects induced by transverse inertia play a crucial role in shaping the propagation of waves in rock dynamics. These effects, which arise from the consideration of rotational inertia and shear deformation, lead to frequency-dependent phase and group velocities, the existence of multiple wave modes, and complex resonance phenomena. Understanding and accurately modeling these effects is essential for a wide range of applications in rock engineering, from blast vibration prediction to seismic hazard assessment.

The consideration of dispersion effects induced by transverse inertia represents a significant advancement in the field of rock dynamics, bridging the gap between classical wave theory and the complex behavior observed in real rock formations. By accounting for these effects, engineers and researchers can develop more accurate models and predictions, leading to safer and more efficient designs in a wide range of geotechnical and geological applications.

2.2 Rock Stress Wave Theory

2.2.1 Coaxial Collision of Two Elastic Bars

The coaxial collision of two elastic bars represents a foundational principle in rock stress wave theory, providing essential insights into wave propagation phenomena in rock dynamics. When the two elastic bars collide coaxially, the compressive stress wave generates at the collision point, and they propagate to the two elastic bars from the collision point, then reflects the stretching wave at the free end of the elastic bar, and finally forms a complex wave interaction mode in the system. The study of this process is vital for understanding stress wave behavior in rock masses.

The propagation process of the longitudinal wave in the elastic bars is controlled by the one-dimensional wave equation:

$$\frac{\partial^2 u}{\partial t^2} = c^2 \frac{\partial^2 u}{\partial x^2} \tag{2.36}$$

where u is the displacement, t is time, x is the spatial coordinate, and c is the wave speed in the material.

For two elastic bars of equal cross-sectional area and material properties, the stress wave generated at the impact interface can be described by:

$$\sigma(t) = \frac{1}{2}\rho c v_0 \left[H(t) - H\left(t - \frac{2L}{c}\right) \right] \tag{2.37}$$

where σ is the stress, ρ is the density of the bar material, v_0 is the initial velocity of the striking bar, L is the length of each bar, c is the wave speed in the material, and $H(t)$ is the Heaviside step function.

The transmission and reflection of waves at the interface between the colliding bars depend on their relative impedances. The transmission coefficient (T) and reflection coefficient (R) for stress waves at the interface are given by:

$$T = \frac{2Z_2}{Z_1 + Z_2} \tag{2.38}$$

$$R = \frac{Z_2 - Z_1}{Z_1 + Z_2} \tag{2.39}$$

where Z_1 and Z_2 are the impedances of the first and second bar, respectively.

The impedance is defined as:

$$Z = \rho c A \tag{2.40}$$

where ρ is the density of the bar material, c is the wave speed in the material, and A is the cross-sectional area of the bar.

In rock dynamics, the coaxial collision model of two elastic bars can be extended to analyze wave transmission across rock interfaces and discontinuities, where impedance mismatches significantly influence wave propagation characteristics. This principle finds critical applications in various geological engineering contexts.

For instance, in hydraulic fracturing operations, the stress waves generated by the fluid pressure interact with the rock mass in a manner analogous to the collision of elastic bars. The transmission and reflection of these waves at rock-fluid interfaces play a crucial role in fracture propagation and rock fragmentation. Similarly, the propagation of seismic waves in the layered rock structures and the stress waves generated in the rock blasting also propagate in a way similar to the waves in the coaxial collision model. In addition, the study of coaxial collision also provides implications for the design of a dynamic rock testing equipment, such as the Split Hopkinson Pressure Bar (SHPB).

In conclusion, the study of coaxial collision of elastic bars provides a foundational understanding of stress wave propagation in rock dynamics. It offers valuable insights into wave generation, transmission, and reflection, which are crucial for analyzing and predicting rock behavior under dynamic loading conditions. As research in rock dynamics continues to advance, the principles derived from coaxial collision studies will remain fundamental in developing more sophisticated models and techniques.

2.2.2 Interaction of Two Elastic Waves

The interaction of elastic waves fundamentally governs rock stress wave theory, with applications in seismic exploration, rock blasting, and underground excavation. When elastic waves meet in rock media, they undergo superposition, interference, and scattering, phenomena that determine wave propagation patterns and energy dissipation characteristics. Understanding these interactions is essential for accurately predicting wave propagation patterns and energy dissipation in rock masses.

Superposition of Elastic Waves

The principle of superposition states that when two or more waves meet at a point, the resultant displacement is the algebraic sum of the individual wave displacements. This principle is fundamental to understanding the interaction of elastic waves in rocks. For two waves with displacements $u_1(x,t)$ and $u_2(x,t)$, the resultant displacement $u(x,t)$ at any point x and time t is given by:

$$u(x, t) = u_1(x, t) + u_2(x, t) \tag{2.41}$$

This superposition can lead to constructive or destructive interference, depending on the phase relationship between the interacting waves.

Interference of Elastic Waves

Interference occurs when two or more coherent waves superpose, resulting in a new wave pattern. In rock masses, interference can significantly affect the distribution of stress and energy. The interference pattern depends on the relative amplitudes, frequencies, and phases of the interacting waves.

Constructive interference occurs when the peaks of the interacting waves align, resulting in an amplification of the wave amplitude. This can lead to localized stress concentrations in the rock mass. The condition for constructive interference is:

$$\Delta\phi = 2n\pi, \text{ where } n = 0, 1, 2, \ldots \tag{2.42}$$

Here, $\Delta\phi$ represents the phase difference between the two waves.

Conversely, destructive interference occurs when the peak of one wave aligns with the trough of another, resulting in a reduction or cancellation of the wave amplitude. The condition for destructive interference is:

$$\Delta\phi = (2n+1)\pi, \text{ where } n = 0, 1, 2, \ldots \tag{2.43}$$

In rock dynamics, understanding these interference patterns is crucial for predicting stress distributions and potential failure zones in rock masses subjected to multiple sources of dynamic loading.

Scattering of Elastic Waves

When elastic waves encounter discontinuities, inhomogeneities, or boundaries within a rock mass, scattering occurs. This phenomenon leads to the redistribution of wave energy in various directions. Scattering is particularly important in fractured or jointed rock masses, where the presence of discontinuities can significantly alter the propagation pattern of stress waves.

The scattering of elastic waves in rocks can be described using various mathematical models, depending on the nature of the scatterer and the wavelength of the incident wave. For instance, the Rayleigh scattering model is applicable when the scatterer size is much smaller than the wavelength, while the Mie scattering model is used when the scatterer size is comparable to the wavelength.

The intensity of scattered waves I_s in relation to the incident wave intensity I_i can be expressed as:

$$\frac{I_s}{I_i} = f(\theta, \lambda, a) \tag{2.44}$$

where θ is the scattering angle, λ is the wavelength, and a is the characteristic size of the scatterer. The function f depends on the specific scattering model used.

Mode Conversion

When elastic waves interact with interfaces or discontinuities in rock masses, mode conversion can occur. This phenomenon involves the transformation of one wave type into another. For example, an incident P-wave (compressional wave) can generate both reflected and transmitted P-waves and S-waves (shear waves) at an interface. The angles of reflection and transmission for these converted waves are governed by Snell's law:

$$\frac{\sin\theta_1}{V_1} = \frac{\sin\theta_2}{V_2} \tag{2.45}$$

where θ_1 and θ_2 are the angles of incidence and transmission (or reflection), respectively, and V_1 and V_2 are the wave velocities in the incident and transmitted medium.

Mode conversion is particularly important in understanding the complex wave field generated in rock masses during dynamic events such as earthquakes or explosions. It contributes to the redistribution of energy between different wave types and can significantly influence the overall stress distribution in the rock mass.

Nonlinear Wave Interactions

In rock masses subjected to high-intensity stress waves, such as those generated by large explosions or strong earthquakes, nonlinear wave interactions can occur. These interactions are characterized by phenomena such as harmonic generation, wave mixing, and soliton formation. Nonlinear effects become significant when the wave amplitude is large enough to induce changes in the elastic properties of the rock material.

The nonlinear wave equation for one-dimensional propagation reflects the dependence of wave velocity on strain:

$$\frac{\partial^2 u}{\partial t^2} = c^2(1 + \beta\varepsilon)\frac{\partial^2 u}{\partial x^2} \tag{2.46}$$

where u is the displacement, c is the linear wave velocity, β is the nonlinearity parameter, ε is the strain and x is the spatial coordinate variable in the direction of wave propagation.

Understanding nonlinear wave interactions is crucial for accurately predicting rock behavior under extreme loading conditions and for designing structures to withstand such events.

In conclusion, the interaction of elastic waves in rock masses is a complex phenomenon that plays a crucial role in rock dynamics. The interaction of elastic waves in rock masses governs rock dynamics through superposition, interference, scattering, reflection, transmission, mode conversion, and nonlinear interactions. Understanding these wave phenomena is fundamental for predicting rock mass response to dynamic loading and optimizing rock engineering applications.

2.3 Rock Dynamic Strength Theory

2.3.1 Rock Stress–Strain Relationship

The stress–strain relationship under dynamic loading conditions fundamentally characterizes rock dynamic strength theory, differing significantly from static conditions due to strain rate effects, wave propagation phenomena, and inertial effects. These factors influence rock deformation and failure mechanisms during rapid loading events such as blasting, impacts, or seismic activities. A key distinguishing feature is the increase in rock strength and stiffness under high strain rates, quantified through the dynamic increase factor (DIF)–the ratio of dynamic to static strength. Experimental data demonstrates that DIF increases proportionally with strain rate, expressed mathematically as:

$$\text{DIF} = \frac{f_d}{f_s} \tag{2.47}$$

where f_d is the dynamic strength and f_s is the static strength of the rock.

Experimental observations have shown that the DIF for rocks can vary significantly depending on the rock type, confining pressure, and the range of strain rates considered.

The mechanical behavior of rocks exhibits notable strain rate dependence, which can be attributed to several interrelated mechanisms. Fundamentally, micro-crack propagation plays a crucial role in this phenomenon, as high strain rates restrict the time available for micro-cracks to propagate and coalesce, thereby resulting in higher apparent strength. This behavior is further complicated by inertial effects, where rapid loading induces resistive inertial forces that oppose deformation. Additionally, certain rocks demonstrate viscous behavior under high strain rate conditions, which manifests as increased resistance to deformation. The strain rate dependence is also influenced by thermal effects, as rapid loading can generate localized heating that potentially alters the material properties and subsequently affects the stress–strain response.

To capture the effects of dynamic loading on rock materials, various mathematical models have been proposed. The Johnson-Holmquist (JH) model is widely used and incorporates strain rate effects into the constitutive equations for rock materials. It expresses normalized equivalent stress (σ) as a function of pressure (P), damage (D), and strain rate ($\dot{\varepsilon}^*$), with material constants A, B, C, M, and N:

$$\sigma^* = \left[A(P^* + T^*)^N\right](1 - D) + B(P^*)^M \left[1 + C\ln(\dot{\varepsilon}^*)\right] \tag{2.48}$$

where T^* is the normalized maximum tensile hydrostatic pressure, and the asterisks denote normalized quantities.

Another key aspect of the dynamic stress–strain relationship is the observed nonlinearity at high strain rates, particularly in the pre-peak region. This nonlinearity results from the complex interaction of elastic wave propagation, plastic deformation, and damage accumulation within the rock material.

To address this nonlinearity, the Unified Strength Theory (UST) incorporates both hydrostatic pressure dependence and the intermediate principal stress effect. The UST framework describes the nonlinear behavior of rocks under dynamic loading, using the first and second stress tensor invariants (I_1, J_2) and the Lode angle (θ):

$$F(I_1, J_2, \theta) = \alpha I_1 + \lambda \sqrt{J_2} \left[(1 + \sin\phi)\cos\theta - \frac{1 - \sin\phi}{3}\left(\cos\theta + \sqrt{3}\sin\theta\right) \right]$$
$$- c\cos\phi = 0 \qquad (2.49)$$

where α and λ are material parameters, ϕ is the internal friction angle, and c is the cohesion.

Confining pressure significantly influences rock dynamic strength and deformation characteristics, with higher confinement increasing both dynamic strength and ductility while amplifying strain rate effects. The dynamic Hoek–Brown criterion extends the static model to incorporate these combined strain rate and confinement effects:

$$\sigma_1 = \sigma_3 + \sigma_{ci}\left(m_b \frac{\sigma_3}{\sigma_{ci}} + s\right)^a \left[1 + D\log_{10}\left(\frac{\dot{\varepsilon}}{\dot{\varepsilon}_0}\right)\right] \qquad (2.50)$$

where σ_1 and σ_3 are the major and minor principal stresses, respectively, σ_{ci} is the uniaxial compressive strength of the intact rock, m_b, s, and a are material constants, D is a strain rate sensitivity parameter, $\dot{\varepsilon}$ is the strain rate, and $\dot{\varepsilon}_0$ is a reference strain rate.

Discontinuities (joints, fractures, and bedding planes) significantly alter wave propagation and mechanical response in rock masses under dynamic loading. The equivalent viscoelastic model addresses this through a continuum approach integrating both intact rock and discontinuity properties. The stress–strain relationship is expressed as:

$$\sigma(t) = E_\infty \varepsilon(t) + \int_0^t E(t-\tau)\frac{d\varepsilon(\tau)}{d\tau}d\tau \qquad (2.51)$$

where $\sigma(t)$ is the stress at time t, $\varepsilon(t)$ is the strain at time t, E_∞ is the long-term elastic modulus, and $E(t)$ is the time-dependent relaxation modulus that accounts for the viscoelastic behavior of the rock mass.

Another important aspect of the dynamic stress–strain relationship in rocks is the energy dissipation and damage accumulation process. During dynamic loading, rocks experience rapid deformation and failure, with energy dissipating through microcrack propagation, friction, and heat generation. These dissipation mechanisms are fundamental to dynamic strength theories and rock behavior prediction under extreme

2.3 Rock Dynamic Strength Theory

loads. The energy-horizon theory of dynamic fragmentation (EHTDF) was developed to correlate rock dynamic strength with fragmentation energy requirements.

$$\sigma_d = \left(\frac{24\rho c^2 \Gamma}{\dot{\varepsilon}}\right)^{\frac{1}{3}} \quad (2.52)$$

where σ_d is the dynamic strength, ρ is the material density, c is the elastic wave speed, Γ is the specific fracture energy, and $\dot{\varepsilon}$ is the strain rate.

Qi et al. [10] enhanced the EHTDF by integrating energy-based criteria with static size effect laws to predict dynamic fragmentation scales. This framework enables analysis of rock failure characteristics under various loading conditions. Under high-energy impact loading, rapid deformation induces adiabatic heating, significantly affecting the rock's mechanical properties.

To account for temperature effects in the dynamic stress–strain relationship, researchers have developed thermo-mechanical coupling models that incorporate both mechanical and thermal processes. One such model is the thermo-mechanical coupled damage model, which expresses the stress–strain relationship as:

$$\sigma_{ij} = (1-D)C_{ijkl}\left(\varepsilon_{kl} - \varepsilon_{kl}^p - \varepsilon_{kl}^T\right) \quad (2.53)$$

where σ_{ij} is the stress tensor, D is the damage variable, C_{ijkl} is the elastic stiffness tensor, ε_{kl} is the total strain tensor, ε_{kl}^p is the plastic strain tensor, and ε_{kl}^T is the thermal strain tensor.

The damage variable D evolves with mechanical and thermal loading, simulating coupled thermo-mechanical effects on rock behavior. Rocks under complex loading paths (e.g., underground excavations, seismic events) exhibit distinct stress–strain responses compared to uniaxial or triaxial conditions. A path-dependent constitutive model incorporating isotropic and kinematic hardening was developed to capture these effects:

$$d\sigma = D\varepsilon_p : d\varepsilon \quad (2.54)$$

where $d\sigma$ is the stress increment, $d\varepsilon$ is the strain increment, and $D\varepsilon_p$ is the elasto-plastic tangent stiffness tensor that evolves based on the loading history and stress path.

Fluid-saturated rocks display complex dynamic behavior due to solid–fluid interactions, manifesting as pore pressure buildup, hydraulic fracturing, and wave attenuation. Coupled hydro-mechanical models, notably the dynamic Biot's theory, extend static poroelasticity to dynamic loading conditions. The stress–strain relationship is expressed as:

$$\sigma_{ij} = 2G\varepsilon_{ij} + \lambda\varepsilon_{kk} - \alpha p \delta_{ij} \quad (2.55)$$

where σ_{ij} is the total stress tensor, G is the shear modulus, λ is Lamé's constant, ε_{ij} is the strain tensor, α is the Biot coefficient, p is the pore pressure, and δ_{ij} is the Kronecker delta.

The pore pressure evolution in this model is governed by a coupled diffusion equation that accounts for both fluid flow and solid deformation:

$$\frac{\partial p}{\partial t} = \frac{k}{\eta}\nabla^2 p + \alpha\frac{\partial \varepsilon_{kk}}{\partial t} \qquad (2.56)$$

where t is the time variable that controls the dynamic evolution of pore pressure, ε_{kk} is the volume strain, which characterizes the overall volume change of the solid skeleton, k is the permeability, η is the fluid viscosity, and ∇^2 is the Laplacian operator.

Rock microstructural imperfections (microcracks, voids, inclusions) influence crack initiation and propagation under dynamic loading, affecting failure patterns. The statistical crack mechanical model (SCRAM) was developed to relate macroscopic stress–strain response to microcrack evolution under dynamic conditions.

$$\sigma = E_0(1-D)\varepsilon \qquad (2.57)$$

where σ is the macroscopic stress, E_0 is the initial elastic modulus, D is the damage variable representing the microcrack density, and ε is the macroscopic strain.

The damage variable D evolves with microcrack initiation and propagation thresholds, simulating dynamic failure processes. Rock anisotropy, originating from geological formation (e.g., sedimentary bedding planes, metamorphic foliation), influences wave propagation and mechanical response. An anisotropic damage model expressing the directional stress–strain relationship was developed as:

$$\sigma_{ij} = (1-D)_{ijkl} C_{klmn} \varepsilon_{mn} \qquad (2.58)$$

where σ_{ij} is the stress tensor, ε_{mn} is the strain tensor, C_{klmn} is the fourth-order elastic stiffness tensor, and $(1-D)_{ijkl}$ is the fourth-order damage effect tensor that accounts for the anisotropic evolution of damage under dynamic loading.

Loading rate history, particularly during multiple dynamic events or cyclic loading, influences rock behavior through fatigue damage accumulation and strain hardening/softening effects. A rate-dependent damage model incorporating loading history through a time-dependent damage variable was developed:

$$D = 1 - exp\left(-\int_0^t f(\sigma, \dot{\varepsilon})\, dt\right) \qquad (2.59)$$

where D is the damage variable, $f(\sigma, \dot{\varepsilon})$ is a function that depends on the stress state and strain rate, and t is the loading time.

This model simulates complex loading scenarios, including cyclic and repeated dynamic events, by capturing loading history effects on rock stress–strain response. The dynamic behavior is influenced by multi-scale discontinuities (microcracks to faults), which affect wave propagation and mechanical response. To address

2.3 Rock Dynamic Strength Theory

this, researchers developed multi-scale models integrating intact rock and discontinuity properties, hierarchical approach combining continuum and discrete element methods:

$$\sigma = (1-\xi)\sigma_c + \xi\sigma_d \qquad (2.60)$$

where σ is the macroscopic stress, σ_c is the stress in the continuum phase (intact rock), σ_d is the stress in the discontinuity phase, and ξ is a weighting factor that depends on the volume fraction of discontinuities.

This multi-scale approach enables simulation of dynamic stress–strain behaviors in rock masses across discontinuity scales. Under high strain-rate loading conditions, rocks exhibit localized deformation through shear banding, significantly affecting stress–strain response and failure characteristics. The Continuous Surface Cap Model (CSCM) incorporates strain softening and localization phenomena through a cap-type yield surface and damage formulation to account for these effects:

$$F(I_1, J_2, J_3, \kappa) = J_2 - [X(\kappa) - L(I_1)]^2 R^2 + [X(\kappa) - L(I_1)]^2 = 0 \qquad (2.61)$$

where F is the yield function, I_1 is the first invariant of the stress tensor, J_2 and J_3 are the second and third invariants of the deviatoric stress tensor, κ is the hardening parameter, $X(\kappa)$ is the intersection of the yield surface with the I_1 axis, $L(I_1)$ is the compaction surface, and R is the ratio of major to minor axes of the ellipse.

This model simulates strain localization and shear banding in rocks under dynamic loading, reflecting experimental failure mechanisms. Pre-existing damage from prior loading events or environmental factors significantly alters dynamic stress–strain responses compared to pristine rocks. The unified viscoelastic-plastic-damage model incorporates initial damage states and their evolution under dynamic loading, expressing the stress–strain relationship as:

$$\sigma_{ij} = (1 - D_0 - D) C_{ijkl} \left(\varepsilon_{kl} - \varepsilon_{kl}^p - \varepsilon_{kl}^{ve} \right) \qquad (2.62)$$

where σ_{ij} is the stress tensor, D_0 is the initial damage variable representing pre-existing damage, D is the evolving damage variable, C_{ijkl} is the elastic stiffness tensor, ε_{kl} is the total strain tensor, ε_{kl}^p is the plastic strain tensor, and ε_{kl}^{ve} is the viscoelastic strain tensor.

This model simulates dynamic stress–strain behavior in rocks with pre-existing damage for engineering applications. The dynamic stress–strain relationship in rocks is governed by multiple factors: strain rate effects, wave propagation, inertial effects, confining pressure, temperature, loading path, fluid saturation, microstructural features, anisotropy, loading history, discontinuities, strain localization, and pre-existing damage. Advanced constitutive models incorporating elasticity, plasticity, damage mechanics, and wave propagation theory have enhanced our ability to predict rock behavior under dynamic loading conditions. However, the complexity of rock materials and diverse loading conditions necessitate continued research to refine existing models and develop new approaches, validated against experimental data across various rock types and loading conditions.

2.3.2 Dynamic Strength Damage Criteria for Rocks

Dynamic strength, unlike static strength, exhibits strain rate dependency and determines rock material resistance under dynamic loading. This section examines dynamic strength damage criteria for rocks.

The DIF characterizes dynamic rock strength as the ratio of dynamic to static strength. Bukit Timah granite. The applications of Mohr–Coulomb and Hoek–Brown criteria to brittle rock dynamic strength using Bukit Timah granite were studied. Their findings showed dynamic compressive strength increases with loading rate and confining pressure. While Mohr–Coulomb criterion accurately represents rock strength at low confining pressures with rate-dependent cohesion, Hoek–Brown criterion better describes behavior at higher confining pressures. The dynamic Mohr–Coulomb criterion can be expressed as:

$$\tau = c_d + \sigma_n tan\varphi_d \tag{2.63}$$

where τ is the shear strength, c_d is the dynamic cohesion, σ_n is the normal stress, and φ_d is the dynamic internal friction angle.

Zhao et al. [11] observed that under dynamic loading conditions, the changes in rock strength primarily result from variations in the cohesion component c_d, while the internal angle of friction remains relatively unchanged. The dynamic Hoek–Brown criterion, on the other hand, is represented by:

$$\sigma_1 = \sigma_3 + \sigma_c(m_d\sigma_3/\sigma_c + 1.0)^{0.5} \tag{2.64}$$

where σ_1 and σ_3 are the major and minor principal stresses, respectively, σ_c is the uniaxial compressive strength, and m_d is the dynamic material constant.

The study indicated that the parameter m_d is unaffected by loading rate changes, making the Hoek–Brown criterion more versatile for dynamic strength prediction across a wide range of confining pressures [11].

Liao et al. [12] developed a unified strength theory for geomaterials' dynamic strength parameters, incorporating deviatoric stress, effective consolidation pressure, and dynamic equilibrium line effects. The dynamic equilibrium line equation is expressed as:

$$q'_u = m + n\sigma'_u \tag{2.65}$$

where q'_u is the unified strength deviatoric stress, σ'_u is the effective consolidation pressure, and m and n are coefficients derived from dynamic strength parameters. These coefficients can be used to determine the effective dynamic internal friction angle ($\varphi_d\,'$) and cohesion ($c_d\,'$):

2.3 Rock Dynamic Strength Theory

$$\varphi'_d = sin^{(-1)}\left(\frac{n}{n+2}\right) \quad (2.66)$$

$$c'_d = m\left(\frac{1 - sin\varphi'_d}{2cos\varphi'_d}\right) \quad (2.67)$$

Liao et al.'s [12] unified strength theory offers a physically clear and simplified approach to determine geomaterials' dynamic strength parameters, particularly useful for predicting earth structure failure during earthquakes.

Zhang et al. [13] reviewed dynamic experimental techniques and rock mechanical behavior, demonstrating that rock dynamic strengths increase significantly above critical loading rates, transitioning from simple fracturing to multiple fragmentation and pulverization. To describe this strain rate effect, researchers developed semi-empirical equations of the form:

$$\sigma_d = \sigma_s\left(1 + Alog\left(\frac{\dot{\varepsilon}}{\dot{\varepsilon}_0}\right)\right) \quad (2.68)$$

where σ_d is the dynamic strength, σ_s is the static strength, $\dot{\varepsilon}$ is the strain rate, $\dot{\varepsilon}_0$ is a reference strain rate (typically 10^{-5} s^{-1}), and A is a material-dependent constant.

This logarithmic relationship between strength and strain rate has been widely observed in various rock types.

At very high strain rates, Zhang et al. [13] noted that Johnson-Holmquist and micromechanics-based models (sliding crack and statistical crack mechanical models) more accurately simulate rock behavior under dynamic loading by incorporating thermal activation, Stefan effect, and dynamic fragmentation mechanisms.

Xie et al. [14] developed the True Triaxial Electromagnetic Hopkinson Bar (TEHB) System for precise triaxial dynamic impact testing under in-situ conditions. The TEHB system calculates dynamic stress $\sigma(t)$, strain $\varepsilon(t)$, and strain rate $\dot{\varepsilon}(t)$ using one-dimensional stress wave theory:

$$\sigma(t) = \frac{1}{2}\left(\frac{A}{A_s}\right)E\left(\varepsilon_{L_{inc}} + \varepsilon_{R_{inc}} + \varepsilon_{L_{refl}} + \varepsilon_{R_{refl}}\right) \quad (2.69)$$

$$\varepsilon(t) = \frac{C}{L_s}\int_0^t\left(\varepsilon_{L_{inc}} + \varepsilon_{R_{inc}} - \varepsilon_{L_{refl}} - \varepsilon_{R_{refl}}\right)dt \quad (2.70)$$

$$\dot{\varepsilon}(t) = \frac{C}{L_s}\left(\varepsilon_{L_{inc}} + \varepsilon_{R_{inc}} - \varepsilon_{L_{refl}} - \varepsilon_{R_{refl}}\right) \quad (2.71)$$

where A and A_s are the cross-sectional areas of the bar and specimen, respectively, E is the elastic modulus of the bar, C is the wave speed in the bar, L_s is the specimen length, and ε_{Linc}, ε_{Rinc}, ε_{Lrefl}, and ε_{Rrefl} are the incident and reflected strain waves in the left and right bars.

The TEHB system enables 3D dynamic compression, tension, and fracturing tests, facilitating the development of comprehensive dynamic strength criteria under complex stress states.

Zhou et al. [15] proposed an energy-based dynamic strength criterion incorporating strain rate effects, analyzing energy transformation from mechanical work to strain and thermal energy. The critical energy release rate G_{cd} was developed to account for strain rate effects.

$$G_{cd} \propto \frac{\sigma_{cd}^2}{E_d} \quad (2.72)$$

where σ_{cd} is the dynamic uniaxial compressive strength and E_d is the dynamic Young's modulus.

This energy-based criterion provides a more physically meaningful framework for predicting the yield and failure characteristics of rock materials under dynamic loads, addressing the limitations of traditional strength-based criteria.

An elastoplastic damage dynamic model incorporating the Unified Strength Theory (UST) and equivalent stress history to predict dynamic enhancement effects was developed. The dynamic yield strength is given by:

$$F = f(\sigma_1, \sigma_2, \sigma_3, \kappa) - \kappa = 0 \quad (2.73)$$

where f is a function of principal stresses (σ_1, σ_2, σ_3) based on the UST, and κ determines the strength degradation and hardening behavior of the rock material based on damage and strain rate effects.

This model offers a more comprehensive understanding of rock responses under dynamic loads without relying on explicit strain or stress rate-dependent functions.

The diversity of dynamic strength damage criteria reflects rock's complex behavior under dynamic loading. While traditional Mohr–Coulomb and Hoek–Brown criteria have been adapted for dynamic conditions, emerging models incorporate energy principles and damage mechanics, addressing strain rate effects, stress states, and energy transformation processes. Future dynamic strength criteria are expected to integrate multiphysics approaches, including temperature effects, fluid interactions, and microstructural evolution, with applications in mining, tunneling, and earthquake engineering [11–16].

2.3.3 Rock Dynamic Strength Versus Strain Rate

The relationship between rock dynamic strength and strain rate governs rock materials from quasi-static to high-strain-rate dynamic loading. This behavior has profound implications for engineering applications, including mining, tunneling, earthquake engineering, and protective structures design.

2.3 Rock Dynamic Strength Theory

In this section, we will explore the complex relationship between rock dynamic strength and strain rate, examining the underlying mechanisms, experimental observations, theoretical models, and practical implications of this phenomenon.

Mechanisms of Strain Rate Effects on Rock Strength

Rocks exhibit strain rate-dependent mechanical properties, particularly strength, which typically increases with loading rate. This rate dependency manifests across loading conditions from quasi-static (strain rates < 10^{-5} s^{-1}) to dynamic (strain rates > 10^0 s^{-1}), and a gradual transition between different strain rates.

The strain rate sensitivity is governed by multiple physical mechanisms. At high strain rates, inertial forces drive stress wave propagation and localized deformation, and the time-dependence of crack growth and condensation is also affected by the loading rate. In saturated rocks, pore fluid pressurization varies across strain rate regimes. Rapid deformation can generate localized heating effects, while high-rate loading may activate distinct microscale deformation mechanisms compared to quasi-static conditions.

Mechanism dominance varies with rock type and strain rate regime, with inertial confinement prominent at high strain rates (>10^3 s^{-1}) and crack propagation kinetics at intermediate rates.

To illustrate the interplay of these mechanisms, an energy-based dynamic strength criterion, which incorporates both the static strength and the additional strength contribution due to dynamic loading, was proposed:

$$\sigma_d = \sigma_s + \Delta\sigma_d \tag{2.74}$$

where σ_d is the dynamic strength, σ_s is the static strength, and $\Delta\sigma_d$ is the dynamic strength increment.

The dynamic strength increment is:

$$\Delta\sigma_d = \sqrt{\frac{2\rho c^2 \Gamma_{\text{eff}}}{l_c}} \tag{2.75}$$

where ρ is the rock density, c is the P-wave velocity, Γ_{eff} is the effective surface energy, and l_c is a characteristic length scale related to the dominant crack size.

This model, incorporating material properties (ρ, c, Γ_{eff}) and loading rate effects (l_c decreases with increasing strain rate), predicts the non-linear relationship between dynamic strength and strain rate across rock types.

Experimental Techniques for Measuring Rock Dynamic Strength

Accurately measuring rock dynamic strength across a wide range of strain rates presents significant challenges due to the transient nature of high-rate loading and the complexities of rock behavior.

Rock dynamic strength measurement at different strain rates require the use of different experimental techniques. Rock dynamic strength measurement across varying strain rates requires specialized experimental techniques for different strain rate ranges. Conventional uniaxial or triaxial compression quasi-static tests can be applied for strain rates up to 10^{-3} s^{-1}; Servo-hydraulic testing machines and specialized loading frames can be applied for strain rates in the range of 10^{-3} to 10^0 s^{-1}; Split Hopkinson Pressure Bar (SHPB) testing can be applied for strain rates in the range of 10^1 to 10^3 s^{-1}; and, for strain rates exceeding 10^4 s^{-1}, a plate impact test using an air gun or dynamite is required.

The Split Hopkinson Pressure Bar (SHPB) technique deserves special attention due to its widespread use in rock dynamics research. In a typical SHPB setup, a cylindrical rock specimen is sandwiched between two long bars (incident and transmitted bars). A striker bar impacts the free end of the incident bar, generating a stress wave that propagates through the system. By analyzing the incident, reflected, and transmitted waves recorded by strain gauges on the bars, researchers can determine the stress–strain behavior of the rock specimen at high strain rates.

Dynamic stress ($\sigma(t)$), strain ($\varepsilon(t)$), and strain rate ($d\varepsilon/dt$) are calculated using the following equations [13]:

$$\sigma(t) = \frac{A_b}{A_s} E_b \varepsilon_i(t) \tag{2.76}$$

$$\varepsilon(t) = -\frac{2C_0}{L_s} \int_0^t \varepsilon_r(\tau)\, d\tau \tag{2.77}$$

$$\frac{d\varepsilon}{dt} = -\frac{2C_0}{L_s} \varepsilon_t(t) \tag{2.78}$$

where A_b and A_s are the cross-sectional areas of the bar and specimen, respectively, E_b is the Young's modulus of the bar material, C_0 is the wave speed in the bar, L_s is the specimen length, and $\varepsilon_i(t)$, $\varepsilon_r(t)$, and $\varepsilon_t(t)$ are the incident, reflected, and transmitted strain pulses, respectively. While it is very difficult to achieve stress equilibrium and uniform deformation in performing SHPB tests at high strain rates in brittle rock samples, these problems can be solved using pulse shaping techniques to modify the incident wave shape and end-friction reduction methods.

Modeling Approaches for Rock Dynamic Strength Versus Strain Rate

Accurately modeling the relationship between rock dynamic strength and strain rate is essential for predicting rock behavior in various engineering applications. Several modeling approaches have been developed to capture this complex relationship, ranging from empirical correlations to physically-based constitutive models.

Empirical models use power-law, logarithmic, and exponential functions fitted to experimental data. The Johnson–Cook model, adapted from metals, incorporates strain hardening, strain rate sensitivity, and thermal softening, while its variant, the Holmquist-Johnson–Cook (HJC) model, addresses brittle materials by including pressure-dependent strength and damage evolution.

2.3 Rock Dynamic Strength Theory

Advanced approaches include coupled elastoplastic damage models combining plasticity theory with damage mechanics, micromechanics-based models linking macro-behavior to microscale processes, and energy-based models. Statistical models address strength variability with strain rate.

A modified Johnson–Cook model to describe the rate-dependent strength of sandstone under biaxial compression was proposed:

$$\sigma = [A + B(\varepsilon_p)^n][1 + C\ln(\dot{\varepsilon}^*)][1 - (T^*)^m][1 + D(P^*)^q] \quad (2.79)$$

where σ is the stress, A, B, C, D, n, m, and q are material parameters, ε_p is plastic strain, $\dot{\varepsilon}^*$ is dimensionless strain rate, T_* is homologous temperature, and P_* is normalized pressure.

A constitutive model of statistical damage in the weathered granite that can accurately predict the rate-dependent behavior in SHPB tests under dynamic load, considering the effect of strain rate on the elastic modulus and damage evolution process, was developed:

$$\sigma = E(\dot{\varepsilon})(1 - D) \quad (2.80)$$

where σ is the stress, $E(\dot{\varepsilon})$ is the strain rate-dependent elastic modulus, and D is a Weibull-distributed damage variable.

An energy-based model for dynamic rock fragmentation was proposed:

$$s_{min} = \left[\frac{K\Gamma}{\rho\dot{\varepsilon}^2}\right]^{\frac{1}{3}} \quad (2.81)$$

where s_{min} is minimum fragment size, K is material constant, Γ is specific fracture energy, ρ is rock density, and $\dot{\varepsilon}$ is strain rate.

While these models have shown success in capturing various aspects of rock dynamic behavior, it is important to note that no single model can accurately predict the full range of rock responses across all loading conditions and rock types. The choice of modeling approach often depends on the specific application, available data, and computational resources.

2.4 Rock Fracture Mechanics Theory

2.4.1 Linear Elastic Fracture Mechanics

Linear elastic fracture mechanics (LEFM) forms the foundation of modern fracture mechanics and plays a crucial role in understanding the behavior of brittle materials, including rocks, under various loading conditions. This section delves into the principles and key concepts of LEFM in rock mechanics, providing a comprehensive overview of its theoretical underpinnings and practical implications.

Historical Development and Fundamental Concepts

The development of LEFM can be traced back to the pioneering work of A.A. Griffith in the early twentieth century [17]. Griffith's seminal paper, "The Phenomena of Rupture and Flow in Solids," published in 1920, laid the groundwork for understanding the discrepancy between the theoretical strength of materials and their observed strength in practice. Griffith proposed that this discrepancy could be attributed to the presence of microscopic flaws or cracks within materials, which act as stress concentrators and significantly reduce the material's resistance to fracture.

Griffith's theory was based on an energy balance approach, which posited that crack propagation occurs when the energy release rate due to crack extension exceeds the energy required to create new crack surfaces.

While Griffith's original work focused primarily on brittle materials like glass, subsequent researchers extended and refined his ideas to apply to a broader range of materials, including metals and rocks. Notable contributions came from Irwin, who introduced the concept of stress intensity factors in the 1950s, and Orowan, who modified Griffith's theory to account for plastic deformation at the crack tip [18].

The fundamental premise of LEFM is that the stress field near a crack tip in an elastic material can be characterized by a single parameter, known as the stress intensity factor (K). This parameter depends on the applied load, crack geometry, and specimen configuration. The stress intensity factor approach provides a convenient means of quantifying the severity of a crack and predicting its propensity for propagation.

Stress Intensity Factors and Fracture Modes

In LEFM, three basic modes of crack tip displacement are recognized [18]:

1. Mode I (Opening mode): The crack faces move apart in a direction normal to the crack plane.
2. Mode II (Sliding mode): The crack faces slide relative to each other in a direction perpendicular to the crack front.
3. Mode III (Tearing mode): The crack faces move relative to each other parallel to the crack front.

For each of these modes, a corresponding stress intensity factor (KI, KII, KIII) can be defined. The general form of the stress intensity factor is given by:

2.4 Rock Fracture Mechanics Theory

$$K = Y\sigma\sqrt{\pi a} \tag{2.82}$$

where Y is a dimensionless geometric factor that depends on the crack and specimen geometry, σ is the applied stress, and a is the crack length.

In rock mechanics, Mode I fracture is often the most critical, as it represents the opening of tensile cracks. The Mode I stress intensity factor (KI) is particularly important in predicting the onset of crack propagation. For a given material, there exists a critical stress intensity factor, denoted as KIC, which represents the material's resistance to crack propagation under Mode I loading. This property is also known as the fracture toughness of the material. For rocks, K_{IC} is typically determined through laboratory testing, such as the chevron notched short rod (CNSR) method or the cracked chevron notched Brazilian disc (CCNBD) method [18].

The fracture toughness of rocks can vary significantly depending on factors such as mineralogy, grain size, porosity, and confining pressure. For most rocks, K_{IC} typically ranges from 0.5 to 3.0 MPa$\sqrt{m}$. However, some very brittle rocks may have lower values, while some tougher rocks may have higher values.

In anisotropic rocks, such as those with bedding planes or foliation, the fracture toughness can vary with orientation. For example, in a recent study on Mode I fracture growth in anisotropic rocks [19], it was found that the fracture toughness varied significantly with orientation relative to the bedding planes in sedimentary rocks like the Mont Terri Opalinus Clay.

The concept of stress intensity factors allows for the characterization of the stress state near a crack tip. The stress field in the vicinity of a crack tip can be expressed in terms of the stress intensity factor as follows:

$$\sigma_{ij} = \frac{K}{\sqrt{2\pi r}} f_{ij}(\theta) + \text{higher order terms} \tag{2.83}$$

where σ_{ij} represents the stress components, r and θ are polar coordinates centered at the crack tip, and $f_{ij}(\theta)$ are known functions that depend on the mode of loading.

Energy Approach

While the stress intensity factor approach provides a convenient means of characterizing the stress state near a crack tip, the energy approach offers an alternative perspective on fracture mechanics. The energy approach, rooted in Griffith's original work, considers the energy balance of a cracked body during fracture propagation.

The energy release rate, G, is defined as the rate of change in potential energy with respect to crack area for a linear elastic material. It can be expressed as:

$$G = -\frac{d\Pi}{dA} \tag{2.84}$$

where Π is the potential energy of the system, and A is the crack area.

For linear elastic materials, there is a direct relationship between the energy release rate and the stress intensity factor:

$$G = \frac{K^2}{E} \quad \text{(for plane stress)} \tag{2.85}$$

$$G = \frac{K^2}{E}\left(1 - v^2\right) \quad \text{(for plane strain)} \tag{2.86}$$

where E is the Young's modulus, and v is Poisson's ratio.

Limitations and Extensions of LEFM for Rocks

LEFM has established itself as a powerful tool in rock mechanics; however, its application to rock materials presents several significant limitations and challenges that warrant careful consideration. A primary concern is the non-linear stress–strain behavior exhibited by rocks, particularly under high confining pressures or in the presence of discontinuities, which can result in the development of a substantial process zone near the crack tip, thereby violating LEFM's small-scale yielding assumption. Furthermore, the inherent heterogeneity and anisotropy of rocks, characterized by features such as bedding planes and foliation, can generate complex fracture patterns that deviate from LEFM predictions [19]. This complexity is compounded by notable size effects, where rock specimens demonstrate size-dependent fracture behavior not accounted for in classical LEFM, leading to disparities between laboratory-scale measurements and field-scale observations. Additionally, practical scenarios often involve complex stress states resulting in mixed-mode fracturing, and while LEFM can be extended to such conditions, accurate prediction of fracture propagation remains challenging [20]. Finally, rocks exhibit time-dependent deformation and fracture behavior, such as subcritical crack growth, which falls outside the scope of classical LEFM's framework.

To address the limitations of LEFM in rock materials, researchers have developed several significant extensions and modifications. Among these advances, Elastic–plastic fracture mechanics (EPFM) represents a fundamental extension of fracture mechanics concepts to materials exhibiting substantial plastic deformation before fracture. While EPFM was initially developed for metals, its principles have proven applicable to rocks under specific conditions.

A more specialized approach involves cohesive zone models, which introduce a process zone ahead of the crack tip where progressive damage occurs. This methodology enables more accurate predictions of fracture propagation by better capturing the non-linear behavior of rocks near the crack tip [21]. Complementing this approach, micromechanical models incorporate the effects of rock microstructure, such as grain boundaries and microcracks, into fracture mechanics predictions, thereby providing valuable insights into how rock heterogeneity influences fracture behavior [22].

Recognizing the inherent variability in rock properties, researchers have developed statistical methods to account for the probabilistic nature of rock fracture. These statistical approaches enhance engineering design by quantifying uncertainty in fracture predictions [23]. Furthermore, in applications where fluid pressure plays a crucial role, such as hydraulic fracturing, coupled hydro-mechanical models have

2.4 Rock Fracture Mechanics Theory

been developed to account for the complex interactions between fluid flow and rock deformation [24].

Numerical Modeling of Rock Fracture Using LEFM Principles

The application of LEFM principles in numerical modeling has greatly enhanced our ability to simulate and predict rock fracture behavior in complex geological settings. Various numerical methods have been developed to incorporate LEFM concepts into computational models, allowing for the analysis of fracture initiation, propagation, and interaction in rock masses. Some of the most commonly used numerical approaches include:

1. Finite Element Method (FEM): Traditional FEM approaches have been extended to incorporate LEFM principles through the use of special crack tip elements and remeshing techniques. While these methods can provide accurate solutions for simple crack geometries, they can become computationally expensive for complex fracture networks.
2. Extended Finite Element Method (XFEM): This method allows for the modeling of crack propagation without the need for remeshing by enriching the finite element approximation with discontinuous functions. XFEM has gained popularity in rock mechanics due to its ability to handle complex fracture geometries and mixed-mode loading conditions [21].
3. Boundary Element Method (BEM): BEM is particularly well-suited for modeling linear elastic fracture problems, as it only requires discretization of the boundary and crack surfaces. This can lead to significant computational efficiency, especially for problems involving multiple cracks or complex geometries.
4. Discrete Element Method (DEM): While not strictly based on LEFM principles, DEM has been widely used to model fracture propagation in rock masses by representing the material as an assembly of discrete particles or blocks. This approach can naturally capture the discontinuous nature of fractured rock masses and has been particularly useful in modeling dynamic fracture processes [22].
5. Hybrid methods: Various hybrid approaches have been developed to combine the strengths of different numerical methods. For example, coupled FEM-DEM models have been used to simulate fracture propagation in rocks under complex loading conditions [21].

2.4.2 Elasto-Plastic Fracture Mechanics

Elasto-plastic fracture mechanics (EPFM) is an extension of LEFM that accounts for the nonlinear behavior of materials under high stresses, particularly in the vicinity of crack tips. While LEFM provides a good approximation for brittle materials or situations where plastic deformation is limited, many rocks exhibit significant plastic deformation before failure, especially under high confining pressures or at elevated temperatures. EPFM addresses these limitations by incorporating plasticity and nonlinear material behavior into fracture analysis, making it particularly relevant for understanding rock fracture mechanics in complex geological settings.

The development of EPFM has been crucial in bridging the gap between the idealized brittle fracture models of LEFM and the complex, often ductile, behavior observed in many rock types under various loading conditions. This approach is especially important in rock dynamics, where understanding the initiation, propagation, and coalescence of cracks under dynamic loading is essential for predicting rock behavior in scenarios such as earthquakes, explosions, or rapid excavation processes.

Fundamental Concepts of Elasto-Plastic Fracture Mechanics

Elasto-plastic fracture mechanics builds upon the foundations of LEFM but extends the analysis to account for plastic deformation near the crack tip. In EPFM, the stress intensity factor (K) used in LEFM is no longer sufficient to characterize the stress state at the crack tip due to the presence of a plastic zone. Instead, EPFM introduces new parameters to describe the fracture behavior of materials exhibiting significant plastic deformation.

The two primary approaches in EPFM are: The J-integral approach, and the Crack Tip Opening Displacement (CTOD) approach.

Both methods aim to characterize the energy available for crack propagation in the presence of plastic deformation, providing a more accurate representation of fracture behavior in materials that deviate from purely elastic responses.

1. The J-integral Approach

The J-integral is a path-independent contour integral that characterizes the energy release rate associated with crack advance in a nonlinear elastic material. It is defined as:

$$J = \int_{\Gamma} (W\, dy - T_i \frac{\partial u_i}{\partial x}\, ds) \tag{2.87}$$

where W is the strain energy density, T_i are components of the traction vector, u_i are the displacement vector components, ds is an element of arc length along the contour Γ.

The J-integral represents the change in potential energy for a virtual crack extension and can be interpreted as a nonlinear energy release rate. In linear elastic materials, J is equivalent to G, the energy release rate used in LEFM. However, the J-integral's applicability extends to nonlinear elastic materials and can be used to characterize fracture in elasto-plastic materials under certain conditions.

The J-integral's path independence makes it a powerful tool for analyzing fracture in complex geometries and loading conditions. It provides a measure of the intensity of deformation at the crack tip that is applicable even when significant plastic deformation occurs. In rock mechanics, the J-integral approach has been particularly useful in analyzing fracture behavior in ductile rocks or under high confining pressures where plastic deformation becomes significant.

2.4 Rock Fracture Mechanics Theory

2. **The Crack Tip Opening Displacement (CTOD) Approach**

The Crack Tip Opening Displacement (CTOD) approach, developed by Wells [25], provides another method for characterizing fracture in materials exhibiting plastic deformation. CTOD is defined as the distance between the crack faces measured at the original crack tip location. As plastic deformation occurs, the initially sharp crack tip blunts, and the CTOD provides a measure of this blunting.

The CTOD can be related to the J-integral and the stress intensity factor K under small-scale yielding conditions:

$$\text{CTOD} = \frac{J}{m\sigma_y} \tag{2.88}$$

where σ_y is the yield stress, m is a dimensionless constant that depends on the stress state and material properties.

The CTOD approach is particularly useful in situations where the plastic zone size is significant compared to the crack length or specimen dimensions. It provides a physical measure of crack tip deformation that can be directly related to the material's fracture resistance.

In rock mechanics, the CTOD approach has been applied to analyze fracture behavior in rocks exhibiting significant plastic deformation, such as certain sedimentary rocks or rocks under high confining pressures. It provides insights into the ductile-to-brittle transition in rock fracture and helps in understanding the role of microstructural features in controlling fracture resistance.

Plastic Zone Analysis in Rock Fracture

One of the key aspects of EPFM is the analysis of the plastic zone that forms around a crack tip under load. In rocks, this plastic zone can take various forms depending on the material properties, confining pressure, and loading conditions. Understanding the size and shape of the plastic zone is crucial for accurately predicting fracture behavior and determining the applicability of different fracture mechanics approaches.

Irwin [25] proposed a simple model for estimating the size of the plastic zone under plane stress conditions:

$$r_p = \frac{1}{2\pi}\left(\frac{K}{\sigma_y}\right)^2 \tag{2.89}$$

where r_p is the plastic zone size, K is the stress intensity factor, σ_y is the yield stress.

This model assumes a circular plastic zone and provides a first-order approximation of the plastic zone size. However, in rocks, the plastic zone shape can be more complex due to anisotropy, heterogeneity, and pressure-dependent yield behavior.

The Dugdale model, also known as the strip-yield model, provides an alternative approach to estimating the plastic zone size. It assumes that all plastic deformation is concentrated in a narrow strip ahead of the crack tip. The plastic zone size in this model is given by:

$$r_p = \frac{\pi}{8}\left(\frac{K}{\sigma_y}\right)^2 \qquad (2.90)$$

The Dugdale model is particularly useful for analyzing fracture in materials with significant plasticity, such as some sedimentary rocks or rocks under high confining pressures.

In rocks, the yield behavior is often pressure-dependent, which affects the shape and size of the plastic zone. Models incorporating pressure-dependent yield criteria, such as the Mohr–Coulomb or Drucker-Prager criteria, provide more accurate representations of plastic zone development in rocks. These models typically result in asymmetric plastic zones, with the size and shape depending on the confining pressure and material properties.

For example, using the Mohr–Coulomb criterion, the plastic zone size can be estimated as:

$$r_p = \frac{K^2}{2\pi (c\cos\phi + \sigma_m \sin\phi)^2} \qquad (2.91)$$

where c is the cohesion, φ is the friction angle, σ_m is the mean stress.

This model captures the pressure-dependent nature of rock yielding and provides a more realistic representation of plastic zone development in rocks under various stress states.

Integration of Elasto-Plastic Fracture Mechanics with Other Rock Mechanics Approaches

The full potential of EPFM in rock mechanics is realized when it is integrated with other advanced approaches in rock mechanics and engineering. This integration allows for a more comprehensive understanding of rock behavior across various scales and loading conditions. Some key areas of integration include.

Integrating EPFM with micromechanical modeling approaches allows for a more detailed understanding of how microscale deformation and damage processes contribute to macroscale fracture behavior. This integration is particularly important for rocks, where the heterogeneous microstructure plays a crucial role in determining overall mechanical behavior.

Zhu and Tang [26] developed a numerical model that combines EPFM principles with micromechanical damage models to simulate the progressive failure of heterogeneous rocks. Their approach allows for the explicit consideration of microstructural features, such as grain boundaries and pre-existing microcracks, in predicting the development of macroscale fractures under complex loading conditions.

The integration of EPFM with continuum damage mechanics (CDM) provides a powerful framework for modeling the progressive degradation of rock properties

2.4 Rock Fracture Mechanics Theory

leading to fracture. This combined approach allows for the consideration of both localized fracture processes and distributed damage accumulation.

Shojaei et al. [27] proposed a coupled elastoplastic-damage model for simulating the behavior of porous rocks under various loading conditions. Their model incorporates EPFM concepts to describe localized fracture propagation while using CDM principles to account for distributed damage accumulation. This integrated approach provides a more comprehensive description of rock failure processes, particularly in situations involving complex loading histories or heterogeneous materials.

For saturated or partially saturated rocks, the integration of EPFM with poromechanics is crucial for understanding the coupled effects of fluid pressure and mechanical deformation on fracture behavior. This integration is particularly important for applications in reservoir engineering, CO_2 sequestration, and geothermal energy extraction.

Segura and Carol [28] developed a coupled hydro-mechanical model that incorporates EPFM principles to simulate fluid-driven fracture propagation in porous media. Their approach allows for the consideration of both fluid pressure effects and plastic deformation in predicting fracture growth and interaction with pre-existing discontinuities.

The integration of EPFM with thermomechanical models is essential for understanding fracture behavior in rocks subjected to significant temperature changes or gradients. This integration is particularly relevant for applications in geothermal energy, nuclear waste disposal, and fire-induced spalling in underground openings.

Chen et al. [29] proposed a thermo-mechanical model that incorporates EPFM concepts to simulate thermally induced fracturing in rocks. Their approach accounts for the coupled effects of thermal expansion, plastic deformation, and fracture propagation, providing insights into the complex failure mechanisms observed in rocks subjected to extreme temperature conditions.

Integrating EPFM with dynamic fracture mechanics principles allows for the analysis of rock fracture under high-rate loading conditions, such as those encountered during earthquakes, blasting, or impact events. This integration is crucial for understanding the time-dependent aspects of rock fracture and the influence of wave propagation on fracture processes.

Camacho and Ortiz [30] developed a computational framework for dynamic fracture that incorporates elasto-plastic effects and cohesive zone models. While their work focused primarily on brittle materials, the principles they developed have been extended to analyze dynamic fracture in rocks exhibiting significant plastic deformation.

Conclusion

Elasto-plastic fracture mechanics has emerged as a powerful tool for analyzing and predicting fracture behavior in rocks exhibiting significant plastic deformation. By extending the principles of linear elastic fracture mechanics to account for nonlinear material behavior, EPFM provides a more realistic framework for understanding rock fracture processes across a wide range of loading conditions and scales.

The integration of EPFM with experimental techniques, numerical methods, and other advanced approaches in rock mechanics has led to significant advancements in our understanding of rock behavior. From improving predictions of excavation stability to optimizing hydraulic fracturing treatments, EPFM has found numerous practical applications in rock engineering and geosciences.

However, challenges remain in fully capturing the complex behavior of rocks, particularly with respect to heterogeneity, anisotropy, and time-dependent effects. Future research directions in EPFM for rocks should focus on developing multiscale models that can efficiently bridge microscale deformation mechanisms with macroscale fracture behavior, incorporating environmental effects and dynamic loading conditions, and integrating EPFM principles with other advanced modeling approaches.

As our understanding of rock fracture mechanics continues to evolve, EPFM will undoubtedly play a crucial role in addressing the complex challenges faced in rock engineering and geosciences. By providing a more comprehensive and physically based approach to analyzing rock fracture, EPFM contributes to the development of safer, more efficient, and more sustainable solutions for a wide range of geological and engineering applications.

2.4.3 Rock Failure Criteria

Fracture mechanics theory plays a crucial role in understanding the behavior of rocks under various loading conditions, particularly in predicting crack initiation, propagation, and ultimate failure. This section delves into the failure criteria that form the foundation for analyzing and modeling rock fracture processes.

While the parameters discussed above provide valuable insights into the fracture process, practical rock engineering often requires simpler failure criteria that can be easily applied in design. Several failure criteria have been proposed for rocks, incorporating both fracture mechanics concepts and empirical observations. Some of the most widely used criteria are:

1. Mohr–Coulomb Criterion:
   The Mohr–Coulomb criterion is one of the oldest and most widely used failure criteria in rock mechanics. It relates the shear stress (τ) at failure to the normal stress (σ_n) acting on the failure plane:

   $$\tau = c + \sigma_n \tan(\phi) \tag{2.92}$$

   where c is the cohesion and φ is the angle of internal friction. While this criterion does not explicitly account for fracture mechanics parameters, it has been widely used due to its simplicity and the ease of obtaining the required parameters from standard laboratory tests.
2. Hoek–Brown Criterion:

2.4 Rock Fracture Mechanics Theory

The Hoek-Brown criterion is an empirical failure criterion that has gained widespread acceptance in rock engineering. The generalized form of the criterion is [23]:

$$\sigma_1 = \sigma_3 + \sigma_{ci}\left(m_b \frac{\sigma_3}{\sigma_{ci}} + s\right)^a \tag{2.93}$$

where σ_1 and σ_3 are the major and minor principal stresses at failure, σ_{ci} is the uniaxial compressive strength of the intact rock, and m_b, s, and a are material constants that depend on the geological strength index (GSI), disturbance factor (D), and intact rock parameter (m_i).

The Hoek–Brown criterion has the advantage of being applicable to both intact rock and rock masses, and it can be related to fracture mechanics parameters through empirical correlations.

3. Griffith Criterion:
   The Griffith criterion, based on the work of A.A. Griffith [17], relates the onset of fracture to the energy balance between the surface energy created by a new crack and the elastic strain energy released. For a brittle material under uniaxial tension, the Griffith criterion predicts that fracture will occur when:

$$\sigma_f = \sqrt{\frac{2E\gamma}{\pi a}} \tag{2.94}$$

   where σ_f is the fracture stress, E is Young's modulus, γ is the surface energy per unit area, and a is the length of the critical flaw.

   While the original Griffith criterion was developed for tensile fracture, it has been extended to compressive stress states and forms the basis for many fracture mechanics-based failure criteria in rock mechanics.

4. Modified Griffith Criterion:
   Recognizing that the original Griffith criterion often underestimates the compressive strength of rocks, McClintock and Walsh proposed a modified version that accounts for crack closure under compression:

$$(\sigma_1 - \sigma_3)^2 = 8T_0(\sigma_1 + \sigma_3) \tag{2.95}$$

   where T_0 is the uniaxial tensile strength of the rock.

5. Strain Energy Density Criterion:
   The strain energy density criterion, proposed by Sih, suggests that crack growth occurs in the direction of minimum strain energy density. The criterion states that fracture occurs when the strain energy density factor S reaches a critical value S_c:

$$S = a_{11}K_I^2 + 2a_{12}K_I K_{II} + a_{22}K_{II}^2 + a_{33}K_{III}^2 \tag{2.96}$$

where a_{ij} are functions of the elastic properties and the angle θ from the crack plane. This criterion has the advantage of being applicable to mixed-mode loading conditions.

6. Maximum Tangential Stress Criterion:
   Also known as the σ-criterion, this criterion proposes that a crack will propagate in the direction perpendicular to the maximum tangential stress at the crack tip. Fracture occurs when the maximum tangential stress reaches a critical value. For Mode I loading, this criterion predicts that fracture will occur when:

$$K_I = K_{IC} \tag{2.97}$$

   For mixed-mode loading, the criterion becomes more complex but can still be expressed in terms of the stress intensity factors.

7. Maximum Energy Release Rate Criterion:
   This criterion, also known as the G-criterion, states that a crack will propagate in the direction that maximizes the energy release rate. Fracture occurs when the maximum energy release rate reaches a critical value G_c. For pure Mode I loading, this criterion is equivalent to the K_{IC} criterion. For mixed-mode loading, it can be expressed as:

$$\frac{(K_I^2 + K_{II}^2)(1 - \nu^2)}{E} + \frac{K_{III}^2(1 + \nu)}{E} = G_c \tag{2.98}$$

8. Cohesive Zone Model:
   The cohesive zone model, introduced by Dugdale and Barenblatt, provides a way to model the fracture process zone ahead of a crack tip. In this model, the material separation is resisted by cohesive tractions that follow a prescribed traction-separation law. Failure occurs when the separation reaches a critical value.

   The cohesive zone model has been particularly useful in modeling quasi-brittle materials like rocks, where there is a significant process zone ahead of the crack tip. It allows for the incorporation of nonlinear material behavior and can capture phenomena like crack bridging and R-curve behavior.

9. Continuum Damage Mechanics Approach:
   Continuum damage mechanics provides a framework for modeling the progressive degradation of material properties due to microcracking and other damage mechanisms. In this approach, a damage variable D is introduced, which ranges from 0 (undamaged material) to 1 (completely damaged material). The constitutive equations are modified to account for the effect of damage:

$$\sigma = (1 - D)C : \epsilon \tag{2.99}$$

where C is the elastic stiffness tensor and ε is the strain tensor.

Failure criteria in continuum damage mechanics are often expressed in terms of a critical damage value D_c, at which macroscopic failure is assumed to occur.

10. Phase-field Models:
    Recent advancements in computational methods have led to the development of phase-field models for fracture. These models represent cracks as diffuse interfaces characterized by a scalar field variable. The evolution of this field variable is governed by the minimization of a total energy functional that includes both elastic and fracture energy terms.

A recent study by Wang et al. [20] proposed a phase-field model for mixed-mode fracture based on a unified tensile fracture criterion. This model can capture complex fracture patterns and transitions between different fracture modes, making it particularly useful for modeling fracture in heterogeneous and anisotropic rocks.

The phase-field approach has several advantages, including the ability to handle complex crack geometries and topological changes without explicit tracking of crack surfaces. However, it requires careful calibration of model parameters and can be computationally intensive for large-scale problems.

2.5 Rock Dynamic Damage Mechanics Theory

2.5.1 Fatigue Damage of Rocks Under Stress Wave Action

Stress waves generated by cyclic loading, such as those induced by earthquakes, blasting, or machinery vibrations, can cause significant fatigue damage in rock masses over time. This section examines the mechanisms and characteristics of fatigue damage in rocks subjected to repeated stress wave action, based on recent experimental studies and theoretical developments.

The phenomenon of stress wave propagation in rock materials presents a complex mechanism of progressive damage accumulation through cyclic loading and unloading processes. This dynamic process manifests primarily through the initiation, propagation, and eventual coalescence of microcracks within the rock mass. The evolution of such fatigue damage is governed by several interconnected factors that warrant careful consideration in rock mechanics analyses.

Foremost among these factors are the characteristics of the stress waves themselves. The amplitude, frequency, and duration of these waves play crucial roles in determining both the rate and extent of damage accumulation. Higher amplitude waves generally accelerate the damage process, while sustained exposure to even low-amplitude waves can induce substantial fatigue effects over extended periods [27, 28]. Furthermore, the intrinsic properties of the rock mass significantly influence its response to cyclic loading. Specifically, the mineral composition, porosity, and presence of pre-existing discontinuities determine the rock's susceptibility to fatigue damage, with higher porosity and pre-existing microcracks typically indicating greater vulnerability to such effects [29, 31].

The temporal aspects of loading history constitute another critical dimension in understanding fatigue damage evolution. The cumulative impact of multiple stress wave events, including both the sequence of loading cycles and the intervals between them, substantially influences the rate of damage accumulation [32, 33]. Additionally, environmental conditions play a significant role in modulating the fatigue damage process. Factors such as temperature variations, moisture content, and confining pressure can markedly affect how rocks respond to cyclic stress waves [34, 35].

Recent experimental investigations have significantly advanced the understanding of fatigue damage mechanisms in rocks subjected to stress wave loading. Through systematic observation and analysis, researchers have identified distinct patterns in the progression of fatigue damage across various rock types. The damage evolution typically manifests in three characteristic stages: an initial phase of rapid damage accumulation, followed by a period of steady-state damage growth, and culminating in an accelerated damage phase leading to ultimate failure. This consistent pattern has been documented across multiple rock types under cyclic loading conditions [36, 37].

Advanced high-resolution imaging techniques have enabled detailed visualization of microcrack network development during the fatigue process. These observations reveal a progressive evolution from isolated microcracks to increasingly complex, interconnected networks. The transformation of discrete microcracks into integrated systems ultimately culminates in macroscopic failure, providing crucial insights into the mechanical behavior of rocks under repeated loading [38, 39]. This microscale analysis has been complemented by studies of energy dissipation mechanisms during cyclic loading. Researchers have identified significant correlations between damage accumulation and changes in hysteresis loop characteristics, while concurrent acoustic emission patterns serve as valuable indicators of progressive fatigue damage [40, 41].

A particularly significant finding from these experimental studies concerns the identification of specific stress thresholds. Below these critical levels, fatigue damage accumulation becomes minimal or negligible. These thresholds, which vary depending on rock properties and loading conditions, have important implications for engineering design and practice [42, 43]. Such findings provide essential guidance for developing more reliable and efficient design parameters in rock engineering applications.

Significant theoretical advances have been made in modeling the complex phenomena of fatigue damage evolution in rocks under stress wave loading. These modeling approaches span multiple scales and methodologies, each offering unique insights into different aspects of the damage process. Continuum damage mechanics has emerged as a foundational framework, employing both scalar and tensor damage variables to characterize the progressive deterioration of rock properties under cyclic loading conditions [44, 45]. This approach provides a robust mathematical foundation for describing the macroscopic manifestation of damage accumulation.

2.5 Rock Dynamic Damage Mechanics Theory

Complementing these continuum approaches, micromechanics-based models have been developed to bridge the gap between microscale phenomena and macroscopic behavior. These models explicitly account for the heterogeneous nature of rock microstructures, often utilizing statistical methods to represent the distribution and evolution of microscale crack growth processes [46, 47]. The integration of microscale mechanics with macroscopic observations helps to predict damage evolution across multiple scales.

Energy-based methodologies have demonstrated particular effectiveness in modeling fatigue damage under complex loading scenarios. Models based on cumulative energy dissipation or strain energy density have proven successful in predicting rock failure patterns under varied loading histories [48, 49]. Furthermore, researchers have developed rate-dependent formulations to capture the critical influence of loading frequency on fatigue damage accumulation, providing more accurate representations of time-dependent damage processes [50, 51]. These advanced formulations have significantly improved the capability to predict rock behavior under dynamic loading conditions.

In conclusion, understanding the fatigue damage of rocks under stress wave action is crucial for accurate assessment of long-term rock mass behavior in various engineering contexts. Ongoing research continues to refine the knowledge of the underlying mechanisms, improve predictive models, and develop strategies to mitigate fatigue-related risks in rock engineering applications. Future work should focus on bridging the gap between laboratory-scale observations and field-scale predictions, incorporating the effects of complex loading histories and environmental factors, and developing advanced non-destructive techniques for monitoring fatigue damage progression in situ.

2.5.2 Damage Laws of Rocks Under Cyclic Impact

The damage laws of rocks under cyclic impact loading are complex and multifaceted, involving progressive deterioration of mechanical properties, accumulation of microcracks, and eventual macroscopic failure.

Progressive Damage Accumulation

Rocks subjected to cyclic impact loading exhibit a progressive accumulation of damage over multiple loading cycles. Each impact loading cycle initiates new microcracks and propagates existing ones, leading to a gradual increase in crack density within the rock matrix [33, 47]. As damage accumulates, rocks experience a reduction in elastic modulus, strength, and overall load-bearing capacity [32, 36]. The rate of damage accumulation is often nonlinear, with distinct phases of slow initial damage, steady accumulation, and accelerated damage leading to failure [33, 47]. Ren et al. [33] proposed a coupled damage variable that accounts for both compaction-induced and cracking-induced damages under cyclic loading. Their research showed that the damage variable initially decreases due to compaction effects before increasing as cracking dominates, following an inverted S-shaped curve.

Strain Rate and Loading Frequency Effects

The rate of loading and frequency of impacts significantly influence the damage process in rocks. Higher strain rates generally lead to increased dynamic strength and altered failure modes [34, 52], with Liu et al. [52] demonstrating that the dynamic strength of rocks increases linearly with the logarithm of strain rate. The frequency of cyclic loading affects the rate of damage accumulation and the number of cycles to failure, with higher frequencies often resulting in more rapid damage accumulation [36, 47, 53]. The number of cycles a rock can withstand before failure (fatigue life) is inversely related to the loading amplitude and frequency.

Energy Dissipation and Damage Correlation

Energy-based approaches have proven effective in characterizing rock damage under cyclic impact. The dissipated energy per loading cycle typically increases with accumulated damage [36], with Song et al. [54] establishing a linear relationship between dissipated energy and damage variables. Researchers have identified critical energy thresholds beyond which damage accumulation accelerates, leading to rapid failure [36]. Several studies have proposed damage models based on the ratio of dissipated energy to total input energy [33, 36]. The damage laws under cyclic impact are significantly affected by various factors. Rock type and composition play a crucial role, with different rock types exhibiting varying susceptibility to cyclic damage; for instance, layered rocks like sandstone may show anisotropic damage patterns [54]. Pre-existing damage or thermal treatment can significantly alter the damage accumulation process under cyclic loading [55]. Higher confining pressures generally increase rock resistance to cyclic damage by suppressing crack propagation [36]. Additionally, the shape of the loading pulse (e.g., sinusoidal, triangular) can influence the damage accumulation process [56].

Damage Localization and Failure Modes

As cyclic impact damage accumulates, rocks exhibit characteristic patterns of damage localization and failure. Damage tends to concentrate in specific regions, often leading to the formation of shear bands or tensile fracture zones [47, 52]. With increasing loading cycles or strain rates, failure modes may transition from tensile splitting to shear failure and eventually to complete pulverization [52, 57]. The fragmentation patterns resulting from cyclic impact often exhibit fractal characteristics, with the fractal dimension evolving as damage accumulates [33, 57].

Constitutive Modeling of Cyclic Impact Damage

To capture the complex damage laws under cyclic impact, researchers have developed various constitutive models. Statistical damage models often use Weibull distribution functions to describe the statistical nature of microcrack initiation and propagation [33]. Models that relate damage evolution to energy dissipation have shown good agreement with experimental results [36, 58]. For scenarios involving thermal effects, models that couple thermal and mechanical damage have been proposed [59]. Advanced micromechanical models incorporating grain-scale interactions and crack propagation mechanisms provide detailed insights into damage processes [57].

In conclusion, the damage laws of rocks under cyclic impact are governed by a complex interplay of loading conditions, rock properties, and energy dissipation mechanisms. Understanding these laws is crucial for predicting rock behavior in various engineering applications, from mining and tunneling to earthquake engineering and deep geological repositories. Ongoing research continues to refine our understanding of these phenomena, leading to more accurate predictive models and improved engineering designs.

2.5.3 Attenuation of Stress Waves in Rock Mass

The attenuation of stress waves as they propagate through rock masses is a critical phenomenon in rock dynamics, with significant implications for blast design, seismic hazard assessment, and underground excavation stability. This section examines the mechanisms and mathematical descriptions of stress wave attenuation in rock masses, focusing on the key factors that influence this process.

Mechanisms of Stress Wave Attenuation

Stress wave attenuation in rock masses is governed by multiple interconnected mechanisms, including geometric spreading, material damping, scattering and radiation damping, all of which work together to reduce wave energy during propagation. The primary mechanism involves geometric spreading, whereby stress waves propagating outward from their source experience amplitude reduction as their energy becomes distributed across progressively larger volumes. Besides, material damping arises from the internal friction and inelastic behavior of the rock material, leading to energy dissipation as the wave propagates. Furthermore, the presence of discontinuities, joints, and inhomogeneities within the rock mass leads to scattering, causing waves to undergo reflection, refraction, and diffraction, which further disperses their energy. Additionally, radiation damping occurs at material interfaces, where stress waves partially convert to other forms of energy, such as heat, resulting in further energy loss.

Mathematical Description of Attenuation

The attenuation of stress waves in rock masses is commonly described using an exponential decay function:

$$A(r) = A_0 e^{-\alpha r} \qquad (2.100)$$

where:
$A(r)$ is the amplitude of the stress wave at distance r from the source
A_0 is the initial amplitude at the source

α is the attenuation coefficient

r is the distance from the source

The attenuation coefficient α is a complex parameter that incorporates the effects of various attenuation mechanisms and is influenced by factors such as frequency, rock properties, and discontinuity characteristics.

Factors Influencing Attenuation

Stress wave attenuation in rock masses is influenced by a variety of factors. These include the frequency of the waves, with higher frequency waves generally attenuating more rapidly due to increased scattering and material damping. The rock's properties, such as its elastic modulus, density, and internal friction, also affect the rate of attenuation. In addition, the presence, orientation, spacing, and properties of discontinuities like joints and fractures contribute to attenuation by scattering and reflection. The in-situ stress state of the rock mass can influence wave propagation, particularly for shear waves, and the presence of water in rock pores and fractures further alters wave propagation and attenuation characteristics. All these factors interact to determine how stress waves dissipate their energy in the rock mass.

Experimental and Field Measurements

Measuring stress wave attenuation in rock masses mainly includes controlled source experiments, seismic surveys, and laboratory testing. Controlled source experiments use explosives or mechanical impacts to generate stress waves, with amplitude measurements taken at various distances to observe attenuation. Seismic surveys analyze the amplitude decay of seismic waves originating from natural or artificial sources. Additionally, laboratory testing measures attenuation in rock samples under controlled conditions, allowing for the determination of material-specific attenuation properties. These approaches collectively help quantify how stress waves dissipate energy within rock masses.

Implications for Rock Engineering

Understanding stress wave attenuation in rock masses is crucial for rock engineering. It plays a key role in blast design, where optimizing charge weights and timing helps achieve the desired fragmentation while minimizing damage to surrounding rock. In seismic hazard assessment, understanding wave attenuation is vital for evaluating the potential impact of earthquakes on underground structures and surface facilities. It also contributes to excavation stability, helping to assess the dynamic loading on underground openings from blasting or seismic events. Furthermore, it is important in geophysical exploration, where the interpretation of seismic data aids in resource exploration and geological characterization.

Advanced Modeling Approaches

Recent advances in numerical modeling have led to more sophisticated approaches for simulating stress wave attenuation in rock masses. Discrete element methods (DEM) allow for explicit representation of discontinuities and their effect on wave propagation. Finite element methods (FEM), combined with advanced constitutive models, incorporate complex material behavior and damage evolution to better capture attenuation effects. Hybrid continuum-discontinuum models combine the strengths of different modeling approaches, representing both intact rock and discontinuities.

In conclusion, the attenuation of stress waves in rock masses is a complex phenomenon influenced by multiple factors. Understanding and accurately modeling this process is essential for various applications in rock engineering and geophysics. Ongoing research continues to refine our understanding of attenuation mechanisms and improve our ability to predict and account for their effects in practical applications [27–29, 31, 32].

2.6 Conclusion

This chapter presents an in-depth examination of the fundamental theories and applications in rock dynamics, revealing significant findings in several key areas. Rock dynamics theory, founded on Newton's second law, provides the theoretical basis for understanding and predicting rock mass behavior under dynamic loading conditions. The governing equations of elastic waves, ranging from basic wave equations in isotropic, homogeneous media to complex considerations of anisotropy, heterogeneity, and coupled processes, establish a robust framework for analyzing wave propagation in rock dynamics.

The fundamental theory of stress wave propagation, derived from studies of coaxial collision of elastic bars, offers essential insights into wave generation, propagation, and reflection. Principles of superposition, interference, scattering, reflection, transmission, and mode conversion collectively govern stress wave behavior in rocks. Research on the relationship between rock dynamic strength and strain rate has revealed significant implications for various practical applications, from mining blast design to seismic hazard assessment in earthquake engineering.

The integration of Linear Elastic Fracture Mechanics (LEFM) and Elasto-Plastic Fracture Mechanics (EPFM) provides a comprehensive theoretical framework for analyzing and predicting rock fracture behavior. EPFM, in particular, offers a more practical approach to understanding rock fracture processes under various loading conditions and scales by considering nonlinear material behavior and plastic deformation. However, challenges remain in fully capturing complex rock behavior, especially regarding heterogeneity, anisotropy, and time-dependent effects.

Despite these advances, several challenges remain, including the characterization of rock material heterogeneity and anisotropy, time-dependent effects, and multi-scale problems. Future research should focus on: developing multi-scale models that effectively bridge microscale deformation mechanisms with macroscale fracture behavior; incorporating environmental effects and dynamic loading conditions; advancing non-destructive in-situ monitoring techniques; and improving predictive capabilities from laboratory to field scale. Through these efforts, we can provide safer, more efficient, and more sustainable solutions for geological engineering, resource development, and environmental protection.

References

1. Hynd R. Newton's second law with a semiconvex potential. Newton I. Philosophiae naturalis principia mathematica. G. Brookman; 1833. Part Differ Equ Appl. 2022;3(1):11.
2. Terada K, ItoT, Kikuchi N. Characterization of the mechanical behaviors of solid-fluid mixture by the homogenization method. Comput Methods Appl Mech Eng. 1998;153(3-4):223–257.
3. Pourciau B. Newton's interpretation of Newton's Second Law. Arch Hist Exact Sci. 2006;60(2):157–207.
4. Fan LF, Wang M, Du XL. The Method of Characteristics for Stress Wave Propagation in the Rock Mass. Springer, 2025.
5. Wang XM, Zhou YQ. Research on elastodynamic theory based on the famework of energy conservation. Acta Phys Sin. 2023;72(07):262–72 [in Chinese].
6. Zhou YQ, Zhang XM, Liu L, et al. Formulations of the elastodynamic equations in anisotropic and multiphasic porous media from the principle of energy conservation. Prog Theor Exp Phys. 2022; 2022(12):123A01.
7. Rubino JG, Caspari E, Müller TM, et al. Numerical upscaling in 2-D heterogeneous poroelastic rocks: Anisotropic attenuation and dispersion of seismic waves. J Geophys Res Solid Earth. 2016;121(9):6698–6721.
8. He YX, Wang SX, Sun C, et al. Analysis of the frequency dependence characteristics of wave attenuation and velocity dispersion using a poroelastic model with mesoscopic and microscopic heterogeneities. Geophy Prospect. 2021;69(6):1260–81.
9. Hu JY, Sheng DF, Qin FF, et al. Progressive failure characteristics and damage constitutive model of fissured rocks under water–rock coupling. Theor Appl Fract Mec. 2025;135:104765.
10. Qi CZ, Yan FY, Zhao F, et al. On the nature of energy-horizon and determination of length scales in dynamic fragmentation of rocks. Int J Impact Eng. 2022; 166:104242.
11. Zhao J. Applicability of Mohr-Coulomb and Hoek-Brown strength criteria to the dynamic strength of brittle rock. Int J Rock Mech Min Sci. 2000;37(7):1115–21.
12. Liao HJ, Wu JY, Huang FQ. The dynamic strength parameters of geotechnical materials are solved by the uniform strength theory. J Rock Mech Eng. 2003;12:1994–2000 [in Chinese].
13. Zhang QB, Zhao J. A review of dynamic experimental techniques and mechanical behaviour of rock materials. Rock Mech Rock Eng. 2014;47:1411–78.
14. Xie H, Zhu J, Zhou T. Novel three-dimensional rock dynamic tests using the true triaxial electromagnetic Hopkinson bar system. Rock Mech Rock Eng. 2021;54:2079–86.
15. Zhou CT, Xie HP, Zhu JB. Dynamic destruction criterion based on energy theory. J Rock Mech Eng. 2023;42(08):1890–8 [in Chinese].
16. Hu X, Zhang M, Zhang X. A coupled elastoplastic damage dynamic model for rock. Shock Vib. 2021;2021(1):5567019.
17. Griffith AA. The phenomena of rupture and flow in solids. Philos Trans R Soc Lond Ser A Contain Pap Math Phys Char. 1921; 221(582–593):163–98.

References

18. Atkinson BK. Introduction to fracture mechanics and its geophysical applications. Fract Mech Rock. 1987; 1–26.
19. Nejati M, Aminzadeh A, Amann F, et al. Mode I fracture growth in anisotropic rocks: theory and experiment. Int J Solids Struct. 2020;195:74–90.
20. Wang Q, Feng YT, Zhou W, et al. A phase-field model for mixed-mode fracture based on a unified tensile fracture criterion. Comput Methods Appl Mech Eng. 2020;370: 113270.
21. Mohammadnejad M, Liu H, Chan A, et al. An overview on advances in computational fracture mechanics of rock. Geosyst Eng. 2021;24(4):206–29.
22. Cho N, Martin CD, Sego DC. A clumped particle model for rock. Int J Rock Mech Min Sci. 2007;44(7):997–1010.
23. Hoek E, Brown ET. Practical estimates of rock mass strength. Int J Rock Mech Min Sci. 1997;34(8):1165–86.
24. Shen B, Stephansson O, Rinne M. Modelling rock fracturing processes: a fracture mechanics approach using FRACOD. Springer Science & Business Media; 2013.
25. Irwin GR. Analysis of stresses and strains near end of a crack traversing a plate. J Appl Mech. 1956;24(24):361–4.
26. Zhu W, Tang C. Micromechanical model for simulating the fracture process of rock. Rock Mech Rock Eng. 2004;37(1):25–56.
27. Sun G, Wang J, Tang S, et al. A dynamic damage constitutive model and three-dimensional internal crack propagation of rock-like materials in bond-based peridynamic. Theoret Appl Fract Mech. 2024;130: 104320.
28. Li HY, Shi GY. A dynamic material model for rock materials under conditions of high confining pressures and high strain rates. Int J Impact Eng. 2016;89:38–48.
29. Liu WY, Zhu QZ, Zhang J, et al. A micromechanical fatigue damage model for quasi-brittle rocks subjected to triaxial compressive cyclic loads. Comput Geotech. 2023;163: 105747.
30. Camacho GT, Ortiz M. Computational modelling of impact damage in brittle materials. Int J Solids Struct. 1996;33(20–22):2899–938.
31. Lu D, Wang G, Du X, et al. A nonlinear dynamic uniaxial strength criterion that considers the ultimate dynamic strength of concrete. Int J Impact Eng. 2017;103:124–37.
32. Liu XR, Kou MM, Lu YM, et al. An experimental investigation on the shear mechanism of fatigue damage in rock joints under pre-peak cyclic loading condition. Int J Fatigue. 2018;106:175–84.
33. Ren C, Yu J, Liu X, et al. Cyclic constitutive equations of rock with coupled damage induced by compaction and cracking. Int J Min Sci Technol. 2022;32(5):1153–65.
34. Liu XS, Ning JG, Tan YL, et al. Damage constitutive model based on energy dissipation for intact rock subjected to cyclic loading. Int J Rock Mech Min Sci. 2016;85:27–32.
35. Gao R, Wang H, Cao R, et al. Damage deterioration mechanism and damage constitutive modelling of marble after cyclic impact loading. J Market Res. 2024;29:1293–304.
36. Yu X, Tan Y, Song W, et al. Damage evolution of rock-encased-backfill structure under stepwise cyclic triaxial loading. J Rock Mech Geotech Eng. 2024;16(2):597–615.
37. Ahmadi MH, Molladavoodi H. Rock failure analysis under dynamic loading based on a micromechanical damage model. Civil Eng J. 2018;4(11):2801–12.
38. Zheng Q, Liu E, Sun P, et al. Dynamic and damage properties of artificial jointed rock samples subjected to cyclic triaxial loading at various frequencies. Int J Rock Mech Min Sci. 2020;128: 104243.
39. Liu K, Wang T. Dynamic mechanical behaviours of frozen rock under sub-zero temperatures and dynamic loads. Int J Rock Mech Min Sci. 2024;180: 105813.
40. Lu J, Yin T, Guo W, et al. Dynamic response and constitutive model of damaged sandstone after triaxial impact. Eng Fail Anal. 2024;163: 108450.
41. Wang Y, Tang P, Han J, et al. Energy-driven fracture and instability of deeply buried rock under triaxial alternative fatigue loads and multistage unloading conditions: prior fatigue damage effect. Int J Fatigue. 2023;168: 107410.
42. Song H, Zhang H, Fu D, et al. Experimental analysis and characterization of damage evolution in rock under cyclic loading. Int J Rock Mech Min Sci. 2016;88:157–64.

43. Yao W, Yu J, Liu X, et al. Experimental and theoretical investigation of coupled damage of rock under combined disturbance. Int J Rock Mech Min Sci. 2023;164: 105355.
44. Chen Z, Zhan H, Zhou Z, et al. Experimental investigation on progressive damage and failure mechanism of deep hard-rock tunnels subjected to cyclic loading. Theor Appl Fract Mech 2024; 104501.
45. Zhang L, Wang E, Liu Y, et al. Experimental research into the dynamic damage characteristics and failure behavior of rock subjected to incremental repeated impact loads. Eng Geol. 2024;331: 107435.
46. Niu Z, Zhu Z, Que X, et al. Experimental study on damage evolution process, seepage characteristics, and energy response of columnar jointed basalt under true triaxial cyclic loading. Geoenergy Sci Eng. 2024;237: 212774.
47. Ran Q, Chen P, Liang Y, et al. Hardening-damage evolutionary mechanism of sandstone under multi-level cyclic loading. Eng Fract Mech. 2024;307: 110291.
48. Guo Y, Chen X, Wang Z, et al. Identification of mixed mode damage types on rock-concrete interface under cyclic loading. Int J Fatigue. 2023;166: 107273.
49. Xie LX, Yang SQ, Gu JC, et al. JHR constitutive model for rock under dynamic loads. Comput Geotech. 2019;108:161–72.
50. Grigoriev AS, Shilko EV, Skripnyak VA, et al. Kinetic approach to the development of computational dynamic models for brittle solids. Int J Impact Eng. 2019;123:14–25.
51. Zhang Y, Huang M, Jiang Y, et al. Mechanics, damage and energy degradation of rock-concrete interfaces exposed to high temperature during cyclic shear. Constr Build Mater. 2023;405: 133229.
52. Liu H, Zhu W, Yu Y, Xu T. Dynamic mechanical properties of rock under the influence of strain rate: experimental and theoretical analysis. Rock Mech Rock Eng. 2018;51(9):2791–804.
53. Zheng Z, Li RH, Li SJ, et al. A novel dynamic fractional mechanical model for rock fracture under true triaxial static-dynamic combined loading and its engineering application. Rock Mech Rock Eng. 2024;57:9343–69.
54. Song ZP, Cheng Y, Yang TT. Experimental study on fatigue damage evolution mechanism of hard layered sandstone under cyclic loading. Chinese J Geotech Eng. 2024;46(03):490–9.
55. Feng G, Zhang J, Guo J, et al. Numerical study on the shear damage behavior of sandstone under normal disturbance effects. KSCE J Civ Eng. 2024;28(7):2761–77.
56. Yu Y, Yang Y, Liu J, et al. Experimental and constitutive model study on the mechanical properties of a structural plane of a rock mass under dynamic disturbance. Sci Rep. 2022;12(1):21238.
57. Luo H, Gong H, Luo Y, et al. The macroscopic and microscopic fatigue failure mechanisms of high-temperature thermally-damaged granite under cyclic impact loading. Geothermics. 2024;121: 103047.
58. Miao S, Liu Z, Zhao X, et al. Plastic and damage energy dissipation characteristics and damage evolution of Beishan granite under triaxial cyclic loading. Int J Rock Mech Min Sci. 2024;174: 105644.
59. Feng C, Wang Z, Wang J, et al. A thermo-mechanical damage constitutive model for deep rock considering brittleness-ductility transition characteristics. J Central South Univ. 2024;31(7):2379–92.

Open Access This chapter is licensed under the terms of the Creative Commons Attribution-NonCommercial-NoDerivatives 4.0 International License (http://creativecommons.org/licenses/by-nc-nd/4.0/), which permits any noncommercial use, sharing, distribution and reproduction in any medium or format, as long as you give appropriate credit to the original author(s) and the source, provide a link to the Creative Commons license and indicate if you modified the licensed material. You do not have permission under this license to share adapted material derived from this chapter or parts of it.

The images or other third party material in this chapter are included in the chapter's Creative Commons license, unless indicated otherwise in a credit line to the material. If material is not included in the chapter's Creative Commons license and your intended use is not permitted by statutory regulation or exceeds the permitted use, you will need to obtain permission directly from the copyright holder.

Chapter 3
Rock Dynamics Test Device and Test Technique

3.1 Principle and Classification of the Test Device

3.1.1 Background of Test Device Design

The design and development of rock dynamics test devices are driven by the need to understand and characterize the complex behavior of rock materials under dynamic loading conditions. These devices play a crucial role in various fields, including mining engineering, civil engineering, geotechnical engineering, and earthquake engineering. The background of test device design encompasses several key aspects, including the historical development of rock dynamics testing, the specific challenges posed by dynamic loading, and the evolving requirements of modern engineering applications.

Historical Development

The study of rock dynamics has its roots in the early twentieth century, with the initial focus primarily on understanding the behavior of rocks under static loading conditions. However, as engineering projects became more ambitious and complex, particularly in the realms of mining and underground construction, the need to understand rock behavior under dynamic loads became increasingly apparent [1].

The first significant developments in rock dynamics testing occurred in the mid-twentieth century, driven largely by the demands of the mining industry and the need to understand rock bursts and other dynamic failure phenomena. Early test devices were often adaptations of static testing equipment, with modifications to allow for higher loading rates. These early attempts, while groundbreaking, were limited in their ability to accurately simulate the complex stress states and loading conditions encountered in real-world scenarios [2].

As the field progressed, researchers began to develop more specialized equipment designed specifically for dynamic testing. The Split Hopkinson Pressure Bar (SHPB) test, originally developed for metal testing, was adapted for rock mechanics in the 1960s and became one of the most widely used methods for studying the dynamic properties of rocks. This marked a significant step forward in the ability to characterize rock behavior under high strain rates [3].

In the latter part of the twentieth century and into the twenty-first century, advancements in technology, particularly in sensors, data acquisition systems, and control mechanisms, have led to the development of increasingly sophisticated test devices. These modern devices are capable of applying complex loading patterns, simulating a wide range of in-situ conditions, and capturing detailed data on rock response across multiple physical domains [4].

Specific Design Considerations

The choice of loading mechanism depends on the specific requirements of the test, including the desired strain rate, force capacity, and degree of control needed [1]. The rock testing apparatus design requires a sophisticated balance between ensuring representative specimen dimensions and maintaining practical constraints for uniform stress states, while incorporating precisely calibrated boundary conditions to simulate loading scenarios and guarantee reliable results [3–5]. The system integrates strategically positioned high-speed sensors and advanced data acquisition systems, coupled with sophisticated processing algorithms, to capture and analyze dynamic testing responses with exceptional accuracy [2, 4]. Essential safety features are robustly engineered to safeguard both personnel and equipment from high-energy dynamic loading and potential specimen failures, while the apparatus's modular and adaptable architecture accommodates diverse testing configurations for optimal cost-effectiveness [6]. The design is further enhanced by comprehensive environmental control systems that maintain precise temperature, pressure, and fluid conditions throughout the testing process, ensuring experimental integrity and reproducibility [2] (Fig. 3.1).

Emerging Trends in Test Device Design

Several emerging trends are shaping the future of rock dynamics test device design: Modern test devices are being developed at micro and nano scales to investigate fundamental rock deformation and failure mechanisms, while incorporating advanced X-ray CT imaging technologies for real-time 3D visualization of internal deformation processes [2, 4]. These devices are integrating AI and machine learning capabilities for sophisticated testing protocols and enhanced data interpretation [7], evolving to include remote operation and autonomous capabilities for extended experimental runs [8], and being engineered to accurately simulate and measure complex multi-physics interactions within rock masses [2]. furthermore, ongoing development is pushing into ultra-high-speed testing to study phenomena such as rock fragmentation under explosive loading [1], while advancing in-situ field applications to measure rock behavior under authentic environmental conditions that overcome conventional laboratory testing limitations; additionally [7], with growing environmental awareness, there is a trend towards designing more energy-efficient devices that utilize environmentally friendly materials and processes [9].

3.1 Principle and Classification of the Test Device 65

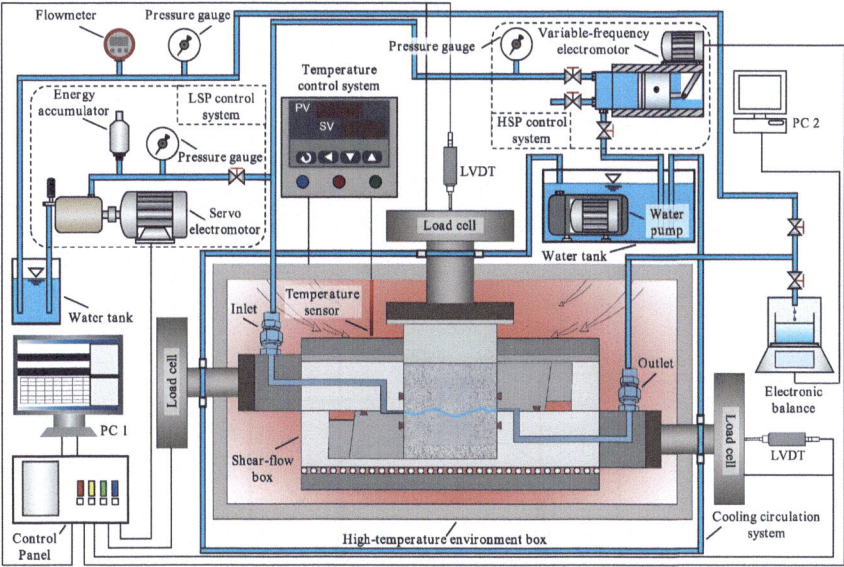

Fig. 3.1 The design is further optimized by advanced environmental control systems that precisely regulate temperature, pressure, and fluid conditions during the testing process. This figure shows the structure diagram of the THM coupling test system [2]

Case Studies in Test Device Design

To further illustrate the background and evolution of rock dynamics test device design, it is instructive to examine several case studies of innovative test devices developed in recent years:

1. Large-Size High-Temperature True Triaxial Hydraulic Fracturing Apparatus

Tan et al. [6] engineered an innovative apparatus to investigate hydraulic fracturing processes in hot dry rock (HDR) geothermal reservoirs. The device effectively simulates the extreme temperature and pressure conditions characteristic of deep geothermal environments while enabling true triaxial loading and hydraulic fracturing experiments. Through this apparatus, researchers examined intricate fracture morphologies that emerge under elevated temperatures and analyzed how various parameters, including injection rates and stress conditions, influence fracture propagation. The experimental results provide valuable insights for optimizing hydraulic fracturing techniques in HDR geothermal energy extraction systems (Fig. 3.2).

2. Thermo-Hydro-Mechanical Coupling Test System for Rock Fractures

Wu et al. [2] developed an advanced experimental system to investigate the thermo-hydro-mechanical (THM) behavior of rock fractures. The apparatus enables comprehensive analysis of the coupled interactions between thermal, hydraulic, and mechanical processes in fractured rock masses—critical knowledge for geothermal energy

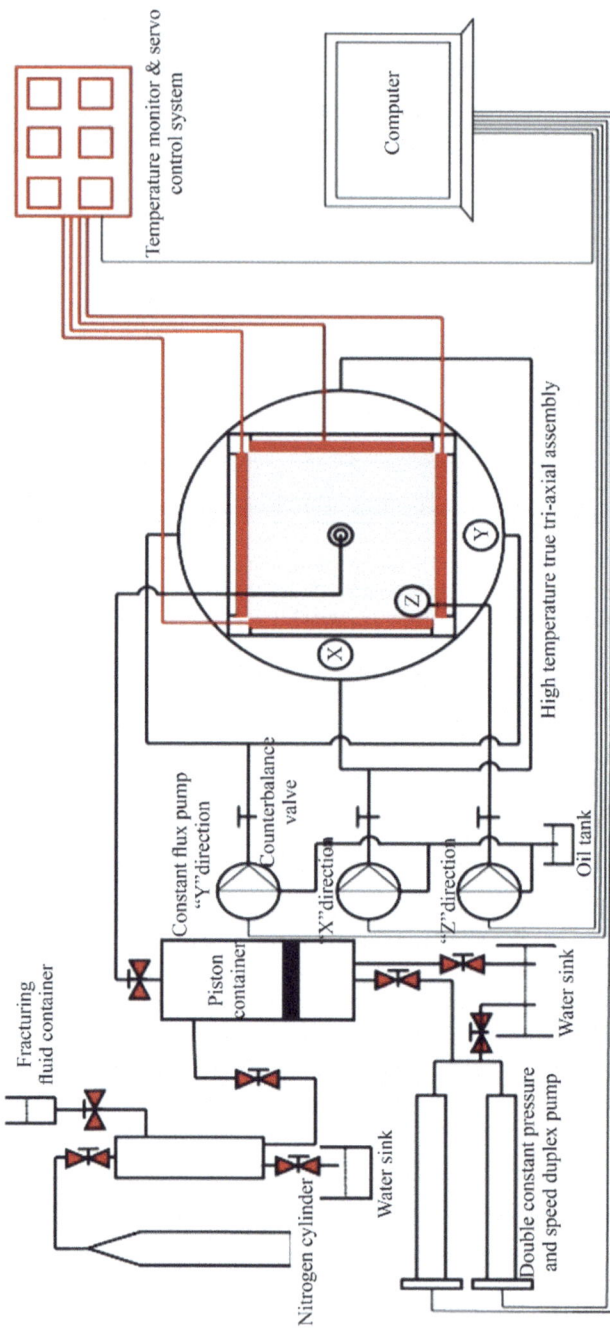

Fig. 3.2 The high-temperature true triaxial apparatus [6]

3.1 Principle and Classification of the Test Device

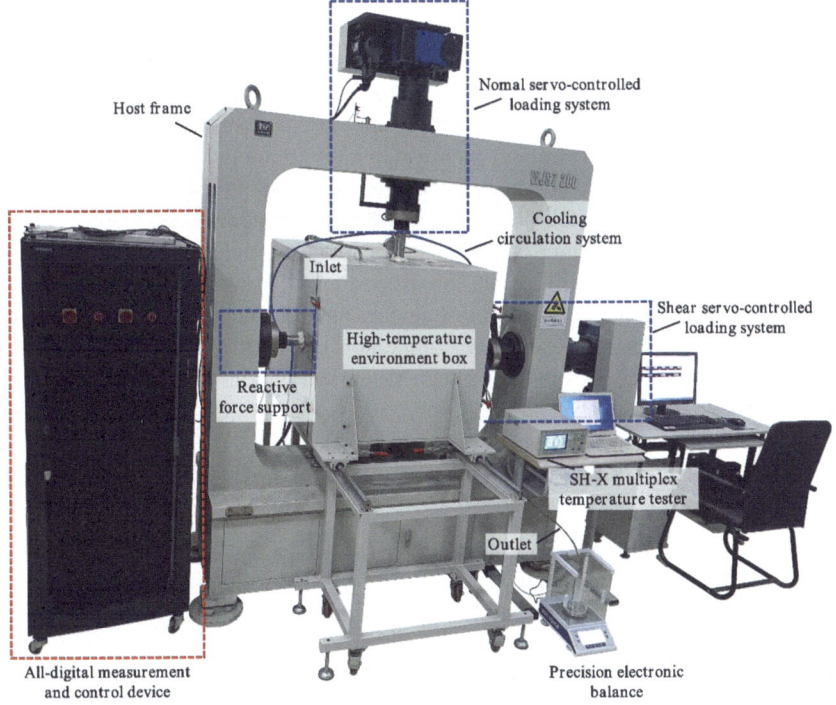

Fig. 3.3 The THM coupling test system for a single rock fracture [2]

extraction and nuclear waste disposal applications. The system facilitates systematic examination of the interrelationships among temperature gradients, fluid pressure variations, and mechanical loading conditions that govern fracture behavior. Through extensive experimentation, the researchers demonstrated the system's analytical capabilities by quantifying complex correlations between thermal expansion, fracture closure mechanisms, and hydraulic conductivity parameters (Fig. 3.3).

3. Large-Scale Three-Dimensional Apparatus for Rockfall Simulation

Xin et al. [9] engineered a large-scale experimental apparatus to investigate rockfall mechanisms in underground engineering applications. The system enables simulation of blocky rock mass behavior at scales approximating actual underground excavations, addressing a critical gap in experimental capabilities. Through this apparatus, researchers systematically analyzed the complete rockfall failure sequence—from initial stability conditions through progressive failure mechanisms—under controlled laboratory conditions. The integrated multi-physics monitoring system facilitated the identification of precursor instability indicators, including variations in natural frequency response and acoustic emission signatures.

4. In-Situ Triaxial Test Method for Rock Masses

Tetsuji et al. [7] established an innovative in-situ triaxial testing methodology for rock masses that overcomes fundamental limitations of conventional laboratory and field-testing approaches. The method enables direct measurement of stress–strain relationships in undisturbed rock specimens within their native geological environment. This novel approach facilitates comprehensive characterization of rock mass deformation and strength properties at multiple depths while maintaining the specimen's original stress state, thereby providing critical data for understanding rock mass behavior in situ. These cases studies illustrate the diverse and innovative approaches being taken in the design of rock dynamics test devices and related decision-making tools. They highlight the trend towards more comprehensive, multi-physics approaches to testing, the importance of simulating realistic in-situ conditions, and the growing role of advanced data analysis and decision-making techniques in rock mechanics research and engineering practice.

3.1.2 Types of Rock Dynamics Testing Devices

Rock dynamics testing devices play a crucial role in understanding the behavior of rocks under dynamic loading conditions. These devices are designed to simulate various dynamic loading scenarios and measure the response of rock specimens to such loads. The development of these testing apparatuses has significantly advanced our understanding of rock mechanics in dynamic environments, which is essential for numerous applications in geotechnical engineering, mining, and civil engineering. In this section, we will discuss the various types of rock dynamics testing devices, their principles of operation, and their specific applications in rock mechanical research.

Split Hopkinson Pressure Bar (SHPB)

SHPB, also known as the Kolsky bar, is one of the most widely used devices for testing the dynamic properties of materials, including rocks [10]. The SHPB has been extensively developed and modified since its inception to accommodate various testing conditions and material types.

1. Basic Principle

The SHPB typically consists of three main components: a striker bar, an incident bar, and a transmitted bar. The basic principle of operation involves generating a stress wave through the impact of the striker bar on the incident bar. This stress wave propagates through the incident bar, interacts with the specimen placed between the incident and transmitted bars, and then continues through the transmitted bar. By analyzing the stress waves in the incident and transmitted bars, researchers can determine the dynamic stress–strain behavior of the specimen.

The governing equations for the SHPB test are based on one-dimensional elastic wave propagation theory. The key equations for calculating stress, strain, and strain rate in the specimen are as follows [10]:

3.1 Principle and Classification of the Test Device

Dynamic stress:

$$\sigma(t) = \frac{A_b}{A_s} E_b \varepsilon_t(t) \tag{3.1}$$

Dynamic strain:

$$\varepsilon(t) = \frac{2C_0}{L_s} \int_0^t [\varepsilon_i(t) - \varepsilon_r(t) - \varepsilon_t(t)] dt \tag{3.2}$$

Strain rate:

$$\dot{\varepsilon}(t) = \frac{2C_0}{L_s} [\varepsilon_i(t) - \varepsilon_r(t) - \varepsilon_t(t)] \tag{3.3}$$

where: A_b and A_s are the cross-sectional areas of the bar and specimen, respectively. E_b is the elastic modulus of the bar material. C_0 is the wave speed in the bar. L_s is the length of the specimen. $\varepsilon_i(t)$, $\varepsilon_r(t)$, and $\varepsilon_t(t)$ are the incident, reflected, and transmitted strain signals, respectively.

2. Modifications and Improvements

Over the years, several modifications have been made to the traditional SHPB to enhance its capabilities and address specific testing requirements for rock mechanics. The introduction of pulse shaping techniques has significantly improved the quality of stress wave propagation in SHPB tests. Pulse shapers, typically made of soft materials like copper or rubber, are placed at the impact end of the incident bar. They help to modify the incident wave shape, resulting in a more gradual rise time and better stress equilibrium in the specimen. This improvement is particularly important for brittle materials like rocks, where achieving stress equilibrium can be challenging. The use of conical-shaped impactors instead of traditional cylindrical strikers has been shown to improve the waveform generated in SHPB tests. This modification helps in achieving a more stable and repeatable loading condition, which is crucial for accurate measurements of rock dynamic properties. The choice of bar material has been expanded to accommodate different testing needs. While traditional SHPB systems often use steel bars, materials such as aluminum alloys, magnesium alloys, and even polymeric materials have been employed for specific testing requirements. The selection of bar material is important for impedance matching with the specimen, which affects the transmission of stress waves and the overall test accuracy. Modified SHPB systems have been developed to conduct tests at elevated temperatures, allowing researchers to investigate the dynamic behavior of rocks under conditions similar to those found in deep geological formations or in proximity to geothermal activities. Some SHPB systems have been adapted to include triaxial confinement capabilities, enabling the study of rock behavior under more complex stress states that better represent in-situ conditions [10] (Figs. 3.4 and 3.5).

3. Applications in Rock Dynamics

The SHPB is widely utilized in rock dynamics research for various applications. Its primary use in rock mechanics is to measure the dynamic compressive strength and stress–strain behavior of rocks under high strain rates, providing essential insights into rock behavior during impact or blast loading conditions. Modified SHPB setups, such as the Brazilian disc test configuration, facilitate the measurement of the dynamic tensile strength of rocks, which is critical for understanding rock fragmentation processes. Additionally, specially designed SHPB configurations enable the analysis of dynamic shear properties of rocks and rock joints, offering valuable information on the behavior of rock masses under dynamic loading. SHPB systems have also been adapted to assess the dynamic fracture toughness of rocks, shedding light on crack propagation under high loading rates. Furthermore, high-temperature SHPB tests allow for the exploration of the combined effects of dynamic loading and elevated temperatures on rock behavior, which is particularly relevant to deep mining and geothermal energy applications [3]. Lastly, SHPB tests on water- or gas-saturated rock specimens provide critical data on the influence of pore fluids on dynamic rock properties, advancing our understanding of reservoir rock behavior [10].

True Triaxial Electromagnetic Hopkinson Bar (TEHB) System

TEHB system represents a significant advancement in rock dynamics testing capabilities. This innovative system allows for three-dimensional dynamic testing of rocks under coupled triaxial dynamic impacts and triaxial in-situ stresses, providing a more comprehensive understanding of rock behavior under complex loading conditions [11].

1. System Components and Testing Principles

The system consists of a six-orthogonal triaxial frame constructed from titanium alloy bars, providing structural support for applying multidirectional loads to rock specimens. It is capable of applying confining pressures of up to 300 MPa along each axis, enabling the simulation of diverse in-situ stress conditions. Six electromagnetic pulse generators produce controllable and adjustable stress pulses, ensuring synchronous, repeatable, and precise generation of multiple identical stress pulses. A sophisticated data acquisition system collects and processes test data with high accuracy. The TEHB system operates on principles similar to the SHPB but incorporates significant enhancements. Symmetrical loading is applied to both sides of the specimen along each axis, allowing stress equilibrium to be achieved more rapidly than in conventional SHPB systems. Additionally, the TEHB system achieves strain rates of up to approximately 10^3 s^{-1}, significantly expanding the range of dynamic loading conditions that can be studied [11].

2. Significance and Applications

The TEHB system represents a significant advancement in rock dynamics testing capabilities. Its ability to simulate complex three-dimensional (3D) loading conditions offers valuable insights into rock behavior under scenarios that closely resemble

3.1 Principle and Classification of the Test Device

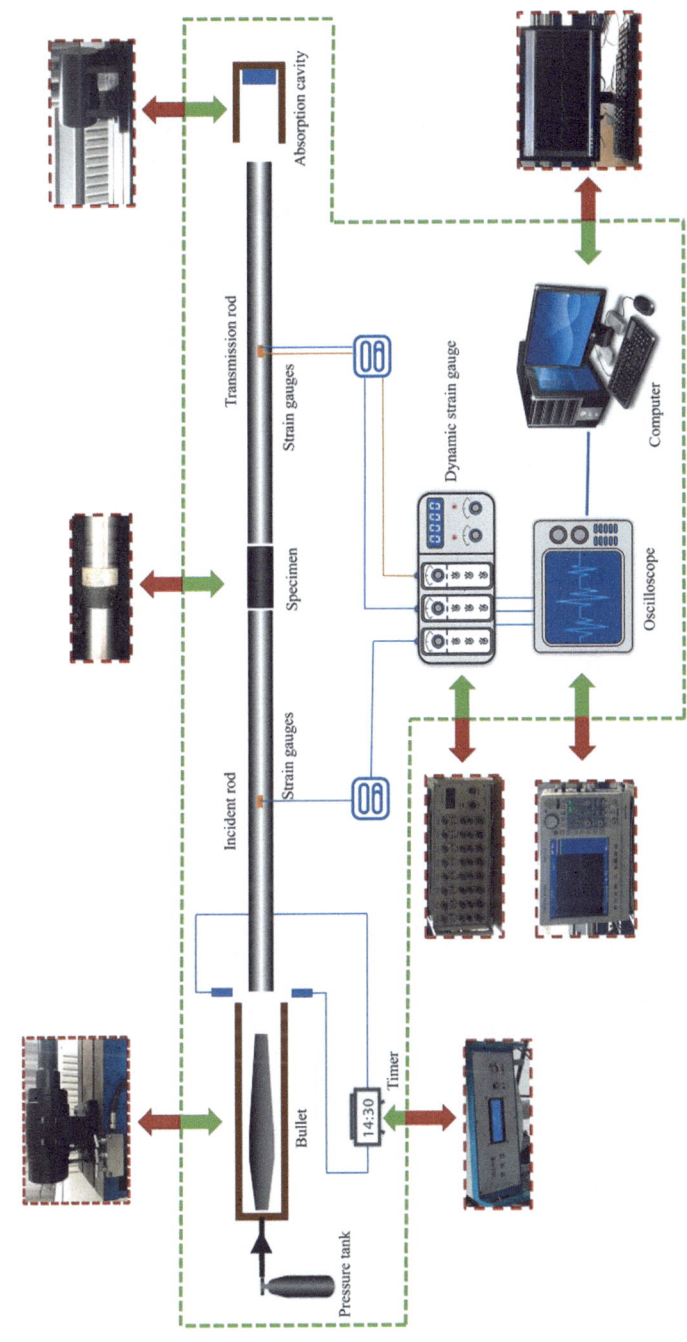

Fig. 3.4 The main components of SHPB test apparatus [10]

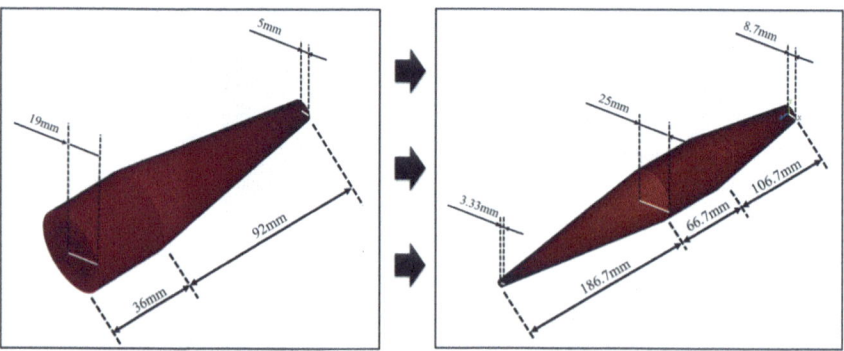

Fig. 3.5 The conical impact projectile of SHPB test apparatus before and after modification [10]

real-world conditions. This system enhances understanding of rock mass behavior during earthquakes, providing critical insights into fault mechanics and seismic hazard assessment. By replicating complex stress states and thermal conditions, the TEHB system contributes to improved understanding and management of geothermal and hydrocarbon reservoirs. The TEHB system fosters the development of more robust theories in three-dimensional rock dynamics, advancing the field of rock mechanics as a whole [11].

Dynamic Shear-Slip Testing Apparatus

The development of a new apparatus for testing shear-slip properties of rock joints subjected to dynamic disturbance represents another significant advancement in rock dynamics testing capabilities. This apparatus is designed to simulate and study the behavior of rock joints under various loading conditions, including both static and dynamic loads [12].

1. Apparatus Design and Capabilities

The apparatus can apply both constant normal loading (CNL) and constant normal stiffness (CNS) boundary conditions, allowing for the simulation of various in-situ stress states [4]. An electromagnetic-driven disturbance generator is used to produce low-frequency dynamic loads (0–15 Hz), simulating various types of dynamic disturbances such as earthquakes or blasting. The apparatus can apply initial shear stress before dynamic disturbance, enabling the study of rock joints in different initial stress states. The system allows for both quasi-static direct shear tests and dynamic shear tests, providing a comprehensive understanding of rock joint behavior under various loading conditions. The apparatus can accommodate specimens up to 300 mm × 300 mm × 50 mm, allowing for testing of larger, more representative rock joint samples. The system can apply a maximum normal pressure of 33 MPa and shear displacement rates from 0.01 to 10 mm/s [12].

3.1 Principle and Classification of the Test Device

2. Key Equations and Parameters

Several key equations and parameters are used in the analysis of rock joint behavior using this apparatus:

Fractal Dimension Calculation:

$$D = \frac{\log_{10} 4}{\log_{10}\left[2\left(1 + \cos\tan^{-1}\left(\frac{2h}{l}\right)\right)\right]} \qquad (3.4)$$

where: D is the fractal dimension. h is the average height of joint asperity. l is the average base length of joint asperity.

Joint Roughness Coefficient (JRC) Calculation:

$$\text{JRC} = 85.2671(D-1)^{0.5679} \qquad (3.5)$$

Slip Triggering Condition (without initial shear stress):

$$F_{sd}(t) > \mu_s F_n \qquad (3.6)$$

Slip Triggering Condition (with initial shear stress):

$$F_{s0} + F_{sd}(t) > \mu_s F_n \qquad (3.7)$$

where: $F_{sd}(t)$ is the disturbance load. μ_s is the static friction coefficient. F_n is the normal load. F_{s0} is the initial shear load.

3. Significance and Applications

The dynamic shear-slip testing apparatus has significant applications in geotechnical engineering, offering critical insights into the behavior of rock joints under both static and dynamic loading conditions. This apparatus is essential for slope stability analysis in seismically active regions, underground excavation design in areas prone to dynamic disturbances, and the development of rock support systems tailored for dynamic loading. It also plays a crucial role in assessing the potential for slip reactivation in faulted rock masses and evaluating rock mass stability in mining operations subject to blast-induced vibrations. By providing a deeper understanding of the complex mechanical behavior of rock joints, the apparatus contributes to the development of more accurate models of rock mass behavior and supports the design of safer and more reliable rock engineering projects.

Comparison of Testing Devices

While each of the rock dynamics testing devices discussed above has its unique capabilities and applications, it is important to consider how they complement each other and contribute to a comprehensive understanding of rock behavior under dynamic

Table 3.1 Comparison of testing devices

	Loading conditions	Strain rates	Specimen size	Measurement capabilities
SHPB	Primarily uniaxial loading, with modifications for triaxial confinement	Typically 10^1~10^4 s^{-1}	Generally smaller specimens due to the need for stress equilibrium	Primarily measures dynamic stress–strain behavior
TEHB	True triaxial loading capabilities, allowing for complex 3D stress states	Up to approximately 10^3 s^{-1}	Can accommodate larger specimens due to its 3D loading capabilities	Measures 3D stress–strain behavior and can incorporate temperature effects
Dynamic shear-slip testing apparatus	Focuses on shear behavior of rock joints under both static and dynamic loads	Lower strain rates, focusing on low-frequency dynamic loads (0–15 Hz)	Can test larger specimens (up to 300 mm × 300 mm × 50 mm)	Focuses on shear strength and slip behavior of rock joints

loading conditions. This section will compare the different devices and discuss how their integration can lead to a more holistic approach to rock dynamics research (Table 3.1).

The various rock dynamics testing devices discussed in this section represent significant advancements in our ability to study and understand rock behavior under dynamic loading conditions. Each device offers unique capabilities and insights, and their integration provides a powerful toolkit for comprehensive rock dynamics research. As these technologies continue to evolve and new techniques emerge, our understanding of rock dynamics will undoubtedly deepen, leading to improved predictions, designs, and safety measures in various rock engineering applications.

3.1.3 Innovations in Device Development

The field of rock dynamics has seen significant advancements in recent years, particularly in the development of sophisticated test devices and techniques. These innovations have enabled researchers to better simulate and study the complex behavior of rocks under various dynamic loading conditions. This section focuses on the key innovations in device development that have revolutionized rock dynamics testing.

True Triaxial Testing Systems

One of the most significant advancements in rock dynamics testing is the development of true triaxial testing systems, which enable the application of independent stresses in three orthogonal directions to replicate the complex stress states experienced by rocks in deep underground environments [13–15]. The true triaxial test system developed by Feng et al. [14] represents a notable breakthrough in this field. This apparatus integrates a high-temperature module, allowing researchers to study the thermal–mechanical properties of hard rocks under conditions characteristic of deep underground environments. The system is composed of four main components: mechanical loading in three axes, thermal loading up to 250 °C, deformation and temperature measurement, and a data acquisition center. Moreover, Xia et al. [16] developed a novel true triaxial shear apparatus capable of testing rocks under high temperature and high stress conditions. This apparatus allows for true triaxial stress states with independent loading in three directions, high-temperature testing up to 300 °C, and both constant normal load (CNL) and constant normal stiffness (CNS) boundary conditions. It supports both short-term stress–strain tests and long-term creep tests, making it versatile for a range of applications. By enabling the investigation of rock behavior under coupled high-temperature and high-pressure conditions, The system provides critical insights into the stability of deep underground engineering projects (Figs. 3.6 and 3.7).

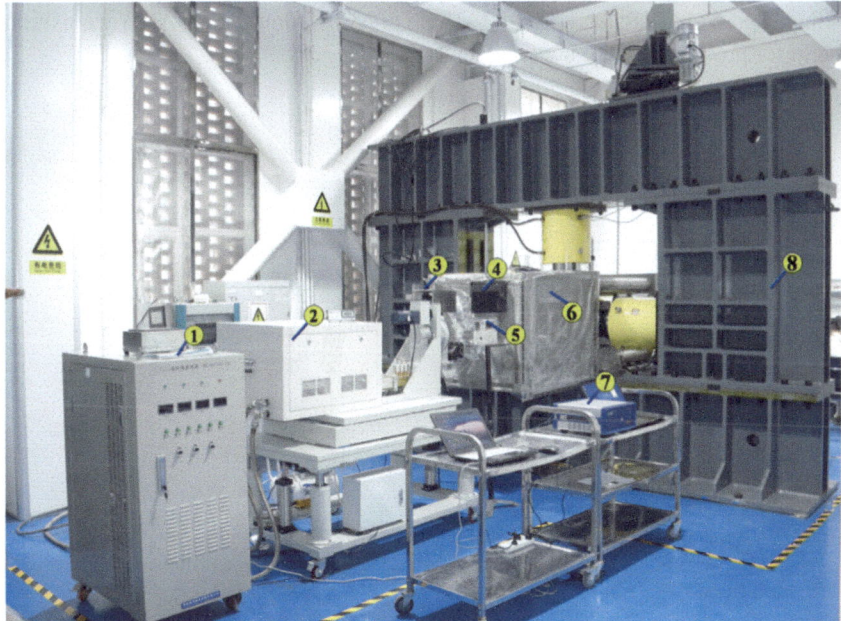

Fig. 3.6 True triaxial system for microwave-induced fracturing of hard rocks [15]

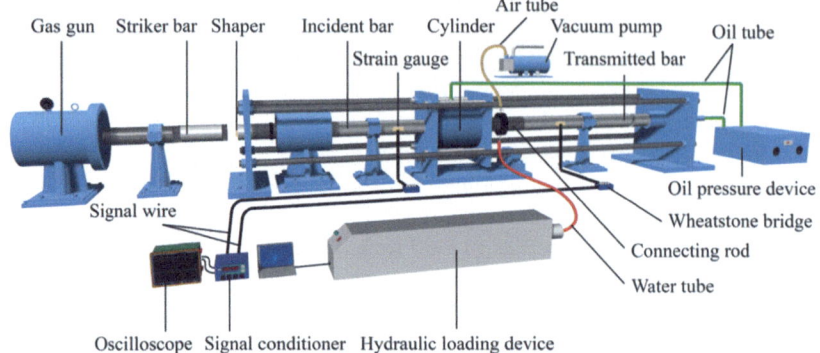

Fig. 3.7 The novel SHPB experimental system with pore pressure loading cell [16]

Dynamic and Static Combined Loading Systems

A notable advancement in rock dynamics testing is the development of systems capable of simultaneously applying dynamic and static loads, effectively replicating real-world conditions. Luo et al. [17] designed a modified triaxial SHPB system that includes a pore pressure loading cell, enabling the study of the dynamic compressive behavior of porous rocks under coupled hydraulic-mechanical loading. The system integrates three essential components: a Split Hopkinson Pressure Bar for dynamic loading, a pore pressure loading cell for fluid pressure control, and a data acquisition system for precise measurement and analysis (Fig. 3.8).

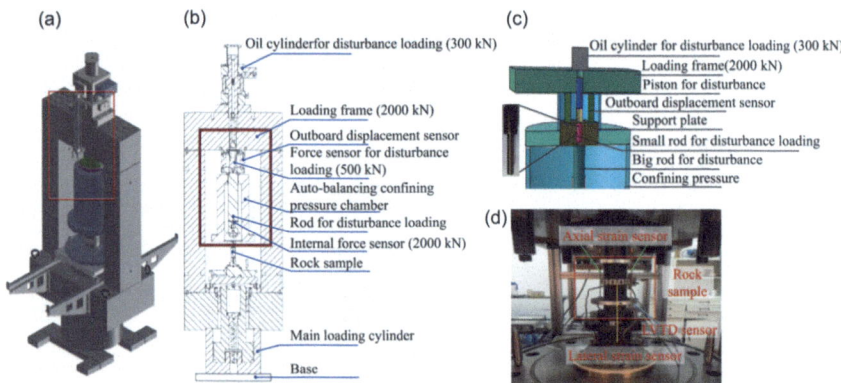

Fig. 3.8 Dynamic disturbance servo triaxial loading device and system. **a** model of rock triaxial instrument; **b** schematic diagram of the instrument; **c** loading device and disturbance rods; **d** rock sample installation [17]

3.1 Principle and Classification of the Test Device

Novel Monitoring and Analysis Techniques

Advancements in monitoring and analysis techniques have greatly enhanced our ability to observe and understand rock behavior during testing.

Zhang et al. [18] developed a novel Digital Image Correlation (DIC) based methodology for crack identification in jointed rock masses. This method allows for quantitative identification of crack types and occurrence times using full-field deformation data. Key innovations include:

The use of a Rate of Effective Variance Change (*REVC*) measure to quantify strain dispersion:

$$REVC = \frac{\partial V_e|_{d,k}}{\partial \varepsilon} \approx \frac{V_e|_{d,k+1} - V_e|_{d,k-1}}{2\Delta \varepsilon} \quad (3.8)$$

where V_{eld} is the effective variance of strain components, ε is the axial strain, and k is the time step.

The introduction of a Rate of Mutation (*RM*) to characterize how fast the *REVC* increases:

$$RM = \frac{REVC_K}{REVC_{K-1}} \quad (3.9)$$

where K is the time step associated with the occurrence of a new type of crack.

This method allows for automatic identification of crack types, captures full-field deformation characteristics, and can identify both crack mechanisms and occurrence times [18] (Fig. 3.9).

Cryogenic Testing Systems

Innovations in cryogenic testing systems have significantly advanced the study of rock behavior under extremely low temperatures, which is essential for certain underground engineering applications. Cha et al. [13] developed a laboratory system specifically designed for investigating cryogenic thermal rock fracturing. These tools include temperature and pressure measurements, acoustic wave transmission, pressure decay tests, and X-ray CT scanning. This system enables controlled cryogenic fracturing experiments under simulated downhole conditions, providing high-quality data that enhance the understanding of cryogenic fracturing mechanisms and their potential applications in underground engineering [13].

Novel Stress Path Testing Methods

Innovations in stress path testing methods have significantly enhanced the ability to simulate complex in-situ stress conditions in laboratory environments. Ma et al. [19]

Fig. 3.9 Schematics of the reference and current subsets [18]

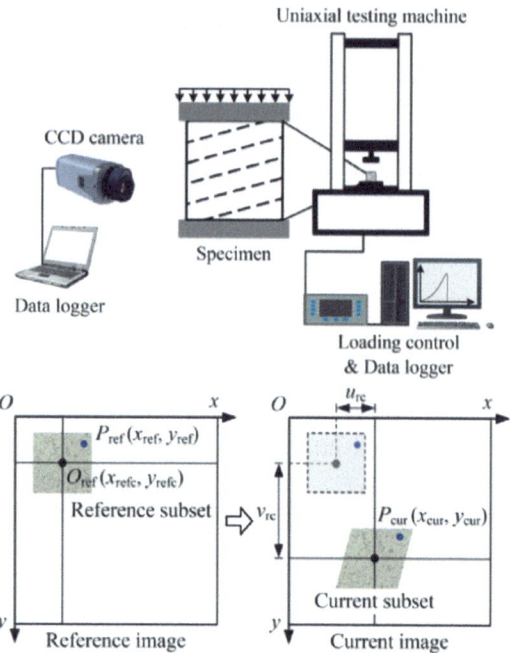

introduced a novel stress path method for true-triaxial testing using a pseudo-triaxial apparatus (PTA).

Verification of a novel stress path method by true-triaxial test

This method leverages the control of the PTA's loading ratio to replicate stress paths equivalent to true-triaxial conditions in generalized stress space (p-q space). The key control equation governing the PTA loading is:

$$d\sigma_3 = \left(\frac{b+1}{b-2}\right)d\sigma_1 \tag{3.10}$$

where σ_1 and σ_3 are the major and minor principal stresses, and b is the intermediate principal stress coefficient.

This innovative approach extends the functionality of standard pseudo-triaxial equipment to perform certain true-triaxial testing operations, offering cost-effective solutions and simplified procedures for complex stress path simulations [19].

In conclusion, these innovations in device development have significantly advanced our ability to study rock dynamics under various conditions. True triaxial testing systems, dynamic and static combined loading systems, novel monitoring and analysis techniques, cryogenic testing systems, and novel stress path testing methods have all contributed to a more comprehensive understanding of rock behavior under complex loading conditions. These advancements are crucial forz addressing challenges in deep underground engineering, waste disposal, and other applications requiring a thorough understanding of rock dynamics [13–17, 19–22].

3.2 Impact Test Technique

3.2.1 High-Strain Rate Loading Methods

In the field of rock dynamics, understanding the behavior of rocks under high-strain rate loading conditions is crucial for various applications, including mining, tunneling, and earthquake engineering. This section focuses on the methods used to apply high-strain rate loads to rock specimens, with particular emphasis on impact test techniques. These methods allow researchers to simulate and study the dynamic response of rocks under conditions similar to those experienced during blasting, rock bursts, and other rapid loading events.

Split Hopkinson Pressure Bar (SHPB) Technique

SHPB technique, also known as the Kolsky bar method, is one of the most widely utilized experimental approaches for investigating the dynamic mechanical properties of materials, including rocks, under high-strain-rate loading conditions. This method is particularly well-suited for studying material behavior at strain rates typically ranging from $10^2 - 10^4$ s^{-1}, making it invaluable for applications where materials experience extreme dynamic loading. By accurately measuring stress–strain relationships under these conditions, the SHPB technique provides critical insights into the deformation and failure mechanisms of rock materials, contributing to a deeper understanding of their behavior under dynamic and impact loading scenarios.

While the standard SHPB setup is primarily designed for compression testing, modifications have been developed to investigate rock behavior under tensile and shear loading conditions. The Brazilian Disc SHPB utilizes a disc-shaped specimen loaded diametrically to generate indirect tensile stress [23]. The dynamic tensile strength (σ_t) is calculated using the formula:

$$\sigma_t = \frac{2P}{\pi D t} \tag{3.11}$$

where P is the applied load, D is the specimen diameter, and t is the specimen thickness (Fig. 3.10).

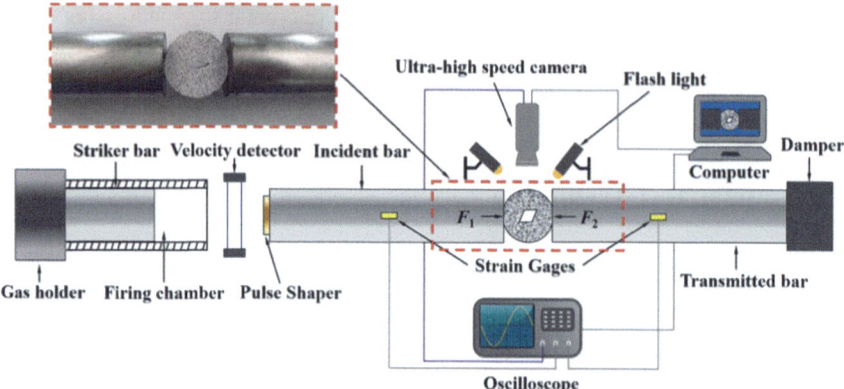

Fig. 3.10 Experimental setup containing an SHPB system, a stain recording system and an ultrahigh-speed camera [24]

3.2.2 Split-Hopkinson Pressure Bar (SHPB) Test

SHPB is a typical testing apparatus for rock dynamics, comprising several essential components. First, the system employs a loading mechanism, typically a gas gun, to accelerate a striker bar that impacts an incident bar, generating stress waves. These stress waves then propagate through a long, slender incident bar to the specimen, and subsequently through the specimen to a transmitted bar. To prevent multiple loading events, a momentum trap is installed at the end of the transmitted bar to absorb residual energy. Meanwhile, strain gauges mounted on both the incident and transmitted bars measure stress wave signals, which are recorded using high-speed oscilloscopes or data acquisition cards. Additionally, the specimen is properly aligned and supported between the incident and transmitted bars using a specialized specimen holder.

Applications in Rock Dynamics Research

The SHPB testing technique has been widely adopted in rock dynamics research, particularly in investigating rock behavior under high-strain-rate loading conditions. One of its primary applications lies in studying the dynamic strength and deformation characteristics of rocks. Extensive research has demonstrated its unique advantages in investigating the effects of strain rate on both compressive and tensile strength of rocks, as well as their deformation behavior under dynamic loading [25].

For instance, research has revealed that the dynamic compressive strength of rocks typically increases with strain rate, often following a power-law relationship. This strain-rate sensitivity underscores the influence of loading rate on rock mechanical properties, making SHPB testing an invaluable tool for understanding rock behavior in scenarios such as blasting, seismic events, and high-speed impacts. Moreover, this

3.2 Impact Test Technique

Fig. 3.11 Rock sample and testing equipment diagram: **a** red sandstone specimens; **b** acoustic instrument; and **c** drying machine [26]

technique provides deep insights into the mechanisms underlying dynamic deformation, including crack initiation, propagation, and energy dissipation processes, thereby offering a comprehensive understanding of rock response under extreme conditions.

1. Effect of Freeze–Thaw Cycles on Dynamic Properties.

In this field, Yu et al. [26] conducted an in-depth investigation into the influence of freeze–thaw cycles on the dynamic mechanical properties of red sandstone using SHPB tests. Their findings revealed that as the number of freeze–thaw cycles increased, both the dynamic compressive strength and elastic modulus of the rock decreased, while the cumulative strain at failure increased. Furthermore, freeze–thaw damage significantly affected the rock's energy dissipation characteristics, specifically manifesting as a decrease in energy utilization efficiency with increasing freeze–thaw cycles (Fig. 3.11).

2. Dynamic Response of Jointed Rock:

Xu et al. [27] used SHPB tests to investigate the dynamic response of granite specimens containing a single pre-existing joint. The presence of the joint significantly influenced wave propagation and energy dissipation in the rock. The dynamic strength of jointed specimens decreased compared to intact rock, with the extent of reduction depending on the joint orientation relative to the loading direction. The study also revealed that the fracture patterns in jointed specimens were strongly influenced by the joint orientation and loading rate (Fig. 3.12).

These case studies demonstrate the versatility and importance of SHPB testing in rock dynamics research, providing valuable insights into various aspects of dynamic rock behavior under different loading conditions and material states.

Fig. 3.12 Rock-like layered samples with different dip angles: **a** 0°, **b** 90°, **c** 22.5°, **d** 45°, **e** 67.5°, **f** top view, **g** front view [27]

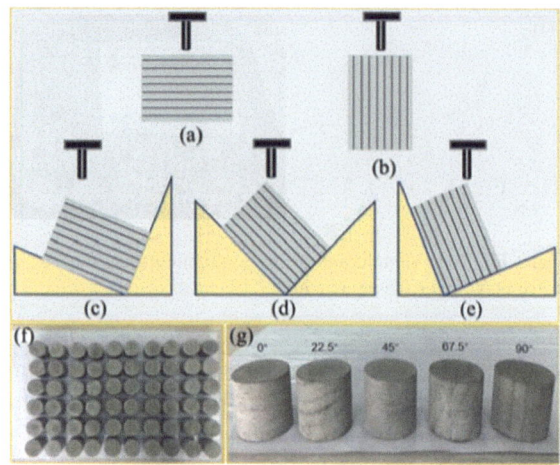

3.2.3 Data Acquisition and Analysis in Impact Tests

In the field of rock dynamics, understanding rock materials' behavior under impact loading conditions relies heavily on accurate data acquisition and analysis. This section presents an overview of the essential techniques, methodologies, and considerations involved in this process.

Data Acquisition Systems

The foundation of reliable impact test results rests upon sophisticated data acquisition systems designed to capture high-speed, transient events occurring within microseconds to milliseconds [28, 29]. At the heart of this measurement process are strain gauges and load cells, typically mounted on the incident and transmitted bars in SHPB systems. These sensors work together to measure elastic deformation and calculate stress wave propagation. The systems integrate high-speed data acquisition capabilities with sampling rates ranging from 1 to 10 MHz, 12-bit analog-to-digital converters (ADCs), and multi-channel inputs for simultaneous recording from multiple sensors.

Signal conditioning maintains data quality through amplification of low-level signals from strain gauges, reduction of high-frequency noise using low-pass filters (with cutoff frequencies between 100 kHz and 1 MHz), and proper impedance matching between sensors and acquisition systems. These methodologies enable accurate characterization of rock behavior under dynamic loading conditions (Fig. 3.13).

Data Processing and Analysis

Once raw data is acquired, it must be processed and analyzed to extract meaningful information about the rock specimen's behavior under impact loading.

3.2 Impact Test Technique

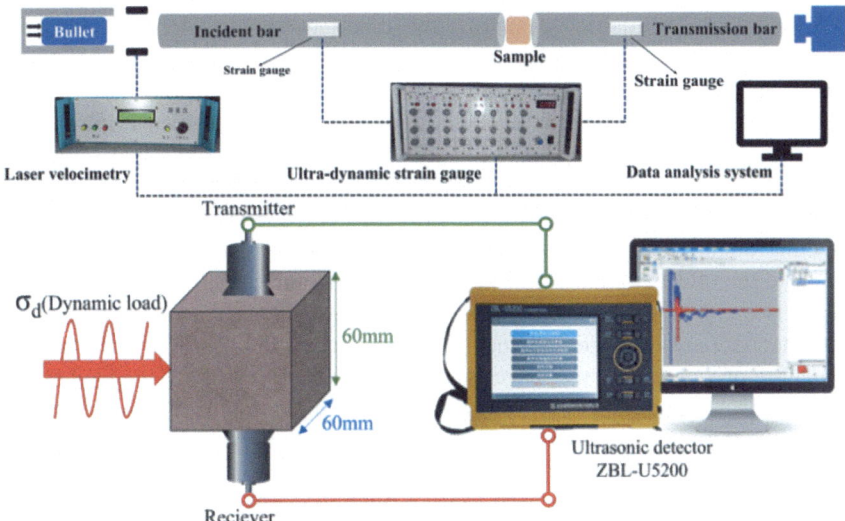

Fig. 3.13 Schematic diagram of ultrasonic testing system [28]

Wave Propagation Analysis

The analysis of stress wave propagation forms the basis for interpreting SHPB test results [30]. Incident, reflected, and transmitted waves are identified and separated from the recorded strain gauge signals [31]. Due to the dispersive nature of wave propagation in bars, corrections are applied to account for the frequency-dependent wave velocity. This correction is particularly important for high-frequency components of the stress waves. The separated waves are time-shifted to the specimen-bar interfaces to synchronize the stress and strain calculations [31, 32].

3.2.4 Application in Resource Exploitation Research

Impact test techniques in resource exploitation research have gained significant attention in recent years due to the growing demand for efficient and sustainable extraction methods. This section examines various impact testing methods and their relevance to resource extraction in deep underground environments.

Chemical Corrosion Effects on Granite Under Impact Loading

In resource exploitation, rocks are frequently exposed to corrosive environments that significantly alter their mechanical properties. Niu et al. [33] investigated the effects of acid corrosion on the dynamic mechanical properties and energy evolution of granite under impact loading. The study treated granite samples with H_2SO_4 solutions at pH values of 3 and 5 for 30 and 60 days, followed by dynamic impact tests using a SHPB system. The findings emphasize the importance of considering both chemical corrosion and dynamic loading when evaluating rock behavior. This integrated approach enables more accurate predictions of rock mass stability, enhanced safety measures, and optimized extraction methods in corrosive environments (Fig. 3.14).

Supercritical CO_2 Impact Fracturing Technology and Applications

Supercritical CO_2 ($ScCO_2$) impact fracturing has emerged as a novel technology, attracting attention for its efficiency and environmental benefits. Research [34] examined the effects of supercritical CO_2 impact fracturing on rock-like materials under various temperatures and pressures. This technology demonstrates significant potential in resource development through creating complex fracture networks to enhance production from tight reservoirs, improving permeability in geothermal energy projects, enhancing in-situ leaching effectiveness for mineral resources, optimizing underground gas storage cavities, and serving as a pre-conditioning technique in hard rock mining to reduce energy consumption and improve productivity. Furthermore, the use of CO_2 as a fracturing fluid offers environmental advantages

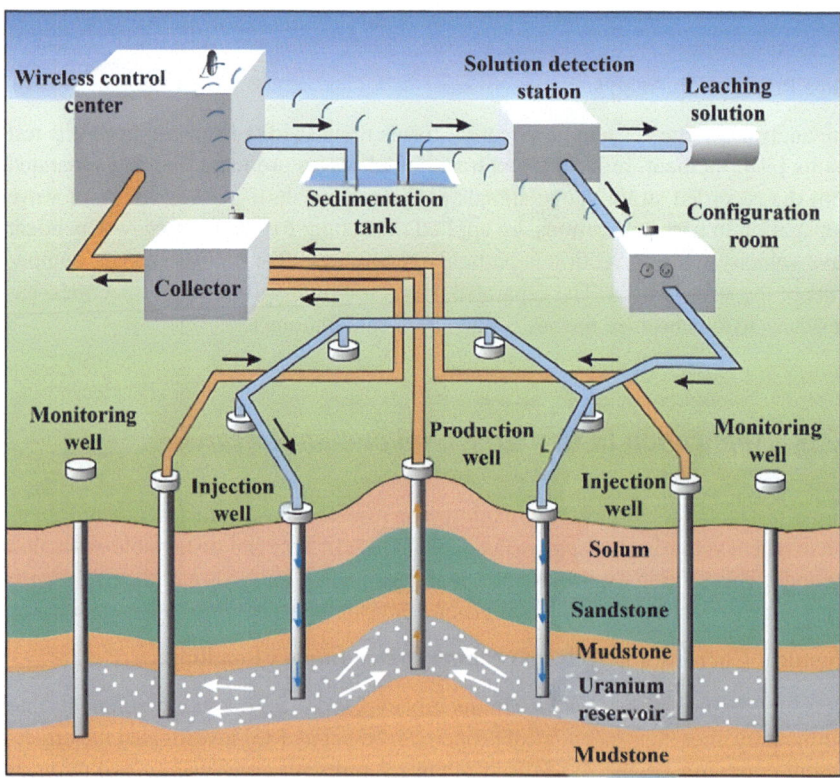

Fig. 3.14 Schematic diagram of in-situ leaching mining of sandstone-type uranium deposits [33]

3.3 Dynamic and Static Combined Loading Test Technique

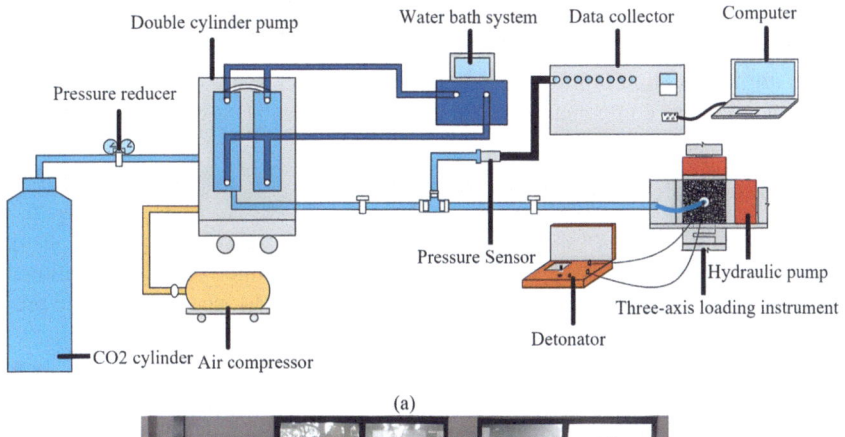

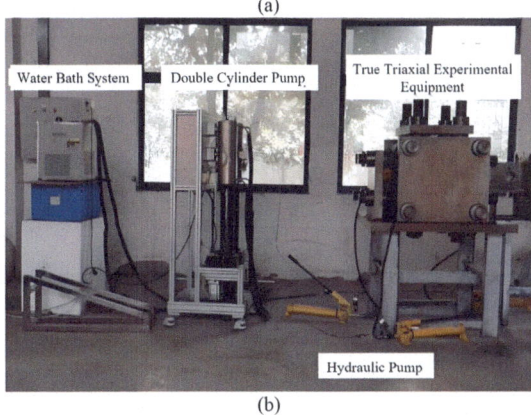

Fig. 3.15 True triaxial experimental setup for supercritical CO_2 thermal impact fracturing: **a** schematic diagram of the test system, and **b** physical diagram of the test system [34]

by reducing water consumption while supporting carbon capture and utilization strategies (Fig. 3.15).

3.3 Dynamic and Static Combined Loading Test Technique

3.3.1 Design Methods

The development of mechanical testing techniques for rocks under coupled static-dynamic loading conditions is essential for understanding rock behavior in complex stress environments, such as those found in deep mining and underground construction. These techniques are designed to replicate in-situ stress states while incorporating dynamic loading to investigate critical phenomena such as failure mechanisms, stress wave propagation, and strain rate effects. This section explores advancements in traditional SHPB systems, the introduction of true triaxial dynamic testing devices, and the implementation of innovative loading techniques, emphasizing their underlying principles, key components, and the role of instrumentation in ensuring accurate and reliable results.

Split Hopkinson Pressure Bar (SHPB) Based Designs

SHPB system, also known as the Kolsky bar, serves as the foundation for many dynamic and static combined loading test designs. By modifying the traditional SHPB setup, researchers have developed several methods to incorporate static pre-stress conditions.

1. Axial Pre-Stress SHPB Systems

One of the most common approaches to combining static and dynamic loads is the incorporation of an axial pre-stress mechanism into the SHPB system. This design typically involves the addition of a hydraulic loading device or a mechanical screw system to apply static axial stress to the rock specimen before the dynamic impact [35, 36]. An example is presented by Yang et al. [37], where they developed a modified SHPB system capable of applying axial static loads up to 13 MPa before dynamic impact. Their design incorporated a hydraulic loading device and a transfer flange to maintain static load during dynamic testing (Fig. 3.16).

2. True Triaxial Pre-Stress SHPB Systems

To better replicate the dynamic fracture behaviors under high in-situ stress conditions in deep mining, researchers have developed a modified SHPB system with coupled static-dynamic loading capabilities. These systems allow for pre-application of confining pressures before dynamic loading to simulate realistic excavation scenarios. Key components of such systems include standard SHPB elements, an axial pre-compression mechanism for static stress application, advanced data acquisition units for capturing strain and stress signals, and cylindrical rock specimen holders designed to withstand high pressure. This configuration enables the study of stress wave propagation, strain rate effects, and energy dissipation in rocks under true triaxial stress conditions, providing critical insights into failure mechanisms (Fig. 3.17).

True Triaxial Dynamic Testing Systems

While SHPB-based systems are widely used, some researchers have developed standalone true triaxial dynamic testing systems that offer greater flexibility in loading paths and specimen sizes.

These systems typically use cubical rock specimens and allow for independent control of stresses in three orthogonal directions, with one or more axes capable of applying dynamic loads. Key components of cubical specimen true triaxial systems include: Rigid loading frame; Three pairs of loading platens (one pair for each axis); Static loading actuators (hydraulic or mechanical); Dynamic loading actuators

3.3 Dynamic and Static Combined Loading Test Technique

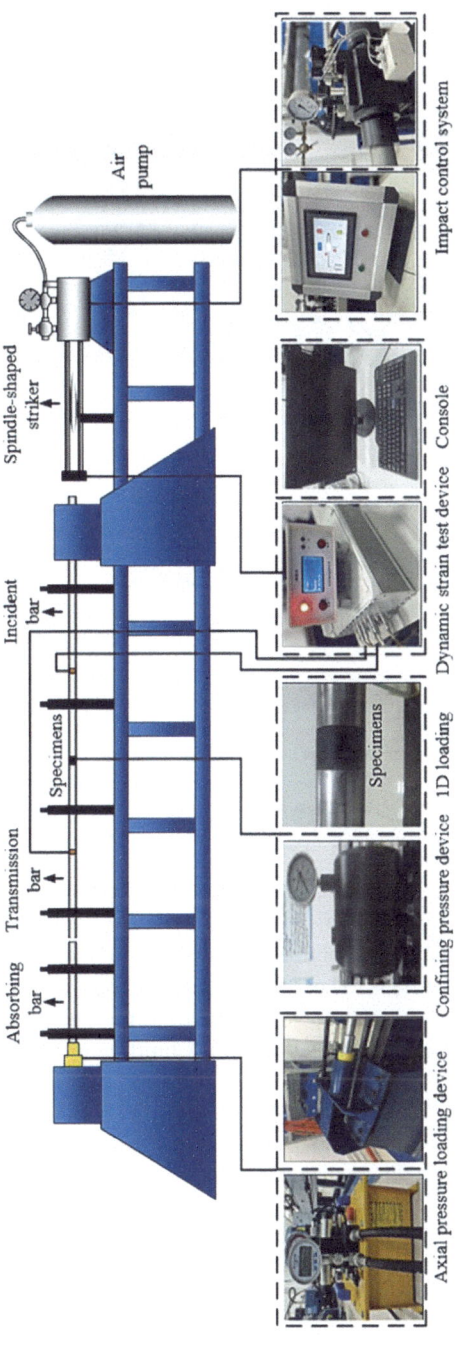

Fig. 3.16 Axial pre-stress SHPB systems [37]

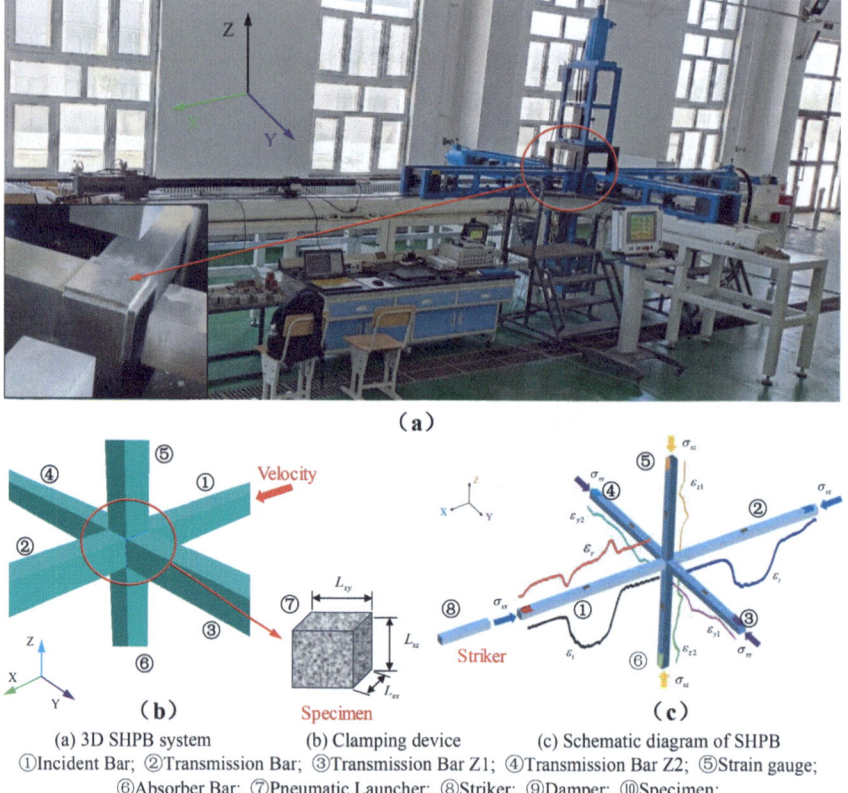

(a) 3D SHPB system   (b) Clamping device   (c) Schematic diagram of SHPB
①Incident Bar; ②Transmission Bar; ③Transmission Bar Z1; ④Transmission Bar Z2; ⑤Strain gauge;
⑥Absorber Bar; ⑦Pneumatic Launcher; ⑧Striker; ⑨Damper; ⑩Specimen;

Fig. 3.17 True triaxial pre-stress SHPB systems [38]

(e.g., pneumatic, hydraulic, or electromagnetic); Multi-axial load cells and displacement sensors; High-speed data acquisition system. He et al. [39] designed a true triaxial testing system capable of applying slight dynamic disturbances to rock specimens under high static stresses. Their system successfully reproduced rockburst phenomena under laboratory conditions, highlighting the significance of coupled static-dynamic loading in understanding deep rock failure mechanisms (Fig. 3.18).

Novel Combined Loading Techniques

In addition to the more established testing methods, researchers are continually developing novel techniques to better simulate the complex loading conditions encountered in deep rock environments.

1. Coupled Thermo-Mechanical Loading Systems

Recognizing the importance of temperature effects in deep rock masses, some researchers have developed systems that combine thermal loading with static and dynamic mechanical loads. The thermo-mechanical loading system represents a

3.3 Dynamic and Static Combined Loading Test Technique

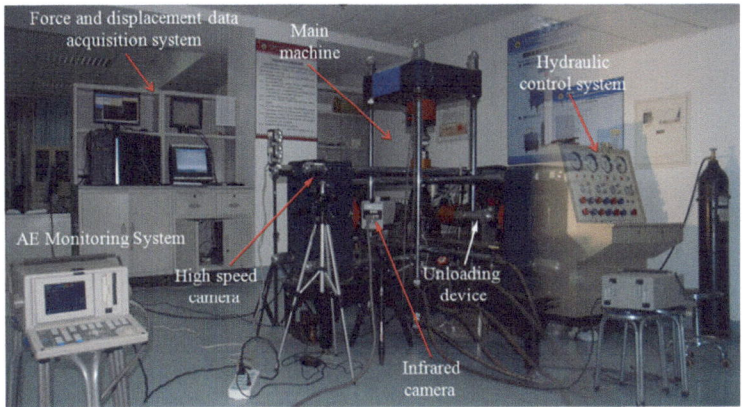

Fig. 3.18 Rockburst testing system [39]

sophisticated apparatus for investigating rock behavior under coupled thermal and mechanical conditions. At its core, the system employs a comprehensive temperature control system comprising heating elements and cooling systems that enable precise thermal regulation. Essential to maintaining controlled experimental conditions is the thermal insulation, which prevents unwanted heat dissipation and ensures temperature stability. Temperature sensors are strategically positioned throughout the system to provide real-time monitoring and feedback of thermal conditions. The mechanical aspects are addressed through specialized loading components capable of both static and dynamic force application. Finally, all instrumentation must be specifically designed to withstand elevated temperatures, utilizing thermal-resistant materials and construction to maintain accuracy and reliability throughout testing procedures. This integrated configuration enables researchers to effectively study the complex interplay between thermal and mechanical effects on rock behavior (Fig. 3.19).

2. Electromagnetic Dynamic Loading Systems

Some researchers have explored the use of electromagnetic forces to apply dynamic loads to pre-stressed rock specimens, offering potential advantages in load control and waveform generation. The electromagnetic dynamic loading system, which is crucial for studying rock dynamic behavior, consists of several essential components. The system requires electromagnetic coils for generating electromagnetic force fields, coupled with a power supply and control system to regulate and maintain stable loading conditions. To prevent electromagnetic interference, both the loading frame and specimen holder must be constructed from non-magnetic materials. Additionally, the system incorporates a static loading mechanism for applying initial stress. Furthermore, all testing instrumentation must be made of non-metallic or non-magnetic materials to ensure measurement accuracy. The proper configuration

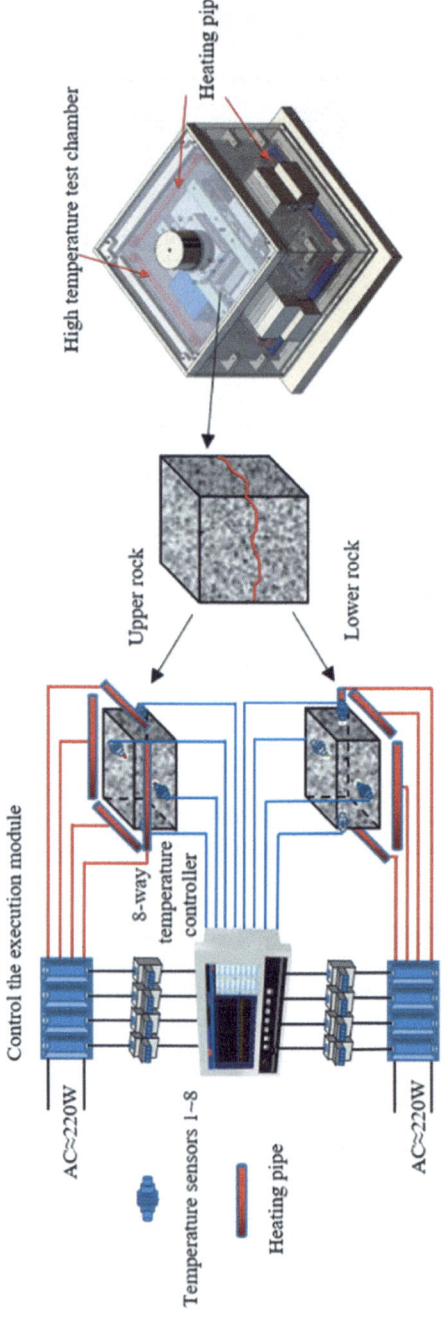

Fig. 3.19 Schematic diagram showing the control system used to heat the rock sample uniformly for long periods [14]

3.3 Dynamic and Static Combined Loading Test Technique

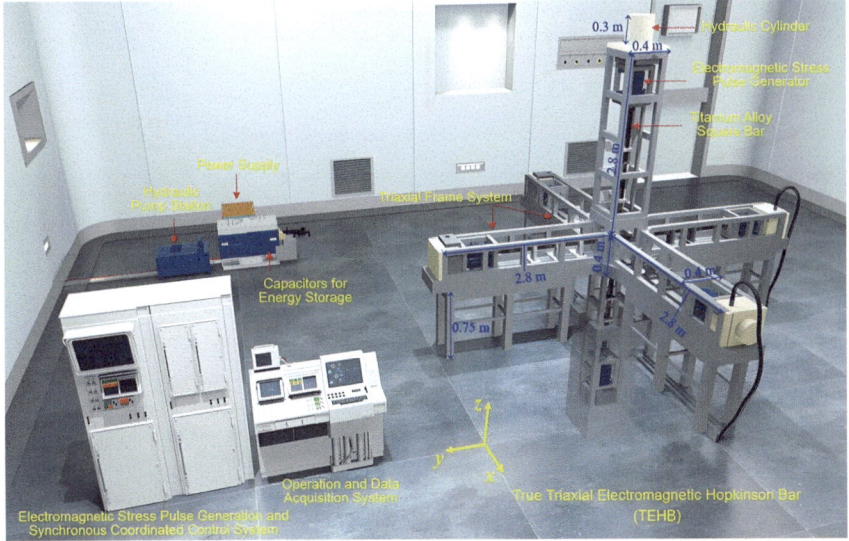

Fig. 3.20 The schematic diagram of the TEHB [11]

and coordinated operation of these components are key to successfully conducting electromagnetic dynamic loading experiments (Fig. 3.20).

Instrumentation and Data Acquisition

The design of instrumentation and data acquisition systems is crucial for the success of combined loading tests on rocks. These systems must be capable of measuring a wide range of parameters under challenging conditions.

1. Stress and Strain Measurement

Stress and strain measurement is a critical aspect of material science and structural engineering, as it helps to evaluate the mechanical behavior of materials under various loads and conditions. The process involves the use of precise instruments and techniques to quantify the amount of stress (internal resistance) and strain (deformation) a material undergoes when subjected to external forces (Fig. 3.21).

Design considerations for stress and strain measurement include selecting sensors with appropriate range, resolution, and response time for the expected loading conditions, and optimizing sensor placement to capture relevant data while minimizing interference with the specimen or loading system. Additionally, temperature compensation is crucial to account for temperature effects on sensor readings, especially in thermo-mechanical tests. Robust calibration procedures must be developed to handle complex loading conditions, and precise data synchronization is essential to ensure accurate timing alignment between different measurement channels.

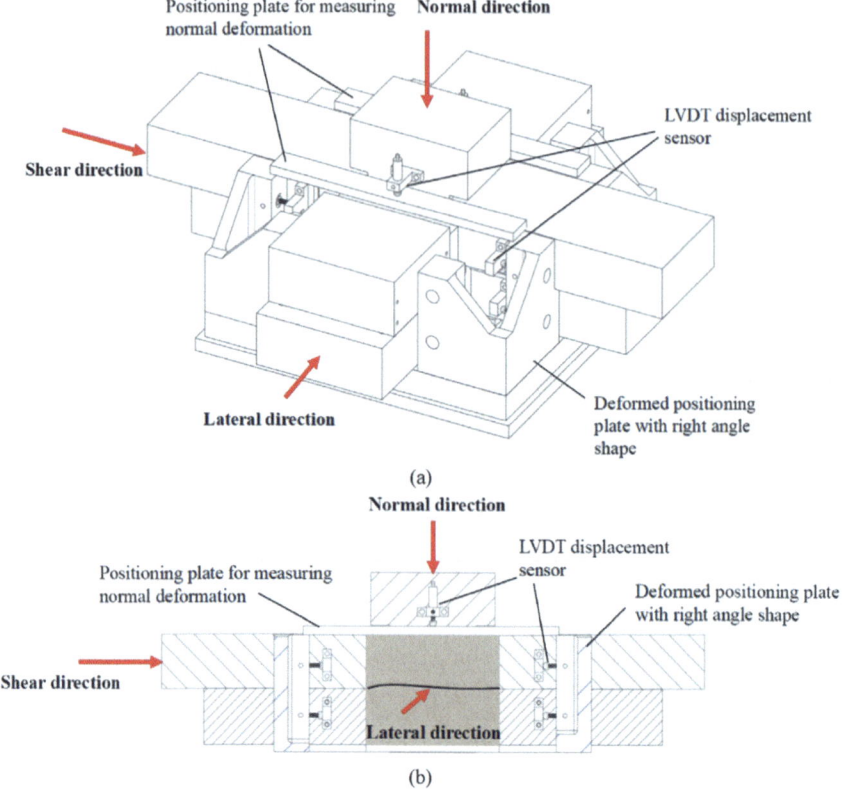

Fig. 3.21 Method used to measure the volumetric deformation of the rock sample during true-triaxial testing: **a** schematic diagram showing the locations of the LVDTs; and **b** side-on profile of the equipment showing the installed LVDTs [14]

2. Data Acquisition and Control Systems

The heart of any combined loading test system is its data acquisition and control system, which must handle multiple input and output channels with high speed and precision. Key components of data acquisition and control systems include: Multi-channel analog-to-digital converters; High-speed data buses (e.g., PCI Express, USB 3.0); Real-time control processors; Data storage systems; User interface software.

Design considerations for data acquisition and control systems include ensuring a sufficient sampling rate to capture rapid dynamic events and accommodating all necessary sensors and control outputs through an appropriate channel count. Precise synchronization across all channels and with external events is critical, along with real-time processing capabilities for on-the-fly data analysis and control decisions. Reliable data capture and storage must be ensured to maintain data integrity, even during high-speed operation. Additionally, the system should be designed with flexibility to adapt to different test configurations and support future upgrades.

Calibration and Validation Procedures

The accuracy and reliability of combined loading test systems depend heavily on proper calibration and validation procedures.

1. Static Load Calibration

The calibration of static loading components is a critical process that ensures the accuracy and reliability of experimental results in rock dynamics. This typically involves three key steps: the calibration of load cells using traceable standards to verify force measurements; the calibration of pressure sensors for confining pressure systems to ensure precise pressure control; and the calibration of displacement sensors to guarantee accurate measurement of deformation. These procedures form the foundation for achieving high-precision experimental setups in rock mechanics studies.

2. Dynamic Load Calibration

The calibration of dynamic loading components presents unique challenges compared to static systems, due to the high strain rates and rapid load changes involved. Key calibration procedures include the precise calibration of strain gauges on Hopkinson bars to ensure accurate strain measurement; wave propagation checks in SHPB systems to validate the consistency and accuracy of stress wave transmission; and the calibration of dynamic load cells to accurately capture transient loading forces. These steps are essential for ensuring the reliability and accuracy of dynamic loading experiments in rock mechanics.

3. System Validation

The validation of combined loading test systems is a critical step in ensuring their reliability and accuracy. This process typically involves testing standard materials with well-documented and known properties to verify system performance. Additionally, results are compared with those obtained from established testing methods to ensure consistency and reliability. Round-robin testing between multiple laboratories is often conducted to assess inter-laboratory variability and reproducibility. Furthermore, comparisons with analytical solutions for simple loading cases are performed to validate the fundamental accuracy of the system. These validation steps collectively ensure the robustness and credibility of the testing system.

3.3.2 Mechanical Testing Techniques

In the field of rock dynamics, understanding the behavior of rock materials under combined static and dynamic loading conditions is crucial for various engineering applications, including deep mining, underground excavation, and seismic protection. This section provides a comprehensive overview of the mechanical testing techniques used to investigate rock properties under coupled static-dynamic loads.

Split Hopkinson Pressure Bar (SHPB) with Static Pre-stress

1. System Components and Operation

The basic SHPB system consists of three main components: a striker bar, an incident bar, and a transmitted bar. The rock specimen is placed between the incident and transmitted bars. In the modified version for coupled static-dynamic loading, a static loading system is integrated to apply pre-stress to the specimen before the dynamic impact [7, 9].

The operation of a rock dynamics testing system involves a sequence of interconnected steps. First, a controlled axial pre-stress is applied to the rock specimen using a hydraulic loading system or a screw-driven mechanism to establish a predefined static load. Dynamic loading is then initiated as the striker bar impacts the incident bar, generating a stress wave that propagates through the system. This wave travels through the incident bar, interacting with the specimen, where part of the wave is reflected at the specimen-incident bar interface and the remainder is transmitted through the specimen to the transmitted bar. Strain gauges mounted on the incident and transmitted bars measure the incident, reflected, and transmitted waves, providing critical data for analysis. Using one-dimensional wave propagation theory, these strain signals are processed to calculate the stress, strain, and strain rate in the specimen, enabling a comprehensive assessment of its dynamic mechanical behavior under combined static and dynamic loading conditions (Fig. 3.22).

2. Advantages and Limitations

The dynamic testing system offers several advantages and limitations that define its applicability and performance. Advantages include the ability to simulate complex loading conditions encountered in real-world scenarios, making it highly relevant for practical applications. The system can achieve high strain rates, typically in the range of 10^2 to 10^4 s^{-1}, allowing for the study of material behavior under extreme conditions. Additionally, data analysis is relatively straightforward, leveraging one-dimensional wave propagation theory. The system is also versatile, capable of testing

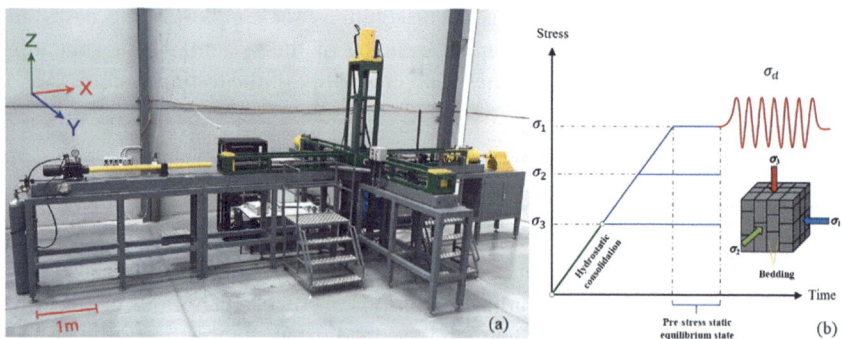

Fig. 3.22 **a** Triaxial Hopkinson Bar (Tri-HB) system and **b** the stress path to apply true triaxial static pre-stresses followed by the dynamic loading in red [40]

3.3 Dynamic and Static Combined Loading Test Technique

a wide range of materials and accommodating various loading conditions. However, there are limitations to consider. Achieving stress equilibrium at very high strain rates can be challenging, potentially affecting the accuracy of the results. The test duration is inherently limited by the finite length of the bars, which restricts the time available for observations. Maintaining a constant strain rate throughout the test can also be difficult, leading to variations in the loading conditions. Furthermore, end effects and friction at the specimen-bar interfaces may introduce additional complexities, potentially influencing the test results.

True Triaxial Static-Dynamic Loading System

While the modified SHPB system provides valuable insights into material behavior under uniaxial or confined compression, many real-world scenarios involve complex three-dimensional stress states. True triaxial static-dynamic loading systems have been developed to address this need, allowing researchers to investigate rock behavior under more realistic stress conditions [41].

True Triaxial Testing

1. System Components and Operation

A typical true triaxial static-dynamic loading system consists of key components for applying and measuring multi-directional loads on rock specimens. It includes a rigid loading frame capable of independent loading in three orthogonal directions, hydraulic or servo-controlled static loading actuators, and a dynamic loading system, such as a gas gun or drop weight, for impact testing. The specimen is held in a specially designed containment cell, allowing three-dimensional loading. Instrumentation, including load cells, displacement transducers, and strain gauges, measures forces and deformations, while a high-speed data acquisition system captures and analyzes test data.

The operation of a true triaxial static-dynamic loading system involves a structured sequence of steps to simulate realistic in-situ stress conditions and analyze the rock specimen's behavior under combined static and dynamic loads. First, a cubic or prismatic rock specimen is carefully prepared and instrumented as needed to ensure accurate measurements. Controlled static loads are then applied in three orthogonal directions to replicate the desired in-situ stress state. Once the static loads are stabilized, a dynamic load is introduced in one or more directions to simulate transient loading conditions. Throughout the test, forces, displacements, and strains are recorded using high-speed data acquisition systems to capture the specimen's response in real time. Finally, the collected data is processed during post-test analysis to determine stress–strain relationships, failure criteria, and other critical mechanical parameters, providing valuable insights into the rock's behavior under complex loading conditions.

2. Advantages and Limitations

The true triaxial static-dynamic loading system offers several advantages. It allows independent loading in three orthogonal directions, accurately simulating real triaxial stress states. With both static and dynamic loading capabilities, it enables the study of material behavior under complex conditions. Its specially designed specimen containment ensures a realistic three-dimensional loading environment, while high-precision instruments like load cells and displacement transducers, combined with high-speed data acquisition systems, provide accurate and reliable test data. This system is versatile and widely used in fields such as rock mechanics and structural engineering.

However, the system also has certain limitations. Its complex structure requires specialized expertise for operation and maintenance, and its advanced components, such as servo controllers and high-speed acquisition systems, significantly increase costs. Additionally, it is typically limited to testing small specimens, which may not fully represent large-scale structural behaviors. Non-uniform stress distributions during dynamic loading can complicate result interpretation, and the time required for test preparation and data analysis can reduce overall efficiency.

Cyclic Loading Systems

1. System Components and Operation

Cyclic loading plays a crucial role in understanding rock behavior in scenarios such as earthquake loading, machine vibrations, and repetitive mining operations. Several testing systems have been developed to investigate rock properties under cyclic static-dynamic loading conditions.

Cyclic triaxial systems are adaptations of conventional triaxial testing apparatus, designed to apply cyclic axial loads while maintaining constant confining pressure. These systems are particularly useful for studying the behavior of rock under conditions similar to those experienced during earthquakes or other cyclic loading scenarios.

A cyclic triaxial system is composed of key components designed to accurately simulate and measure soil or rock behavior under cyclic loading conditions. At its core is the triaxial cell, a pressure chamber that houses the specimen and applies confining pressure. The system includes a servo-controlled axial loading system to apply both static and cyclic axial loads, along with a confining pressure system to maintain consistent lateral pressure. For saturated tests or studies involving pore fluid effects, a pore pressure control system is integrated. Precise measurement is ensured by advanced instrumentation, such as load cells, displacement transducers, and pore pressure sensors. All parameters are managed and recorded by a data acquisition and control system, enabling efficient operation and reliable data collection (Fig. 3.23).

2. Advantages and Limitations

Cyclic loading systems offer several advantages, including the ability to realistically simulate real-world conditions such as earthquakes, traffic, or wave loads, making them invaluable for geotechnical and structural studies. They provide versatile testing capabilities, supporting static, dynamic, and cyclic loading with options for confining and pore pressure control, enabling detailed studies of saturated or unsaturated specimens. Advanced instrumentation ensures high precision in measuring forces,

3.3 Dynamic and Static Combined Loading Test Technique

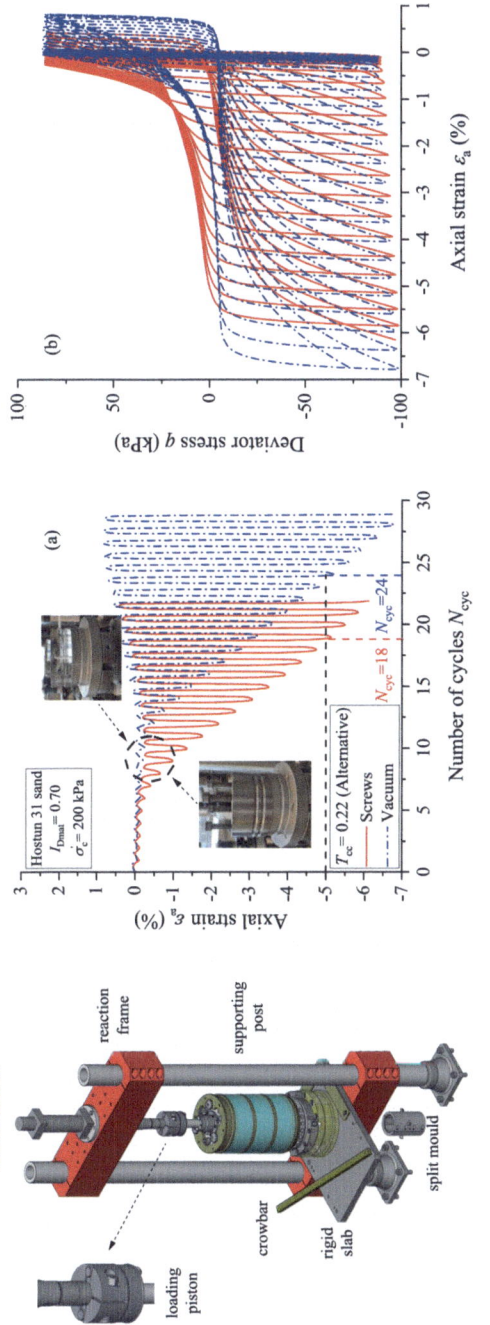

Fig. 3.23 Schematic view of the reaction frame and the confining chamber and testing results [42]

displacements, strains, and pore pressures, enhancing understanding of material behavior under repeated loading, such as fatigue and stiffness degradation.

However, these systems have limitations, including high costs due to advanced components, complexity requiring specialized expertise for operation, and time-intensive test preparation and data analysis. Additionally, specimen size limitations may reduce representativeness for large-scale structures, and non-uniform stress distributions during cyclic loading can complicate result interpretation.

Impact Testing

Impact testing is crucial for understanding rock behavior under high-strain-rate loading conditions, such as those encountered in blasting, rockbursts, and high-speed drilling operations. Several specialized testing systems have been developed to investigate rock properties under impact loading.

1. System Components and Operation

Drop weight impact testing systems are simple yet highly effective tools for investigating rock behavior under intermediate strain rates, typically ranging from $1 \sim 10^2$ s^{-1}. These systems consist of a guided drop weight, which is raised to a specific height and released to impact the specimen, and an anvil, a rigid base that supports the specimen during impact. The instrumented tup, located at the impacting end of the drop weight, is equipped with a load cell to measure the impact force. Additionally, displacement or velocity sensors monitor the motion of the drop weight, while a high-speed data acquisition system records rapid changes in force and displacement during the impact event, enabling precise analysis of the material's dynamic response (Fig. 3.24).

The testing procedure for drop-weight impact tests involves a systematic process to evaluate the dynamic response and energy absorption characteristics of rock specimens. First, the specimen is securely positioned on the anvil to ensure stability and

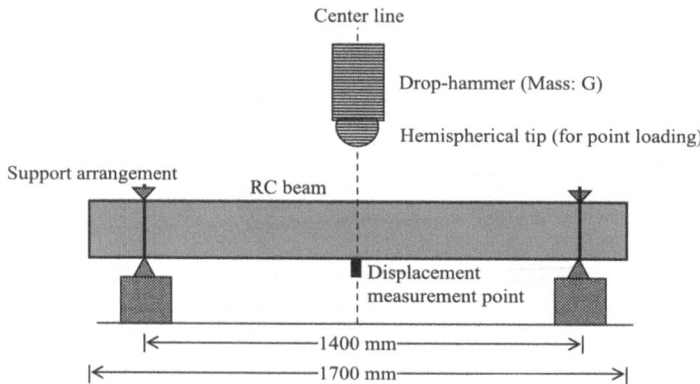

Fig. 3.24 Schematic diagram of drop hammer impact test setup [43]

3.3 Dynamic and Static Combined Loading Test Technique

minimize movement during the test. The drop weight is then raised to a predetermined height, carefully selected to achieve the desired impact energy. Once positioned, the drop weight is released to freely fall, generating a high-energy impact on the specimen. During the impact event, force–time and displacement–time data are recorded using sensors integrated with a high-speed data acquisition system. Finally, the recorded data are analyzed to calculate critical parameters such as the impact energy, peak force, and energy absorption characteristics of the specimen, providing valuable insights into its dynamic mechanical behavior.

2. Advantages and Limitations

Drop weight systems offer several advantages. They are relatively simple and cost-effective while effectively simulating intermediate strain rate conditions, making them ideal for studying material behavior under real-world impact scenarios like rockfalls or blasts. These systems provide valuable insights into dynamic response and failure mechanisms, enabling the measurement of key parameters such as impact energy, peak force, and energy absorption. Additionally, high-speed data acquisition ensures detailed force, displacement, and velocity measurements, offering high-resolution data for comprehensive analysis.

However, impact testing systems also have limitations. They offer less precise control over loading conditions compared to servo-controlled systems, and results can vary significantly depending on specimen size, making generalization to larger structures challenging. These systems are typically designed for single-event impacts, limiting their ability to study fatigue or repeated impact effects. Furthermore, non-uniform stress distribution during impact can complicate material behavior interpretation, and the complex datasets generated require specialized expertise for accurate analysis and understanding.

Biaxial loading systems

In many real-world scenarios, rocks are subjected to complex loading conditions that involve combinations of compression, tension, and shear. Several specialized testing systems have been developed to simulate these complex loading conditions in laboratory settings.

1. System Components and Operation

Biaxial loading systems are designed to apply loads or displacements in two perpendicular directions, providing valuable insights into rock behavior under plane stress conditions. These systems consist of a rigid loading frame capable of applying loads in two orthogonal directions and hydraulic or electromechanical actuators to deliver controlled forces or displacements. Precise measurement is ensured by load cells and displacement transducers, which track applied forces and resulting deformations. Specimen grips or platens are used to securely transfer loads to the specimen while minimizing stress concentrations. A control and data acquisition system manages the testing process and records data, ensuring accurate and efficient operation (Fig. 3.25).

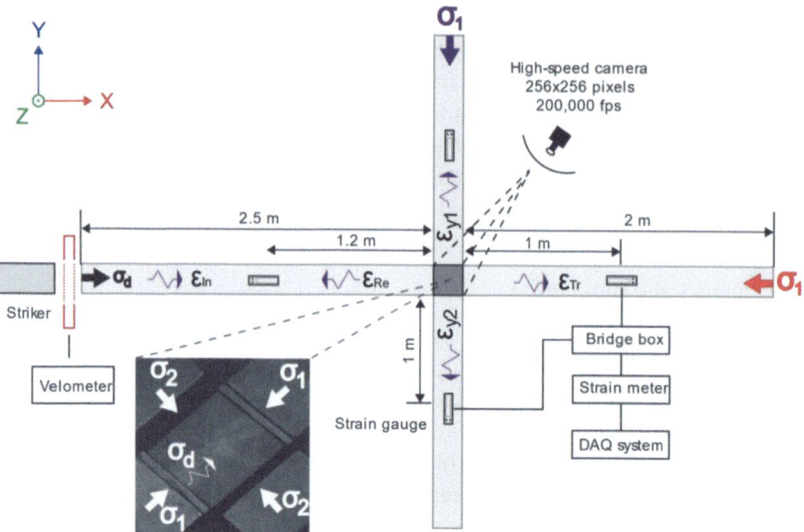

Fig. 3.25 Simplified 2D view of the biaxial configuration of a Triaxial Hopkinson bar (Tri-HB) system [44]

2. Advantages and Limitations

Biaxial loading systems offer several advantages, including the ability to simulate plane stress conditions, providing precise control of loads or displacements in two perpendicular directions, and enabling detailed analysis of stress–strain relationships and failure mechanisms. They are versatile, suitable for various materials, and equipped with advanced instrumentation for accurate measurement of forces and deformations. However, these systems also have limitations, such as their inability to replicate three-dimensional stress states, the complexity of setup and calibration, high costs for acquisition and maintenance, and restrictions on specimen size. Additionally, interpreting data from biaxial stress states can be more challenging compared to simpler testing methods.

3.4 Temperature–Pressure Coupling Test Technique

3.4.1 *Thermal Effects on Rock Properties*

The study of thermal effects on rock properties is crucial for understanding the behavior of rocks in various engineering applications, including geothermal energy extraction, underground nuclear waste storage, and deep mining operations. This

3.4 Temperature–Pressure Coupling Test Technique

section examines the complex interplay between temperature changes and rock properties, focusing on the latest research findings and experimental techniques used to investigate these phenomena.

Temperature-induced changes in rock properties can be broadly categorized into physical, mechanical, and hydraulic effects. These changes are often interrelated and can significantly impact the overall behavior of rock masses under varying thermal conditions.

Physical Property Changes

Thermal loading significantly affects the physical and mechanical properties of rocks through thermal expansion, cracking, and changes in wave propagation. As rocks are heated, differential thermal expansion of their mineral constituents generates stresses at grain boundaries, often leading to microcrack formation. These microcracks weaken the rock's structure and, with repeated thermal cycles, propagate and coalesce, causing extensive damage such as spalling or structural failure.

The extent of these changes is influenced by factors like temperature range, frequency of thermal cycles, and mineral composition. Rocks with minerals that have varying thermal expansion coefficients are more susceptible to cracking. At higher temperatures, chemical changes, such as mineral oxidation or deoxidation, further degrade mechanical properties.

Furthermore, wave propagation characteristics, particularly P-wave velocity, are indicators of rock integrity under thermal stress. P-wave velocity typically decreases with rising temperatures due to microcrack formation, propagation, and changes in mineral elasticity. These effects are especially pronounced beyond certain temperature thresholds, as demonstrated in studies of various rock types [45].

In summary, the interplay between thermal expansion and grain boundary stresses plays a critical role in rock degradation under thermal conditions. Understanding these processes is essential for applications in geothermal energy, underground waste storage, and high-temperature engineering.

Mechanical Property Changes

1. Compressive Strength

The uniaxial compressive strength (UCS) of rocks generally decreases with increasing temperature, although the exact relationship can be non-linear and rock-type dependent. In the study on Tournemire shale [46], the researchers observed a significant decrease in compressive strength with increasing temperature. However, the study on clay-rich Hawkesbury sandstone [47] revealed a more complex behavior, where the compressive strength initially increased from 25 to 600 °C (after an initial reduction at 200 °C), followed by a decrease at temperatures above 600 °C. This non-monotonic behavior highlights the importance of considering the specific mineralogical composition and microstructure of rocks when predicting their thermal response.

2. Young's Modulus

The elastic modulus or Young's modulus of rocks typically decreases with increasing temperature. This reduction in stiffness is primarily due to the formation and propagation of microcracks, as well as changes in the elastic properties of individual minerals. The study on granite [48] showed a dramatic decrease in elastic modulus after the first thermal cycle, from 16.53 GPa to 4.84 GPa, representing a 70.7% reduction. Further thermal cycling led to continued, albeit slower, degradation of the elastic modulus.

3. Poisson's Ratio

The effect of temperature on Poisson's ratio is less consistent across different rock types. In some cases, Poisson's ratio may increase with temperature, while in others, it may decrease or show non-monotonic behavior. The study on Tournemire shale [46] reported that Poisson's ratio generally decreased with increasing confining pressure but increased with temperature.

4. Failure Characteristics

Temperature changes can significantly alter the failure mode and crack propagation patterns in rocks. The study on granite [48] observed a transition from brittle to more ductile behavior with thermal cycling. Untreated samples failed by axial splitting in a brittle manner, while samples subjected to thermal cycling exhibited conical shear failure. This transition was accompanied by changes in acoustic emission (AE) behavior, with thermally cycled samples showing more continuous AE activity occurring mainly after peak stress (Fig. 3.26).

The failure envelope of rocks in the p-q plane (mean stress versus deviatoric stress) also changes with temperature. The study on Tournemire shale [46] found that failure surfaces became more nonlinear with increasing temperature. Additionally,

Fig. 3.26 Fracture patterns of granite specimens after the thermal cycle treatments in the uniaxial compressive test: **a** 0 thermal cycle; **b** 1 thermal cycle; **c** 40 thermal cycles; **d** 100 thermal cycles [48]

3.4 Temperature–Pressure Coupling Test Technique

the intrinsic cohesion of the shale was observed to be very low, and the frictional coefficient decreased significantly with temperature increase.

Hydraulic Property Changes

Temperature changes can have a profound impact on the hydraulic properties of rocks, particularly permeability and porosity. These changes are closely linked to the thermal-induced microstructural alterations and can significantly affect fluid flow in geothermal reservoirs and other subsurface applications.

1. Permeability

The permeability of rocks generally increases with temperature due to thermal cracking and mineral alterations. However, the relationship between temperature and permeability can be complex and non-monotonic, especially under coupled thermo-hydro-mechanical (THM) conditions.

2. Porosity

Temperature changes generally lead to an increase in rock porosity due to thermal cracking and mineral decomposition. The study on Tournemire shale [46] reported a significant increase in porosity with temperature, particularly above 80 °C. The porosity increased from 8.3% at 20 °C to 10.4% at 120 °C.

The marble study [49] provided detailed insights into pore structure changes under different thermal treatments. Under THM conditions, pore volume and connectivity decreased at 300 °C compared to 200 °C and 400 °C, indicating pore closure. This contrasted with HTO (heat treatment only) conditions, where pore volume and connectivity increased monotonically with temperature. Interestingly, below 400 °C, HTO treatment produced higher porosity than THM treatment, while above 500 °C, THM treatment resulted in higher porosity.

3. Pore Size Distribution

Temperature changes can also alter the pore size distribution in rocks. The proportion of larger pores (meso- and macropores) in the marble increased with temperature, especially above 500 °C [49]. This shift in pore size distribution can have significant implications for fluid flow and storage capacity in high-temperature environments.

Microstructural Changes

The thermal effects on rock properties are fundamentally linked to microstructural changes occurring at the grain and pore levels. Several studies have employed advanced imaging techniques to visualize and quantify these changes.

1. Thermal Cracking

Thermal cracking is a primary mechanism driving changes in rock properties at elevated temperatures. The study on clay-rich sandstone [47] used scanning electron microscopy (SEM) to observe the evolution of thermal cracks with increasing temperature:

At 25–200 °C: Kaolinite structure transformed from hexagonal to fiber-like.

At 200–600 °C: Further transformation to tangled fiber structure.
Above 600 °C: Thermal degradation of fiber structure (Fig. 3.27).

The granite study [48] also observed progressive development of thermal cracks along grain boundaries and through grains using SEM imaging. By 80 thermal cycles, loosened mineral grains were observed spalling from the structure.

2. Mineral Transformations

Temperature elevation triggered systematic mineralogical transformations in the studied sandstone [47]: quartz underwent polymorphic transformation to tridymite at temperatures exceeding 870 °C, kaolinite experienced dehydroxylation above 500 °C with subsequent conversion to mullite at elevated temperatures, while siderite exhibited substantial thermal degradation at 1000 °C, collectively altering the rock's mineralogical composition.

3. Fractal Analysis

Fractal dimension analysis has been employed to characterize the complexity of pore and fracture networks in thermally treated rocks. The marble study [49] found that the fractal dimension of pores/fractures increased continuously with temperature under HTO conditions, while it fluctuated at 200 °C under THM conditions. This analysis provides insights into the scale-invariant properties of thermally induced microstructural changes.

Implications for Engineering Applications:

Thermal effects on rock properties have significant implications across various engineering applications: in geothermal energy extraction, temperature-enhanced permeability and porosity facilitate fluid circulation and heat extraction, though thermal cracking and mineral alterations require careful reservoir management; for underground nuclear waste storage, heat from radioactive decay necessitates thorough evaluation of non-monotonic mechanical and hydraulic property changes in host rocks; in deep mining operations, geothermal gradients increasingly influence rock strength and deformability, demanding comprehensive stability assessments; for underground coal gasification, extreme temperature effects on clay-rich sandstone critically impact reservoir integrity; while in thermal energy storage applications, despite notable thermal degradation, properly engineered granite demonstrates potential viability as a storage medium.

The thermal effects on rock properties are complex and multifaceted, involving interrelated changes in physical, mechanical, and hydraulic characteristics. These changes are driven by microstructural alterations, including thermal cracking, mineral transformations, and pore structure evolution. The non-monotonic behavior observed in some properties underscores the importance of considering specific mineralogical compositions and coupling effects when predicting rock behavior at elevated temperatures.

3.4 Temperature–Pressure Coupling Test Technique

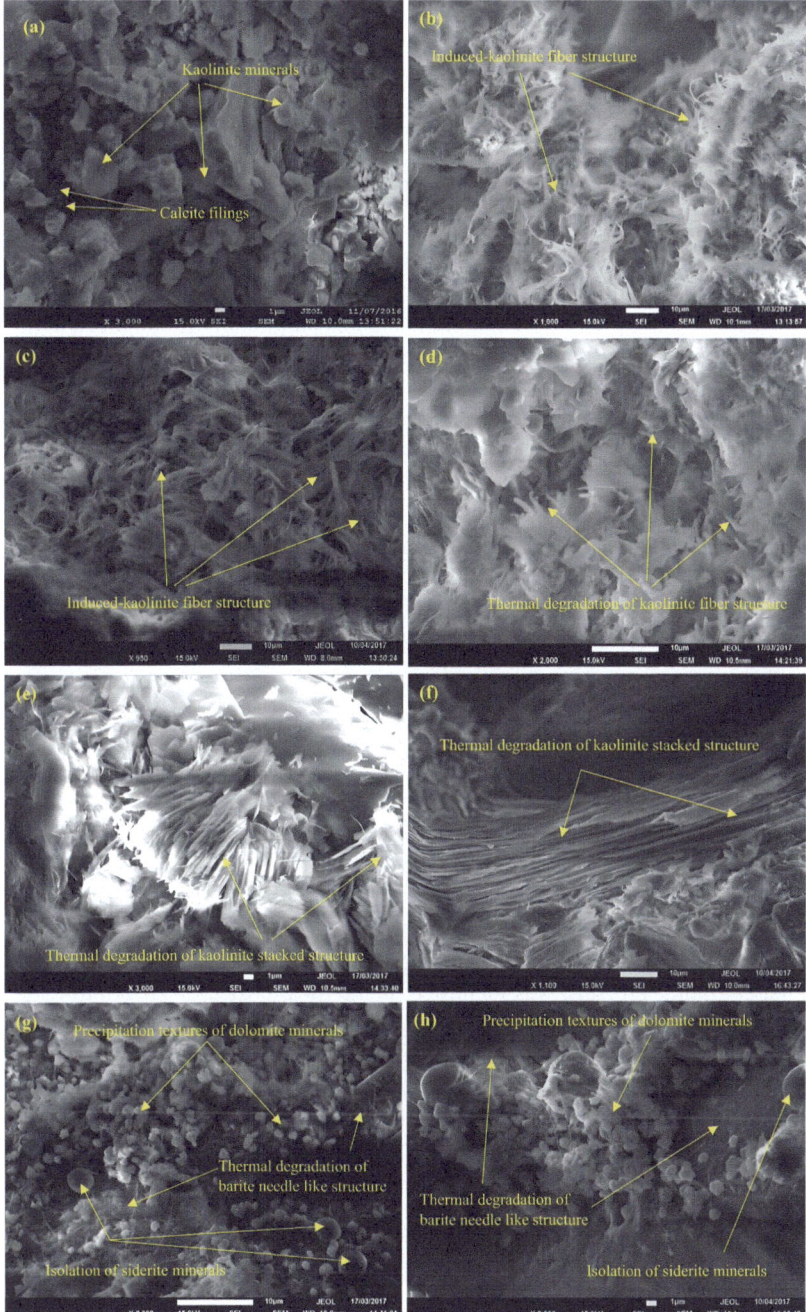

Fig. 3.27 Mineral structures in sample at different temperature [47]

3.4.2 Pressure Effects on Rock Properties

The Influence of Pressure on Rock Properties: Insights from Experimental Studies and Theoretical Analyses

1. Pressure Effects on Hydraulic Properties of Rock Fractures

A novel thermo-hydro-mechanical (THM) coupling testing system was employed to investigate the influence of confining pressure on the hydraulic aperture of granite fractures. Results indicate that hydraulic aperture decreases significantly with increasing confining pressure. This relationship is particularly pronounced at lower pressure ranges, where hydraulic aperture rapidly decreases as confining pressure rises. For instance, under constant normal load (CNL) conditions, a confining pressure increase can reduce hydraulic aperture by up to 84.55%. This reduction is primarily attributed to the closure of microcracks and pores on the rock matrix and fracture surfaces.

Additionally, the effect of confining pressure on hydraulic aperture is nonlinear. At higher pressures, the rate of decrease in hydraulic aperture slows, indicating that the degree of fracture closure diminishes with increasing pressure. This nonlinear behavior likely arises from the closure of larger and more compliant fractures and pores at low pressures, while smaller and stiffer fractures resist closure at higher pressures.

2. Pressure Effects on Mechanical Behavior of Rock

Studies on deep sandstone under thermo-hydro-mechanical coupling conditions demonstrate that confining pressure significantly influences the strength and deformation characteristics of rock. Using orthogonal experimental design, the effects of variations in confining pressure, axial pressure, and pore water pressure on rock behavior were analyzed. Results reveal that higher confining pressures substantially enhance the peak strength of sandstone by suppressing the initiation and propagation of internal microcracks, thereby improving the overall rock strength. Among all factors, confining pressure was identified as the most critical in determining peak strength, surpassing the influence of temperature or impact air pressure.

Confining pressure also plays a key role in rock deformation. As confining pressure increases, rock deformation typically decreases due to enhanced resistance to compression. However, this relationship is more complex than that observed for peak strength, as axial pressure and temperature also significantly affect deformation behavior.

3. Pressure Effects on Wave Velocity Characteristics of Rock

Research on the P-wave response of thermally dry rocks, such as granite and sandstone, under conditions of high temperature and high confining pressure, reveals significant variations in P-wave velocity with changing confining pressure [50]. This behavior is primarily attributed to the closure of microcracks and pores, which increases the continuity of the rock medium and enhances wave propagation efficiency. The relationship between P-wave velocity and confining pressure is nonlinear, exhibiting three distinct stages:

At the low-pressure stage, P-wave velocity increases rapidly due to the closure of larger, more compliant fractures and pores. This stage is characterized by significant changes in wave propagation as the rock structure becomes increasingly compact. In the high-pressure stage, the rate of velocity increase slows down as smaller and stiffer fractures and pores close gradually, resulting in a more stable medium. Finally, at the ultra-high-pressure stage, a slight decline in P-wave velocity is observed, likely caused by the formation of new complex fractures under extreme stress conditions.

For granite, P-wave velocity demonstrates a rapid increase of 17.42% within the 20–50 MPa pressure range, with nonlinear growth continuing up to approximately 80 MPa. Beyond this point, a slight decline occurs, reflecting the onset of microstructural changes. Similarly, sandstone shows a 28.6% linear increase in P-wave velocity within 65 MPa, followed by a 1.94% decrease at higher pressures. These trends highlight the influence of both pore and fracture closure, as well as the potential initiation of new fractures, on wave propagation in rocks under extreme conditions.

This study provides critical insights into the behavior of P-waves in thermally dry rocks under high confining pressures, which has important implications for understanding subsurface stress conditions, seismic wave propagation, and dynamic material properties in geological engineering and exploration.

4. Pressure Effects on Thermal Expansion Behavior of Rock:

Studies using a high-temperature, high-pressure triaxial compression system reveal that the thermal expansion coefficient of granite under 25 MPa confining pressure ranges from 0.2×10^{-5} to 1.25^{-5} °C. This indicates that pressure influences thermal expansion behavior, which is critical for understanding coupled thermo-mechanical behavior in deep geological environments.

Furthermore, granite samples tested under high confining pressures typically fail in shear mode, exhibiting X-shaped fractures. This failure pattern is characteristic of rocks under high confining pressures, where the development of a single dominant fracture is inhibited, leading to a more dispersed failure mode (Fig. 3.28).

Fig. 3.28 Failure of granite specimen at 300 °C; **a** shear plane **b** full view [51]

Pressure exerts profound and complex effects on the hydraulic, mechanical, wave velocity, and thermodynamic properties of rock. These effects are often nonlinear, exhibiting significant thresholds or critical points. Additionally, pressure interacts with environmental factors such as temperature and pore fluid pressure, leading to coupled effects that are essential for understanding rock behavior in deep underground environments.

These pressure-related effects have critical implications for engineering applications such as deep mining, geothermal energy extraction, underground energy storage, and nuclear waste disposal. Future research should focus on the coupled effects of pressure, temperature, and fluids on rock properties under extreme conditions.

3.4.3 Coupled Effects on Rock Mechanics

The study of coupled effects on rock mechanics is crucial for understanding the behavior of rock masses under complex environmental conditions, particularly in deep underground settings where high temperatures and pressures prevail. This section explores the intricate interplay between thermal, hydraulic, mechanical, and chemical processes in rocks, with a focus on how these coupled effects influence rock properties, strength, and failure mechanisms.

Thermal–Mechanical Coupling

The interaction between thermal and mechanical processes in rocks plays a significant role in determining their behavior under various environmental conditions. This coupling is particularly important in applications such as geothermal energy extraction, nuclear waste disposal, and deep mining operations.

1. Thermal Expansion and Contraction

Thermal expansion and contraction of rocks play a crucial role in the development of thermal stresses and subsequent cracking. The thermal expansion coefficient (α) of rocks is a key parameter in understanding their thermal–mechanical behavior. For granite, Ma et al. [52] reported that the thermal expansion coefficient ranged from 0.2×10^{-5} to $1.25 \times 10^{-5}\,°C$ under 25 MPa confining pressure. This variation in thermal expansion with temperature and pressure highlights the complexity of thermal–mechanical coupling in rocks.

2. Residual Strain as a Measure of Thermal Damage

Using residual strain after a heating–cooling cycle as a quantitative measure of thermal damage provides valuable insights into the permanent changes in rock structure caused by thermal loading. This method offers a practical and effective

3.4 Temperature–Pressure Coupling Test Technique

approach to evaluating thermal damage in rocks, which is essential for predicting their long-term behavior and stability in high-temperature environments.

3. Influence of Confining Pressure on Thermal Effects

The interplay between temperature and confining pressure plays a critical role in determining rock behavior. In granite, peak strength initially increases with temperature up to a threshold, after which it decreases. This behavior underscores the complex interaction between thermal and mechanical processes in rocks. Confining pressure enhances the initial strengthening effect of temperature, likely by facilitating crack closure and increasing friction between mineral grains. However, at higher temperatures, thermal damage becomes the dominant factor, reducing strength regardless of confining pressure.

4. Temperature Effects on Fracture Mechanisms

The temperature not only affects rock strength but also influences the fracture mechanisms. The study on granite [52] observed: At low temperatures, brittle failure characteristics were dominant. As temperature increased, especially above 200 °C, granite exhibited increasingly plastic behavior. Stress–strain curves showed increasing deformation ability with higher horizontal stress and temperature.

These changes in fracture behavior have important implications for the stability of rock structures in high-temperature environments. The transition from brittle to more ductile behavior with increasing temperature suggests that different failure criteria may be needed for rocks at elevated temperatures.

5. Mineral-Specific Thermal Effects

Different mineral components within rocks respond differently to thermal loading, contributing to the overall thermal–mechanical behavior. The study on granite [52] reported:

With increasing temperature, different minerals exhibited distinct mechanical responses: quartz demonstrated a transition from river-pattern to step-pattern cleavage, feldspar developed conjugate intragranular fracturing, while biotite exhibited enhanced fracture toughness (Figs. 3.29, 3.30 and 3.31).

Hydraulic-Mechanical Coupling

The interaction between fluid flow and mechanical deformation in rocks is a critical aspect of rock mechanics, particularly in applications such as hydrocarbon extraction, geothermal energy production, and underground waste storage (Fig. 3.32).

The interaction between fluid pressure and rock deformation plays a critical role in fracture propagation and stability. A study on the mechanical properties of a single rock fracture under thermal-hydro-mechanical coupling [3] revealed:

Thermal cycling exhibited distinct responses under different boundary conditions: under Constant Normal Load (CNL), heating induced thermal expansion with normal displacement up to 0.840 mm and decreased hydraulic aperture, while under Constant Normal Displacement (CND), thermal stress accumulated up to 3 times the initial

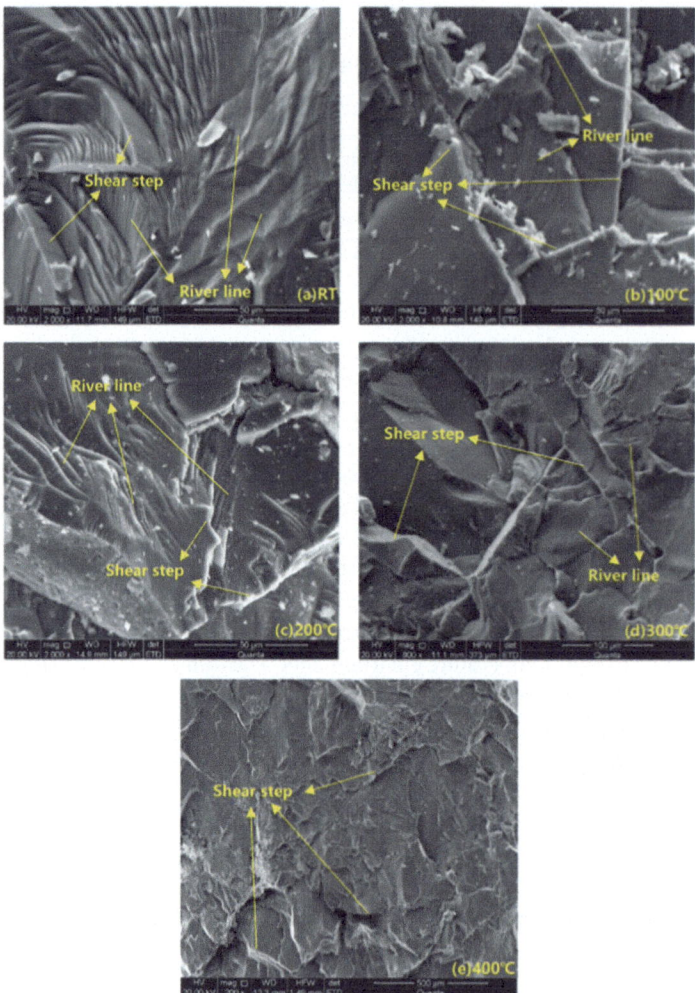

Fig. 3.29 Effect of temperature on fracture morphology of quartz particles [52]

magnitude with concurrent aperture reduction; in both conditions, subsequent cooling led to partial recovery of fracture properties.

The thermal–mechanical-hydraulic coupling in fractured rocks demonstrated significant temperature dependence: heating to 120 °C resulted in a 15.92% increase in water dynamic viscosity and an 84.55% reduction in hydraulic aperture, while subsequent cooling-induced shrinkage led to a substantial hydraulic aperture recovery of up to 241.72%, emphasizing the complex interplay between thermal effects, mechanical deformation, and fluid flow properties.

3.4 Temperature–Pressure Coupling Test Technique

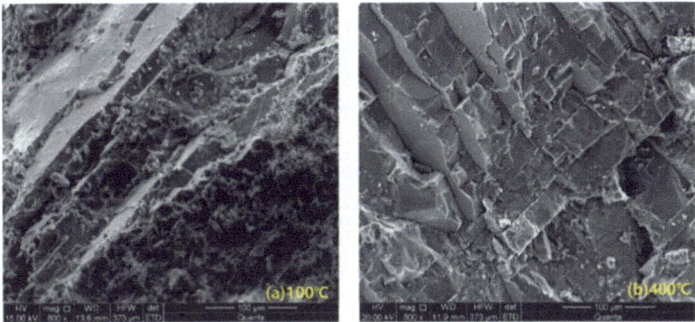

Fig. 3.30 Effect of temperature on fracture morphology of feldspar particles [52]

These findings demonstrate the significant impact of temperature on fracture hydraulic properties, which is crucial for understanding fluid flow in geothermal reservoirs and other high-temperature subsurface environments.

Chemical–Mechanical Coupling

The interaction between chemical processes and mechanical behavior in rocks adds another layer of complexity to the understanding of rock mechanics in various geological and engineering contexts. This section explores the effects of chemical reactions on rock properties and the resulting impacts on mechanical behavior (Fig. 3.33).

1. Chemical Effects on Rock Strength

Chemical reactions involving water can profoundly modify rock mechanical properties through complex geochemical processes, with significant implications for long-term geological applications such as CO_2 storage and nuclear waste repositories. These transformative interactions primarily occur through three key mechanisms: mineral dissolution, which can weaken rock structure by reducing intergranular cohesion; mineral precipitation, a process that simultaneously presents the potential to enhance rock strength by filling pores and microcracks while risking structural expansion and cracking when precipitated minerals exceed the volume of dissolved components; and ion exchange, which can fundamentally alter the mechanical characteristics of clay minerals, thereby influencing the overall rock strength and deformation behavior. Understanding these intricate chemical–mechanical interactions is crucial for predicting the long-term stability and integrity of geological media subjected to complex environmental conditions.

2. Chemical Effects on Fracture Propagation

Chemical processes play a critical role in fracture propagation dynamics within rock masses, with profound implications for understanding both anthropogenic hydraulic fracturing and natural fracture network evolution. Two key mechanisms fundamentally drive this phenomenon: subcritical crack growth, where chemical reactions at

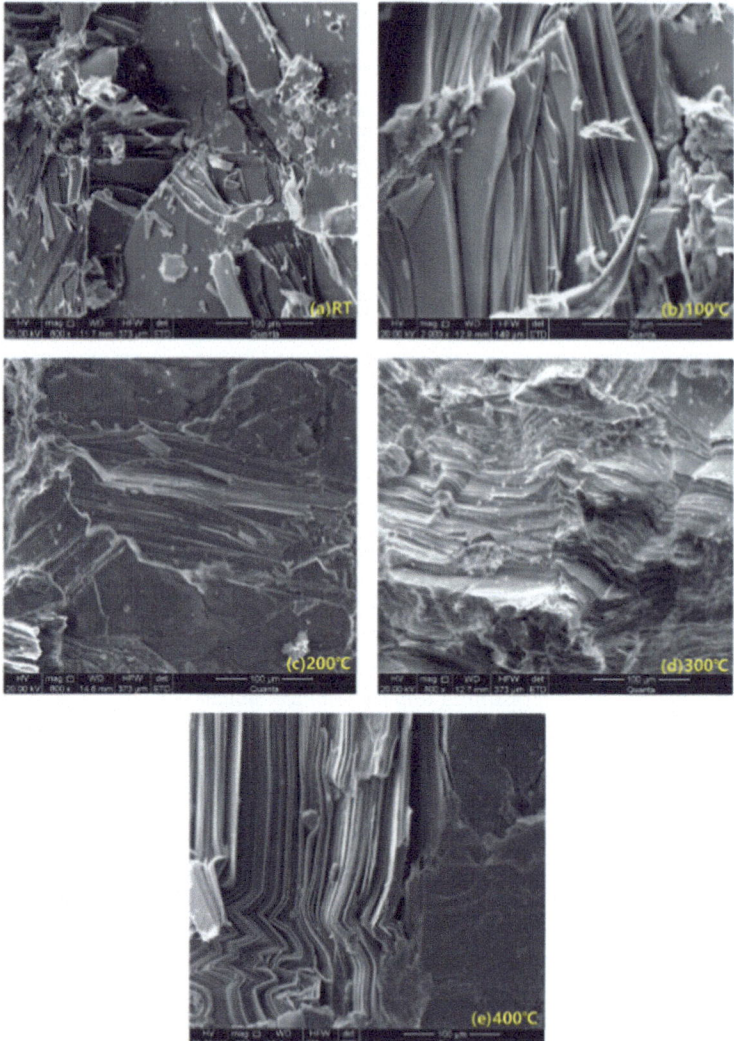

Fig. 3.31 Effect of temperature on fracture morphology of biotite particles [52]

crack tips enable crack advancement at stress intensities below traditional mechanical fracture thresholds, and fluid-rock interactions, which can dynamically modify local stress states and rock mechanical properties through complex geochemical exchanges.

3. Chemo-Mechanical Effects on Permeability

Chemical reactions and mechanical deformation interact in ways that can significantly alter rock permeability, a critical factor in understanding fluid dynamics

3.4 Temperature–Pressure Coupling Test Technique

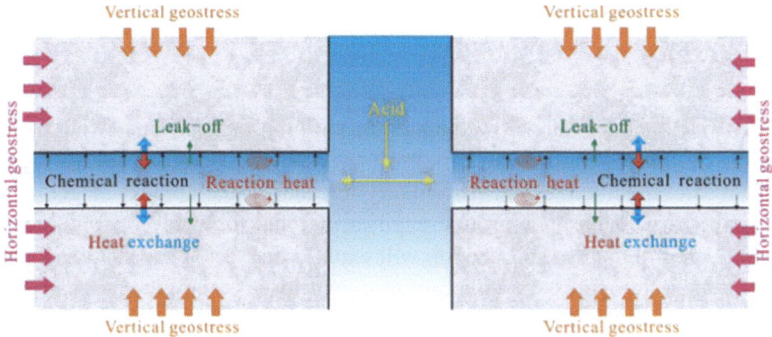

Fig. 3.32 THMC coupling during acid fracturing of carbonate geothermal reservoirs with heterogeneous fractures [53]

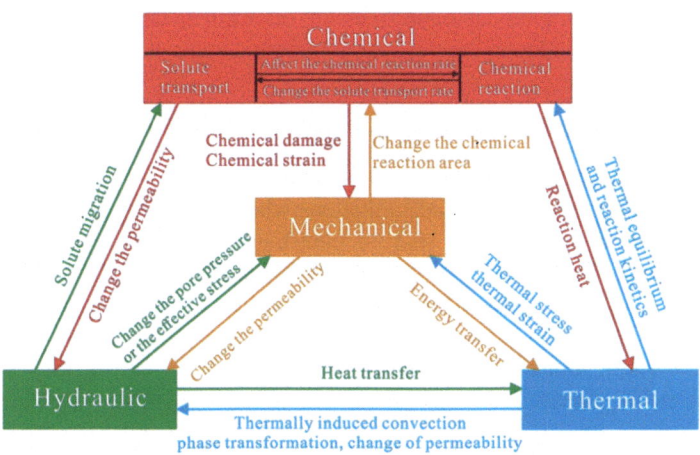

Fig. 3.33 Schematic diagram of full coupling of THMC [53]

within geological formations over extended periods. The key processes driving these changes include dissolution-induced compaction, where the dissolution of load-bearing minerals leads to rock compaction and subsequent permeability reduction, and precipitation-induced fracturing, in which mineral precipitation within pores and fractures generates internal stresses that can initiate new fractures or propagate existing ones. These complex chemical–mechanical interactions fundamentally reshape the internal structure and fluid transmission capabilities of rock masses, highlighting the dynamic nature of geological systems.

3.4.4 Challenges in High-Temperature and High-Pressure Testing

Rock dynamics testing under high-temperature and high-pressure conditions presents numerous challenges that researchers must overcome to obtain accurate and reliable data. These challenges are particularly relevant in the context of geothermal energy extraction, deep underground engineering, and the study of rock behavior in extreme environments. This section will explore the various difficulties encountered in temperature–pressure coupling tests and discuss the innovative solutions developed to address these issues.

Sealing and Pressure Maintenance

One of the primary challenges in high-temperature and high-pressure testing is maintaining an effective seal around the rock sample while simultaneously withstanding extreme conditions. Traditional sealing methods often fail under the combined effects of high temperature and pressure, leading to inaccurate measurements and potential safety hazards.

1. Conventional Sealing Limitations

Conventional heat-shrinkable tubing, commonly used in rock mechanics testing, has been shown to have significant limitations when exposed to high-temperature and high-pressure conditions. A study demonstrated that traditional high-voltage tubing (ethylene–vinyl acetate copolymer -based) failed at relatively low conditions of 55.01 °C and 33.26 MPa. Similarly, dual-wall tubing with an ethylene–vinyl acetate copolymer outer wall and hot-melt adhesive inner wall failed at 58.62 °C and 74.46 MPa. These failure points are far below the conditions required for simulating deep rock environments, which can exceed 150 °C and 140 MPa [54] (Fig. 3.34).

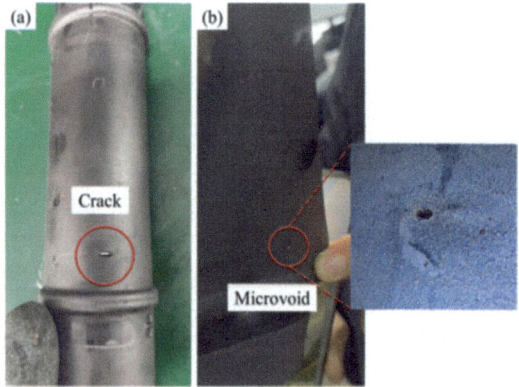

Fig. 3.34 Failure of heat-shrinkable tubing under different temperature and pressure conditions: **a** high-voltage heat-shrinkable tubing, and **b** dual-wall heat-shrinkable tubing [54]

3.4 Temperature–Pressure Coupling Test Technique

Conventional sealing materials exhibit failure mechanisms under extreme conditions that stem from multiple complex interactions. These include thermal degradation, where high temperatures induce polymer chain breakdown, resulting in substantial loss of mechanical strength and sealing capabilities; pressure-induced deformation, characterized by ultrahigh pressures causing excessive material distortion that potentially leads to rupture or complete loss of sealing effectiveness; and material incompatibility, wherein certain sealing materials undergo chemical reactions with rock samples or testing fluids at elevated temperatures, fundamentally compromising their structural integrity and functional performance. These interconnected failure modes underscore the critical challenges in developing robust sealing solutions for high-stress, high-temperature geological and engineering applications.

2. Innovative Sealing Solutions

To address the limitations of conventional sealing materials, researchers have developed novel polymer heat-shrinkable tubing capable of withstanding high-temperature and ultrahigh-pressure conditions. The study [54] introduced a polymer composite tubing that maintained excellent sealing performance at 140 MPa and 150 °C. This innovative tubing is composed of: 50% acrylic rubber; 30% polyethylene; 15% flame retardant; 1% antioxidant; 4% coloring agent.

The novel polymer tubing achieves its exceptional performance through meticulously engineered material composition, adeptly balancing heat resistance and mechanical strength. Its superior characteristics stem from the synergistic interaction between acrylic rubber and polyethylene, the significant enhancement of thermal stability by flame retardants and antioxidants, and the material's remarkable adaptability to pressure-induced deformation. This composite material optimizes thermomechanical properties at the molecular scale while simultaneously ensuring sealing performance stability at the macroscopic level.

3. Rock Surface Effects on Sealing Performance

An important consideration in high-temperature and high-pressure testing is the impact of rock surface characteristics on sealing performance. The study [54] revealed that rock surface irregularities significantly affect the deformation of heat-shrinkable tubing, potentially leading to localized stress concentrations and failure.

To quantify this effect, a failure criterion was developed based on 3D laser scanning of rock surface characteristics:

$$T_{max} = \left(\sum_{i=1}^{n} N_{ij} N_{i+1j} - r_j \theta \right) \frac{2\pi}{\theta} + \varepsilon_{r,max}^{h} < 1.27 D_0 \qquad (3.12)$$

where: T_{max} is the maximum radial deformation. $\varepsilon_{r,max}^{h}$ is the maximum radial rock deformation. D_0 is the initial tubing circumference. $N_{i,j}$ represents the ith data point of the jth interface.

This criterion provides a valuable tool for predicting tubing performance based on rock surface characteristics and experimental conditions. It highlights the importance of considering rock surface irregularities when designing sealing solutions for high-temperature and high-pressure tests.

Temperature Control and Uniformity

Achieving and maintaining uniform temperature distribution throughout the rock sample and testing apparatus is another significant challenge in high-temperature and high-pressure testing. Non-uniform temperature distribution can lead to inaccurate measurements, thermal stress gradients, and potential damage to the testing equipment.

1. Temperature Gradients and Their Effects

Rock samples subjected to temperature gradients experience complex mechanical property transformations that profoundly influence their structural integrity and failure mechanisms. The study by investigating the mechanical behavior of Australian Strathbogie granite under geothermal reservoir conditions revealed intricate temperature-dependent responses [55]. Strength variations demonstrated a nonlinear relationship between deviatoric stress at failure and temperature, initially increasing up to 200°C and subsequently decreasing at 300°C, a phenomenon attributed to the competing mechanisms of thermal cracking and mineral strengthening. Elastic properties exhibited similarly nuanced changes, with elastic modulus and Poisson's ratio increasing up to 100°C before experiencing progressive decreases and fluctuations at elevated temperatures. Microstructural analyses using SEM further illuminated these complex interactions, revealing that thermal cracking becomes increasingly pronounced above 300°C, particularly affecting larger crystal structures exceeding 0.5 mm in size (Fig. 3.35).

These observations underscore the importance of achieving uniform temperature distribution throughout the sample to ensure accurate and representative measurements of rock properties.

2. Innovative Temperature Control Systems

To address the challenges of temperature control and uniformity, researchers have developed advanced heating systems for high-temperature and high-pressure testing apparatus. The study on the development of a thermo-hydro-mechanical coupling test system introduced a specially designed high-temperature environment box with the following features: a double circulation heating system, which improves air flow and ensures more uniform heating throughout the test area; a temperature range capable of maintaining temperatures up to 300°C with an accuracy of $\pm$ 4°C; and a heat insulation system carefully designed to prevent temperature transmission to sensitive components of the testing apparatus [3]. The effectiveness of this temperature control system is demonstrated by its ability to maintain stable and uniform temperatures during long-term creep tests lasting over 2000 h.

3.4 Temperature–Pressure Coupling Test Technique

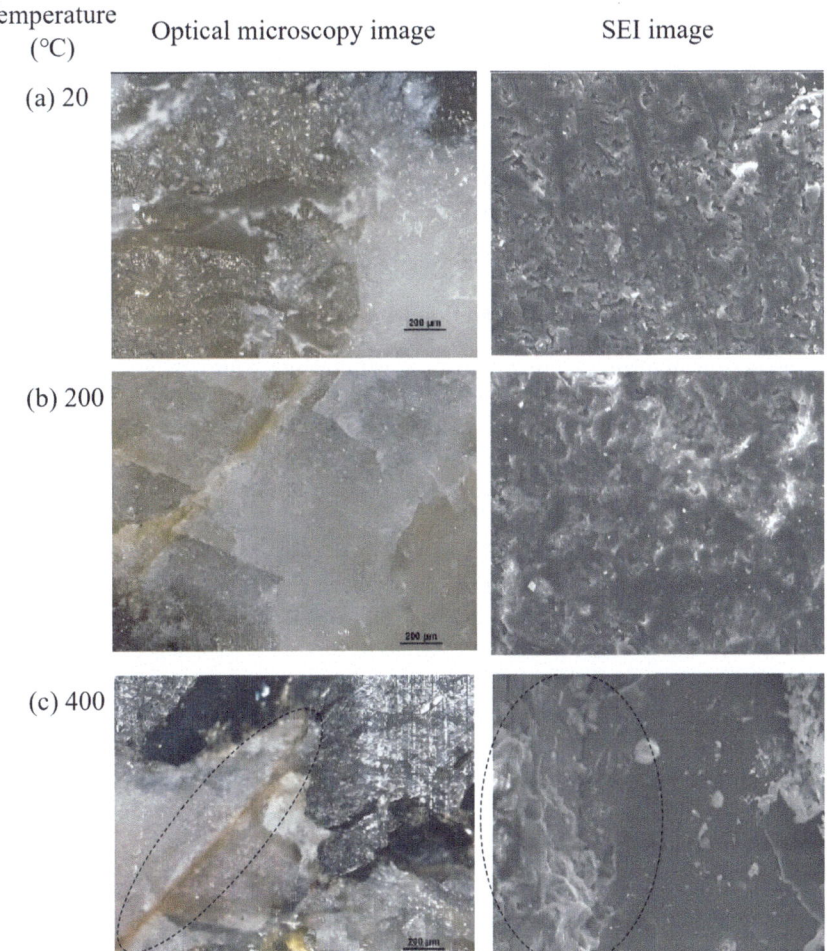

Fig. 3.35 Optical microscopic and SEM images of thin sections of Strathbogie granite **a** at room temperature; **b** pre-heated to 100 °C **c** pre-heated to 400 °C [55]

3. Temperature-Dependent Material Properties

Accurate temperature control is crucial for understanding the temperature-dependent behavior of rocks under high-pressure conditions. The study on the mechanical properties of granite under real-time high temperature and three-dimensional stress conditions revealed several important temperature-dependent phenomena: peak strength variations, where the peak strength of granite first increased up to a temperature threshold (which depended on the applied horizontal stress) and then decreased [52]. The maximum strength increase compared to room temperature ranged from 23.1% to 71.6%, depending on the horizontal stress level. Additionally, cohesion and internal friction angle changes were observed, with the internal friction angle remaining relatively stable around 46°, while cohesion increased with temperature up to a threshold and then decreased. In addition, different mineral components within the granite exhibited varying responses to high temperatures. For example, quartz showed a transition from river pattern cleavage to step pattern cleavage with increasing temperature, while biotite underwent toughness enhancement at high temperatures.

These findings highlight the complexity of temperature-dependent rock behavior and underscore the need for precise and uniform temperature control in high-temperature and high-pressure testing.

Strain Measurement and Deformation Monitoring

Accurate measurement of strain and deformation under high-temperature and high-pressure conditions presents significant technical challenges. Traditional strain measurement techniques may be inadequate or unreliable in extreme environments, necessitating the development of specialized instrumentation and methodologies.

1. Limitations of Conventional Strain Gauges

Conventional strain gauges, widely used in rock mechanics testing under ambient conditions, encounter significant limitations when exposed to high temperatures and pressures. These limitations encompass multiple critical aspects: temperature sensitivity, where many strain gauges are prone to thermal expansion effects that can obscure the actual strain in rock samples; adhesive degradation, as the bonding materials may deteriorate or fail at elevated temperatures, resulting in inaccurate measurements or complete data loss; pressure effects, wherein high pressures can cause direct deformation of the strain gauge, potentially generating erroneous readings; and durability concerns, as the extreme environment of high-temperature and high-pressure tests can dramatically reduce the operational lifespan of conventional strain gauges, rendering them unsuitable for long-term experimental investigations.

2. Advanced Strain Measurement Techniques

Conventional strain gauges face significant limitations in high-temperature and high-pressure testing environments, prompting researchers to develop and implement advanced strain measurement techniques. These innovations provide more accurate, reliable, and versatile solutions for studying the mechanical behavior of rock materials under extreme conditions.

High-temperature strain gauges are specifically designed to endure elevated temperatures by utilizing materials with low thermal expansion coefficients and high-temperature-resistant adhesives. These gauges are tailored to maintain measurement accuracy in conditions where conventional gauges would fail.

Non-contact optical methods, such as DIC, offer a significant advantage by enabling strain measurement without direct contact with the rock sample. By using appropriate viewing windows and cooling systems, these methods can be adapted for high-temperature environments, making them a versatile tool for extreme testing conditions.

3.4 Temperature–Pressure Coupling Test Technique

Fiber optic sensors, particularly Fiber Bragg Grating (FBG) sensors, have gained prominence for their ability to measure both strain and temperature simultaneously. These sensors are highly resistant to electromagnetic interference and can be embedded within or attached to rock samples, allowing for continuous and reliable monitoring throughout the testing process.

Acoustic emission monitoring, though not a direct strain measurement technique, provides valuable insights into the behavior of rock samples. By detecting and analyzing acoustic signals generated during crack initiation and propagation, this method complements strain measurements by offering critical information about damage mechanisms under high-temperature and high-pressure conditions.

These advanced techniques significantly enhance the capability to monitor and analyze the deformation and failure of rock materials under extreme conditions, contributing to a deeper understanding of rock mechanics and improving the reliability of experimental studies in rock dynamics and related fields.

Fluid Flow and Permeability Measurement

Many high-temperature and high-pressure rock mechanics tests involve the study of fluid flow and permeability, particularly in the context of geothermal energy extraction and hydrocarbon reservoir characterization. Measuring fluid flow and permeability under extreme conditions presents unique challenges that must be addressed to obtain accurate and reliable data.

1. Temperature and Pressure Effects on Fluid Properties

Fluid flow measurements at high temperatures and pressures present significant challenges due to substantial changes in fluid properties under extreme conditions. The development of a thermo-hydro-mechanical coupling test system emphasized the critical importance of understanding temperature-dependent fluid properties [3]. Viscosity changes emerged as a key concern, with dynamic water viscosity increasing significantly with temperature—specifically, heating to approximately 120°C resulted in up to a 15.92% increase in dynamic viscosity. Density variations also play a crucial role, as fluid density changes with temperature and pressure, directly impacting flow behavior and permeability measurements. Additionally, phase changes at extreme temperatures and pressures can trigger dramatic transformations, such as transitions from liquid to supercritical fluid states, which fundamentally alter fluid behavior and characteristics.

2. Fracture Flow Behavior

In many rock types, especially those relevant to geothermal energy extraction, fluid flow primarily occurs through fractures. The behavior of these fractures under high-temperature and high-pressure conditions significantly impacts overall fluid flow and permeability. The modified cubic law was employed to depict fluid flow in a single rock fracture:

$$b = \sqrt[3]{\frac{12q\mu L}{D \cdot \Delta P}} \qquad (3.13)$$

where: b is the equivalent hydraulic aperture (m); q is the flow rate (m^3/s); μ is the dynamic viscosity of water (Pa·s); L is the flow distance (m); D is the fracture width perpendicular to flow direction (m); ΔP is the pressure difference between inlet and outlet (Pa).

3. Advanced Flow Measurement Techniques

Accurate fluid flow measurement under extreme conditions of high temperature and high pressure poses significant challenges. To overcome these limitations, researchers have developed advanced techniques that enable precise monitoring and visualization of fluid behavior within rock samples, providing critical insights into subsurface processes.

High-temperature flow meters are specifically designed to operate under extreme conditions, ensuring accurate measurement of flow rates. These devices are engineered with materials capable of withstanding high thermal and pressure environments, making them essential tools for fluid flow experiments in challenging settings.

Tracer tests using heat-resistant tracers provide detailed information about fluid flow paths and residence times within rock samples. This technique is particularly valuable for understanding the movement of fluids in porous and fractured media under extreme testing conditions.

Acoustic methods, including acoustic emissions monitoring, offer indirect but insightful measurements of fluid behavior. These techniques can detect fluid-induced fracturing and track changes in flow paths, offering real-time information about the dynamic interactions between fluids and rock structures.

Computed tomography (CT) scanning, when integrated with specialized pressure vessels, allows for real-time visualization of fluid flow within rock samples under high-temperature and high-pressure conditions. This technique provides a three-dimensional perspective on fluid distribution and movement, making it a powerful tool for studying flow dynamics in complex rock systems.

These advanced flow measurement techniques enable researchers to better understand fluid-rock interactions under extreme conditions, enhancing the accuracy and reliability of experiments in fields such as reservoir engineering, geothermal energy exploration, and rock mechanics.

4. Permeability Evolution

Understanding how permeability evolves under high-temperature and high-pressure conditions is crucial for many applications, including geothermal reservoir management. To accurately measure and interpret permeability evolution under high-temperature and high-pressure conditions, researchers must develop experimental protocols that allow for permeability measurements at various stages of thermal and pressure loading, implement numerical models that can account for the competing effects of temperature and pressure on pore and fracture networks, and consider the time-dependent aspects of permeability evolution, particularly for long-term experiments or when simulating reservoir conditions over extended periods.

3.4 Temperature–Pressure Coupling Test Technique

Time-Dependent Behavior and Long-Term Testing

Understanding the time-dependent behavior of rocks under high-temperature and high-pressure conditions is crucial for many engineering applications, including geothermal energy extraction and deep geological disposal of radioactive waste. However, conducting long-term tests under extreme conditions presents numerous challenges.

1. Creep and Stress Relaxation

Rocks often exhibit creep (time-dependent deformation under constant stress) and stress relaxation (time-dependent stress reduction under constant strain) behaviors, which can be significantly affected by high temperatures and pressures. The study on the development of a thermo-hydro-mechanical coupling test system demonstrated the capability to maintain creep conditions for over 2000 h, highlighting the importance of long-term testing capabilities [52].

Under high-temperature and high-pressure conditions, studying creep and stress relaxation involves complex experimental challenges centered on three critical aspects: maintaining consistent environmental stability, mitigating measurement uncertainties, and precisely differentiating thermal and mechanical deformation mechanisms. The technical difficulty lies in preserving constant temperature, pressure, and auxiliary parameters over extended periods, which demands sophisticated experimental apparatus. Long-term tests are particularly vulnerable to sensor drift and instrumental degradation, necessitating advanced calibration techniques and specialized instrumentation. Furthermore, the nuanced distinction between thermal-induced dimensional changes and genuine time-dependent mechanical deformation presents a significant analytical challenge, especially when temperature fluctuations introduce additional complexity to the experimental system.

2. Fatigue and Cyclic Loading

In many geotechnical applications, rocks are subjected to cyclic loading conditions, which can lead to fatigue damage over time. Studying fatigue behavior under high-temperature and high-pressure conditions adds another layer of complexity to rock mechanics testing. Challenges in this area include: Designing loading systems capable of applying cyclic loads at high frequencies under extreme conditions; Monitoring and characterizing progressive damage accumulation during long-term cyclic loading tests; Developing predictive models for fatigue life that account for the combined effects of temperature, pressure, and cyclic loading.

3. Equipment Durability and Reliability

Conducting long-term tests under high-temperature and high-pressure conditions places extreme demands on testing equipment. Ensuring the durability and reliability of all components over extended periods is a significant challenge. Key considerations include: Material selection for pressure vessels, seals, and other critical components; Implementing redundant safety systems to prevent catastrophic failures during long-term tests; Developing maintenance and inspection protocols for long-term testing equipment.

High-temperature and high-pressure rock mechanics testing presents a wide array of challenges that span multiple disciplines, including materials science, instrumentation, data analysis, and modeling. Addressing these challenges requires innovative approaches, advanced technologies, and interdisciplinary collaboration. As researchers continue to develop new techniques and methodologies, our understanding of rock behavior under extreme conditions will improve, leading to more reliable and efficient solutions for geotechnical engineering problems in deep underground environments.

3.5 Advanced Test Techniques and Methods

3.5.1 Micro-scale Rock Dynamics Testing

In recent years, the field of rock dynamics has seen significant advancements in testing techniques, particularly at the micro-scale level. These developments have been driven by the need for a more comprehensive understanding of rock behavior under dynamic loading conditions, especially in complex geological environments. Micro-scale rock dynamics testing offers unique insights into the fundamental mechanisms governing rock deformation, fracture, and failure processes, which are crucial for various geotechnical and geological applications, including mining, tunneling, and geological storage of hazardous materials.

This section explores the cutting-edge techniques and methodologies employed in micro-scale rock dynamics testing, focusing on their principles, applications, and limitations. We will discuss various advanced methods, including acoustic emission monitoring, X-ray micro-computed tomography (micro-CT) and 3D printing.

Acoustic Emission and X-ray Micro-CT for Crack Propagation Monitoring

One of the most promising advancements in micro-scale rock dynamics testing is the integration of acoustic emission (AE) monitoring with X-ray micro-computed tomography (micro-CT). This innovative approach enables researchers to correlate acoustic signals with real-time visual observations of crack initiation and propagation, offering unprecedented insights into the fracture mechanics of rocks under dynamic loading conditions. By combining these techniques, researchers can integrate multiple data sources and apply advanced processing methods to extract meaningful information from micro-scale rock dynamics tests (Fig. 3.36).

This integrated approach has several significant implications for the field of rock mechanics. It enhances the understanding of fracture processes by linking acoustic signals to visual crack development, which is essential for improving predictive models of rock behavior under dynamic loading. It also allows for the quantification of microstructural effects, providing direct observations of how variations in rock microstructures influence their dynamic responses. This information supports the development of more accurate constitutive models that account for microstructural

3.5 Advanced Test Techniques and Methods

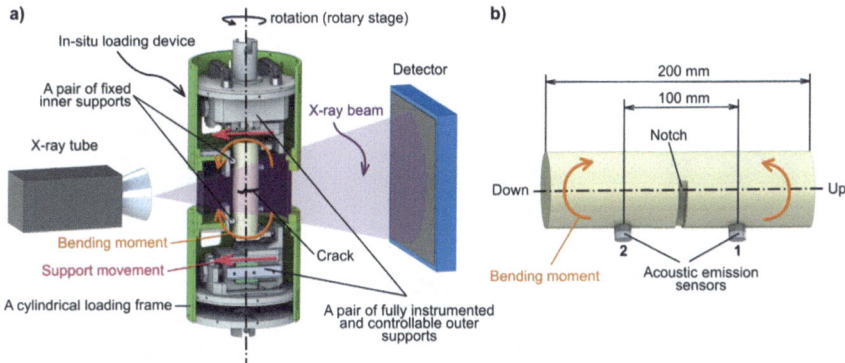

Fig. 3.36 The micro-scale rock dynamics testing experiment using acoustic emission monitoring and X-ray micro-computed tomography [56]

variability. Furthermore, the AE-micro-CT method improves the characterization of the fracture process zone (FPZ), offering detailed insights into energy dissipation mechanisms in rocks during dynamic loading. Finally, the comprehensive experimental data generated through this approach serves as a critical resource for validating and refining numerical models of rock fracture under dynamic conditions.

Despite its many advantages, the combined AE and X-ray computed tomography (XCT) approach for micro-scale rock dynamics testing has limitations that warrant attention in future research. One challenge lies in the resolution of current XCT imaging, which, at 21.79 μm voxel size in this study, may fail to capture the finest details of crack initiation and early propagation. Advancements in spatial resolution, while maintaining the capability to image larger sample sizes, are essential for more precise observations.

Another limitation is temporal resolution, as the interrupted loading procedures required for XCT scanning may overlook critical transient phenomena. Developing faster XCT scanning techniques or exploring alternative imaging methods that enable continuous monitoring could address this issue and provide a more comprehensive understanding of dynamic fracture processes. Future efforts should aim to overcome these challenges, ensuring the AE-XCT method reaches its full potential in advancing rock mechanics research.

Application of 3D Printing in Micro-Scale Rock Dynamics Testing

An emerging technology that holds great promise for advancing micro-scale rock dynamics testing is 3D printing. This additive manufacturing technique allows for the creation of rock-like specimens with precisely controlled internal structures, enabling researchers to isolate and study specific aspects of rock behavior under dynamic loading conditions.

A comprehensive review highlights the potential applications and current limitations of 3D printing in rock mechanics research [57, 58]. The technology has been applied to five major areas in rock mechanics: Preparation of rock specimens

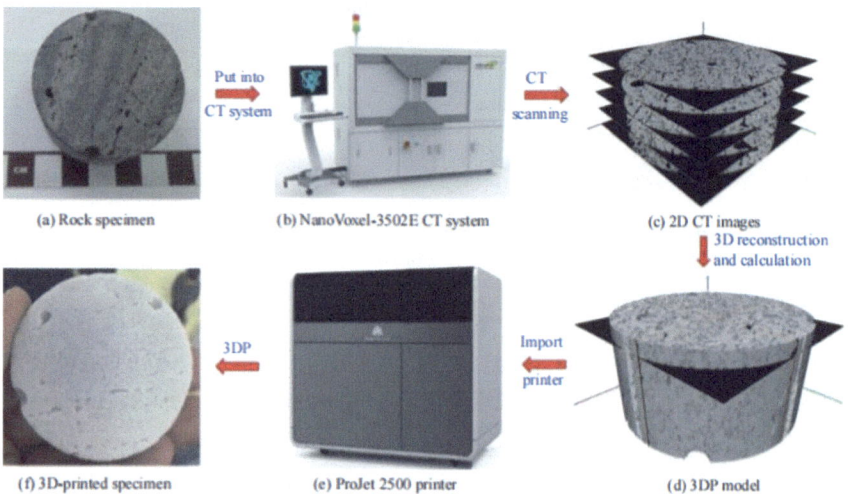

Fig. 3.37 General procedures for preparing specimens using 3D-printed technologies coupled with CT scanning technology [57]

(including pre-flawed specimens); Preparation of joints; Preparation of geophysical models; Reconstruction of complex rock structures; Bridging experimental testing and numerical simulation.

1. 3D Printed Rock Specimens for Dynamic Testing

One of the most significant advantages of 3D printing in micro-scale rock dynamics testing is the ability to create specimens with controlled internal structures and pre-existing flaws. This capability allows researchers to systematically study the influence of specific structural features on dynamic rock behavior (Fig. 3.37).

A variety of 3D-printed materials have been employed to fabricate rock-like specimens, including polymers, resin, gypsum, sand, ceramics, and geological materials. Each material offers distinct mechanical properties, making their selection dependent on the specific aspects of rock behavior being studied. For instance, resin-based specimens closely mimic brittle rocks in terms of fracture behavior, while gypsum and sand-based specimens are suitable for simulating weaker rocks like sandstone. Additionally, rock-like geological materials are particularly effective for replicating layered rocks and complex heterogeneous structures. It is worth noting that most 3D-printed specimens typically exhibit higher ductility, lower strength, and lower stiffness compared to natural rocks. However, these differences can be advantageous in certain scenarios, such as studying slow crack propagation processes that occur too rapidly in natural rocks for detailed observation.

3D printing excels in creating specimens with controlled pre-existing flaws, making it invaluable for studying crack initiation and propagation under dynamic loading. This technique offers several key advantages: precise control over flaw

3.5 Advanced Test Techniques and Methods

geometry, orientation, and distribution; the ability to produce complex three-dimensional flaw networks; and reproducibility for statistically significant testing. Common materials for pre-flawed specimens include transparent resin (enabling visual observation of crack growth), gypsum, sand powder, and polymers. Research indicates that crack initiation and propagation in 3D-printed flawed specimens closely resemble those in natural rocks, validating the method as a robust tool for fundamental fracture mechanics studies (Fig. 3.38).

The integration of 3D printing and micro-CT imaging offers transformative opportunities for micro-scale rock dynamics research. This approach allows researchers to create 3D-printed specimens based on micro-CT scans of natural rocks, preserving their intricate internal structures, while enabling dynamic tests to be conducted on these specimens with micro-CT imaging used to monitor internal deformation and fracture processes. Furthermore, the models can be iteratively refined based on experimental observations, creating a continuous cycle of improvement [59]. This process provides an unprecedented level of control and insight into rock dynamics, advancing the field in ways previously unattainable.

Despite its potential, the application of 3D printing in micro-scale rock dynamics testing faces several challenges that must be addressed to fully realize its capabilities. One major limitation lies in the material properties of current 3D-printable materials, which do not fully replicate the strength and brittleness of natural rocks. Additionally, the resolution of current 3D printers is insufficient to capture the intricate details of rock microstructures, particularly for fine-grained rocks. Limited multi-material printing capabilities further hinder the recreation of the complex mineralogical compositions inherent to natural rocks. Moreover, the restricted size of specimens

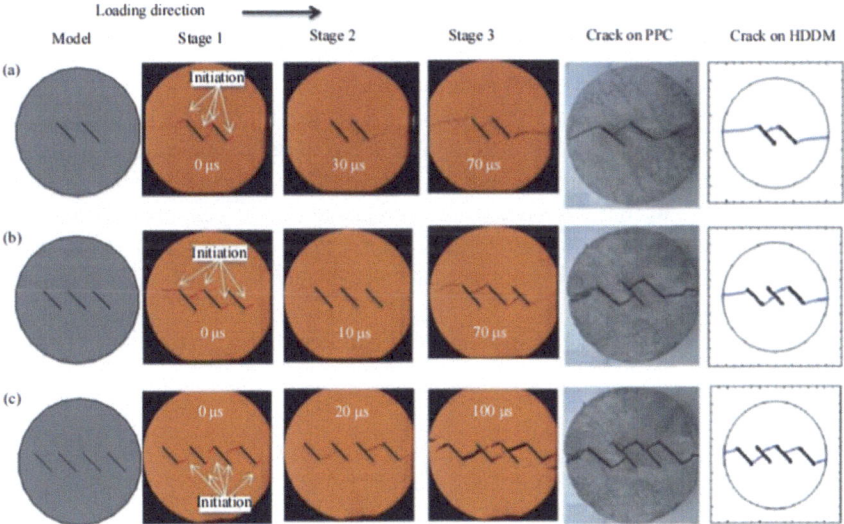

Fig. 3.38 Specimens with pre-existing flaws and their cracks evolution [57]

due to current printer dimensions poses challenges for studying scale effects in rock dynamics.

3.5.2 Real-Time Monitoring and Imaging Techniques

In recent years, the field of rock dynamics has witnessed significant advancements in real-time monitoring and imaging techniques. These innovations have revolutionized our ability to observe, analyze, and understand the complex behaviors of rock masses under dynamic loading conditions. This section explores the cutting-edge technologies and methodologies that enable researchers and engineers to gain unprecedented insights into rock dynamics processes.

Acoustic Emission (AE) Monitoring

AE monitoring has emerged as a powerful tool for real-time analysis of rock damage and fracture propagation. This technique relies on the detection and analysis of elastic waves generated by the rapid release of energy from localized sources within materials under stress [60–62].

AE monitoring in rock dynamics involves several key components that work together to detect and analyze micro-fracture events within the rock. The primary sensor used is typically a piezoelectric transducer, which detects high-frequency elastic waves generated by these microfractures. A high-speed data acquisition system is employed to record and process the AE signals, usually operating at sampling rates in the MHz range to capture the high-frequency content of the events. Once the data is collected, advanced signal processing techniques are applied to extract meaningful information, such as event location, magnitude estimation, and frequency content analysis. These components work in unison to provide detailed insights into the fracture processes occurring within the rock.

One of the key advantages of AE monitoring is its ability to locate the sources of micro-fracture events. This is generally achieved through triangulation methods, which rely on arrival time differences between multiple sensors to pinpoint the location of the AE events [63]. To improve the accuracy of source location, error analysis techniques, such as the use of error ellipsoids, can be incorporated. These techniques help refine the location estimates and enhance the precision of the monitoring process.

The relative magnitude of AE events can be estimated using the following formula [61]:

$$M_r = \frac{2}{3}\log_{10}\frac{1}{N}\sum_{i=1}^{N}r_iA_i \quad (3.14)$$

where M_r is the relative magnitude, N is the number of signals, r_i is the source-to-sensor distance, A_i is the absolute value of the first peak amplitude.

3.5 Advanced Test Techniques and Methods

This magnitude estimation allows researchers to quantify the energy release associated with different stages of rock damage and failure.

Recent research has demonstrated the effectiveness of AE monitoring in studying fracture propagation in rocks with pre-existing flaws. AE monitoring has shown high sensitivity to damage initiation, enabling the detection of damage in rocks at stress levels well below the macroscopic failure threshold. This early detection capability is crucial for understanding rock behavior before complete failure occurs. Furthermore, AE signals exhibit a near-linear correlation with tensile damage accumulation in rocks, making them a useful tool for monitoring damage progression under tensile stress. However, the relationship between AE signals and shear damage accumulation is more complex and non-linear, requiring additional analysis to interpret accurately. AE events also tend to cluster around flaw tips, providing insights into the stress concentrations and damage initiation processes that occur in fractured rocks. This clustering behavior helps researchers better understand the conditions that lead to crack propagation and fracture growth.

Digital Image Correlation (DIC)

DIC has emerged as a powerful non-contact technique for measuring full-field deformation and strain in rock samples during dynamic loading tests. This method offers high spatial resolution and sensitivity, making it ideal for studying the complex strain fields associated with crack initiation and propagation in rocks. The principles behind DIC involve several key steps: First, the surface of the rock sample is prepared with a high-contrast speckle pattern to facilitate tracking. Then, high-resolution digital cameras capture a series of images of the sample surface as it undergoes loading. Advanced algorithms process these images, tracking the movement of surface features between successive images to calculate displacement fields. These displacement gradients are then used to compute full-field strain maps across the sample surface, providing detailed insights into the deformation behavior (Fig. 3.39).

Recent studies have demonstrated the effectiveness of DIC in analyzing rock fracture behavior, particularly in understanding the initiation and propagation of cracks. One of the primary applications of DIC in rock fracture studies is its ability to precisely track the initiation and growth of cracks on the rock surface, offering valuable information about the underlying failure mechanisms. The technique also reveals areas of strain concentration that often precede visible crack formation, helping researchers identify critical regions of damage. Furthermore, DIC allows

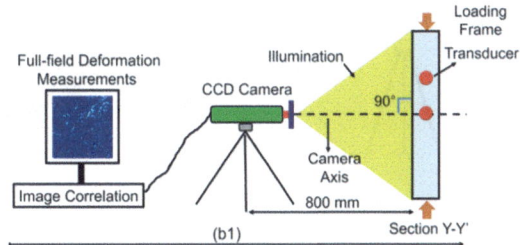

Fig. 3.39 The experimental set-up used for monitoring the progression of damage in the rock specimens using DIC [64]

for the quantification of non-elastic strains, including apparent tensile and shear strains, which are crucial for understanding the accumulation of damage in rocks and the progression toward failure.

Infrared Thermography

Infrared thermography has emerged as a valuable non-contact technique for monitoring thermal changes associated with rock deformation and failure. This method enables researchers to visualize and quantify temperature distributions on rock surfaces during dynamic loading tests, providing insights into energy dissipation processes and damage evolution. The application of infrared thermography in rock dynamics involves several critical components. A high-sensitivity infrared camera captures thermal radiation emitted from the rock surface during loading. The camera is calibrated to convert the detected radiation into accurate temperature measurements. After acquiring the thermal images, advanced image processing techniques are applied to extract meaningful temperature data and patterns, allowing researchers to study the thermal behavior of the rock as it undergoes deformation and failure (Fig. 3.40).

Recent research has shown that infrared thermography can detect subtle thermal changes associated with different stages of rock damage. These changes offer valuable information about the progression of rock failure. During the initial compaction phase, a slight temperature increase is often observed, which is linked to pore closure and microcrack formation. During the elastic deformation stage, temperature changes are typically minimal. However, as cracks begin to grow stably, localized temperature increases can be observed along crack paths due to friction and plastic deformation. In the unstable crack development phase, rapid temperature increases are often seen

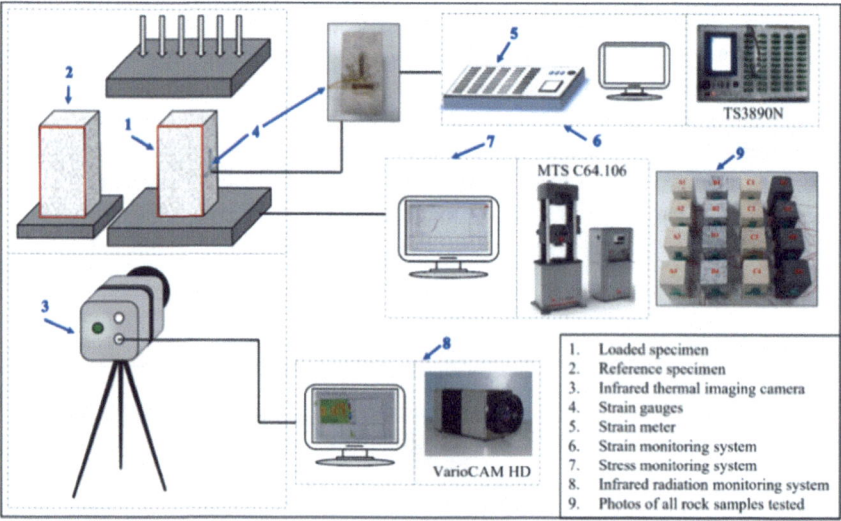

Fig. 3.40 Equipment physical diagram, test schematic diagram, and test rock sample photos [65]

3.5 Advanced Test Techniques and Methods

immediately before and during macroscopic failure, reflecting the release of stored elastic energy.

To extract meaningful insights from infrared thermography data, several quantitative analysis techniques have been developed. The Average Infrared Radiation Temperature (AIRT) provides a global measure of the thermal state of the rock sample, which can be used to track overall damage evolution. Additionally, spatial temperature gradients on the rock surface can reveal areas of stress concentration and incipient failure, while temporal temperature evolution at specific points or regions of interest offers further insights into local damage processes.

Recent advancements have seen the integration of infrared thermography with machine learning techniques for automated damage state identification. This approach involves preprocessing infrared images to remove noise and background thermal effects, followed by feature extraction to capture relevant details such as temperature matrices or statistical descriptors. A convolutional neural network (CNN) is then trained to classify rock damage states based on these features. Once trained, the model can be applied in real-time to identify damage states during dynamic loading tests. This machine learning integration has shown promising results, with reported accuracies exceeding 99% in classifying rock damage states based on infrared thermography data.

Advantages of infrared thermography in rock dynamics include its non-contact nature, which avoids interfering with the rock sample or loading apparatus, and its ability to provide full-field information through spatial temperature distributions across the entire visible surface. Modern infrared cameras also have high sensitivity, allowing them to detect minute temperature changes that are associated with early-stage damage. However, there are limitations, such as the fact that infrared thermography is restricted to surface measurements, and it is sensitive to environmental factors like ambient temperature variations in surface emissivity. Additionally, interpreting the observed thermal patterns in terms of specific damage mechanisms can be complex and may require integration with other techniques. Despite these challenges, infrared thermography provides a unique perspective on energy dissipation and damage processes in rocks under dynamic loading, complementing other real-time monitoring techniques.

Fiber-Optic Strain Sensing

Fiber-optic strain sensing technology has recently emerged as a powerful tool for real-time monitoring of rock deformation and fracture processes. This technique is particularly valued for its high spatial resolution, sensitivity, and ability to perform distributed measurements along the entire length of the fiber. The application of fiber-optic strain sensing in rock dynamics typically involves the use of a specially designed optical fiber, which is either embedded in or attached to the rock sample. A coherent light source, often a laser, sends light pulses through the fiber, and the backscattered light is analyzed using an Optical Frequency Domain Reflectometer (OFDR). This device measures strain along the fiber by processing the data from the reflected light. Advanced algorithms are then used to calculate the distributed strain measurements (Fig. 3.41).

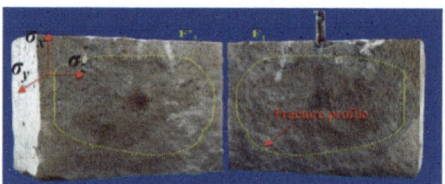

Fig. 6. Post-fracture observation of Sample #1 with fracture patterns.

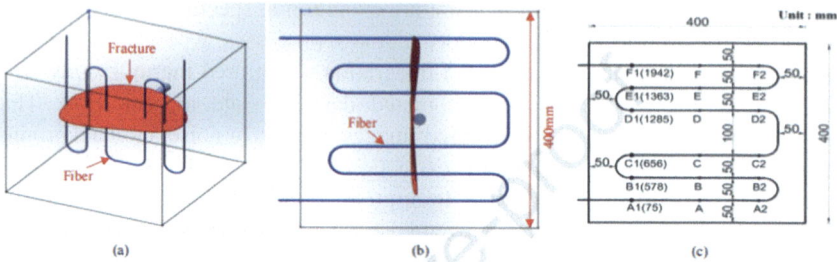

Fig. 3.41 Schematic diagram of fracture reconstruction: **a** 3D model of the sample, **b** Top view of the fracture, and **c** Schematic of the optical fiber path coordinates [66]

One of the key advantages of fiber-optic sensing is its ability to provide continuous, distributed strain measurements along the entire length of the fiber. This capability enables detailed mapping of strain fields within rock samples or across large-scale structures. Recent studies have shown that fiber-optic strain sensing systems can achieve spatial resolutions as fine as 1.28 mm, with strain accuracies of ± 1 μm/m and sampling rates up to 120 Hz. This level of precision makes it particularly effective for monitoring dynamic events such as rock fractures.

In terms of applications, fiber-optic strain sensing has shown particular promise in monitoring hydraulic fracture propagation in rocks. The technique is able to precisely detect the moment of fracture initiation by observing distinct changes in the strain field. Additionally, it allows for real-time tracking of fracture propagation by analyzing the progression of strain along the fiber. Fiber-optic systems can also estimate fracture width by measuring the magnitude of strain once the fracture contacts the fiber. When fibers are placed parallel to the fracture height expansion direction, the extent of axial strain can be used to determine the fracture height.

There are several advantages to using fiber-optic strain sensing in rock dynamics. Its high spatial resolution allows for millimeter-scale measurements along the entire fiber length, providing detailed strain data. The high sensitivity of modern systems allows for the detection of strain changes as small as 1 μm/m. Unlike traditional point sensors, fiber-optic systems provide continuous measurements along the fiber, making them ideal for monitoring large or complex structures. Additionally, the multiplexing capability of fiber-optic systems allows multiple fibers to be interrogated simultaneously, enabling 3D strain field mapping.

However, there are also limitations associated with fiber-optic strain sensing. One of the main challenges is the proper installation of the fibers in rock samples or field settings, which can be technically difficult. Furthermore, measurements are limited to the specific paths where the fibers are installed, meaning that the system cannot capture strain data from areas outside of the fiber's path. The interpretation of the measured strain patterns can also be complex, particularly when attempting to relate them to the intricate 3D geometries of rock fractures.

Ultrasonic Imaging Techniques

Ultrasonic imaging has become a powerful non-destructive technique for visualizing and analyzing the internal structures and defects within rock samples. Recent advancements in ultrasonic technology have significantly enhanced its ability to provide real-time monitoring of dynamic processes within rocks, offering valuable insights into fracture initiation, propagation, and the evolution of weak structural planes.

The principles of ultrasonic imaging in rock dynamics involve several key components. Ultrasonic transducers generate and receive high-frequency sound waves, which are used to explore the internal structure of the rock. A coupling medium is often required to facilitate the transmission of these ultrasonic waves between the transducer and the rock sample. Additionally, a high-speed data acquisition system is used to capture and process the reflected ultrasonic signals, which form the basis for subsequent analysis (Fig. 3.42).

One promising development in ultrasonic imaging is the use of multi-frequency techniques for analyzing weak structural planes in rocks. This approach typically involves scanning the rock at multiple frequencies, ranging from 2 to 5 MHz, to capture detailed information about internal structures. Multi-angle scanning is also employed to improve spatial resolution and enhance defect detection. By analyzing the attenuation characteristics of the reflected ultrasonic waves, researchers can infer the properties of the rock's internal structures. Moreover, gradient feature analysis allows for the detection and characterization of weak structural planes, offering insights into areas of potential failure within the rock.

To extract meaningful information from ultrasonic images, several quantitative analysis techniques have been developed. One method involves the normal vector sum of gradient features, which analyzes the orientation and magnitude of gradients in difference images to detect weak structural planes. Reflection attenuation features

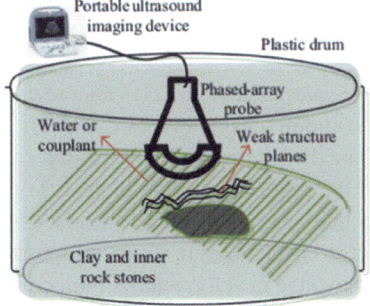

Fig. 3.42 Schematic diagram of multiple frequency ultrasonic imaging experiment system [67]

are another key analysis method, which examines changes in the attenuation of reflected ultrasonic waves to infer the properties of internal structures. Additionally, overall gray value changes in ultrasonic images can provide valuable insights into the evolution of internal structures over time, aiding in the understanding of how cracks and fractures develop under dynamic loading.

Recent advancements in ultrasonic imaging technology have enabled real-time monitoring of rock internal structures during physical model tests. This capability allows for the non-destructive visualization of cracks, fractures, and weak planes within rock samples as they evolve during dynamic loading. Researchers can track these internal structural changes over time by continuously capturing and analyzing ultrasonic images. Furthermore, the technique allows for both qualitative and quantitative analysis of weak structural planes, providing information on their morphological characteristics and spatial distribution.

The advantages of ultrasonic imaging in rock dynamics include its non-destructive nature, which allows for internal visualization without damaging the rock sample. Modern systems are capable of real-time image capture and processing, providing immediate feedback during dynamic tests. The use of multi-frequency and multi-angle scanning provides high-resolution images, offering detailed insights into internal rock structures. Advanced image processing techniques also enable quantitative analysis of these internal features, adding depth to the interpretation of the data.

However, there are limitations to ultrasonic imaging. The penetration depth of ultrasonic waves is limited, particularly in highly attenuating materials, which may restrict the technique's effectiveness in certain rock types. There is also a trade-off between resolution and penetration; higher frequencies offer better resolution but less penetration, requiring a balance between these factors. Additionally, the interpretation of complex ultrasonic images can be challenging and may require expertise and integration with other techniques.

Emerging Techniques and Future Prospects

The field of rock dynamics is evolving rapidly, driven by innovations in technology that provide deeper insights into rock behavior under dynamic loading conditions. As these advancements unfold, several emerging techniques hold promise for enhancing our ability to monitor, analyze, and predict rock behavior, offering more accurate models and real-time monitoring solutions.

1. Nano-Scale Sensing Technologies

The continued progress in nanotechnology is opening new possibilities for ultra-high resolution sensing in rock dynamics. Nano-sensors, for example, are being developed to be distributed throughout rock samples, providing highly detailed measurements of internal structures, which were previously difficult to access. Quantum sensing technologies, capable of ultra-sensitive measurements, are also being explored for detecting changes in strain, temperature, and other key physical parameters. Another innovative development is molecular tagging, which involves introducing specially

3.5 Advanced Test Techniques and Methods

designed molecular tags into rock pores and fractures to track fluid flow and chemical processes during dynamic loading.

2. 4D Imaging Techniques

The advent of 4D imaging (3D + time) is significantly improving the ability to visualize dynamic processes in rocks. Techniques such as 4D X-ray micro-tomography allow for real-time, high-speed X-ray imaging to capture 3D structural changes as they occur under dynamic loading. Similarly, 4D acoustic imaging uses advanced ultrasonic systems to generate time-resolved 3D images of internal rock structures, offering unprecedented detail in fracture development and material response. Furthermore, the development of 4D electromagnetic imaging is enabling real-time 3D mapping of rock properties such as porosity and fluid content, which are crucial for understanding how rocks behave under various loading conditions.

3. Advanced Data Analytics and Artificial Intelligence

With the increasing complexity and volume of data generated by multi-modal monitoring systems, advanced data analytics techniques are becoming essential. Deep learning algorithms are being developed to automatically extract meaningful features from complex, multi-dimensional datasets, reducing the need for manual intervention. AI-driven predictive models are also being created to forecast rock behavior based on real-time monitoring data, allowing researchers to predict failure events and other critical behaviors. Additionally, intelligent systems for automated experiment control are being developed to adjust experimental parameters in real-time, optimizing experimental conditions based on ongoing data analysis.

4. Virtual and Augmented Reality Interfaces

Immersive technologies such as virtual reality (VR) and augmented reality (AR) are transforming the way rock dynamics data is visualized and interpreted. VR laboratories, for example, allow researchers to interact with 3D representations of rock samples and monitor dynamic processes in a fully immersive environment. AR is being developed for field applications, where real-time monitoring data can be overlaid on physical rock outcrops or engineering structures, aiding in decision-making during field investigations. Furthermore, the integration of haptic feedback systems is allowing researchers to "feel" rock properties and stress distributions in virtual environments, providing a more intuitive understanding of the data.

5. Integration with Multi-Scale Modeling

A key challenge in rock dynamics is bridging the gap between microscopic and macroscopic behaviors. The integration of real-time monitoring techniques with multi-scale modeling approaches is beginning to provide solutions. Real-time model updating systems are being developed to continuously update numerical models based on real-time monitoring data, allowing for a dynamic simulation of rock behavior. Multi-physics coupling is another exciting development, where mechanical, thermal, hydraulic, and chemical processes are integrated into real-time models of rock behavior. Additionally, scale-bridging frameworks are being created to seamlessly transition between different scales, from micro-fractures to large-scale rock masses, providing more comprehensive insights into the rock's response to loading.

6. In-Situ Monitoring in Extreme Environments

Advancements in sensor technology and data transmission are enabling real-time monitoring in increasingly challenging environments. For instance, robust monitoring systems are being developed for deep underground environments, capable of operating under high pressure and temperature conditions. Similarly, there are growing efforts to create monitoring systems for offshore and subsea applications, such as those used for underwater rock structures or seafloor installations. The design of monitoring systems for extraterrestrial applications is also underway, aiming to study rock behavior on other planetary bodies, such as the Moon or Mars, as part of space exploration and resource utilization initiatives.

7. Bioengineered Sensing Systems

The emerging field of bioengineering is presenting unique opportunities for rock monitoring. Microbial sensors are being engineered to detect specific conditions within rock pores and fractures, offering an innovative approach to environmental monitoring. Additionally, bio-inspired sensing networks are being developed, mimicking biological neural networks to create distributed systems that can respond to changes in rock behavior. Furthermore, self-healing sensor systems are being designed, inspired by biological processes, which could adapt to changing conditions in the rock or recover from damage over time.

These emerging technologies are set to revolutionize the way we understand and monitor rock dynamics. By integrating these advancements with sophisticated modeling and analysis tools, the field is poised to make significant strides in rock engineering, geohazard mitigation, and resource extraction. The future holds exciting possibilities for real-time monitoring, enhanced predictive modeling, and deeper insights into the complex behaviors of rocks under dynamic loading conditions.

3.5.3 Integration of AI and Machine Learning in Data Analysis

The integration of artificial intelligence (AI) and machine learning (ML) techniques into rock dynamics data analysis has revolutionized the field, offering unprecedented capabilities in predicting, interpreting, and optimizing various aspects of rock behavior under dynamic loading conditions. This section explores the cutting-edge applications of AI and ML in rock dynamics testing, highlighting their potential to enhance experimental design, data processing, and result interpretation.

One of the most significant contributions of AI and ML to rock dynamics testing lies in their ability to predict complex rock properties and behaviors based on limited

3.5 Advanced Test Techniques and Methods

input data. This capability is particularly valuable in scenarios where extensive laboratory testing may be impractical or cost-prohibitive. This high level of accuracy underscores the potential of AI-driven approaches to complement and, in some cases, even surpass conventional predictive methods in rock dynamics analysis.

The integration of AI and ML techniques has facilitated the development of advanced data fusion approaches, enabling researchers to extract more comprehensive insights from diverse data sources. Liu et al. [68] proposed a methodology combining DIC strain nephograms and AE data with deep learning techniques to analyze rock damage evolution during compression tests. By employing a dual-model approach, including a Transformer + UNet hybrid model for lithology identification and an optimized ResNet-18 model for determining rock weathering degrees, they achieved high accuracy in predicting rock strength values, with an average error of only 9.33% compared to laboratory test results. This sophisticated data fusion approach demonstrates the potential of AI and ML to enhance our understanding of complex rock behavior under dynamic loading conditions by integrating multiple data streams and analytical techniques.

Moreover, AI and ML techniques have shown remarkable potential in integrating diverse data sources to enhance our understanding of rock behavior under dynamic loading. Xu et al. [69] developed an improved Incremental Transformer (ZITS) algorithm to repair distorted or missing strain nephograms from DIC during uniaxial compression tests. This innovative approach resulted in an 8–10% increase in information completeness, leading to more accurate damage analysis that aligned better with actual rock failure processes compared to conventional DIC or acoustic emission (AE) methods. The integration of image processing techniques with AI algorithms showcases the potential for enhancing data quality and interpretation in rock dynamics experiments.

The application of AI and ML extends beyond simple property prediction to more complex tasks such as constitutive modeling and fracture behavior analysis. Gong et al. [70] employed a combination of linear and non-linear machine learning algorithms, including LASSO, Ridge Regression, Elastic Net, Artificial Neural Networks (ANN), and Multivariate Adaptive Regression Splines (MARS), to estimate Young's modulus and evaluate rock fracturability. Their study revealed that non-linear regression methods (ANN and MARS) outperformed linear regularized methods in predicting Young's modulus, with ANN achieving the lowest root mean square error (RMSE). This demonstrates the capability of AI techniques to capture intricate non-linear relationships in rock mechanical properties, which are often challenging to model using traditional approaches.

The application of AI and ML techniques in rock dynamics testing can also optimize experimental design and improve testing methodologies. For instance, Zhao et al. [71] developed a hybrid model combining Extreme Gradient Boosting (XGBoost) with an Artificial Bee Colony (ABC) optimization algorithm to predict the UCS of rock materials in deep mines. This XGBoost-ABC model demonstrated superior performance compared to other machine learning approaches, achieving R2 values of 0.98 and 0.93 for training and testing datasets, respectively. The successful implementation of such hybrid models illustrates the potential for AI-driven optimization in rock dynamics testing procedures, potentially leading to more efficient and accurate experimental designs.

The application of AI and ML in rock dynamics testing also extends to the realm of microstructural analysis and its relationship to macroscopic rock properties. Gong et al. [70] incorporated microstructural information from thin section images into their machine learning models to enhance the understanding of fracture behavior. Their analysis revealed important correlations between microstructural features, particularly clay content and laminations, and rock mechanical behavior and fracturability. This integration of microstructural data with AI-driven analysis techniques opens new avenues for developing more comprehensive and accurate predictive models in rock dynamics, bridging the gap between microscale rock characteristics and macroscale dynamic behavior (Fig. 3.43).

Despite the significant advancements and potential benefits offered by AI and ML techniques in rock dynamics testing, several challenges and limitations must be addressed. One primary concern is the risk of overfitting, particularly when dealing with limited datasets. To mitigate this issue, researchers have employed various strategies, such as cross-validation techniques and ensemble learning approaches. For instance, Wang et al. [72] used tenfold cross-validation to evaluate their ensemble model's performance, while Zhao et al. [71] split their dataset into training (86 sets) and testing (20 sets) subsets to ensure robust model validation.

Another challenge lies in the interpretability of complex AI and ML models, particularly deep learning architectures. While these models often demonstrate superior predictive performance, their internal decision-making processes can be opaque,

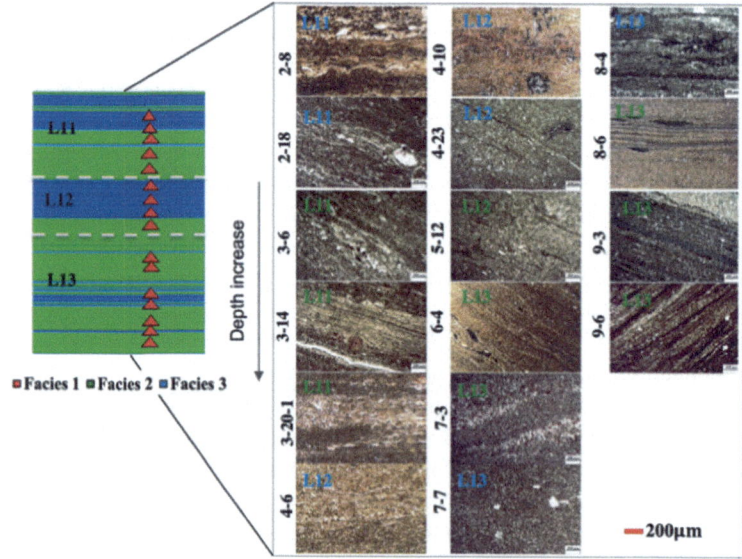

Fig. 3.43 The microstructure information relevant to the geomechanical facies [70]

making it difficult for researchers to gain insights into the underlying physical mechanisms governing rock behavior under dynamic loading. To address this issue, researchers are increasingly turning to explainable AI techniques. For example, Sun et al. [73] employed SHAP (Shapley additive explanations) analysis to interpret their model's predictions, revealing the relative importance of different features in determining rock properties and behaviors.

The quality and quantity of available data also pose significant challenges in the application of AI and ML techniques to rock dynamics testing. Many studies rely on relatively small datasets, which can limit the generalizability of the developed models. To overcome this limitation, researchers are exploring various strategies, including data augmentation techniques and the integration of numerically simulated data with experimental results. For instance, Wang et al. [72] used mesoscopic finite element numerical simulations to generate a large training dataset of rock mechanical properties, complementing limited experimental data.

The integration of AI and ML techniques into rock dynamics testing has already demonstrated significant potential for enhancing our ability to predict, analyze, and interpret complex rock behavior under dynamic loading conditions. As these technologies continue to evolve and mature, they are likely to play an increasingly central role in advancing our understanding of rock dynamics and improving the design and safety of rock engineering projects subjected to dynamic loads. However, it is crucial to recognize that AI and ML approaches should complement, rather than replace, traditional experimental and theoretical methods in rock dynamics research. The most promising path forward lies in the synergistic integration of AI-driven techniques with established principles of rock mechanics and experimental expertise, leading to more comprehensive and accurate characterizations of rock behavior under dynamic loading conditions [70, 74, 75].

3.6 Conclusion

This chapter provides a comprehensive overview of the development and advancements in rock dynamics testing devices and techniques, emphasizing their critical role in rock mechanics research and engineering applications. The development of these devices addresses the need to understand the complex behavior of rock materials under dynamic loading conditions, significantly contributing to advancements in mining, civil, geotechnical, and earthquake engineering. From their historical evolution to modern engineering demands, these testing devices reflect technological progress and respond to the challenges of dynamic rock behavior research.

Recent years have seen significant advancements, particularly in the application of the SHPB system for studying rock behavior under high-strain rate loading. The SHPB system has been instrumental in improving the understanding of dynamic rock responses, providing valuable insights for sustainable resource extraction, mining safety, and tunneling under challenging geological conditions.

Furthermore, advancements in testing techniques under static-dynamic coupled loading conditions have led to breakthroughs in understanding complex rock behaviors. Innovations such as modifications to SHPB systems, novel loading techniques, high-performance sensing technologies, and advanced data analysis methods have enabled researchers to explore intricate rock phenomena more effectively. These advancements improve safety and efficiency in rock engineering projects while opening new pathways for innovation.

In the study of rock behavior under high-temperature and high-pressure conditions, this chapter highlights the coupled thermal–mechanical effects on the physical, mechanical, and hydraulic properties of rocks. The findings reveal that mechanisms like thermal expansion, microcrack formation, and mineral transformations significantly influence rock strength, elastic modulus, and failure characteristics. These insights are critical for applications such as geothermal energy extraction, underground waste storage, and deep mining, where understanding the interplay of thermal, hydraulic, and mechanical processes is essential.

Lastly, the integration of AI, ML, and advanced real-time monitoring techniques has revolutionized rock dynamics research. Innovations such as acoustic emission monitoring, digital image correlation, fiber-optic sensing, and ultrasonic imaging have provided unprecedented insights into fracture and deformation processes. By combining these technologies with traditional experimental methods and advanced data analysis, researchers have achieved more accurate and comprehensive models of rock dynamics. These tools are essential for addressing challenges in sustainable resource extraction, infrastructure development, and geohazard mitigation.

In conclusion, this chapter systematically analyzes the development of testing devices, technological innovations, and cutting-edge methodologies in rock dynamics. These contributions not only provide a solid foundation for theoretical research and engineering applications but also point toward new directions for future exploration in the field.

References

1. Li B, Gong X, Wang G, Qiao JX. Benchmark experiment on shear behavior of ice-filled planar rock joints using a novel direct shear testing apparatus. Int J Rock Mech Min Sci. 2024;178: 105757.
2. Wu RY, Chen SW, Wang GB, Zhu YH, Lan XD. Development and application of a thermo-hydro-mechanical coupling test system for a single rock fracture. Measurement. 2025;242: 116197.
3. Saltiel S, Groebner N, Sawi T, McCarthy C. Characterization of seismicity from different glacial bed types: machine learning classification of laboratory stick-slip acoustic emissions. Ann Glaciol. 2024;65(e17):1–8.
4. Rong G, Chen ZH, Li BW, Liu RT, Xu LD. Development of a laboratory test apparatus for simulating core drilling under high in situ stress. Geotech Test J. 2024;47:504–20.
5. Xin G, Yang G, Li F, Liu H. A large-scale three-dimensional apparatus to study failure mechanisms of rockfalls in underground engineering contexts. 2024; 24(7):2068.
6. Tan P, Pang H, Jin Y, Zhou Z. Experiments and analysis of hydraulic fracturing in hot dry rock geothermal reservoirs using an improved large-size high-temperature true triaxial apparatus. Nat Gas Indus B. 2024;11:83–94.

References

7. Tetsuji O, Kazuo T, Hitoshi O, Yoshinori T, Takayasu H, Toshifumi T, Hideo K, Tomohiro N, Susumu K. Development of in-situ triaxial test for rock masses. Int J JCRM. 2006;2:7–12.
8. Ji YG, Wang MY, Li JH, Deng SX, Li ZH, Xu TH, Gao F. A laboratory method to simulate seismic waves induced by underground explosions. J Rock Mech Geotech Eng. 2022;14:1514–30.
9. Aghajari AM, Namin FS. U-HRMES: decision theory-based model for appropriate mining equipment selection in underground hard rock stopes. Expert Syst Appl. 2024;246: 123108.
10. Zou SZ, Gao YT, Yang ZR, Yang C, Qian LY, Zhou Y. Development of the split-Hopkinson pressure bar and its application in testing the dynamic mechanical properties of quasi-brittle materials: a review. J Market Res. 2024;33:9463–83.
11. Xie HP, Zhu JB, Zhou T, Zhao J. Novel three-dimensional rock dynamic tests using the true triaxial electromagnetic hopkinson bar system. Rock Mech Rock Eng. 2021;54(4):2079–86.
12. Yuan W, Li JC, Zou CJ, Zhao J. A new apparatus for testing shear-slip properties of rock joint subjected to dynamic disturbance. Exp Mech. 2024;64(5):745–59.
13. Cha MS, Alqahtani NB, Yin XL, Kneafsey TJ, Yao BW, Wu YS. Laboratory system for studying cryogenic thermal rock fracturing for well stimulation. J Petrol Sci Eng. 2017;156:780–9.
14. Zhao J, Feng XT, Wang JR, Hu L, Guo Y. A multifunctional shear apparatus for rocks subjected to true triaxial stress and high temperature in real-time. J Rock Mech Geotechn Eng. 2024;16(9):3524–43.
15. Feng XT, Zhang JY, Yang CX, Tian J, Lin F, Li SP, Su XX. A novel true triaxial test system for microwave-induced fracturing of hard rocks. J Rock Mech Geotech Eng. 2021;13(5):961–71.
16. Zhao GL, Li X, Xu Y, Xia KW. A modified triaxial split Hopkinson pressure bar (SHPB) system for quantifying the dynamic compressive response of porous rocks subjected to coupled hydraulic-mechanical loading. Geomech Geophys Geo-Energy Geo-Resourc. 2022;8(1):1–15.
17. Luo JZ, Cai YY, Yu J, Zhang JZ, Zhu YL, Wei Y. Development and applications of the quasi-dynamic triaxial apparatus for deep rocks. Deep Underground Sci Eng. 2023;3(1):70–90.
18. Zhang K, Zhang K, Liu WL, Xie JB. A novel DIC-based methodology for crack identification in a jointed rock mass. Mater Des. 2023;230: 111944.
19. Ma Z, Li X, Lv L. Verification of a novel stress path method by true-triaxial test. Sci Rep. 2024;14(1):6110.
20. Han K, Pyrak-Nolte LJ, Bobet A. Direct shear testing apparatus for saturated rock joints. Geotech Test J. 2024;47(6):1229–42.
21. Wang C, Liu ZB, Zhou HY, Wang KX, Shen WQ. A novel true triaxial test device with a high-temperature module for thermal-mechanical property characterization of hard rocks. Eur J Environ Civ Eng. 2022;27(4):1697–714.
22. Jiang Y, Tanabashi Y, Xiao J, Nagaie K. An improved shear-flow test apparatus and its application to deep underground construction. Int J Rock Mech Min Sci. 2004;41(3):385–6.
23. Zhang ZY, Li CJ, Ge JJ. Dynamic Brazilian tests of yellow sandstone under coupled static and dynamic loads. J Vibroeng. 2024;26(3):673–89.
24. Liu W, Li YL, Zhang ZQ, Yang Li Y, Luo Yi, Yue ZW. Loading rate effect on mixed-mode I+II fracture of V-notched Brazilian disc rock specimen under impact loads. Eng Fract Mech 2023; 291:109581.
25. Gao GY, Zhang KH, Wang P, Xu Y, Zhou H, Wang CH. Dynamic strength and fracturing behavior of persistent fractured granite under dynamic loading. Bull Eng Geol Env. 2024;83(6):218.
26. Yu Y, Qin CJ, Qiang L, Feng GL, Zeng JJ, Lu YY. Experimental study on dynamic failure behavior of red sandstone after freeze-thaw cycles. Constr Build Mater. 2024;451: 138582.
27. Xu X, Jing HW, Yin Q, Wu JG, Guzev MA, Jin JW. Dynamic compressive mechanical properties of rock-like material with bedding planes subject to different impact loads. KSCE J Civ Eng. 2024;28(6):2409–19.
28. Zhang L, Wang EY, Liu YB, Yue WT, Chen D. Experimental research into the dynamic damage characteristics and failure behavior of rock subjected to incremental repeated impact loads. Eng Geol. 2024;331: 107435.

29. Liu LW, Li HB, Zhang GK, Fu SY. Dynamic strength and full-field cracking behaviours of pre-cracked rocks under impact loads. Int J Mech Sci. 2024;268: 109049.
30. Cao CH, Ding HN, Zou BP. Dynamic impact properties of deep sandstone under thermal-hydraulic-mechanical coupling loads. J Mt Sci. 2024;21(6):2113–29.
31. Ji DL, Sai VP, Zhao HB, Yue ZR. Dynamic response and constitutive model for coal-rock composite material subjected to impact loading. Eng Fract Mech. 2024;312: 110616.
32. Deng S, Yan YT, Wang X, Ma ZJ. Dynamic mechanical behavior and failure characteristics of jointed rock-like specimen under impact load. Eng Fail Anal. 2025;167: 108975.
33. Niu QH, Hu MW, He JB, Zhang B, Su XB, Zhao LX, Pan JN, Wang ZZ, Du ZG, Wei Yb. The chemical damage of sandstone after sulfuric acid-rock reactions with different duration times and its influence on the impact mechanical behaviour. Heliyon 2023; 9(12):e22346.
34. Hu SB, Yan ZY, Zhu C, He MC, Pang SG. Impact fracturing of rock-like material using carbon dioxide under different temperatures and pressures. J Rock Mech Geotech Eng 2024. (In press)
35. Li D, Gao FH, Han ZY, Zhu QQ. Experimental evaluation on rock failure mechanism with combined flaws in a connected geometry under coupled static-dynamic loads. Soil Dyn Earthq Eng. 2020;132: 106088.
36. Li XB, Zhou ZL, Lok TS, Hong L, Yin TB. Innovative testing technique of rock subjected to coupled static and dynamic loads. Int J Rock Mech Min Sci. 2008;45:739–48.
37. Yang S, Ning JG, Zhang XL, Wang J, Shi XS, Qu XZ. Determination of critical energy for coal impact fracture under coupled static-dynamic loading. Eng Fail Anal. 2024;160: 108222.
38. Wei J, Liao HL, Li N, Liang HJ, Chen KF, Yan H, Fan YG, Zhao XH. Effect of the three-dimensional static pre-stress on the dynamic behaviours of conglomerate: true triaxial Hopkinson pressure bar tests. Geoenergy Sci Eng. 2023;227: 211810.
39. He MC, RibeiroeSousa L, Miranda T, Zhu GL. Rockburst laboratory tests database—Application of data mining techniques. Eng Geol 2015; 185:116–30.
40. Khadivi B, Masoumi H, Heidarpour A, Zhang QB, Zhao J. Mechanical characterization of intact rock under polyaxial static-dynamic stress states. Int J Rock Mech Min Sci. 2025;185: 105977.
41. Feng F, Xie ZW, Chen SJ, Zhao XD, Yan ZK, Shi JL. Investigation on the failure characteristics of fissured sandstone under true triaxial unloading and dynamic disturbance condition. Chinese J Rock Mech Eng 2025. (Online) [in Chinese].
42. Zhu ZH, Zhang F, Peng QY, Chabot B, Dupla JC, Canou J, Cumunel G, Foerster E. Development of an auto compensation system in cyclic triaxial apparatus for liquefaction analysis. Soil Dyn Earthquake Eng 2021; 144:106707.
43. Sarkar S, Chakraborty S, Nayak S. Identification of optimum reinforcement detailing using tuned CDP parameters in RC beam under drop-weight impact. Eng Fail Anal. 2023;146: 107116.
44. Robbiano F, Liu K, Zhang QB, Orellana LF. Dynamic mechanical properties of veined rocks under biaxial confinement. Int J Rock Mech Min Sci 2023; 170: 105538.
45. Zhang WQ, Sun Q, Hao SQ, Geng JS, Lv C. Experimental study on the variation of physical and mechanical properties of rock after high temperature treatment. Appl Therm Eng. 2016;98:1297–304.
46. Masri M, Sibai M, Shao JF, Mainguy M. Experimental investigation of the effect of temperature on the mechanical behavior of Tournemire shale. Int J Rock Mech Min Sci. 2014;70:185–91.
47. Rathnaweera TD, Ranjith PG, Gu X, Perera MS A, Kumari WGP. Experimental investigation of thermomechanical behaviour of clay-rich sandstone at extreme temperatures followed by cooling treatments. Int J Rock Mech Min Sci 2018; 107:208–23.
48. Li BY, Ju F, Xiao M, Ning P. Mechanical stability of granite as thermal energy storage material: an experimental investigation. Eng Fract Mech. 2019;211:61–9.
49. Meng T, Zhang ZJ, Taherdangkoo R, Zhao GH, Butscher C. Temperature-dependent evolution of permeability and pore structure of marble under a high-temperature thermo-hydro-mechanical coupling environment. Acta Geotechnica 2024; 1–22.
50. Deng S, Zhang YX, Yan XP, Yang S, Li C, Peng HP. Experimental study on P-wave response characteristics of hot dry rock under high temperature and high confining pressure. Geomech Energy Environ. 2023;36: 100514.

References

51. Zhao YS, Wan ZJ, Feng ZJ, Yang D, Zhang Y, Fang Q. Triaxial compression system for rock testing under high temperature and high pressure. Int J Rock Mech Min Sci 2012; 52:132–8.
52. Ma X, Wang GL, Hu DW, Liu YG, Zhou H, Liu F. Mechanical properties of granite under real-time high temperature and three-dimensional stress. Int J Rock Mech Min Sci. 2020;136: 104521.
53. Zhang NL, Luo ZF, Chen ZX, Liu FS, Liu PL, Chen WY, Lin W, Zhao LQ. Thermal–hydraulic–mechanical–chemical coupled processes and their numerical simulation: a comprehensive review. Acta Geotech. 2023;18(12):6253–74.
54. Gao H, Xie HP, Zhang ZT, Zhang R, Gao MZ, Li YH, Chen L, Xie HX. Sealing performances of polymer heat-shrinkable tubing for deep rocks under high-temperature and ultrahigh-pressure condition. J Rock Mech Geotech Eng 2024. (In Press)
55. Kumari WGP, Ranjith PG, Perera MSA, Shao S, Chen BK, Lashin A, Al Arif N, Rathnaweera TD. Mechanical behaviour of Australian Strathbogie granite under in-situ stress and temperature conditions: an application to geothermal energy extraction. Geothermics 2017; 65:44–59.
56. Kytýř D, Koudelka P, Drozdenko D, Vavro M, Fíla T, Rada V, Vavro L, Máthis K, Souček K. Acoustic emission and 4D X-ray micro-tomography for monitoring crack propagation in rocks. Int J Rock Mech Min Sci. 2024;183: 105917.
57. Gao YT, Wu TH, Zhou Y. Application and prospective of 3D printing in rock mechanics: a review. Int J Miner Metall Mater. 2021;28(1):1–17.
58. Ishola O, Vilcáez J. Augmenting X-ray micro-CT data with MICP data for high resolution pore-scale simulations of flow properties of carbonate rocks. Geoenergy Sci Eng. 2024;239: 212982.
59. Zhu JB, Zhou T, Liao ZY, Sun L, Li XB, Chen R. Replication of internal defects and investigation of mechanical and fracture behaviour of rock using 3D printing and 3D numerical methods in combination with X-ray computerized tomography. Int J Rock Mech Min Sci. 2018;106:198–212.
60. Gu C, Geng JS, Sun Q, Zhang YL, Hu JJ. Effect of water on granite deterioration under microwave radiation based on real-time AE monitoring. Rock Mech Rock Eng. 2024;57(12):11399–412.
61. Xiong QQ, Lin Q, Gao Y, Han YH, Hampton JC. Laboratory visualization of damage asymmetry formation of rock fracture via acoustic emission and digital imaging correlation. J Rock Mech Geotech Eng. 2024;16(11):4480–90.
62. Li S, Lin H, Cao RH, Wang YX, Zhao YL. Mechanical behavior of rock-like specimen containing hole-joint combined flaw under uniaxial loading: findings from DIC and AE monitoring. J Market Res. 2023;26:3426–49.
63. Lin Q, Yuan HN, Biolzi L, Labuz JF. Opening and mixed mode fracture processes in a quasi-brittle material via digital imaging. Eng Fract Mech. 2014;131:176–93.
64. Shirole D, Hedayat A, Walton G. Damage monitoring in rock specimens with pre-existing flaws by non-linear ultrasonic waves and digital image correlation. Int J Rock Mech Min Sci. 2021;142: 104758.
65. Gao QQ, Ma LQ, Liu W, Wang H, Ma Q, Wang XZ. Identification of damage states of load-bearing rocks using infrared radiation monitoring methods. Measurement. 2025;239: 115507.
66. Zhang YH, Guo TK, Chen M, Qu ZQ, Hu ZP, Zhang B, Xue LR, Wang YP. Real-time monitoring of rock fracture by true triaxial test using fiber-optic strain monitoring in adjacent wells. J Rock Mech Geotech Eng 2024. (Online).
67. Zou XJ, Chen BR, Song H, Ma ZM, Chen SY. The stability analysis of weak structural planes based on multi-frequency ultrasonic imaging characteristics during rock and soil physical model tests. Environ Earth Sci. 2021;80(23):775.
68. Liu LBC, Song ZP, Zhou P, He XH, Zhao L. AI-based rock strength assessment from tunnel face images using hybrid neural networks. Sci Rep. 2024;14(1):17512.
69. Xu MZ, Qi XY, Geng DD. Application of improved and efficient image repair algorithm in rock damage experimental research. Sci Rep. 2024;14(1):14849.

70. Gong YW, El-Monier I, Mehana M. Machine learning and data fusion approach for elastic rock properties estimation and fracturability evaluation. Energy AI. 2024;16: 100335.
71. Zhao JJ, Li DY, Jiang JT, Luo PK. Uniaxial compressive strength prediction for rock material in deep mine using boosting-based machine learning methods and optimization algorithms. CMES Comp Model Eng Sci. 2024;140(1):275–304.
72. Wang HJ, Zhang C, Zhou B, Xue SF, Jia P, Zhu XX. Prediction of triaxial mechanical properties of rocks based on mesoscopic finite element numerical simulation and multi-objective machine learning. J King Saud Univ Sci. 2023;35(7): 102846.
73. Sun JL, Zhang R, Zhang AL, Wang XZ, Wang JX, Ren L, Zhang ZT, Zhang ZL. Rock strength prediction based on machine learning: a study from prediction model to mechanism explanation. Measurement. 2024;238: 115373.
74. Liu HC, Su HZ, Sun LZ, Dias-da-Costa D. State-of-the-art review on the use of AI-enhanced computational mechanics in geotechnical engineering. Artif Intell Rev. 2024;57(8):196.
75. Mahetaji M, Brahma J. A critical review of rock failure Criteria: a scope of Machine learning approach. Eng Fail Anal. 2024;159: 107998.

Open Access This chapter is licensed under the terms of the Creative Commons Attribution-NonCommercial-NoDerivatives 4.0 International License (http://creativecommons.org/licenses/by-nc-nd/4.0/), which permits any noncommercial use, sharing, distribution and reproduction in any medium or format, as long as you give appropriate credit to the original author(s) and the source, provide a link to the Creative Commons license and indicate if you modified the licensed material. You do not have permission under this license to share adapted material derived from this chapter or parts of it.

The images or other third party material in this chapter are included in the chapter's Creative Commons license, unless indicated otherwise in a credit line to the material. If material is not included in the chapter's Creative Commons license and your intended use is not permitted by statutory regulation or exceeds the permitted use, you will need to obtain permission directly from the copyright holder.

Chapter 4
Rock Dynamic Properties

4.1 Basic Concept and Theory of Impact Load

4.1.1 Definition and Characteristics of Impact Load

Impact loading is a fundamental concept in rock dynamics, playing a crucial role in various geological and engineering processes. Understanding the definition and characteristics of impact load is essential for analyzing and predicting the behavior of rock materials under dynamic conditions. This section provides a comprehensive overview of impact load, its definition, and key characteristics, drawing from recent research in rock dynamics and related fields.

Definition of Impact Load

Impact load can be broadly defined as a force applied to a material or structure over a very short duration, typically resulting in a rapid transfer of energy [1–3]. In the context of rock dynamics, impact loads are of particular interest due to their prevalence in natural phenomena (e.g., rockfalls, meteorite impacts) and engineering applications (e.g., blasting, tunneling, and mining operations).

Characteristics of Impact Load

Impact load in rock mechanics represents a complex phenomenon characterized by distinct physical attributes and mechanical responses. The fundamental characteristics begin with the rapid application of force, which typically manifests as a high-magnitude load concentrated in a localized area. This loading scenario exhibits a pronounced time-dependent nature, which can be quantified through detailed load-time history analysis and impulse measurements.

1. Load-Time History

In the analysis of impact loads, the load-time history emerges as the most fundamental characteristic, as it provides a comprehensive description of force variations throughout the duration of an impact event. This temporal evolution of applied forces follows a distinctive three-phase pattern. The initial rising phase is characterized by a rapid increase in force magnitude from zero to its maximum value. This is followed by the achievement of peak force, which represents the maximum force magnitude experienced during the impact event. The final decay phase then occurs, during which the applied force diminishes relatively quickly until it returns to zero. Understanding these distinct phases is crucial for characterizing the dynamic response of rock materials under impact conditions.

2. Impulse

The impulse of an impact load is a critical characteristic that quantifies the total change in momentum imparted to the rock during the impact event.

The impulse provides a measure of the overall effect of the impact, considering both the magnitude of the force and its duration. Two impact loads with different force–time histories may have the same impulse, potentially leading to similar overall effects on the rock material [1, 3].

3. Energy Transfer

Impact events in rock mechanics are characterized by several crucial energy transfer mechanisms. The primary consideration in analyzing these events is the distribution of energy, which manifests in two fundamental components. The first component is kinetic energy, which is directly associated with the motion of the impacting body. This kinetic component transitions into the second key element–strain energy, which becomes stored within the rock material through deformation processes. The interplay between these two energy forms is essential for understanding the complete mechanics of rock impact scenarios.

Understanding energy transfer is critical for assessing the potential for damage or failure in the impacted rock material. The partition of energy between kinetic and strain components, as well as energy dissipation mechanisms (e.g., fracture, plastic deformation), significantly influences the rock's response to impact loading [2, 4].

4. Strain Rate

One of the most distinguishing characteristics of impact load in rock dynamics is the high strain rates it induces in the material. Impact loading generates significantly elevated strain rates in rock materials, typically ranging from 10^2 to 10^4 s^{-1}, which substantially exceeds those observed in quasi-static loading conditions by several orders of magnitude [1, 3]. These elevated strain rates fundamentally alter rock behavior through multiple mechanisms. Most notably, rock materials demonstrate pronounced strain rate sensitivity, manifesting as enhanced strength and stiffness characteristics under dynamic loading conditions. Furthermore, the high strain rates can trigger a transformation in failure mechanisms, potentially inducing a transition from ductile to brittle behavior. Additionally, these dynamic loading conditions necessitate consideration of stress wave propagation effects, which become critically important in understanding and predicting the rock's mechanical response under impact scenarios.

4.1 Basic Concept and Theory of Impact Load

5. Contact Mechanics

The interaction between the impactor and the rock surface during an impact event is governed by contact mechanics. This aspect of impact loading is particularly important for understanding localized effects and energy transfer mechanisms. Key concepts in contact mechanics relevant to rock impact include:

(a) Hertzian contact theory: Describes the stress distribution and deformation at the contact point for elastic materials.
(b) Coefficient of restitution: Characterizes the elasticity of the collision and energy dissipation.
(c) Contact stiffness: Relates the contact force to the indentation depth.

For a spherical impactor on a flat rock surface, the Hertzian contact force can be expressed as:

$$F = \frac{4}{3} E^* \sqrt{R} \delta^{\frac{3}{2}} \qquad (4.1)$$

Where:
F is the contact force
E^* is the effective elastic modulus of the impactor-rock system
R is the radius of the impactor
δ is the indentation depth

The effective elastic modulus E^* is given by:

$$\frac{1}{E^*} = \frac{1 - v_1^2}{E_1} + \frac{1 - v_2^2}{E_2} \qquad (4.2)$$

Where:
E_1, v_1 are the elastic modulus and Poisson's ratio of the impactor
E_2, v_2 are the elastic modulus and Poisson's ratio of the rock

Understanding these contact mechanics principles is crucial for accurately modeling the initial stages of impact and the resulting stress distribution in the rock material [5].

6. Dynamic Fracture and Damage

The response of rock materials to impact loading is characterized by distinctive dynamic fracture and damage mechanisms that differ substantially from quasi-static conditions. When subjected to impact loads, rocks demonstrate complex fracture behavior marked by several key phenomena. Most notably, rocks exhibit increased fracture toughness under dynamic loading conditions, enhancing their resistance to crack propagation. The high-intensity nature of impact loads frequently triggers the simultaneous initiation of multiple cracks throughout the material. Furthermore, these dynamically-induced cracks demonstrate a greater propensity for branching and bifurcation compared to their quasi-static counterparts. Perhaps most significantly, impact-induced cracks can achieve remarkably high propagation velocities, approaching the material's Rayleigh wave speed–a characteristic that fundamentally distinguishes dynamic fracture from quasi-static crack growth [1, 3, 4].

The dynamic stress intensity factor, which characterizes the stress state near a crack tip under dynamic loading, can be expressed as:

$$KId(t) = k(v) \cdot KI(t) \tag{4.3}$$

where:
$KId(t)$ is the dynamic stress intensity factor
$KI(t)$ is the equivalent static stress intensity factor
$k(v)$ is a velocity-dependent factor
v is the crack propagation velocity

The function $k(v)$ accounts for inertial effects and typically increases with crack velocity, leading to higher apparent fracture toughness under dynamic conditions [1–4].

7. Scale Effects

Impact loading on rocks exhibits notable scale-dependent characteristics, which have significant implications for both experimental research and practical engineering applications. The manifestation of these scale effects occurs through several interrelated mechanisms. Primarily, larger rock specimens demonstrate reduced strength under impact loading, primarily due to the increased probability of containing critical flaws within their volume. Furthermore, as specimen size increases, the relative significance of stress wave propagation and reflection phenomena becomes more pronounced, leading to more complex dynamic responses. This scaling relationship also influences the spatial distribution of strain rates, with larger specimens typically experiencing more heterogeneous strain rate patterns during impact events. Additionally, the transition in specimen size can alter the dominant energy dissipation mechanisms, fundamentally affecting the overall dynamic response behavior of the rock mass [2–4].

8. Thermal Effects

While often overlooked, thermal effects can play a significant role in the characteristics of impact loading on rocks, particularly for high-energy impacts. Key aspects of thermal effects in impact loading include:

Adiabatic heating: The rapid deformation during impact can lead to localized temperature increases due to adiabatic heating. Thermal softening: Elevated temperatures can cause thermal softening of the rock material, affecting its mechanical response. Thermo-mechanical coupling: The interaction between thermal and mechanical effects can lead to complex material behavior under impact. Phase transitions: In extreme cases, high-energy impacts can induce phase transitions in rock minerals.

4.1 Basic Concept and Theory of Impact Load

The temperature rise due to adiabatic heating during impact can be estimated using the following equation:

$$\Delta T = \frac{\beta}{\rho C_p} \int \sigma d\varepsilon \tag{4.4}$$

Where:
ΔT is the temperature rise
β is the Taylor-Quinney coefficient (typically 0.8–0.9 for rocks)
ρ is the density of the rock
C_p is the specific heat capacity
σ is the stress
ε is the strain

Understanding and accounting for these thermal effects is crucial for accurately characterizing the behavior of rocks under high-energy impact loading conditions [2].

The definition and characteristics of impact loading in rock dynamics encompass a wide range of phenomena, from the fundamental physics of stress wave propagation to complex failure mechanisms and engineering applications. The high strain rates, rapid energy transfer, and unique failure modes associated with impact loading distinguish it from static and quasi-static loading conditions. Understanding these characteristics is crucial for accurately predicting and analyzing the behavior of rock materials under dynamic conditions, with implications spanning from small-scale laboratory experiments to large-scale geological processes and engineering projects.

4.1.2 Theory and Types of Stress Wave Propagation

Stress wave propagation are fundamental concepts in rock dynamics, playing crucial roles in understanding the behavior of rock materials under dynamic conditions. This section delves into the various types of stress waves that propagate through rock materials, providing a comprehensive overview of these essential aspects of rock dynamics.

Theory of Stress Wave Propagation

Stress wave propagation is the fundamental mechanism by which the effects of impact loading are transmitted through rock materials. Understanding the theory of stress wave propagation is crucial for analyzing the dynamic response of rocks and predicting their behavior under various loading conditions [6–9].

1. Basic Principles of Stress Wave Propagation

When a rock is subjected to impact loading, the resulting deformation and motion are not uniform throughout the material. Instead, there is a wave-like propagation of strain and velocity changes, which can be described as stress waves [8, 10]. The key principles governing stress wave propagation in rocks include:

Wave equation: The propagation of stress waves in elastic media is governed by the wave equation, which relates the displacement of particles to time and position. For a one-dimensional case, the wave equation can be expressed as:

$$\frac{\partial^2 u}{\partial t^2} = \frac{E}{\rho} \cdot \frac{\partial^2 u}{\partial x^2} \tag{4.5}$$

Where:
u is the displacement
x is the position
t is time
E is the elastic modulus
ρ is the density of the material

Wave reflection and transmission: When stress waves encounter interfaces between different materials or free surfaces, they undergo reflection and transmission. The behavior of waves at interfaces depends on the relative wave impedances of the materials involved. One-dimensional stress wave theory provides a simplified yet powerful framework for analyzing wave propagation in long, slender objects such as drill rods or rock bolts [7, 8, 10].

Types of Stress Waves in Rock Materials

Stress waves in rock materials can be classified into several types based on their propagation characteristics and the particle motion they induce. Understanding these different types of waves is crucial for analyzing the dynamic response of rock masses under various loading conditions [8, 9, 11].

1. Body Waves

Body waves, which propagate through the interior of rock masses, consist of two distinct types: P-waves and S-waves, each characterized by unique properties and behaviors. P-waves, also known as Primary or Longitudinal waves, represent the fastest-traveling waves in rock materials, with velocities typically ranging from 3000 to 7000 m/s. These waves are distinguished by particle motion that occurs parallel to the direction of wave propagation, resulting in compression and dilation of the rock material. In contrast, S-waves (Secondary or Shear waves) exhibit markedly different characteristics, with particle motion occurring perpendicular to the wave propagation direction. S-waves propagate more slowly than P-waves, typically achieving velocities of only 60–70% of P-wave velocity, and are notably characterized by their inability to propagate through fluids while causing shear deformation in the rock material.

The ratio of P-wave to S-wave velocity is often used as an indicator of rock properties and can provide insights into rock mass quality and fracture density [7].

2. Surface Waves

Surface waves, which play a crucial role in seismology and near-surface geophysics, propagate along the free surface or interfaces of rock masses. Among these, Rayleigh waves represent a significant type of surface wave, characterized by distinctive properties that influence their behavior and impact. These waves exhibit elliptical particle motion occurring in a vertical plane as they travel along the free surface of the medium. A notable characteristic of Rayleigh waves is their velocity, which is slightly lower than that of S-waves. Furthermore, their amplitude demonstrates a predictable decay pattern, decreasing exponentially with depth from the surface [7–9].

3. Shock Waves

Extreme loading conditions, such as those encountered during explosions or hypervelocity impacts, can generate shock waves within rock materials, representing a distinct class of wave phenomena that demands specialized analysis techniques. These shock waves exhibit several unique characteristics that distinguish them from conventional elastic waves. Most notably, they produce discontinuous changes in pressure, density, and particle velocity across the wave front, while achieving propagation velocities that exceed the material's elastic wave speed. The extreme conditions associated with shock waves induce nonlinear material behavior, manifesting as plastic deformation and phase transformations within the rock mass. Due to these complex phenomena, the analysis of shock waves in rocks necessitates specialized analytical approaches, particularly the implementation of Hugoniot curves, which establish the fundamental relationship between particle velocity and shock velocity [7, 8].

The theory and types of stress wave propagation form a fundamental basis for understanding the dynamic behavior of rock materials under impact loading. The complex interplay between material properties, discontinuities, and loading characteristics governs the propagation, attenuation, and dispersion of stress waves in rock masses. Ongoing research in this field continues to refine our understanding of these phenomena, leading to improved prediction and control of rock behavior in various engineering applications.

4.1.3 Effects of Strain Rate on Rock Mechanical Properties

The effect of strain rate on rock mechanical properties is a crucial aspect of rock dynamics that has significant implications for various engineering applications, including mining, tunneling, and earthquake engineering. This section explores the fundamental concepts and theories related to the influence of strain rate on rock behavior, synthesizing findings from multiple studies to provide a comprehensive understanding of this complex phenomenon.

Strain Rate Sensitivity of Rock Strength

The strength of rocks is known to be highly dependent on the rate at which they are loaded. This strain rate sensitivity is observed across a wide range of rock types, from brittle to ductile materials. Generally, as the strain rate increases, the strength of the rock also increases, but the magnitude of this effect varies significantly depending on the rock type and testing conditions [10, 12, 13].

For brittle rocks, such as limestone and sandstone, the increase in strength with strain rate is often quantified using the DIF, which is the ratio of dynamic strength to static strength. Experimental studies have shown that the DIF for compressive strength can range from 4 to 6 for strain rates between 1 and $5.2*10^2$ s^{-1} [10]. The DIF for tensile strength is typically higher, ranging from 1.5 to 3.4 for similar strain rates [10].

The strain rate sensitivity of rock strength can be attributed to multiple fundamental mechanisms that operate simultaneously during deformation [14]. The first key mechanism involves time-dependent crack growth, which exhibits distinct behavior patterns across different loading rates. At low strain rates, cracks have sufficient time to propagate through the rock material, resulting in reduced overall strength. In contrast, when subjected to higher strain rates, the limited time available for crack propagation leads to enhanced apparent strength [15]. This mechanical response is further complicated by inertial effects, which become increasingly significant as strain rates rise. These inertial forces introduce additional confinement and resistance to deformation [16], further contributing to the complex strain rate dependency of rock strength.

Strain Rate Effects on Deformation Characteristics

The influence of strain rate on rock behavior extends beyond strength parameters to significantly affect various deformation characteristics. In terms of elastic properties, the dynamic elastic modulus of rocks typically exhibits an increase with strain rate, though this enhancement is generally less pronounced than the corresponding increase in strength [12]. The relationship between strain rate and Poisson's ratio demonstrates greater complexity, varying across different rock types; while some research has documented a decrease in Poisson's ratio with increasing strain rate [13], other studies have found minimal to no correlation. A more consistent trend appears in the strain at failure parameter, which typically decreases with increasing strain rate in brittle rocks [10, 12], indicating enhanced brittleness under dynamic loading conditions. Furthermore, the stress–strain behavior undergoes notable changes with increasing strain rate, characterized by increasingly linear stress–strain curves and a reduction in the plastic deformation phase [14], particularly evident during the transition from quasi-static to dynamic loading conditions.

Energy Absorption and Dissipation

The energy absorption characteristics of rocks are significantly influenced by strain rate. Several studies have reported an increase in energy absorption capacity with increasing strain rate [10, 14, 17]. This is often quantified using metrics such as the Specific Energy Absorption (SEA) or energy absorption ratio.

For example, research on quartz sandstone under cyclic impact loading has shown that both energy dissipation and energy efficiency increase with the number of impact cycles and with increasing strain rate [17]. This increased energy absorption capacity is attributed to the formation and propagation of microcracks within the rock matrix.

The relationship between energy dissipation and strain rate can be complex and non-linear. Some studies have observed that the energy dissipation ratio increases rapidly at low to medium strain rates but begins to plateau at very high strain rates [14]. This behavior is thought to be related to the transition between different deformation mechanisms as strain rate increases.

The effect of strain rate on rock mechanical properties is a complex phenomenon that involves multiple interacting mechanisms. The strain rate sensitivity varies significantly across different rock types and loading conditions, necessitating careful consideration in both experimental studies and engineering applications. As research in this field continues to advance, our understanding of these complex behaviors will improve, leading to more accurate predictions and safer, more efficient engineering designs in rock mechanics applications.

4.2 Rock Mechanical Behavior Under Impact Load

4.2.1 Dynamic Stress–Strain Relationship of Rocks

The dynamic stress–strain relationship of rocks is a fundamental aspect of rock mechanics that plays a crucial role in understanding and predicting rock behavior under impact loading conditions. This relationship differs significantly from the static stress–strain behavior due to the high strain rates and complex wave propagation phenomena involved in dynamic loading scenarios. In this section, we will delve into the intricacies of the dynamic stress–strain relationship of rocks, exploring its characteristics, influencing factors, and implications for rock engineering applications.

General Characteristics of Dynamic Stress–Strain Curves

The dynamic stress–strain curve of rocks under impact loading exhibits distinct features that set it apart from its static counterpart. These characteristics are essential for understanding rock behavior in various dynamic loading scenarios, such as blasting, and rock bursts.

1. Typical Stages of Dynamic Stress–Strain Curves

A typical dynamic stress–strain curve for rocks can be divided into several stages, each representing different aspects of rock deformation and failure under impact loading.

These stages are:

(a) Compaction Stage: This initial stage occurs as the stress wave propagates through the rock specimen. During this phase, pre-existing microcracks and pores in the rock are closed, resulting in a non-linear increase in stress with strain. The duration and extent of this stage depend on the initial porosity and microstructure of the rock.

(b) Elastic Stage: Following the compaction stage, the rock exhibits linear elastic behavior. This stage is characterized by a constant slope in the stress–strain curve, representing the dynamic elastic modulus of the rock. The elastic stage typically extends to higher stress levels in dynamic loading compared to static loading due to the strain rate effect.

(c) Crack Propagation Loading Stage: As the stress continues to increase, microcracks within the rock begin to propagate and coalesce. This stage is marked by a gradual decrease in the tangent modulus of the stress–strain curve. The onset and progression of this stage are highly dependent on the loading rate and the rock's microstructure.

(d) First Unloading Stage: After reaching the peak stress, the rock enters the first unloading stage. During this phase, the external load begins to decrease, but the strain continues to increase due to the inertial effects and continued crack propagation. This stage is unique to dynamic loading and is not observed in static stress–strain curves.

(e) Second Unloading Stage: In the final stage, both stress and strain decrease as the rock specimen continues to unload. The residual strain at the end of this stage represents the permanent deformation of the rock due to the impact loading (Fig. 4.1).

These stages provide valuable insights into the deformation and failure processes of rocks under dynamic loading conditions. It is important to note that the exact nature and extent of these stages can vary depending on factors such as rock type, loading rate, and confining pressure.

2. Key Points on the Dynamic Stress–Strain Curve

Several critical points on the dynamic stress–strain curve are of particular interest in rock dynamics research: (a) Peak Stress Point: This point represents the maximum stress that the rock can withstand under the given dynamic loading conditions. The

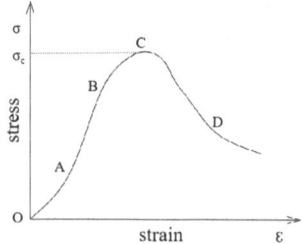

Fig. 4.1 Typical dynamic stress–strain curve

4.2 Rock Mechanical Behavior Under Impact Load

peak stress is a crucial parameter in assessing the dynamic strength of rocks and is significantly influenced by the loading rate. (b) Peak Strain Point: The strain corresponding to the peak stress is known as the peak strain. This point is essential for understanding the deformation capacity of rocks under dynamic loading. (c) Critical Unloading Stress Point: This point marks the transition from the first unloading stage to the second unloading stage. It provides information about the rock's ability to sustain deformation after peak stress has been reached.

The relationship between these key points and their evolution with changing loading conditions offers valuable insights into rock behavior under impact loading.

Factors Influencing Dynamic Stress–Strain Relationships

The dynamic stress–strain relationship of rocks is influenced by various factors, including loading rate, confining pressure, rock type, and temperature. Understanding these influences is crucial for accurately predicting and modeling rock behavior under dynamic loading conditions.

1. Strain Rate Effects

One of the most significant factors influencing the dynamic stress–strain relationship of rocks is the strain rate. Numerous studies have shown that rock strength and deformation characteristics are highly sensitive to the rate of loading [10, 18–20].

(1) Strain Rate Sensitivity of Dynamic Strength

The dynamic strength of rocks generally increases with increasing strain rate. This phenomenon is often quantified using the DIF, defined as the ratio of dynamic strength to static strength:

$$\text{DIF} = \frac{f_d}{f_c} \tag{4.6}$$

Where f_d is the dynamic peak strength and f_c is the static uniaxial compressive strength.

Research has shown that the relationship between DIF and strain rate can often be approximated by logarithmic functions. For example, a study on granite under different confining pressures yielded the following relationships:

$$\text{DIF}_{\sigma_3=2.5} = 0.853\ln(\dot{\varepsilon}) - 2.296 \tag{4.7}$$

$$\text{DIF}_{\sigma_3=5} = 0.997\ln(\dot{\varepsilon}) - 2.720 \tag{4.8}$$

$$\text{DIF}_{\sigma_3=10} = 1.394\ln(\dot{\varepsilon}) - 4.471 \tag{4.9}$$

$$\text{DIF}_{\sigma_3=20} = 0.333\ln(\dot{\varepsilon}) + 0.924 \tag{4.10}$$

Where $\dot{\varepsilon}$ is the strain rate and σ_3 is the confining pressure in MPa.

These relationships demonstrate that the strain rate effect on rock strength is more pronounced at lower confining pressures and becomes less significant as confining pressure increases.

(2) Critical Strain Rate

Research has identified the existence of a critical strain rate for many rock types, typically around 10^2 to 10^3s^{-1} [22]. Below this critical rate, the increase in strength with strain rate is relatively moderate. However, above the critical strain rate, the strength increase becomes much more rapid.

For example, a study on tuff identified a critical strain rate of approximately 76 s^{-1}.

This behavior is attributed to the transition from quasi-static crack growth mechanisms to dynamic fracture processes at high strain rates. Understanding this transition is crucial for accurately modeling rock behavior across a wide range of loading rates.

(3) Strain Rate Effects on Deformation Characteristics

In addition to strength, the strain rate also influences the deformation characteristics of rocks. Generally, as the strain rate increases: the elastic modulus tends to increase, the strain at peak stress (peak strain) often decreases, and the post-peak behavior becomes more brittle.

These changes in deformation characteristics have important implications for energy absorption and fragmentation processes in dynamic rock failure.

2. Confining Pressure Effects

The confining pressure plays a significant role in the dynamic stress–strain behavior of rocks, similar to its effect under static loading conditions. However, the interaction between confining pressure and strain rate effects adds complexity to the dynamic behavior [19].

(1) Influence on Dynamic Strength

Increasing confining pressure generally leads to an increase in the dynamic strength of rocks. This effect can be attributed to the suppression of crack growth and the promotion of plastic deformation mechanisms under higher confining pressures.

Research has shown that the relationship between dynamic strength and confining pressure is often linear, similar to static loading conditions. However, the slope of this relationship (i.e., the internal friction angle) may differ under dynamic loading [19].

(2) Interaction with Strain Rate Effects

The influence of confining pressure on the strain rate sensitivity of rock strength is complex. Some key observations include: the strain rate effect on strength tends to be more pronounced at lower confining pressures. At very high confining pressures, the strain rate sensitivity may decrease significantly. The critical strain rate marking

4.2 Rock Mechanical Behavior Under Impact Load

the transition to rapid strength increase may shift to higher values with increasing confining pressure.

These interactions highlight the importance of considering both confining pressure and strain rate when analyzing and modeling dynamic rock behavior.

(3) Effect on Stress–Strain Curve Shape

Confining pressure also influences the shape of the dynamic stress–strain curve. As confining pressure increases: the initial compaction stage may become less pronounced due to pre-existing crack closure, the elastic stage may extend to higher stress levels, and the post-peak behavior tends to become more ductile, with a more gradual stress drop.

For example, a study on sandstone under dynamic loading showed that at confining pressures of 10–20 MPa, stress–strain curves exhibited yield plateaus near the peak points, indicating reduced brittleness compared to lower confining pressures.

3. Rock Type and Microstructure

The dynamic stress–strain relationship is significantly influenced by the rock type and its microstructural characteristics. Different rock types can exhibit varying sensitivities to strain rate and confining pressure effects.

(1) Influence of Rock Type

Studies have shown that even rocks with similar static strengths can display markedly different dynamic behaviors. For example: Crystalline rocks (e.g., granite, marble) often show higher strain rate sensitivity than sedimentary rocks; rocks with higher porosity generally exhibit more pronounced compaction stages in their dynamic stress–strain curves; the critical strain rate marking the transition to rapid strength increase can vary significantly between rock types.

These differences underscore the importance of rock-specific testing and characterization for accurate dynamic analysis and modeling.

(2) Microstructural Factors

Several microstructural factors influence the dynamic stress–strain behavior of rocks. For example, finer-grained rocks often exhibit higher dynamic strength and more brittle behavior; higher porosity can lead to more pronounced compaction stages and potentially lower dynamic strength.

Understanding these microstructural influences is crucial for developing physics-based models of dynamic rock behavior.

4. Temperature Effects

Temperature can have a significant impact on the dynamic stress–strain relationship of rocks, although this factor has received less attention compared to strain rate and confining pressure effects.

(1) Influence on Dynamic Strength

Generally, lower temperatures lead to higher dynamic strength, similar to the effects observed under static loading conditions. The temperature effect on dynamic strength has been found to be analogous to the strain rate effect, with lower temperatures producing effects similar to higher strain rates [20].

(2) Temperature-Strain Rate Interactions

The interaction between temperature and strain rate effects is complex and can vary depending on the rock type and temperature range.

These interactions highlight the need for further research into the combined effects of temperature and strain rate on rock dynamic behavior, particularly for applications involving extreme temperature conditions (e.g., deep geological repositories, planetary impact studies).

Analytical and Constitutive Models for Dynamic Stress–Strain Relationships

To describe and predict the dynamic stress–strain behavior of rocks, various analytical and constitutive models have been developed. These models aim to capture the key features of dynamic rock behavior while providing a framework for numerical simulations and engineering analyses.

1. Statistical Damage Models

Statistical damage models have been widely used to describe the dynamic stress–strain relationship of rocks. These models are based on the concept that rock failure under dynamic loading is a progressive process involving the accumulation of damage due to microcrack growth and coalescence.

(1) Dynamic Combined Statistical Damage Model.

A comprehensive dynamic combined statistical damage model has been proposed to account for both confining pressure and strain rate effects. The model considers the dynamic axial stress as a superposition of static stress and dynamic inertial stress components.

This model has shown good agreement with experimental data for various rock types under different loading conditions.

2. Visco-Elastic Models

Visco-elastic models attempt to capture the time-dependent aspects of rock behavior under dynamic loading. These models typically combine elastic and viscous elements to represent the instantaneous and time-dependent responses of the rock.

(1) Kelvin-Voigt Model

The Kelvin-Voigt model is a simple Visco-elastic model that has been applied to dynamic rock behavior. It consists of an elastic spring and a viscous damper connected in parallel. The constitutive equation for this model is:

4.2 Rock Mechanical Behavior Under Impact Load

$$\sigma = E\varepsilon + \eta\left(\frac{d\varepsilon}{dt}\right) \quad (4.11)$$

Where E is the elastic modulus and η is the viscosity coefficient.

While this model can capture some aspects of the strain rate dependence of rock behavior, it is often too simplistic to accurately represent the full range of dynamic rock responses.

(2) Maxwell Model

The Maxwell model, consisting of an elastic spring and a viscous damper in series, has also been used to describe dynamic rock behavior. Its constitutive equation is:

$$\frac{d\varepsilon}{dt} = \frac{1}{E}\frac{d\sigma}{dt} + \frac{\sigma}{\eta} \quad (4.12)$$

This model can represent stress relaxation behavior but has limitations in capturing the full dynamic stress–strain response of rocks.

(3) Generalized Visco-elastic Models.

More complex Visco-elastic models, such as the generalized Maxwell model or the Burgers model, have been proposed to better capture the dynamic behavior of rocks. These models incorporate multiple elastic and viscous elements to represent a broader range of time-dependent responses.

3. Rate-Dependent Plasticity Models

Rate-dependent plasticity models extend classical plasticity theory to account for strain rate effects in rock behavior. These models typically modify the yield criterion, flow rule, or hardening law to incorporate strain rate dependence.

(1) Johnson–Cook Model

The Johnson–Cook model is a widely used rate-dependent plasticity model that has been adapted for rock dynamics.

This model can capture the strain rate and temperature dependence of rock strength but may require modifications to accurately represent the full range of rock dynamic behaviors.

(2) Unified Viscoplastic Models

Unified viscoplastic models attempt to describe both the instantaneous and time-dependent aspects of rock behavior within a single framework. These models often use internal state variables to represent the evolving microstructure of the rock under dynamic loading.

4. Micromechanics-Based Models

Micromechanics-based models attempt to link the macroscopic dynamic stress–strain behavior of rocks to underlying microscale processes, such as crack growth and interaction.

This model has shown good agreement with experimental results for various rock types under different loading conditions [18].

5. Empirical and Semi-Empirical Models

In addition to physics-based models, various empirical and semi-empirical models have been developed to describe specific aspects of dynamic rock behavior.

(1) Strain Rate-Dependent Strength Models

Empirical models relating dynamic strength to strain rate are commonly used in rock engineering applications. These models often take the form of power laws or logarithmic functions, such as:

$$\sigma_d = \sigma_s \left(\frac{\dot{\varepsilon}}{\dot{\varepsilon}_0}\right)^\alpha \tag{4.13}$$

Where σ_d is the dynamic strength, σ_s is the static strength, $\dot{\varepsilon}$ is the strain rate, $\dot{\varepsilon}_0$ is a reference strain rate, and α is an empirical parameter.

(2) DIF Models

Models for the DIF as a function of strain rate have been developed for various rock types. These models often incorporate the concept of a critical strain rate, as discussed earlier.

(3) Energy-Based Models

Some researchers have proposed energy-based models to describe dynamic rock failure. These models often relate the energy absorbed by the rock during dynamic loading to parameters such as strain rate, confining pressure, and rock properties.

Experimental Techniques for Measuring Dynamic Stress–Strain Relationships

Accurate measurement of dynamic stress–strain relationships in rocks presents significant challenges due to the high loading rates and short durations involved. Several specialized experimental techniques have been developed to address these challenges.

1. Split Hopkinson Pressure Bar (SHPB) Technique

The SHPB, also known as the Kolsky bar, is the most widely used technique for measuring dynamic stress–strain relationships in rocks at medium to high strain rates (typically 10 to 10^3 s^{-1}) [19] (Fig. 4.2).

The stress, strain, and strain rate in the specimen are calculated using the following equations:

4.2 Rock Mechanical Behavior Under Impact Load

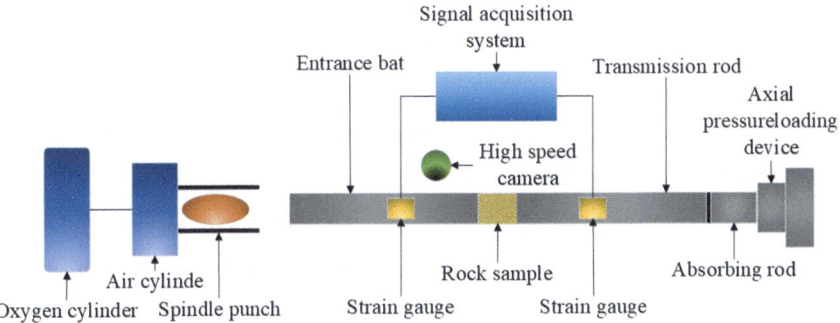

Fig. 4.2 Device of split Hopkinson pressure bar

Stress:

$$\sigma(t) = \left(\frac{A_e E_e}{A_s}\right)\varepsilon_T(t) \tag{4.14}$$

Strain:

$$\varepsilon(t) = -\frac{2c_e}{L_s}\int_0^t \varepsilon_R(\tau)\,d\tau \tag{4.15}$$

Strain rate:

$$\dot{\varepsilon}(t) = -\frac{2c_e}{L_s}\varepsilon_R(t) \tag{4.16}$$

where A_e and A_s are the cross-sectional areas of the bar and specimen, E_e is the elastic modulus of the bar, c_e is the wave speed in the bar, L_s is the specimen length, ε_R and ε_T are the reflected and transmitted strains.

2. Plate Impact Techniques

For very high strain rate testing ($>10^4$ s^{-1}), plate impact techniques are often used. These methods involve launching a flat plate projectile at a rock specimen using a gas gun or powder gun.

3. Drop Weight Impact Tests

Drop weight impact tests provide a simpler alternative for studying rock behavior under low to medium strain rates (typically $< 10^2$ s^{-1}).

4. Advanced Measurement Techniques

Several advanced measurement techniques have been developed to provide more detailed information about the dynamic deformation and failure processes in rocks, such as DIC, Laser Interferometry and Acoustic Emission Monitoring.

Characteristic Features of Dynamic Stress–Strain Curves

The dynamic stress–strain curves of rocks exhibit several characteristic features that distinguish them from static stress–strain behavior. Understanding these features is crucial for interpreting experimental results and developing accurate models of rock behavior under impact loading.

1. Strain Rate Dependence

One of the most prominent features of dynamic stress–strain curves is their dependence on strain rate. As discussed earlier, increasing strain rate generally leads to higher peak strength, increased elastic modulus, reduced strain at peak stress, and more brittle post-peak behavior.

The magnitude of these effects varies depending on rock type, confining pressure, and the range of strain rates considered.

2. Stress Wave Oscillations

Dynamic stress–strain curves often exhibit oscillations, particularly in the early stages of loading. These oscillations are due to stress wave reflections within the specimen and testing apparatus. While these oscillations can complicate data interpretation, they also provide information about wave propagation characteristics in the rock.

3. Apparent Strain Softening

Many rocks exhibit apparent strain softening in their dynamic stress–strain curves, where the stress appears to decrease with increasing strain beyond the peak stress. This behavior is often more pronounced in dynamic tests compared to static tests and can be attributed to several factors, such as inertial effects in the specimen, strain localization and progressive failure, and stress wave interactions. It is important to note that this apparent softening may not always represent the true material behavior and can be influenced by testing conditions and specimen geometry.

4. Rate-Dependent Elastic Limit

The elastic limit or yield point of rocks often shows a strong dependence on strain rate. In many cases, the transition from elastic to inelastic behavior occurs at higher stress levels under dynamic loading compared to static loading. This effect is related to the time-dependent nature of microcrack growth and interaction processes.

5. Energy Absorption Characteristics

Dynamic stress–strain curves provide important information about the energy absorption characteristics of rocks under impact loading. The area under the stress–strain curve represents the energy absorbed per unit volume of the rock. This energy absorption capacity often increases with strain rate, which has important implications for blast resistance and fragmentation processes.

6. Cyclic Loading Effects

Under cyclic impact loading, rocks exhibit characteristic stress–strain behaviors, including degradation of peak strength with increasing impacts, accumulation of residual strain, and changes in the shape of stress–strain loops [22]. These effects result from progressive damage accumulation and fatigue processes in the rock.

Applications of Dynamic Stress–Strain Relationships

Understanding the dynamic stress–strain relationships of rocks is crucial for various engineering and scientific applications. In blast engineering, this knowledge aids in predicting rock fragmentation, designing effective blast patterns, and assessing the stability of nearby structures. In mining, it is essential for predicting and mitigating rockburst hazards, designing support systems for dynamic loading, and optimizing excavation methods in high-stress environments. Additionally, in material science and engineering, it contributes to developing rock-like materials with enhanced dynamic properties, improving concrete and ceramic formulations for impact resistance, and advancing non-destructive testing methods for geomaterials.

Challenges and Future Directions

Despite advances in understanding dynamic stress–strain relationships in rocks, several challenges remain. High strain rate testing requires improved methods to ensure stress equilibrium, mitigate inertial effects, and achieve uniform deformation. Research must expand beyond uniaxial loading to address multiaxial, combined static-dynamic, and cyclic loading conditions. Understanding scale effects is crucial for bridging laboratory tests and field observations, requiring scaling laws and investigations into discontinuities. Advancing microstructure-based models necessitates better insights into crack growth, imaging integration, and multiscale modeling. Developing rate-dependent failure criteria and incorporating them into numerical models is essential. Environmental factors, including temperature, pore fluid effects, and chemical alterations, require further study. Finally, enhancing non-destructive evaluation methods is key for in-situ rock property assessments.

The dynamic stress–strain relationship of rocks is a complex and multifaceted topic with significant implications for various fields of science and engineering. While substantial progress has been made in understanding and modeling this behavior, many challenges remain. Continued research in this area will be essential for advancing our ability to predict and manage rock behavior under the extreme loading conditions encountered in many natural and engineered systems.

4.2.2 Dynamic Failure Mechanisms and Modes of Rocks

Rocks subjected to impact loading exhibit complex dynamic failure mechanisms and modes that differ significantly from their static counterparts. Understanding these mechanisms is crucial for various engineering applications, including mining, tunneling, and protective structures. This section provides a comprehensive review of the current knowledge on dynamic failure mechanisms and modes of rocks under impact load.

Fundamentals of Dynamic Rock Failure

Dynamic rock failure under impact loading is characterized by rapid crack initiation, propagation, and coalescence, leading to material fragmentation. The process is governed by the interplay of stress wave propagation, material properties, and loading conditions [23–25].

1. Stress Wave Propagation

When a rock is subjected to impact loading, stress waves propagate through the material, causing rapid changes in stress and strain states. These waves can be classified into three types:

(a) Longitudinal (P) waves: These are compressive waves that travel parallel to the direction of impact.
(b) Shear (S) waves: These waves propagate perpendicular to the direction of impact and cause shear deformation.
(c) Rayleigh (R) waves: These are surface waves that travel along the free surface of the rock.

The propagation of these waves plays a crucial role in the dynamic failure process. As the waves interact with material boundaries and internal discontinuities, they can lead to stress concentrations and localized failure [23, 24].

2. Strain Rate Effects

One of the most significant factors influencing dynamic rock failure is the strain rate. The strain rate in impact loading scenarios can range from 10 to 10^5 s^{-1}, which is several orders of magnitude higher than in quasi-static loading conditions [25, 26]. This high strain rate has profound effects on rock behavior.

3. Microstructural Influences

The microstructure of rocks, including grain size, porosity, and pre-existing microcracks, significantly influences their dynamic failure behavior [27, 28].

Dynamic Crack Initiation and Propagation

The process of dynamic crack initiation and propagation in rocks under impact loading is fundamentally different from static conditions due to the high strain rates and stress wave interactions involved.

1. Crack Initiation

Under dynamic loading, crack initiation occurs when the local tensile stress exceeds the dynamic tensile strength of the rock.

The critical stress intensity factor for dynamic crack initiation (KId) is generally higher than its static counterpart (KIc) due to the inertial effects and limited time

4.2 Rock Mechanical Behavior Under Impact Load

for stress redistribution. For limestone, the dynamic fracture toughness has been estimated to be 2.4–2.5 MPa·m$^{0.5}$, which is 2–3 times higher than quasi-static values [24].

2. Crack Propagation

Once initiated, cracks propagate rapidly under dynamic loading. The crack propagation process is characterized by high propagation speeds, multiple crack interactions and stress wave interactions.

3. Crack Coalescence and Fragmentation

As multiple cracks propagate under dynamic loading, they interact and coalesce, leading to the formation of larger fractures and ultimately to material fragmentation.

The fragment size distribution resulting from dynamic loading is often described using statistical models such as the Weibull or log-normal distributions.

Failure Modes Under Dynamic Loading

Rocks exhibit a variety of failure modes under dynamic loading, depending on the loading conditions, material properties, and specimen geometry. Understanding these failure modes is crucial for predicting rock behavior in impact scenarios.

1. Tensile Failure

Tensile failure is a common mode of dynamic rock failure, particularly in scenarios involving reflected tensile waves or bending stresses.

2. Compressive Failure

Under dynamic compressive loading, rocks exhibit failure modes that differ from static compression:

The transition between these failure modes can be related to the loading rate and confining pressure. For example, in triaxial compression tests on tuff, the failure mode transitions from brittle tensile splitting to ductile shear as the loading rate decreases or confining pressure increases.

3. Mixed-Mode Failure

In many practical scenarios, rocks experience complex stress states that lead to mixed-mode failure. This is particularly evident in cases involving:

4. Jointed Rock Failure

The presence of joints and discontinuities in rock masses significantly influences their dynamic failure behavior.

Recent studies have shown that joint surface roughness significantly affects the dynamic fracturing behavior of jointed rock. For example, experimental and numerical studies on granite with artificially rough joints revealed that [23]: joint surface roughness influences crack initiation and propagation patterns, and cracks tend to initiate near joint bulges and propagate along the loading direction.

These findings highlight the importance of considering joint properties in predicting the dynamic failure behavior of rock masses.

Size and Boundary Effects

The dynamic failure behavior of rocks is significantly influenced by specimen size and boundary conditions, leading to complex size effects that differ from static loading scenarios.

1. Dynamic Size Effects

Unlike static loading, where larger specimens typically exhibit lower strength (negative size effect), dynamic loading can lead to positive size effects under certain conditions.

These dynamic size effects can be attributed to the interplay between stress wave propagation, crack growth rates, and inertial confinement effects in larger specimens.

2. Boundary Effects

The dynamic failure behavior of rocks is also influenced by boundary conditions, which affect stress wave propagation and reflection.

For example, in triaxial Hopkinson bar tests on sandstone, it was found that the relative contribution of lateral confinement effects to dynamic strength increase was 29.8% for 36 mm diameter specimens and 46.9% for 75 mm diameter specimens. This demonstrates the significant influence of specimen size on confinement effects under dynamic loading.

Influence of Environmental Factors

The dynamic failure behavior of rocks is not only influenced by loading conditions and material properties but also by environmental factors such as temperature, moisture content, and in-situ stress states.

1. Temperature Effects

Temperature can significantly affect the dynamic failure behavior of rocks through its influence on material properties and thermal stress states. For example, Thermal softening: Higher temperatures generally lead to decreased strength and stiffness of rocks, which can alter their dynamic response [25]. Thermal cracking: Rapid temperature changes or high temperatures can induce thermal cracking, creating additional pathways for dynamic crack propagation [25]. Strain rate-temperature coupling: The sensitivity of rock strength to strain rate can be temperature-dependent, with some rocks showing increased rate sensitivity at higher temperatures [25].

2. Moisture Effects

The presence of moisture in rocks significantly influences their dynamic failure behavior through various mechanisms. Pore pressure effects in saturated or partially saturated rocks can cause rapid pressure changes under dynamic loading, altering the effective stress state and failure behavior. Lubrication effects occur as water reduces

4.2 Rock Mechanical Behavior Under Impact Load

friction in microcracks and grain boundaries, potentially modifying crack propagation patterns. Additionally, chemical effects from long-term moisture exposure can alter rock mineral compositions, impacting their mechanical properties and dynamic response.

3. In-Situ Stress Effects

The pre-existing stress state in rock masses, particularly in deep underground environments, can significantly influence their dynamic failure behavior.

Dynamic strength decreased with increasing axial pre-stress σ_1, but increased with increasing lateral pre-stresses σ_2 and σ_3, failure strain generally decreased with increasing confinement, elastic modulus increased significantly from uniaxial to triaxial conditions.

These findings highlight the complex interplay between in-situ stress states and dynamic loading in determining rock failure behavior.

4.2.3 Deformation and Fracture Characteristics at High Strain Rates

Deformation and fracture characteristics of rocks at high strain rates are crucial for understanding their dynamic behavior under impact loading conditions. This section provides a comprehensive review of the current knowledge and recent advancements in this field, drawing from various experimental and numerical studies.

Rock materials exhibit distinct deformation and fracture characteristics when subjected to high strain rate loading compared to quasi-static conditions. These differences are primarily attributed to the rate-dependent nature of rock strength, inertial effects, and the complex interplay between stress wave propagation and material failure mechanisms [3, 29].

One of the key aspects of rock behavior under high strain rates is the increase in dynamic strength. Numerous studies have consistently shown that both the compressive and tensile strengths of rocks increase significantly with increasing strain rate [23, 32]. This phenomenon, often referred to as the strain rate effect, is typically characterized by the DIF, which is the ratio of dynamic strength to static strength.

The deformation behavior of rocks under high strain rates also differs significantly from quasi-static conditions. At high loading rates, rocks tend to exhibit more brittle behavior, characterized by steeper stress–strain curves and reduced plastic deformation before failure [3, 30]. This is partly due to the limited time available for crack growth and coalescence, resulting in a more rapid and localized failure process.

The stress–strain relationship of rocks under dynamic loading can be described using various constitutive models. One commonly used approach is the Holmquist-Johnson–Cook (HJC) model, which incorporates strain rate effects and damage accumulation. The HJC model expresses the normalized equivalent stress (σ^*) as:

$$\sigma^* = \left[A(1-D) + BP^{*N}\right]\left[1 + C\ln(\dot{\varepsilon}^*)\right] \quad (4.17)$$

where A, B, N, and C are material constants, D is the damage parameter, P^* is the normalized pressure, and $\dot{\varepsilon}^*$ is the dimensionless strain rate.

The fracture characteristics of rocks under high strain rates are also significantly influenced by the loading conditions. Duan et al. [29] conducted a comprehensive study on the dynamic fracturing behavior of roughly jointed rock using both experimental and numerical methods. They found that the fracture patterns changed from split failure to fractured or pulverized failure as the loading rate increased. Additionally, the fragment size decreased with increasing loading rate, indicating a more intense fragmentation process under higher strain rates.

The fracture process in rocks under dynamic loading is often characterized by the initiation, propagation, and coalescence of multiple cracks. At high strain rates, the crack propagation velocity can approach the Rayleigh wave speed of the material, leading to complex fracture patterns and multiple fragmentation [3]. The dynamic fracture toughness of rocks also exhibits rate dependence, with higher values observed at higher loading rates [30].

The failure modes of rocks under dynamic loading also exhibit a strong dependence on both strain rate and confining pressure. Under uniaxial conditions and high loading rates, rocks typically fail through brittle tensile splitting, with the number of fracture fragments increasing with loading rate. As the loading rate decreases or confining pressure increases, the failure mode transitions from brittle to ductile shear. This transition in failure mode has important implications for understanding and predicting rock behavior in various engineering applications, such as underground excavations and rock slope stability under seismic loading.

The dynamic properties of fractured rock masses present additional complexities due to the presence of discontinuities. Experimental studies on simulated fractured rock masses have shown that both fracture rate and strain rate significantly influence their dynamic behavior [28]. As strain rate increases, the failure stress of fractured rock specimens increases, regardless of fracture rate. However, the dynamic elastic modulus before failure was found to be independent of strain rate but dependent on fracture rate, increasing as fracture rate decreased. These findings highlight the importance of considering both fracture characteristics and loading conditions when analyzing the dynamic behavior of fractured rock masses.

The study of rock dynamic properties has significant implications for various engineering applications. In mining and tunneling, understanding the dynamic behavior of rocks is crucial for optimizing blasting parameters and designing support systems. For instance, the strain rate-dependent strength and fracture characteristics of rocks influence the extent of the blasting failure zone around boreholes. Research has shown that the blasting failure zone can be divided into three areas: crushing zone, fracture zone, and crack zone, with their respective ranges dependent on both rock properties and loading conditions [32].

In the context of rock burst prevention, the dynamic properties of rocks play a critical role in predicting and mitigating these hazardous events. The strain rate sensitivity of rock strength and the energy dissipation characteristics during dynamic fracturing are key factors in assessing the potential for rock bursts and designing appropriate support measures [3, 31].

The deformation and fracture characteristics of rocks at high strain rates exhibit complex behavior that differs significantly from quasi-static conditions. The strain rate sensitivity of rock strength, the transition in failure modes, the influence of pre-existing discontinuities, and the energy dissipation mechanisms during dynamic fracturing are key aspects that need to be considered in analyzing and predicting rock behavior under impact loading. However, challenges remain in developing comprehensive constitutive models that can accurately capture the multi-physics nature of rock dynamic behavior across a wide range of loading conditions and rock types. Future research directions include further investigation of coupled thermo-hydro-mechanical effects under dynamic loading, advanced imaging techniques for real-time observation of dynamic fracture processes, and the development of multi-scale models that can bridge the gap between microscopic damage mechanisms and macroscopic rock mass behavior.

4.3 Rock Mechanical Properties of Dynamic and Static Combined Loading

4.3.1 Strength Characteristics Under Combined Loading

The mechanical behavior of rocks under combined static and dynamic loading conditions is of paramount importance in understanding and predicting rock response in deep mining and underground excavation scenarios. This section provides a comprehensive review of the strength characteristics of rocks subjected to coupled static-dynamic loads, synthesizing findings from various experimental studies and theoretical analyses.

One of the most significant observations in rock mechanics under combined loading is the complex relationship between static pre-stress and dynamic strength. Experimental evidence consistently shows that the dynamic compressive strength of rocks exhibits a non-monotonic relationship with axial static pre-stress. As the axial pre-stress increases from zero, the dynamic strength initially increases, reaches a maximum, and then decreases rapidly. This phenomenon can be described by the following general relationship:

$$\sigma_d = f(\sigma_s, \dot{\varepsilon}) \tag{4.18}$$

where σ_d is the dynamic strength, σ_s is the static pre-stress, and $\dot{\varepsilon}$ is the strain rate.

The specific form of this function varies depending on the rock type and testing conditions, but a common empirical expression is:

$$\sigma_d = A + B \cdot \sigma_s - C \cdot \sigma_s^2 \tag{4.19}$$

where A, B, and C are material-dependent constants.

Dynamic strength enhancement under static pre-stress occurs through two mechanisms: microcrack closure increasing rock integrity, and strain hardening enhancing dynamic load resistance. These processes jointly contribute to the initial dynamic strength increase.

As static pre-stress approaches the rock's uniaxial compressive strength, the strengthening effect diminishes due to accumulated damage. The maximum dynamic strength occurs at static pre-stress levels of 60–70% of the uniaxial compressive strength [34–36], after which dynamic strength decreases rapidly due to microcrack development.

Rock strength under combined loading exhibits strain rate dependence, increasing logarithmically with strain rate:

$$\sigma_d = \sigma_s \left(1 + D \ln\left(\frac{\dot{\varepsilon}}{\dot{\varepsilon}_0}\right)\right) \quad (4.20)$$

where σ_s is quasi-static strength, D is the material constant, $\dot{\varepsilon}$ is applied strain rate, and $\dot{\varepsilon}_0$ is reference strain rate (10^{-5} s^{-1}).

The interaction between strain rate and static pre-stress affects rock strength, with the DIF varying with both parameters:

$$\text{DIF} = \frac{\sigma_d}{\sigma_s} = f(\dot{\varepsilon}, \sigma_s) \quad (4.21)$$

This relationship can be expressed empirically as:

$$\text{DIF} = 1 + A\left(\frac{\dot{\varepsilon}}{\dot{\varepsilon}_0}\right)^B \left(1 - \frac{\sigma_s}{\sigma_c}\right)^C \quad (4.22)$$

where A, B, and C are fitting parameters, σ_c is uniaxial compressive strength, and $\dot{\varepsilon}_0$ is reference strain rate (Fig. 4.3).

The interaction between static pre-stress and strain rate on rock strength is crucial for understanding rock behavior in deep mining, where rocks experience high in-situ stresses and dynamic loading from blasting, seismic events, or mining-induced stresses [34, 36, 38]. Under combined loading, rocks exhibit maximum energy absorption capacity at an optimal static pre-stress and dynamic load combination [35, 39], coinciding with peak dynamic strength.

The strain energy density (W), defined as the area under the stress–strain curve until failure, quantifies energy absorption under combined loading:

$$W = W_s + W_d \quad (4.23)$$

where W_s represents static pre-stress strain energy density and W_d represents dynamic loading-induced strain energy density.

The relationship between strain energy density and loading conditions mirrors strength behavior, showing initial increase then decrease with rising static pre-stress.

4.3 Rock Mechanical Properties of Dynamic and Static Combined Loading

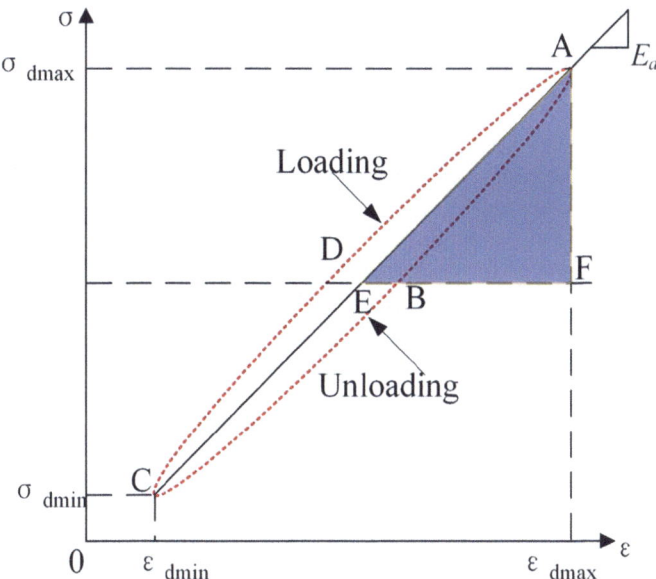

Fig. 4.3 Hysteresis curve of dynamic strain and stress [37]

This is expressed empirically as:

$$W = W_0 \left(1 + \alpha \frac{\sigma_s}{\sigma_c} - \beta \left(\frac{\sigma_s}{\sigma_c}\right)^2\right) \left(\frac{\dot{\varepsilon}}{\dot{\varepsilon}_0}\right)^\gamma \quad (4.24)$$

where W_0 is uniaxial static loading strain energy density, and α, β, γ are fitting parameters.

Under combined loading, rock failure modes transition from tensile splitting to shear failure with increasing static pre-stress [34, 35]. This transition affects fragmentation patterns, progressing from few large fragments at low static pre-stress to numerous smaller fragments at higher levels. The fragmentation process, quantified by fractal dimension (D), increases with both static pre-stress and dynamic load intensity:

$$D = D_0 + k_1 \left(\frac{\sigma_s}{\sigma_c}\right) + k_2 \ln\left(\frac{\dot{\varepsilon}}{\dot{\varepsilon}_0}\right) \quad (4.25)$$

where D_0 is the static loading fractal dimension, and k_1 and k_2 are constants.

Rock burst phenomena in deep mining environments, characterized by violent rock mass failure, are influenced by strength characteristics and energy absorption under combined loading of high static stress and dynamic disturbances [36, 38, 40].

The rock burst proneness can be evaluated using various indices that incorporate both strength and energy considerations. One such index, proposed by researchers based on combined loading experiments, is the dynamic stability index (D_s):

$$D_s = (\varphi_{sa}/\varphi_{sc}) * (\varepsilon_{sa}/\varepsilon_{sc})$$

where φ_{sa} is the energy stored before peak stress, φ_{sc} is the energy consumed after peak stress, ε_{sa} is the strain before peak stress, and ε_{sc} is the total strain after peak stress [40].

This index accounts for both the energy storage/release characteristics and the deformation behavior of the rock under combined loading. Higher values of Ds indicate a greater propensity for unstable failure and rock burst occurrence [40].

Rock strength under combined static-dynamic loading shows non-linear behavior, exhibiting enhancement at moderate pre-stress levels and degradation at high levels. Future research needs constitutive models to capture these behaviors for deep mining and underground structure applications [34–36, 38–41].

4.3.2 Deformation Behavior Under Combined Loading

The deformation behavior of rocks under combined static and dynamic loading is a complex phenomenon that has garnered significant attention in the field of rock dynamics. This section provides a comprehensive review of the current understanding of rock deformation under coupled static-dynamic loading conditions, synthesizing findings from various experimental studies and theoretical models.

Introduction to Combined Loading Conditions

In many practical engineering scenarios, particularly in deep underground excavations and mining operations, rocks are subjected to a combination of static and dynamic loads [36]. The static loads typically arise from in-situ stresses, while dynamic loads can be induced by various sources such as blasting, seismic activity, or mechanical excavation. Understanding the deformation behavior of rocks under these complex loading conditions is crucial for predicting and mitigating geotechnical hazards, as well as for optimizing engineering designs in rock mechanics applications.

Deformation Characteristics Under One-Dimensional Combined Loading

One-dimensional combined loading typically involves the application of axial static pre-stress followed by dynamic impact loading. Several studies have investigated the deformation characteristics of rocks under these conditions [35, 36, 38, 39].

1. Stress–Strain Behavior

Rock stress–strain curves under cyclic impact loading exhibit four phases [36]: compaction (nonlinear, microcrack closure), elastic (linear), crack propagation (nonlinear), and unloading (stress decrease with strain accumulation). Axial pre-stress magnitude influences curve behavior [38, 39], low levels mirror uniaxial dynamic compression, moderate levels accelerate plastic onset, and high levels show predominantly plastic response. Strain accumulation during cycling depends on both pre-stress and dynamic load.

2. Elastic Modulus

The elastic modulus under combined static-dynamic loading shows dual-phase behavior with increasing axial pre-stress: initial increase from microcrack closure, followed by decrease due to damage accumulation [36]. Under combined loading, the elastic modulus falls between pure dynamic and pure static values [38, 39], reflecting pre-stress modification of rock microstructure before dynamic loading. The elastic modulus (E) versus axial pre-stress ratio (η) relationship follows a quadratic function with experimentally determined constants a, b, and c.

$$E = a\eta^2 + b\eta + c \tag{4.26}$$

3. Poisson's Ratio

Poisson's ratio increases with axial pre-stress [38, 39] due to progressive microcrack development, showing inverse correlation with elastic modulus. The Poisson's ratio (v) versus axial pre-stress ratio (η) relationship follows an exponential function with experimentally determined constants α and β.

$$v = \alpha\left(1 - e^{-\beta\eta}\right) \tag{4.27}$$

where α and β are material-specific constants determined through experimental fitting.

4. Strain Rate Effects

Under constant pre-stress, rock dynamic compressive strength (σ_d) increases linearly with strain rate ($\dot{\varepsilon}$), using material constants A and B.

This relationship is often described by a linear function:

$$\sigma_d = A + B \cdot \dot{\varepsilon} \tag{4.28}$$

The strain rate effect peaks at pre-stress levels of 60–70% static strength. The deformation modulus increases with strain rate, indicating higher stiffness at increased loading rates [36].

Deformation Characteristics Under Multi-Dimensional Combined Loading

While one-dimensional combined loading provides valuable insights, many practical scenarios involve more complex stress states. Several studies have investigated rock deformation under multi-dimensional combined loading conditions [34, 36, 38, 39].

1. Two-Dimensional Combined Loading

Under two-dimensional combined loading, rock strength and elastic modulus initially increase then decrease with horizontal static stress [38, 39], peaking at 30–50% of vertical stress for strength. The elastic modulus peaks at slightly lower horizontal stress than strength. Poisson's ratio shows inverse behavior, decreasing then increasing with horizontal stress [38, 39]. Failure modes evolve from tensile splitting (low horizontal stress) through combined shear-tensile failure (moderate stress) to shear failure (high stress).

2. Three-Dimensional Combined Loading

Three-dimensional combined loading incorporates confining pressure with axial stress and dynamic loading. The elastic modulus increases with confining pressure at constant strain rate [36], particularly during elastic deformation [38].

Dynamic compressive strength correlates positively with strain rate and confining pressure [36, 38], quantified through the strength growth factor (DIF) and average strain rate ($\dot{\varepsilon}_{avg}$), expressed as:

$$\text{DIF} = C + D\left(\frac{\dot{\varepsilon}_{avg}}{\dot{\varepsilon}_0}\right)^{\frac{1}{3}} \tag{4.29}$$

where C and D are confining pressure-dependent constants. Strain rate sensitivity of strength decreases with rising confining pressure.

Under constant confinement, deformation modulus decreases with impact cycle count [38], influenced by both confining pressure and axial static stress [38]. These parameters also govern strain rates [38].

Energy Dissipation and Absorption

The deformation behavior of rocks under combined loading is closely related to energy dissipation and absorption processes. Understanding these energy aspects is crucial for explaining phenomena such as rock bursts and for optimizing rock fragmentation in engineering applications.

1. Energy Partitioning

Under combined static-dynamic loading, rock energy partitions into multiple components: elastic strain energy (stored in matrix), plastic dissipation energy (microcrack formation), kinetic energy (fragment motion), and thermal energy (friction and crack propagation). The relative distribution of these components depends on loading conditions and rock properties [35, 36].

2. Effect of Static Pre-stress on Energy Absorption

Static pre-stress significantly affects rock energy absorption under dynamic loading. The energy absorption ratio (absorbed/input energy) peaks when pre-stress approaches the rock's elastic limit [35].

4.3 Rock Mechanical Properties of Dynamic and Static Combined Loading

Pre-stress levels govern energy behavior transitions. As pre-stress increases, rocks shift from energy absorption to energy release during failure [35, 36]. At low pre-stress, rocks absorb dynamic load energy; at high pre-stress, they release stored elastic energy.

3. Energy-Based Failure Criteria

Energy-based criteria effectively predict rock failure under combined loading conditions. The strain energy density criterion incorporates elastic strain energy and damage dissipation energy:

$$W = W_e + W_d \geq W_c \qquad (4.30)$$

where W is total strain energy density, W_e is elastic strain energy density, W_d is damage dissipation energy density, and W_c is the critical failure threshold.

The energy release rate criterion focuses on crack propagation:

$$G \geq G_c \qquad (4.31)$$

where G is energy release rate and G_c is critical energy release rate. These energy approaches provide more comprehensive failure prediction than stress-based criteria.

Microstructural Aspects of Deformation

The macroscopic deformation behavior of rocks under combined loading is ultimately governed by microstructural changes. Several studies have investigated these microstructural aspects using techniques such as acoustic emission monitoring, microscopy, and X-ray computed tomography.

1. Crack Initiation and Propagation

Under combined loading, crack behavior differs from pure static or dynamic conditions. Static pre-stress reduces the crack initiation threshold during subsequent dynamic loading.

Crack orientation depends on stress state: uniaxial pre-stress produces cracks parallel to loading direction, while triaxial pre-stress yields diverse orientations based on principal stress magnitudes. Crack propagation rate increases with both static pre-stress and dynamic load.

2. Damage Evolution

Rock damage under combined loading depends on loading history and stress state. Static pre-stress creates initial damage through microcracks [34, 38], influencing subsequent deformation.

Cyclic dynamic loading causes progressive damage accumulation, with characteristic localization leading to macroscopic failure [38]. This evolution follows a power law:

$$D = 1 - (1 - \alpha\varepsilon)^\beta \tag{4.32}$$

where D is damage variable, ε is strain, and α, β are loading-dependent material parameters [38].

3. Pore Structure Changes

The deformation of rocks under combined loading significantly alters pore structure, affecting mechanical and transport properties. Static pre-stress initially induces pore closure and material stiffening. Progressive loading leads to pore dilation through crack formation and growth, increasing porosity. Under high confining pressure, pore collapse can drive inelastic deformation. These pore structure changes modify rock permeability, substantially impacting fluid flow in geological formations.

Constitutive Modeling

Accurately modeling the deformation behavior of rocks under combined static-dynamic loading is crucial for predicting rock response in various engineering applications. Several constitutive models have been proposed to capture the complex behavior observed in experiments.

1. Viscoelastic-Damage Models

Rock behavior under complex loading can be modeled by combining viscoelastic and damage mechanics, incorporating time-dependent deformation and material degradation. The approach uses spring-dashpot analogues (Maxwell, Kelvin-Voigt, or Burgers models) for viscoelastic behavior, while damage mechanics captures property degradation from microcracking.

The constitutive model combines these mechanisms:

$$\sigma = (1 - D)E_e(\varepsilon - \varepsilon_p - \varepsilon_v) \tag{4.33}$$

where σ is stress, D is damage variable, E_e is elastic modulus, ε is total strain, ε_p is plastic strain, and ε_v is viscous strain. Damage evolution depends on strain and loading rate:

$$\frac{dD}{dt} = f(\varepsilon, \dot{\varepsilon}, D) \tag{4.34}$$

2. Elastoplastic Models with Rate Dependence

An alternative rock behavior model extends elastoplastic theory with rate-dependent effects through modified constitutive relationships. The rate-sensitive yield criterion is:

$$F = J_2 - \left[A + B(\dot{\varepsilon})^n\right] \tag{4.35}$$

where J_2 is the second deviatoric stress invariant, A and B are material constants, $\dot{\varepsilon}$ is strain rate, and n is rate sensitivity parameter.

4.3 Rock Mechanical Properties of Dynamic and Static Combined Loading

The hardening law incorporates strain and strain rate effects:

$$\sigma_y = \sigma_0 + H(\varepsilon_p) + K(\dot{\varepsilon}) \tag{4.36}$$

where σ_y is yield stress, σ_0 is initial yield stress, $H(\varepsilon_p)$ is strain hardening function, and $K(\dot{\varepsilon})$ is strain rate hardening function.

3. Micromechanics-Based Models

Micromechanical models link microscale phenomena to macroscopic rock behavior through crack density evolution modeling under combined loading. Homogenization techniques upscale individual crack and pore behavior to derive macroscopic constitutive relations. Statistical approaches incorporate microstructural feature distributions to predict macroscopic behavior.

4. Numerical Implementation

Implementation of constitutive models in finite element or discrete element methods enables analysis of complex rock deformation problems [38]. Key computational challenges include handling large deformations and material failure, capturing discontinuities and fracture propagation, and balancing computational efficiency with accuracy in complex model implementation.

Implications for Engineering Applications

Combined static-dynamic loading behavior of rocks significantly impacts subsurface engineering applications. In deep mining, high in-situ stresses interact with dynamic loads from blasting and seismic events, directly affecting rockburst prediction and support system design. These loading conditions influence underground construction, particularly tunnel stability and TBM performance optimization. Static pre-stress effects on dynamic rock response inform blasting pattern design and have enabled non-explosive mining techniques utilizing in-situ stress states. For underground works, combined loading analysis ensures storage cavern stability and barrier integrity under long-term static and seismic loads.

Challenges and Future Research Directions

Despite significant advances in understanding rock deformation under combined static-dynamic loading, several challenges and areas for future research remain:

Combined static-dynamic loading of rocks affects subsurface engineering. In deep mining, in-situ stresses and dynamic loads from blasting impact rockburst prediction and support design. These conditions affect tunnel stability and TBM performance. Pre-stress effects on rock response guide blasting design and enable non-explosive mining techniques. In underground works, loading analysis ensures cavern stability under static and seismic loads.

4.3.3 Failure Modes and Mechanisms Under Combined Loading

Rock failure under combined static-dynamic loading impacts geotechnical applications, particularly in deep mining and underground excavation. Understanding failure mechanisms under coupled loads is essential for rockburst prediction and optimizing fragmentation in mining and tunneling operations. This section reviews rock failure behavior under combined loading through experimental, theoretical, and numerical approaches.

Experiment Al Observations of Failure Modes

Experimental studies using various testing apparatuses, particularly modified Split Hopkinson Pressure Bar (SHPB) systems, have provided valuable insights into the failure modes of rocks under combined static-dynamic loading. These observations can be categorized based on the loading conditions and the resulting failure patterns.

1. Failure Modes under One-Dimensional Loading

Under one-dimensional combined loading with static axial pressure preceding dynamic impact, rocks exhibit failure modes varying with load conditions. At low static pre-stress and impact loads, rocks undergo splitting failure from perpendicular tensile stress. Higher impact loads cause transition to fragmentation, shifting from tensile to tensile-shear stress states. Axial static pressure induces shear-compression failure with diagonal fracture planes, while high static pre-stress and dynamic impacts result in explosive failure with violent fragment ejection. Both static prestressing and dynamic impact amplitudes affect these transitions, with higher prestressing levels increasing the percentage of small particles (Fig. 4.4).

2. Failure Modes under Three-Dimensional Loading

Under three-dimensional combined loading simulating deep underground conditions, rocks exhibit failure modes dependent on strain rate and energy levels. At low strain rates and confining pressures, internal shear-compression fracture occurs, forming localized damage zones without overall instability. With increasing strain rates, rocks transition to conical fragmentation due to combined axial impact and confining pressure.

High-energy impact conditions lead to expansion-dominated failure with extensive lateral deformation, occurring when impact energy exceeds the rock's dissipation capacity. At intermediate energy levels, combined tensile-shear and expansion failure emerges from the interplay of stress components. The study shows that the increase of confining pressure inhibits the tensile fracturing and promotes the shear oriented failure mode.

4.3 Rock Mechanical Properties of Dynamic and Static Combined Loading 177

(a) Rock failure under small prestress and impact load.

(b) Rock failure under intermediate prestress and impact load.

(c) Rock failure under very high prestress and impact load or high impact load.

Fig. 4.4 Rock failure patterns under different coupled loads [35]

Failure Mechanisms and Processes

The observed failure modes are the result of complex mechanisms and processes occurring within the rock under combined static-dynamic loading. Understanding these mechanisms is crucial for developing predictive models and designing effective support systems in underground excavations.

1. Progressive Damage Accumulation

Progressive damage accumulation drives rock failure under combined loading [37, 40]. Static pre-stress induces initial microcracks and defects, which determine the rock's response to dynamic loading. Under dynamic impact, pre-existing damage evolves through microcrack growth, coalescence, and new crack initiation. Damage localizes in specific regions, forming macroscopic fractures leading to failure. Both static pre-stress and dynamic load characteristics influence this progression. There are four stages under cyclic impact loads: compaction, elastic deformation, crack propagation and unloading.

2. Energy Dissipation and Release

Energy dissipation and release mechanisms govern rock failure under combined loading [35, 39]. Initially, rocks absorb energy from static pre-stress and dynamic impact as elastic strain energy, with partial dissipation through plastic deformation and microcracking. As damage accumulates, energy dissipation capacity decreases, transitioning from absorption to release. In explosive failures or rockbursts, sudden elastic strain energy release causes rock fragment ejection and underground excavation damage. It was found that the energy absorption ratio reaches a maximum near the elastic limit of the rock, which indicates favorable conditions for the occurrence of rock bursting.

3. Stress Wave Propagation and Interaction

Stress wave propagation and interaction influence rock failure under dynamic loading. At rock boundaries and interfaces, wave reflection and transmission generate stress concentrations and localized damage. Superposition of incident, reflected, and transmitted waves creates complex stress states, particularly during cyclic or multiple impact loading. Wave attenuation occurs through scattering, absorption, and geometric spreading, influenced by rock properties and pre-existing damage. This wave-rock interaction determines damage distribution and failure patterns, exemplified by spalling fractures from reflected tensile waves at free surfaces.

4. Microstructural Evolution

Combined loading alters rock microstructure during failure. Static pre-stress and dynamic loading initiate microcracks from defects, grain boundaries, and stress concentration areas. These microcracks coalesce into larger fractures, transitioning from distributed to localized failure. Additional mechanisms include grain boundary sliding and rotation in granular rocks, and pore collapse with compaction in porous rocks, affecting stress–strain behavior. These microstructural processes inform constitutive models and failure criteria development.

Factors Influencing Failure Modes and Mechanisms

The failure modes and mechanisms of rocks under combined static-dynamic loading are influenced by various factors related to both the loading conditions and the intrinsic properties of the rock. Key factors include:

4.3 Rock Mechanical Properties of Dynamic and Static Combined Loading

1. Static Pre-stress Level

Static pre-stress magnitude affects rock's dynamic response [35, 36, 39, 42]. At 60–70% of uniaxial compressive strength, dynamic strength increases through microcrack closure and grain boundary friction. Above 70–80%, strength decreases from damage accumulation and microcrack development. Higher pre-stress transforms failure from tensile to shear-dominated and increases stored elastic energy, causing violent failures. Peak dynamic strengths at prestress ratios of 0.6–0.7 were observed experimentally, varying with rock type and loading conditions.

2. Dynamic Load Characteristics

Rock failure under combined static-dynamic loading requires further research [36, 37, 39]. Advanced micro-mechanical models need to address static pre-stress, dynamic loading, and multi-scale microstructure interactions. Coupled thermal–hydraulic-chemical processes require study for geothermal and nuclear waste applications.

Future research should examine rock anisotropy, heterogeneity, and time-dependent behavior, especially cyclic loading effects on damage accumulation. New experimental techniques, including true triaxial dynamic systems with in-situ validation, are needed alongside machine learning for failure prediction. Integrating experimental, theoretical, and field studies will improve understanding for deep mining, construction, and energy extraction.

3. Confining Pressure

Confining pressure, simulating three-dimensional stress states in deep underground environments, affects rock failure under combined loading [35, 36]. Increasing confining pressure enhances dynamic compressive strength and ductile failure through tensile crack suppression and increased shear plane friction. Higher confining pressures promote shear-dominated failure over tensile splitting due to stress state changes and different failure mechanisms.

Under higher confining pressures, rocks show greater energy absorption capacity before failure, affecting fragmentation and violent failure potential. Strain rate effects on rock strength decrease under high confining pressures due to shear failure mechanism dominance, critical for predicting rock behavior in deep underground environments.

4. Rock Properties and Microstructure

Rock properties and microstructure determine failure behavior under combined loading [36, 37, 39, 42]. Mechanical properties (uniaxial compressive strength, tensile strength, elastic moduli) control static and dynamic response. Strong, stiff rocks show high failure resistance but risk brittle failure. Porosity and density affect compressibility and wave propagation, with porous rocks exhibiting pore collapse and compaction.

Internal structure affects failure through grain characteristics. Grain size, shape, and arrangement influence fracture behavior and energy dissipation. Fine-grained rocks show higher strength and uniform deformation, while coarse-grained rocks favor intergranular fracturing. Anisotropy from bedding planes, foliation, or mineral orientations creates directional strength and failure patterns. Pre-existing microcracks and joints initiate fractures under dynamic loading. These properties guide predictive modeling and laboratory-to-field scaling.

Theoretical Models and Failure Criteria

To describe and predict the failure behavior of rocks under combined static-dynamic loading, various theoretical models and failure criteria have been proposed. These models aim to capture the complex interplay between static pre-stress, dynamic loading, and rock properties.

1. Damage Mechanics Models

Damage mechanics provides a framework for describing the progressive deterioration of rock strength under combined loading [39]. At the foundation of this framework are continuous damage models, which characterize the evolution of damage as a continuous variable representing the reduction in effective load-bearing area or material stiffness. The evolution of damage under dynamic loading is typically expressed as:

$$\frac{dD}{dt} = f(\sigma, D, \dot{\varepsilon}) \tag{4.37}$$

where D represents the damage variable, σ denotes the stress tensor, and $\dot{\varepsilon}$ indicates the strain rate.

Building upon this foundation, statistical damage models take a probabilistic approach by considering the statistical distribution of microdefects within the rock and their growth under loading. In these models, the overall damage is expressed as a function of the probability of microcrack growth and coalescence. For more complex scenarios, coupled damage-plasticity models integrate damage evolution with plastic deformation to capture the nonlinear behavior of rocks under high-stress conditions. These various damage mechanics approaches are particularly valuable for describing the transition from distributed microcracking to localized failure under combined loading conditions.

2. Energy-Based Criteria

Energy-based failure criteria have been proposed to account for the role of energy absorption and dissipation in rock failure under combined loading [35, 39]. The fundamental concept in this approach is Critical Energy Density, which posits that failure occurs when the total strain energy density within the rock reaches a critical value.

Building on this foundation, more advanced energy-based criteria incorporate energy partitioning principles, considering the distribution of input energy into

4.3 Rock Mechanical Properties of Dynamic and Static Combined Loading

various components, including elastic strain energy, plastic dissipation, and fracture energy. In these sophisticated models, failure prediction relies on analyzing the balance and evolution of these energy components under combined loading conditions. These energy-based approaches prove particularly valuable for understanding and predicting rockburst phenomena, where the sudden release of stored elastic strain energy plays a crucial role.

3. Rate-Dependent Strength Criteria

Various rate-dependent strength criteria have been proposed to account for the strain rate sensitivity of rock strength under dynamic loading [35, 36]. The Johnson-Holmquist model represents a comprehensive approach, expressing dynamic strength (σ_d) as a function of strain rate ($\dot{\varepsilon}$) and confining pressure. This model incorporates static strength (σ_s), reference strain rate ($\dot{\varepsilon}_0$), pressure (P), maximum tensile strength (T), and material constants (A, C, N) through the relationship:

$$\sigma_d = A\left(1 + C\ln\left(\frac{\dot{\varepsilon}}{\dot{\varepsilon}_0}\right)\right)(\sigma_s + T)\left(\frac{P}{T} + 1\right)N \tag{4.38}$$

For applications requiring simpler formulations, alternative models propose a power-law relationship between dynamic strength and strain rate:

$$\sigma_d = \sigma_s \left(\frac{\dot{\varepsilon}}{\dot{\varepsilon}_0}\right)^\alpha \tag{4.39}$$

where α is a material-dependent exponent typically ranging from 0.01 to 0.1 for rocks. These rate-dependent criteria are essential for predicting rock strength and failure behavior across a wide range of loading rates encountered in various engineering applications.

4. Coupled Static-Dynamic Failure Criteria

Several coupled failure criteria address combined static pre-stress and dynamic loading effects. The Modified Mohr–Coulomb Criterion incorporates strain rate effects:

$$\tau = c + \sigma_n \tan\left(\varphi + k\ln\left(\frac{\dot{\varepsilon}}{\dot{\varepsilon}_0}\right)\right) \tag{4.40}$$

where τ is shear strength, c is cohesion, σ_n is normal stress, φ is friction angle, and k is rate-sensitivity parameter.

The Three-Parameter Criterion accounts for static pre-stress, strain rate, and confining pressure:

$$\sigma_d = \sigma_s \left[1 + A\left(\frac{\sigma_0}{\sigma_s}\right)^m\right]\left(\frac{\dot{\varepsilon}}{\dot{\varepsilon}_0}\right)^\alpha \tag{4.41}$$

where σ_0 is static pre-stress, and A, m are fitting parameters. The Unified Strength Theory encompasses intermediate principal stress and failure mode transitions for combined static-dynamic loading in deep underground environments.

5. Catastrophe Theory Models

Catastrophe theory models the sudden instability and failure of rocks under combined loading, particularly rockburst phenomena [43]. The rock system's behavior is described through potential functions and control parameters. For one-dimensional loading, the cusp catastrophe model is expressed as:

$$V = x^4 + ux^2 + vx \tag{4.42}$$

where V is the potential function, x is the system state variable, and u and v are control parameters for static and dynamic loading conditions. For two-dimensional loading, a double cusp catastrophe model captures multiple loading direction interactions. These models reveal conditions for sudden, discontinuous changes in rock behavior under combined loading.

Implications for Engineering Practice

Understanding the failure modes and mechanisms of rocks under combined static-dynamic loading has significant implications for various geotechnical engineering applications, particularly in deep mining and underground construction.

1. Rockburst Prediction and Mitigation

Rock failure analysis under combined loading enables rockburst prediction [36, 39, 43]. Evaluating in-situ stress and mining-induced dynamic disturbance interactions identifies critical rockburst triggers. Microseismic monitoring detects instability precursors through progressive damage assessment. These findings inform yielding support design for energy absorption and guide excavation strategies, including de-stress blasting and pillar sizing for stress control.

2. Blast Design Optimization

Rock fragmentation behavior under combined loading informs blasting optimization in mining and tunneling [35, 39]. Pre-conditioning using static pre-stress enhances fragmentation under dynamic loading, employing methods like hydraulic fracturing before blasting. Stress wave interaction knowledge guides blast timing and sequencing to maximize fragmentation while minimizing surrounding rock damage. Understanding rock energy absorption and dissipation characteristics enables optimal explosive selection and charge configuration.

3. Excavation Design and Stability Analysis

Rock failure analysis under combined loading improves excavation design and stability assessment in deep underground environments [36, 37, 42]. Loading history and stress path considerations enable sophisticated design approaches accounting for stress state evolution during construction. Dynamic loading effects inform load factor development for stability analyses of underground openings under seismic or blast vibrations. Advanced constitutive models incorporating static-dynamic loading effects enhance numerical simulation accuracy for excavation design.

4. In-Situ Testing and Characterization

The complex behavior of rocks under combined loading necessitates advanced in-situ testing and characterization methods [37, 42]. Field-scale testing methods applying combined static and dynamic loads are essential for characterizing rock behavior under realistic conditions. Advanced non-destructive techniques, including acoustic emission monitoring and dynamic wave propagation analysis, enable assessment of rock mass damage state and instability potential under complex loading. Understanding the correlation between laboratory and field-scale failure mechanisms is fundamental for designing large underground excavations.

Future Research Directions

Rock failure under combined static-dynamic loading requires further research [36, 37, 39]. Advanced micro-mechanical models need to address static pre-stress, dynamic loading, and multi-scale microstructure interactions. Coupled thermal–hydraulic-chemical processes require study for geothermal and nuclear waste applications.

Future research should examine rock anisotropy, heterogeneity, and time-dependent behavior, especially cyclic loading effects on damage accumulation. New experimental techniques, including true triaxial dynamic systems with in-situ validation, are needed alongside machine learning for failure prediction. Integrating experimental, theoretical, and field studies will improve understanding for deep mining, construction, and energy extraction.

4.4 Conclusion

This study highlights the critical aspects of rock dynamic properties under impact loading, emphasizing stress wave propagation, strain rate effects, and failure mechanisms. Rocks exhibit significantly different mechanical behaviors under dynamic conditions, with higher strain rates enhancing strength, stiffness, and altering failure modes. The dynamic stress–strain relationship follows distinct phases, influenced by material properties and loading conditions. Energy absorption mechanisms play a crucial role in rock fragmentation and impact resistance, with energy partitioned between kinetic energy, strain energy, and dissipation through fracturing and plastic deformation.

Environmental factors such as temperature, moisture, and confining pressure significantly affect dynamic rock behavior. Elevated temperatures can lead to thermal softening, while confining pressure promotes shear failure over tensile splitting.

Additionally, microstructural factors, including porosity, grain size, and pre-existing fractures, dictate fracture propagation and energy dissipation. Under combined static and dynamic loading, moderate pre-stress enhances impact resistance, whereas high pre-stress levels accelerate failure due to accumulated damage. Understanding the transition between tensile, shear, and mixed-mode failure is essential for accurate rock behavior prediction.

The insights from this study have significant implications for mining, tunnelling, rock blasting, and seismic hazard assessment. A better understanding of dynamic rock behavior aids in optimizing engineering designs and mitigating risks associated with rockbursts and slope failures. Future research should focus on refining constitutive models that integrate strain rate effects, energy dissipation, and microstructural evolution. Advanced experimental techniques, such as high-speed imaging, can enhance real-time rock failure analysis.

Looking ahead, integrating machine learning and AI into rock dynamics can improve predictive modelling and real-time monitoring of extreme loading conditions. Multi-scale modelling approaches that bridge laboratory experiments with field applications will further refine engineering designs, enhancing the resilience of underground structures. By combining experimental, computational, and AI-driven methodologies, rock mechanics can advance toward more precise and adaptable solutions for geotechnical challenges.

References

1. Nie H, Ma W, Wang K, Ren J, Cao J, Dang W. A review of dynamic multiaxial experimental techniques. Rev Sci Instrum. 2020;91(111501):1–20.
2. Łodygowski T, Rusinek A. Constitutive relations under impact loadings: experiments, theoretical and numerical aspects. In: CISM international centre for mechanical sciences. Springer;2014.
3. Xia K, Yao W. Dynamic rock tests using split Hopkinson (Kolsky) bar system: a review. J Rock Mech Geotech Eng. 2015;7(1):27–59.
4. Xu Z, Qian K. Study on the basic problems of dynamics by induction and generalization. Adv Mater Res. 2013;625:12–7.
5. Hertz, H. Über die Berührung fester elastischer Körper. J Für Reine Und Angew Math S. 1881; 156–71.
6. Elías DA, Chiang LE. Dynamic analysis of impact tools by using a method based on stress wave propagation and impulse-momentum principle. J Mech Design 2003; 125(1):131–40.
7. Wang L. Foundations of stress waves. Elsevier;2007.
8. Wang X, Zhang S, Wang C, Shang C, Cao K, Wei P. Investigation into stress wave propagation across interlayers existing in roller compacted concrete (RCC) under impact loadings. Constr Build Mater. 2018;193:13–22.
9. Wang E, Feng J, Zhang Q, Kong X, Liu X. Mechanism of rockburst under stress wave in mining space. J China Coal Soc. 2020;45(1):100–10 [in Chinese].
10. Millon O, Ruiz-Ripoll ML, Hoerth T. Analysis of the behavior of sedimentary rocks under impact loading. Rock Mech Rock Eng 2016; 49:4257–72
11. Chen W. Experimental study on failure mechanism and stress wave propagation of joint rock mass. Master's Thesis, School of Safety Engineering;2022 [in Chinese]

12. Liu S, Xu J. Effect of strain rate on the dynamic compressive mechanical behaviors of rock material subjected to high temperatures. Mech Mater 2015; 82:28–38
13. Liang WG, Zhao YS, Xu SG, Dusseault MB. Effect of strain rate on the mechanical properties of salt rock. Int J Rock Mech Min Sci 2011; 48:161–7
14. Qi C, Qian Q. Physical mechanism of dependence of material strength on strain rate for rock-like material. Chin J Rock Mech Eng. 2003;22(2):177–81 [in Chinese].
15. Zhu W, Wei J, Niu L, Li S, Li S. Numerical simulation on damage and failure mechanism of rock under combined multiple strain rates. Shock Vibrat 2018; 2018:4534250
16. Qi C, Wang H, Liu P, Zhu H. Temporal-spatial properties of deformation and fracture of rock-like materials. J Beijing Univ Civil Eng Arch. 2016;32(3):78–89 [in Chinese].
17. Wang JX, Ma BD, Li SH, Li L, Kou HJ, Sun G. Study on cyclic impact failure characteristics of quartz sandstone under medium strain rate. Blasting 2023; 40(2):0029–13 [in Chinese]
18. Zhou X, Ha Q, Zhang Y, et al. Analysis of deformation localization and the complete stress–strain relation for brittle rock subjected to dynamic compressive loads. Int J Rock Mech Min Sci. 2004;41:311–9.
19. Wang S, Xiong X, Liu Y, et al. Stress–strain relationship of sandstone under confining pressure with repetitive impact. Geomech Geophys Geo-Energy Geo-Resourc. 2021;7:39.
20. Jie-fang J, Xi-bing L, Jun-ran C, et al. Stress strain curve and stress wave characteristics of rock subjected to cyclic impact loadings. Explosion Shock Waves 2013; 33(06):613–9 [in Chinese]
21. Chen J, Zeng B, Zhang J. Dynamic mechanical performance of impact rocks under loading and unloading condition. Chinese J Rock Mech Eng 2024; 43(10):2430–42. [in Chinese]
22. Li XB. Rock dynamics fundamentals and applications. Science Press;1994.
23. Yan Y, Li J, Fukuda D, et al. Experimental and numerical studies on dynamic fracturing behavior of roughly jointed rock. Comput Particle Mech. 2024;11:1715–34.
24. Ahrens TJ, Rubin AM. Impact-induced tensional failure in rock. J Geophys Res. 1993;98(E1):1185–203.
25. Huang C, Mohanty B, Zhu Z. Strain rate effects on dynamic fractures in fine-grained granitic rock. Int J Exp Mech. 2016;52:46–58.
26. Chengzhi Q, Hongsen W, Peng L, et al. Temporal-spatial properties of deformation and fracture of rock-like materials. J Beijing Univ Civil Eng Arch 2016; 32(03):78–88 [in Chinese]
27. Miao Y, Wei C, Niu L. The coupled effect of loading rate and grain size on tensile strength of sandstones under dynamic disturbance. Shock Vib. 2017;13:6989043.
28. Ju Q-H, Wu M-B. Experimental studies of dynamic characteristic of rocks under triaxial compression. Chinese J Geotech Eng 1993;15(03):73–80 [in Chinese]
29. Duan J, Zhou S, Xia C, et al. A dynamic phase field model for predicting rock fracture diversity under impact loading. Int J Impact Eng. 2023;171: 104376.
30. Nie H, Ma W, Wang K, et al. A review of dynamic multiaxial experimental techniques. Rev Sci Instrum 2020; 91:111501
31. Tian H, Li Z, Yin S, et al. Research on infrared radiation response and energy dissipation characteristics of sandstone crushing under impact load. Eng Geol. 2023;322: 107171.
32. Wei D, Chen M, Ye ZW, et al. Study on blasting failure zone based on rate-dependent dynamic characteristics of rock mass. Adv Eng Sci. 2021;53(01):67–74 [in Chinese].
33. Hai-bo Z, Yuan W, Lu-ming W, et al. Experimental research on dynamic characteristics of fractured rock mass. J Hohai Univ (Natural Sciences) 2007; 35(03):309–11. [in Chinese]
34. Li X, Gong F, Tao M, Dong L, Du K, Ma C, Zhou Z, Yin T. Failure mechanism and coupled static-dynamic loading theory in deep hard rock mining: a review. J Rock Mech Geotech Eng. 2017;9(4):767–82.
35. Li X, Zhou Z, Zhao F, Zuo Y, Ma C, Ye Z, Hong L. Mechanical properties of rock under coupled static-dynamic loads. J Rock Mech Geotech Eng. 2009;1(1):41–7.
36. Pan Z, Li K, Niu Y. The research status of dynamic characteristics of rock and the analysis on the mechanism of rock burst under coupled dynamic and static loads. Min Metall 2018;27(02):5–8+14 [in Chinese]
37. Zhou X, Zhang D, Nowamooz H, Jiang C, Ye C. Investigation on damage and failure mechanisms of roadway surrounding rock triggered by dynamic-static combined loads. Rock Mech Rock Eng. 2022;55(9):5639–57.

38. Li XB. Rock dynamics fundamentals and applications. Science Press; 2014[in Chinese]
39. Zuo YJ, Li XB. Failure und fragmentation characteristics of rock under static-dynamic coupling loading. Metallurgical Industry Press;2008 [in Chinese]
40. Yin ZQ, Li XB, Dong LJ, et al. Rockburst characteristics and proneness index undercoupled static and dynamic loads. J Central South Univ Sci Technol. 2014;45(09):3249–56 [in Chinese].
41. Bai J, Dou L, Małkowski P, Li J, Zhou K, Chai Y. Mechanical properties and damage behavior of rock-coal-rock combined samples under coupled static and dynamic loads. Geofluids. 2021;1:3181697.
42. Chen F, Ma DC, Xu JC. Dynamic response and failure behavior of rock under static-dynamic loading. J Central South Univ Technol 2005; 12(3):354–8.
43. Zuo YJ, Li XB, Ma CD, et al. Catastrophic model and testing study on failure of static loading rock system under dynamic loading. Chinese J Rock Mech Eng 2005; (05):741–6.

Open Access This chapter is licensed under the terms of the Creative Commons Attribution-NonCommercial-NoDerivatives 4.0 International License (http://creativecommons.org/licenses/by-nc-nd/4.0/), which permits any noncommercial use, sharing, distribution and reproduction in any medium or format, as long as you give appropriate credit to the original author(s) and the source, provide a link to the Creative Commons license and indicate if you modified the licensed material. You do not have permission under this license to share adapted material derived from this chapter or parts of it.

The images or other third party material in this chapter are included in the chapter's Creative Commons license, unless indicated otherwise in a credit line to the material. If material is not included in the chapter's Creative Commons license and your intended use is not permitted by statutory regulation or exceeds the permitted use, you will need to obtain permission directly from the copyright holder.

Chapter 5
Propagation Characteristics of Stress Wave in Rock

5.1 Basic Principle of Stress Wave

5.1.1 Category of Stress Waves

Stress waves in rocks are complex phenomena that play a crucial role in various geological and engineering processes. Understanding the different categories of stress waves is essential for analyzing and predicting rock behavior under dynamic loading conditions. This section provides a comprehensive overview of the various types of stress waves encountered in rock mechanics, their characteristics, and their significance in rock dynamics.

1. **Body Waves**

Body waves are stress waves that propagate through the interior of a solid medium, such as rock. They are classified into two main types: primary waves (P-waves) and secondary waves (S-waves) [1].

(1) P-waves

P-waves, also known as compressional waves or longitudinal waves, are the fastest-traveling seismic waves. They are characterized by particle motion parallel to the direction of wave propagation, causing alternating compression and dilation of the rock material [1]. P-waves propagate through both solid and fluid media, enabling rock property analysis across various geological settings [1]. These waves create particle oscillations parallel to their propagation direction, generating material compression and rarefaction. Their superior velocity makes them the first arrivals in seismic recordings [1]. Upon encountering rock layer interfaces or discontinuities, P-waves undergo reflection and refraction following Snell's law [1]. During propagation, they experience attenuation influenced by frequency, rock properties, and travel distance [2] (Fig. 5.1).

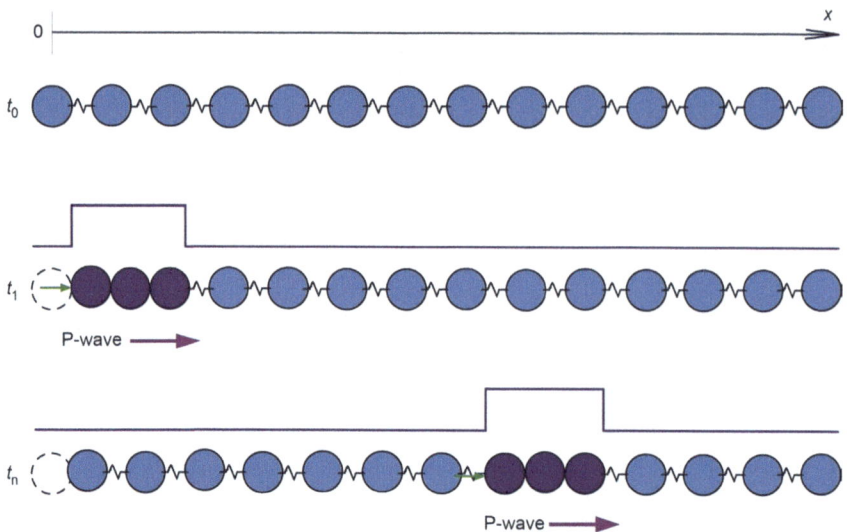

Fig. 5.1 A P-wave particle motion [1]

(2) S-waves

S-waves, also referred to as shear waves or transverse waves, are the second type of body waves. They are characterized by particle motion perpendicular to the direction of wave propagation, causing shearing deformation in the rock material [1]. S-waves exclusively propagate through solid media due to their reliance on shear stress support, facilitating distinction between solid and fluid-filled regions in rock formations [1]. Characterized by perpendicular particle oscillation to wave propagation direction, these waves arrive after P-waves due to lower velocity [1]. S-waves are classified into horizontally (SH) and vertically (SV) polarized waves based on particle motion relative to Earth's surface [3]. Like P-waves, they undergo reflection and refraction at rock interfaces [1], but exhibit higher attenuation, enhancing their sensitivity to rock properties and discontinuities [2] (Fig. 5.2).

2. **Surface Waves**

Surface waves are stress waves that propagate along the surface or interface of a medium. In rock mechanics, two main types of surface waves are of particular interest: Rayleigh waves and Love waves [4].

(1) Rayleigh Waves

Rayleigh waves, named after Lord Rayleigh who first described them mathematically in 1885, are surface waves that travel along the free surface of a solid medium. They result from the interaction between P-waves and SV-waves at the surface [5]. Rayleigh waves are characterized by elliptical particle motion in a vertical plane

5.1 Basic Principle of Stress Wave 189

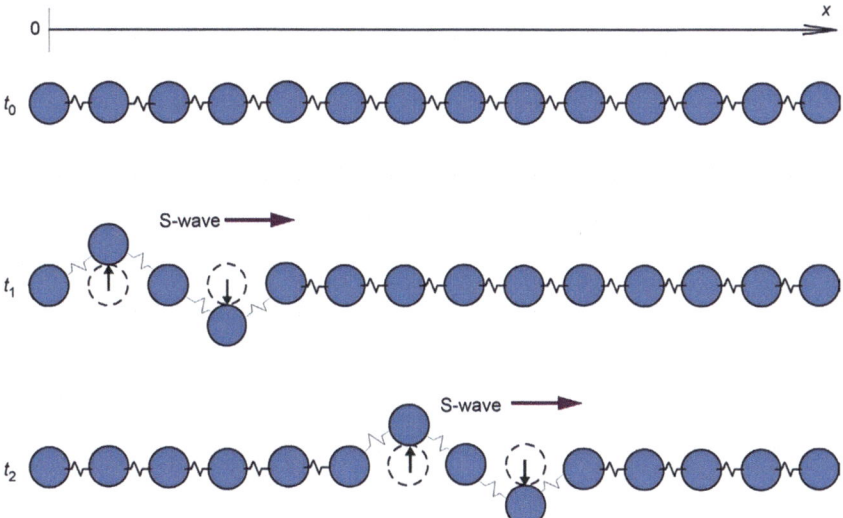

Fig. 5.2 S-wave particle motion [1]

parallel to wave propagation, exhibiting retrograde motion at the surface that transitions to prograde with depth [4]. Their amplitude decreases exponentially with depth, concentrating energy within one wavelength of the surface [4]. In layered or heterogeneous media, these waves display dispersive behavior, with frequency-dependent velocities [4]. This unique set of characteristics makes Rayleigh waves particularly valuable for near-surface geophysical investigations, including rock elastic property determination, subsurface anomaly detection, and rock slope stability assessment [4] (Figs. 5.3 and 5.4).

(2) Love Waves

Love waves, named after A.E.H. Love who first described them mathematically in 1911, are horizontally polarized surface waves that propagate in layered media.

Fig. 5.3 Rayleigh surface wave particle motion [4]

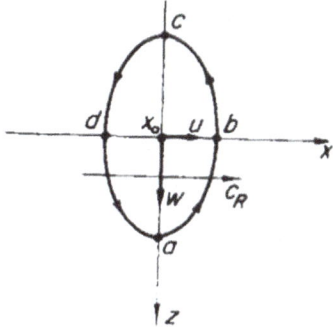

Fig. 5.4 Schematics of Rayleigh surface wave [3]

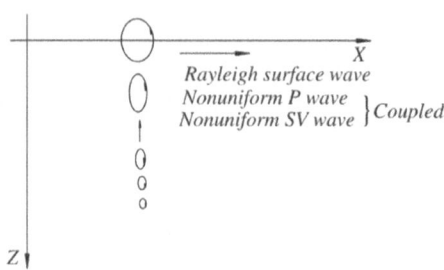

They result from the trapping of SH-waves in a low-velocity surface layer overlying a higher-velocity substrate [4]. Love waves are characterized by horizontal particle motion perpendicular to their propagation direction [4]. While their amplitude decreases with depth, they demonstrate greater penetration capability than Rayleigh waves [4]. These waves exhibit inherent dispersive behavior, with velocity variations across different frequencies. Such characteristics make Love waves valuable tools in seismological and geophysical exploration for investigating Earth's crustal and upper mantle structure and properties.

3. **Guided Waves**

Guided waves are stress waves that propagate along a specific path or within a confined geometry, such as along a borehole or within a rock layer. These waves are of particular interest in rock mechanics and geophysical exploration due to their ability to provide information about rock properties and structures.

(1) Tube Waves

Tube waves, also known as Stoneley waves or borehole guided waves, are cylindrical waves that propagate along the interface between a fluid-filled borehole and the surrounding rock formation. Tube waves are distinguished by their combined radial and axial particle motion, maintaining continuity across fluid–solid interfaces. These waves exhibit dispersive behavior, particularly pronounced at low frequencies and when borehole diameter approaches wavelength scale. Their attenuation characteristics are governed by multiple factors, including formation permeability, fluid viscosity, and borehole conditions. These properties make tube waves valuable tools in borehole geophysical applications.

(2) Flexural Waves

Flexural waves are bending waves that propagate along elongated structures, such as drill strings or rock bolts in underground excavations. They are of particular interest in rock mechanics due to their role in dynamic loading of support systems and their potential for causing damage to structures [6]. Flexural waves are characterized by combined transverse displacement and cross-sectional rotation [6]. These waves exhibit pronounced dispersive behavior, with velocity variations across frequencies [6]. Their attenuation is influenced by multiple factors, including material damping,

5.1 Basic Principle of Stress Wave

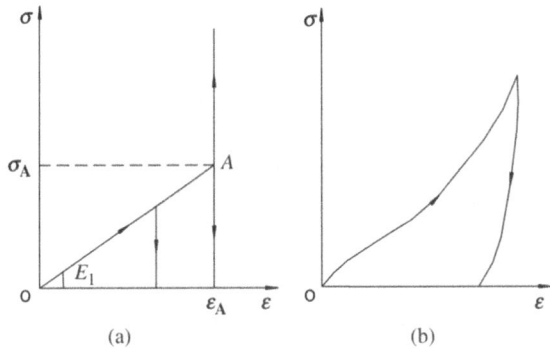

Fig. 5.5 Two stress–strain models that may result in shock waves: **a** linear hardening plastio loading-rigid unloading; **b** stress–strain relation for soils. Note that **a** can be regarded as a simplified model of **b** [3]

radiation into surrounding media, and geometric spreading [6]. These distinctive properties make flexural waves significant in rock mechanics applications.

4. **Shock Waves**

Shock waves are a special category of stress waves characterized by a nearly discontinuous change in pressure, temperature, and density of the medium. In rock mechanics, shock waves are of particular interest in the context of explosive loading, impact events, and high-energy seismic events [7, 8]. Shock waves are distinguished by an abrupt, near-discontinuous pressure increase across the shock front [7, 8]. Their propagation through rock typically induces irreversible deformation, including fracturing, comminution, and phase transformations [7, 8]. The behavior of materials under shock loading is characterized by Hugoniot relations, correlating shock velocity, particle velocity, pressure, and specific volume. As these waves propagate through rock, they undergo rapid attenuation, eventually degenerating into a series of elastic waves [7, 8], making them particularly relevant for studying explosive loading effects in rock blasting applications (Fig. 5.5).

5. **Seismic Waves**

Seismic waves are stress waves generated by earthquakes, explosions, or other sources of ground motion. While they share many characteristics with the previously discussed wave types, seismic waves are often considered separately due to their unique properties and significance in geophysics and earthquake engineering.

(1) Body Waves in Seismology

In seismology, body waves (P-waves and S-waves) exhibit complex behaviors within Earth's large-scale structures. These waves generate travel time curves through their varied arrival times, revealing internal Earth structure. The interaction between seismic waves and Earth's core creates distinctive shadow zones, while wave conversion occurs at layer interfaces, producing additional seismogram phases. Furthermore, seismic wave velocities demonstrate directional variation (anisotropy) due to factors such as aligned cracks and mineral orientation, providing crucial insights into the Earth's crustal and mantle structure and stress conditions.

(2) Surface Waves in Seismology

In seismology, surface waves (Rayleigh and Love waves) demonstrate distinctive characteristics in large-scale Earth structures. Their dispersive nature enables velocity structure inference of Earth's crust and upper mantle through surface wave tomography. These waves propagate in both fundamental and higher modes, offering additional insights into Earth's structure. The group velocity, representing energy propagation speed, differs from phase velocity and serves as a crucial parameter in seismological studies. Notably, surface waves exhibit larger amplitudes than body waves at considerable distances from the source, making them instrumental in earthquake hazard assessment and structural Earth studies.

(3) Normal Modes

Normal modes, or free oscillations, manifest as standing waves when Earth vibrates at specific frequencies following major seismic events. These eigen vibrations exhibit a discrete frequency spectrum, with the fundamental mode having a period of approximately 54 min, and are categorized into spheroidal (radial and tangential motion) and toroidal (tangential motion only) modes. The frequencies can split due to Earth's rotation, ellipticity, and lateral heterogeneities, enabling analysis of Earth's large-scale structure, earthquake source mechanisms, long-period seismic wave propagation, and rotational-gravitational properties.

6. **Acoustic Emission Waves**

Acoustic emission (AE) waves are stress waves generated by rapid energy release from localized sources within materials, particularly significant in monitoring rock failure processes. These waves operate in frequencies from kHz to MHz, depending on source mechanisms and rock properties, and originate from various processes including crack propagation, slip along discontinuities, grain boundary sliding, and phase transformations. Both P-waves and S-waves are generated during AE events, with amplitudes influenced by source mechanism and orientation, though they experience significant attenuation during propagation. The Kaiser effect, where AE activity remains suppressed until previous maximum stress levels are exceeded, provides crucial information about rock stress history. AE monitoring finds extensive applications in predicting laboratory rock failure, assessing underground stability, evaluating structural damage, and studying hydraulic fracturing processes.

7. **Blast Waves**

Blast waves are a special category of stress waves generated by explosions in or near rock masses. They are of particular interest in rock blasting, mining, and tunneling applications. Blast waves exhibit distinctive characteristics across propagation distances, transitioning from high-amplitude, nonlinear behavior with shock fronts in the near-field to elastic wave behavior in the far-field [3]. The energy transfer efficiency from explosive to rock mass is influenced by explosive properties, borehole parameters, stemming characteristics, and rock properties. These waves undergo complex reflection and refraction patterns at interfaces [1], and induce damage through multiple mechanisms including near-field crushing, radial crack formation, spalling, and gas pressurization of existing fractures [3, 8]. Understanding blast wave behavior is essential for optimizing rock fragmentation, controlling excavation damage, and managing environmental impacts in mining and construction applications [7, 8].

8. **Thermoelastic Waves**

Thermoelastic waves are stress waves generated by rapid temperature changes in a material. In rock mechanics, thermoelastic waves are of interest in various contexts, including thermal fracturing, geothermal energy extraction, and the study of meteorite impacts. Thermoelastic waves are characterized by the coupled interaction between temperature changes and mechanical deformation. These waves can be generated through various mechanisms, including rapid surface temperature changes, frictional fault slip, and electromagnetic radiation absorption, and undergo attenuation through both mechanical and thermal dissipation mechanisms. Their applications span crucial areas in rock mechanics, including geothermal reservoir thermal fracturing, rock cutting and drilling operations, meteorite impact analysis, and thermal loading assessment in nuclear waste repositories.

9. **Poroelastic Waves**

Poroelastic waves are stress waves that propagate in fluid-saturated porous media, such as saturated rocks. These waves are described by Biot's theory of poroelasticity and are of particular interest in hydrogeology, petroleum engineering, and the study of fluid-induced seismicity. Poroelastic waves in fluid-saturated porous media are characterized by three distinct body wave types according to Biot's theory: fast P-wave (similar to elastic media P-wave), slow P-wave (highly attenuated, diffusive wave from fluid–solid relative motion), and S-wave (modified from elastic media). These waves exhibit frequency-dependent dispersion and attenuation due to fluid–solid phase interactions, with behavior governed by two characteristic frequencies: Biot's characteristic frequency and squirt flow frequency, which depend on porosity, Young's modulus, and pore size. Applications of poroelastic wave theory are fundamental in hydrocarbon exploration, fluid-induced seismicity studies, saturated rock stability analysis, and understanding earthquake mechanisms.

10. **Nonlinear Waves**

Nonlinear waves in rock mechanics refer to stress waves that exhibit behavior not described by linear elasticity theory. These waves are of particular interest in situations involving large strains, high strain rates, or wave propagation through discontinuous or heterogeneous media [2]. Nonlinear waves exhibit several distinctive characteristics that significantly influence their behavior in rock dynamics. These waves demonstrate amplitude-dependent velocity, which leads to wave steepening and shock formation phenomena. During propagation, they generate higher harmonics, facilitating energy transfer to higher frequencies. Under specific conditions, they can form stable, localized wave packets known as solitons, which propagate while maintaining their shape and velocity. These fundamental properties are crucial for understanding various geomechanical processes, including high-amplitude stress wave propagation in rock blasting and impacts, wave behavior in jointed rock masses, seismic response during major earthquakes, and the development of advanced numerical models for complex geological media [2].

5.1.2 Wave Equations of Elastic Stress Waves

Wave equations of elastic stress waves form the fundamental basis for understanding the propagation of stress waves in rocks. These equations describe the motion and deformation of elastic media under dynamic loading conditions. In this section, we will delve into the derivation and analysis of wave equations for various types of elastic waves, including longitudinal (P) waves, transverse (S) waves, and surface waves. We will also explore the implications of these equations for wave propagation in rocks and discuss their limitations and extensions.

1. **General Wave Equation**

The general wave equation for elastic stress waves can be derived from the principles of continuum mechanics, specifically Newton's second law of motion and Hooke's law for linear elastic materials. Consider a small element of an elastic medium subjected to dynamic forces. The equation of motion for this element can be expressed as:

$$\nabla \cdot \sigma + \rho f = \rho \frac{\partial^2 u}{\partial t^2} \tag{5.1}$$

where σ is the stress tensor, ρ is the density of the medium, f is the body force vector, u is the displacement vector, t is time. For a linear elastic, isotropic material, the stress–strain relationship (Hooke's law) can be written as:

$$\sigma = \lambda(\nabla \cdot \mathbf{u})\mathbf{I} + 2\mu \mathbf{e} \tag{5.2}$$

where λ and μ are Lamé parameters, $\mathbf{I}$ is the identity tensor, e is the strain tensor. Combining these equations and assuming no body forces, we obtain the general wave equation for elastic waves:

$$(\lambda + \mu)\nabla(\nabla \cdot \mathbf{u}) + \mu \nabla^2 \mathbf{u} = \rho \frac{\partial^2 \mathbf{u}}{\partial t^2} \tag{5.3}$$

This vector equation represents the propagation of elastic waves in a three-dimensional, isotropic, linear elastic medium [3].

5.1 Basic Principle of Stress Wave

2. Longitudinal (P) Wave Equation

To derive the wave equation for P-waves, we consider the curl of the displacement field to be zero ($\nabla \times \mathbf{u} = 0$). This condition allows us to express the displacement as the gradient of a scalar potential φ:

$$\mathbf{u} = \nabla \phi \tag{5.4}$$

Substituting this into the general wave equation and simplifying, we obtain the wave equation for P-waves:

$$\nabla^2 \phi = \frac{1}{c_p^2} \frac{\partial^2 \phi}{\partial t^2} \tag{5.5}$$

where c_p is the P-wave velocity given by:

$$c_p = \sqrt{\frac{\lambda + 2\mu}{\rho}} \tag{5.6}$$

This equation describes the propagation of P-waves in an elastic medium. The P-wave velocity depends on the elastic properties (λ and μ) and density (ρ) of the rock [3].

3. Transverse (S) Wave Equation

To derive the wave equation for S-waves, we consider the divergence of the displacement field to be zero ($\nabla \mathbf{u} = 0$). This condition allows us to express the displacement as the curl of a vector potential ψ:

$$\mathbf{u} = \nabla \times \boldsymbol{\psi} \tag{5.7}$$

Substituting this into the general wave equation and simplifying, we obtain the wave equation for S-waves:

$$\nabla^2 \psi = \frac{1}{c_s^2} \frac{\partial^2 \psi}{\partial t^2} \tag{5.8}$$

where c_s is the S-wave velocity given by:

$$c_s = \sqrt{\frac{\mu}{\rho}} \tag{5.9}$$

This equation describes the propagation of S-waves in an elastic medium. The S-wave velocity depends only on the shear modulus (μ) and density (ρ) of the rock [3].

4. Surface Wave Equations

(1) Rayleigh Wave Equation

Rayleigh waves involve both longitudinal and vertical transverse particle motions. The displacement field for Rayleigh waves can be expressed as:

$$\mathbf{u} = \nabla\phi + \nabla \times (0, \psi, 0) \qquad (5.10)$$

where φ and ψ are scalar potentials satisfying the wave equations for P-waves and S-waves.

Where c is the Rayleigh wave velocity. This equation can be solved numerically to determine the Rayleigh wave velocity, which is typically slightly less than the S-wave velocity [3, 4].

(2) Love Wave Equation

Love waves involve horizontal transverse particle motion and can only exist in a layered medium. The displacement field for Love waves in a layer over a half-space can be expressed as:

$$\mathbf{u} = (0, v(z)\exp[i(kx - \omega t)], 0) \qquad (5.11)$$

where $v(z)$ satisfies the equation:

$$\frac{d^2v}{dz^2} + \left(\frac{\omega^2}{\beta^2(z)} - k^2\right)v = 0 \qquad (5.12)$$

Here, $\beta(z)$ is the S-wave velocity as a function of depth, ω is the angular frequency, and k is the wavenumber. The solutions for $v(z)$ must satisfy continuity conditions at the layer interface and decay with depth in the half-space [4].

5. Wave Equations in Cylindrical Coordinates

For problems involving wave propagation in cylindrical structures, such as boreholes or cylindrical rock samples, it is often convenient to express the wave equations in cylindrical coordinates (r, θ, z).

(1) P-wave Equation in Cylindrical Coordinates

The P-wave equation in cylindrical coordinates can be written as:

$$\left(\frac{\partial^2}{\partial r^2} + \frac{1}{r}\frac{\partial}{\partial r} + \frac{1}{r^2}\frac{\partial^2}{\partial \theta^2} + \frac{\partial^2}{\partial z^2}\right)\phi = \frac{1}{c_p^2}\frac{\partial^2\phi}{\partial t^2} \qquad (5.13)$$

5.1 Basic Principle of Stress Wave

Fig. 5.6 Schematic diagram of cylindrical coordinate system [4]

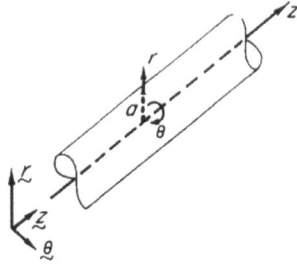

(2) S-wave Equation in Cylindrical Coordinates

The S-wave equation in cylindrical coordinates can be expressed as:

$$\left(\frac{\partial^2}{\partial r^2} + \frac{1}{r}\frac{\partial}{\partial r} + \frac{1}{r^2}\frac{\partial^2}{\partial \theta^2} + \frac{\partial^2}{\partial z^2} - \frac{1}{r^2}\right)\psi = \frac{1}{c_s^2}\frac{\partial^2 \psi}{\partial t^2} \tag{5.14}$$

These equations are particularly useful for analyzing wave propagation in cylindrical rock samples or borehole environments [2, 9] (Fig. 5.6).

6. Wave Equations for Anisotropic Media

Many rocks exhibit anisotropic behavior due to layering, fractures, or preferred mineral orientations. In anisotropic media, wave propagation characteristics depend on the direction of wave travel. The wave equations for anisotropic media are more complex than those for isotropic media.

The general wave equation for anisotropic media can be written as:

$$c_{ijkl}\frac{\partial^2 u_k}{\partial x_j \partial x_l} = \rho \frac{\partial^2 u_i}{\partial t^2} \tag{5.15}$$

where c_{ijkl} is the fourth-order elastic stiffness tensor, which has up to 21 independent components for the most general anisotropic material. For specific types of anisotropy, such as transverse isotropy or orthotropy, the number of independent components is reduced.

7. Dispersion Relations

Dispersion relations describe the relationship between frequency and wavenumber for different wave modes. They are essential for understanding wave propagation characteristics, including phase velocity and group velocity.

(1) P-wave Dispersion Relation

For P-waves in an isotropic medium, the dispersion relation is:

$$\omega = c_p |k| \tag{5.16}$$

where ω is the angular frequency and k is the wave vector. This linear relationship indicates that P-waves in an ideal elastic medium are non-dispersive, meaning all frequencies travel at the same velocity [4].

(2). S-wave Dispersion Relation

Similarly, for S-waves in an isotropic medium, the dispersion relation is:

$$\omega = c_s |k| \tag{5.17}$$

This linear relationship also indicates that S-waves in an ideal elastic medium are non-dispersive [4].

(3). Rayleigh Wave Dispersion Relation

For Rayleigh waves, the dispersion relation is more complex and can be expressed implicitly as:

$$(2 - \frac{c^2}{c_s^2})^2 = 4\sqrt{\left(1 - \frac{c^2}{c_p^2}\right)\left(1 - \frac{c^2}{c_s^2}\right)} \tag{5.18}$$

where $c = \omega/k$ is the phase velocity. This equation must be solved numerically to obtain the dispersion relation [3, 4].

(4). Love Wave Dispersion Relation

The dispersion relation for Love waves in a layer over a half-space is given implicitly by:

$$\tan\left(kH\sqrt{\frac{1}{\beta_1^2} - \frac{1}{c^2}}\right) = \frac{\mu_2\sqrt{\frac{1}{\beta_2^2} - \frac{1}{c^2}}}{\mu_1\sqrt{\frac{1}{\beta_1^2} - \frac{1}{c^2}}} \tag{5.19}$$

where H is the layer thickness, β_1 and β_2 are the S-wave velocities in the layer and half-space, respectively, and μ_1 and μ_2 are the corresponding shear moduli. This equation must be solved numerically to obtain the dispersion relation [4].

8. Energy Considerations

The energy associated with elastic waves is an important aspect of wave propagation. The total energy density of an elastic wave consists of kinetic energy and potential (strain) energy.

(1). Energy Density

The instantaneous energy density (E) of an elastic wave can be expressed as:

$$E = \frac{1}{2}\rho\left(\frac{\partial u}{\partial t}\right)^2 + \frac{1}{2}\sigma_{ij}\varepsilon_{ij} \tag{5.20}$$

5.1 Basic Principle of Stress Wave

where the first term represents kinetic energy density and the second term represents strain energy density.

(2). Energy Flux

The energy flux vector (P), also known as the Poynting vector, represents the rate of energy flow per unit area and is given by:

$$P = -\sigma \cdot \left(\frac{\partial u}{\partial t}\right) \tag{5.21}$$

This vector describes the direction and magnitude of energy propagation in the elastic medium [4].

9. **Attenuation and Absorption**

In real rocks, elastic waves experience attenuation due to various mechanisms, including geometric spreading, scattering, and intrinsic absorption. To account for attenuation, the wave equations can be modified by introducing complex elastic moduli or quality factors.

(1). Viscoelastic Wave Equation

A simple model for attenuation is the Kelvin-Voigt viscoelastic model, which modifies the stress–strain relationship to include a viscous term:

$$\sigma = \lambda(\nabla \cdot \mathbf{u})\mathbf{I} + 2\mu\mathbf{e} + \eta\frac{\partial \mathbf{e}}{\partial t} \tag{5.22}$$

where η is the viscosity tensor. This leads to a modified wave equation that accounts for attenuation [1, 6].

(2). Quality Factor

The quality factor (Q) is often used to characterize attenuation in rocks. It is defined as the ratio of stored energy to dissipated energy per cycle. The wave equation can be modified to include Q by using complex elastic moduli:

$$M^* = M(1 + i/Q) \tag{5.23}$$

where M^* is the complex modulus and M is the real part of the modulus [1, 6].

10. **Numerical Solutions of Wave Equations**

Due to the complexity of wave propagation in real rock masses, numerical methods are often employed to solve the wave equations. Common numerical techniques include:

(1). Finite Difference Method (FDM)

The FDM discretizes the wave equation in both space and time, replacing derivatives with finite differences. This method is particularly well-suited for simple geometries and homogeneous media [1, 3].

(2). Finite Element Method (FEM)

The FEM divides the domain into small elements and approximates the solution within each element using basis functions. This method is versatile and can handle complex geometries and heterogeneous media [1, 3].

(3). Spectral Element Method (SEM)

The SEM combines the flexibility of the FEM with the accuracy of spectral methods. It is particularly effective for large-scale wave propagation problems in complex media [1, 3].

11. Limitations and Extensions of Elastic Wave Equations

While the elastic wave equations provide a powerful framework for understanding stress wave propagation in rocks, they have several limitations:

(1). Small Strain Assumption

The linear elastic wave equations assume small strains and rotations. For large deformations, nonlinear elasticity theory must be considered [1, 6].

(2). Material Heterogeneity

Real rocks are often heterogeneous at various scales. Wave equations for heterogeneous media can be derived using effective medium theory or by explicitly modeling the heterogeneity [1, 3].

(3). Fractures and Discontinuities

The presence of fractures and discontinuities in rocks can significantly affect wave propagation. Extended wave equations that incorporate fracture mechanics or discrete element methods can be used to address this limitation [1, 3].

(4). Poroelasticity

Many rocks are porous and fluid-filled. Biot's theory of poroelasticity extends the elastic wave equations to account for fluid–solid interactions in porous media [1, 6].

(5). Thermoelasticity

Temperature changes can induce thermal stresses and affect wave propagation. Thermoelastic wave equations incorporate the coupling between elastic deformation and temperature changes [1, 6].

5.1 Basic Principle of Stress Wave

12. **Experimental Validation of Wave Equations**

Experimental validation of wave equations in rocks involves various techniques:

(1). Ultrasonic Testing

Ultrasonic pulse transmission tests are commonly used to measure P-wave and S-wave velocities in rock samples. These measurements can be compared with predictions from wave equations [1, 5].

(2). Seismic Surveys

Field-scale seismic surveys provide data on wave propagation in rock masses. These data can be used to validate and calibrate wave propagation models based on the wave equations [1, 5].

(3). Acoustic Emission Testing

Acoustic emission testing monitors the elastic waves generated by microcracking in rocks under stress. These measurements can provide insights into the applicability of wave equations in damaged or fracturing rocks [1, 3].

13. **Applications of Wave Equations in Rock Engineering**

The wave equations of elastic stress waves serve as a fundamental theoretical framework in various rock engineering applications. These equations are instrumental in seismic exploration for energy resources, earthquake hazard assessment and structural design, optimization of rock blasting operations, non-destructive ultrasonic testing of rock properties, and prediction of dynamic responses in underground excavation. While these equations have inherent limitations in modeling complex, heterogeneous, or highly fractured rock masses, they remain essential tools in rock dynamics research and engineering practice. Current research efforts focus on refining these equations to better account for the complexities of real rock behavior under dynamic loading conditions [2–4, 6, 9].

5.1.3 Interaction of Two Elastic Stress Waves

The interaction of elastic stress waves represents a fundamental phenomenon in rock dynamics, encompassing various essential processes in geological and engineering applications. These waves, classified as body waves (P-waves and S-waves) and surface waves (Rayleigh and Love waves), propagate through rock media while preserving elastic deformation characteristics. When such waves intersect, they undergo superposition, resulting in phenomena including constructive and destructive interference, wave focusing, and scattering. Understanding these interaction mechanisms is crucial for accurate analysis and prediction of stress wave propagation in rock masses, with significant implications for rock blasting, seismic exploration, and earthquake studies [1] (Fig. 5.7).

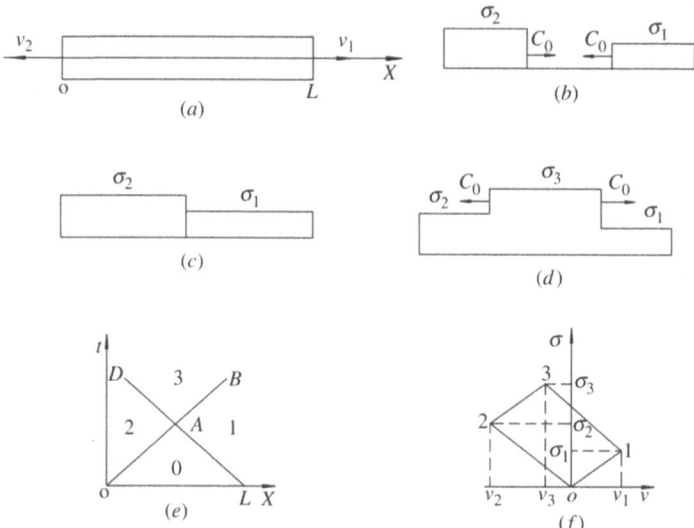

Fig. 5.7 Interaction of two head-on propagating strong discontinuous elastic waves [5]

The study of stress wave interactions in rock is rooted in the fundamental principles of elastodynamics and wave propagation theory. The governing equations for elastic wave propagation in a homogeneous, isotropic medium are derived from Newton's second law of motion and Hooke's law [3, 4]. The interaction of stress waves in rock masses exhibits complex patterns of constructive and destructive interference, with potential applications in enhancing rock fragmentation and reducing explosive consumption [8]. However, the effectiveness of these interactions is constrained by precise timing requirements and complicated by factors including non-linear rock behavior, crack propagation, material heterogeneity, and wave attenuation [1, 8]. Advanced numerical methods, particularly FEM and FDM, have been instrumental in modeling these interactions, as demonstrated by Aasen et al.'s LS-DYNA simulations, which revealed limited interaction zones and significant intermediate hole effects. The implications extend beyond blasting to various applications including seismic exploration, earthquake engineering, non-destructive testing, underground excavation, and hydraulic fracturing [3, 7]. Current research focuses on developing advanced constitutive models, multi-scale modeling approaches, experimental techniques, machine learning applications, and methods for analyzing coupled processes in anisotropic and fractured media [8]. These advancements are essential for bridging the gap between idealized models and complex natural rock mass behavior, ultimately improving applications from rock fragmentation optimization to seismic hazard assessment.

5.2 Stress Wave Propagation Under Different Boundary Conditions

5.2.1 Stress Wave Propagation on a Free Surface

Stress wave propagation on a free surface is a critical aspect of rock dynamics that plays a significant role in various geological and engineering applications. The behavior of stress waves at free surfaces is particularly important because it influences the distribution of energy, the formation of surface waves, and the potential for damage or failure in rock structures. This section will provide a comprehensive examination of stress wave propagation on free surfaces in rock, encompassing theoretical foundations, numerical modeling approaches, and practical implications.

1. **Theoretical Foundations of Stress Wave Propagation on Free Surfaces**

(1). Basic Principles of Stress Wave Reflection at Free Surfaces

When a stress wave encounters a free surface in rock, it undergoes reflection and mode conversion due to the boundary conditions imposed by the surface. The reflection of stress waves at free surfaces is governed by the principle of stress-free boundary conditions, which states that normal and shear stresses must vanish at the free surface [1] (Fig. 5.8).

For P-waves (longitudinal waves) incident on a free surface, the reflection process results in both reflected P-waves and converted SV-waves (vertically polarized shear waves) The amplitudes of these reflected and converted waves depend on the angle of incidence and the elastic properties of the rock medium [10, 11]. For SV-waves incident on a free surface, a similar process occurs, resulting in reflected SV-waves and converted P-waves. In contrast, SH-waves (horizontally polarized shear waves) incident on a free surface undergo total reflection without mode conversion.

(2). Surface Wave Generation

One of the most significant consequences of stress wave interaction with free surfaces is the generation of surface waves, particularly Rayleigh waves. Rayleigh waves are a type of surface wave that propagates along the free surface of an elastic solid, with particle motion following an elliptical path in a plane perpendicular to the surface [12].

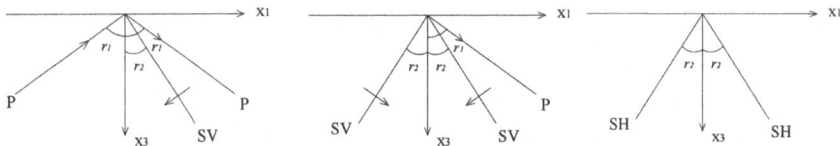

Fig. 5.8 The waves reflected at free surface [10]

The velocity of Rayleigh waves (c_R) in an elastic half-space is slightly less than the shear wave velocity (c_s) and can be approximated by [1]:

$$c_R \approx 0.9554 c_s \tag{5.24}$$

For a Poisson's ratio of 0.25, which is typical for many rocks. The generation of Rayleigh waves at a free surface can be understood as a result of the interference between reflected P-waves and SV-waves. The phase velocity of Rayleigh waves is frequency-independent in a homogeneous half-space, but becomes dispersive in layered or heterogeneous media [13]. In addition to Rayleigh waves, other types of surface waves, such as Love waves, can exist in layered media. Love waves are horizontally polarized shear waves that propagate in a surface layer over a half-space with a higher shear wave velocity [1].

(3). Stress Wave Attenuation and Dispersion at Free Surfaces

As stress waves propagate along a free surface, they experience attenuation and dispersion due to various mechanisms. Attenuation refers to the decay of wave amplitude with distance, while dispersion involves the separation of wave components with different frequencies due to frequency-dependent velocities [1].

For surface waves, geometrical spreading is a significant factor in attenuation. The amplitude of Rayleigh waves decreases as $r^{(-1/2)}$ in two dimensions and as $r^{(-1)}$ in three dimensions, where r is the distance from the source. This is in contrast to body waves, which attenuate as $r^{(-1)}$ in two dimensions and $r^{(-2)}$ in three dimensions. In addition to geometrical spreading, material damping contributes to wave attenuation. Dispersion in surface waves can arise from various sources, including material heterogeneity, layered structures, and frequency-dependent material properties.

(4). Spall Fracture at Free Surfaces

Spall fracture is a dynamic fracture phenomenon that occurs when tensile stresses generated by reflected stress waves exceed the material's tensile strength. This process is particularly relevant for free surfaces in rock subjected to high-intensity stress waves, such as those generated by explosions or impacts [12] (Fig. 5.9).

2. Numerical Modeling of Stress Wave Propagation on Free Surfaces

(1). Finite Difference Methods

Finite difference methods are widely used for simulating stress wave propagation in rock, including the interaction with free surfaces. These methods discretize the spatial and temporal domains, approximating the derivatives in the wave equations using finite differences [14, 15].

One effective approach for modeling stress wave propagation on irregular free surfaces is the use of a boundary-conforming grid method combined with a finite difference scheme. This method involves transforming the governing equations from Cartesian to curvilinear coordinates, allowing for accurate representation of complex topographies [16].

5.2 Stress Wave Propagation Under Different Boundary Conditions

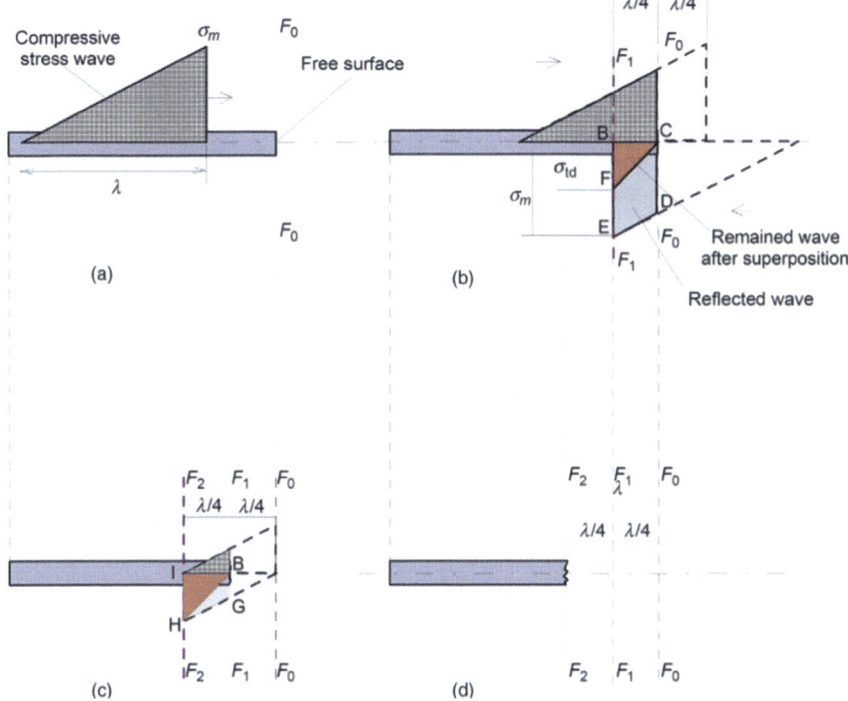

Fig. 5.9 Spalling caused by a triangular wave. **a** The compressive wave travels to the free surface; **b** the 1/4 length of the compressive wave is reflected back from the free surface, and the first spalling happens; **c** the 1/2 length of the remained compressive wave is reflected from the new free surface, and the second spalling occurs; **d** after the second spalling, there is no compressive wave remained [1]

(2). Finite Element Methods

Finite element methods offer another powerful approach for modeling stress wave propagation on free surfaces, particularly for complex geometries and heterogeneous media. These methods divide the domain into smaller elements and approximate the solution within each element using shape functions [15].

(3). Spectral Element Methods

Spectral element methods combine the geometric flexibility of finite element methods with the high-order accuracy of spectral methods. These methods are particularly effective for modeling wave propagation in complex media with free surfaces. In spectral element methods, the domain is divided into elements, and the solution within each element is approximated using high-order polynomials (typically Legendre or Chebyshev polynomials). The weak form of the wave equation is similar to that used in finite element methods, but the integrals are evaluated using Gauss-Lobatto-Legendre quadrature for improved accuracy [11].

(4). Boundary Element Methods

Boundary element methods are well-suited for problems involving stress wave propagation on free surfaces, as they only require discretization of the boundary rather than the entire domain. This makes them particularly efficient for problems with infinite or semi-infinite domains.

(5). Meshless Methods

Meshless methods, such as the Element-Free Galerkin method or the Meshless Local Petrov–Galerkin method, offer an alternative approach for modeling stress wave propagation on free surfaces. These methods do not require a predefined mesh, making them particularly suitable for problems involving large deformations or fracture [14].

In meshless methods, the approximation of the field variables is constructed using a set of nodes distributed over the domain. The weak form of the governing equations is similar to that used in finite element methods, but the shape functions are constructed using techniques such as moving least squares approximation or radial basis functions [14].

(6). Hybrid Methods

Hybrid methods combine different numerical techniques to leverage their respective strengths. For example, a coupled finite element–boundary element method can be used to model stress wave propagation in a bounded domain with an unbounded free surface. This approach allows for efficient modeling of the near-field response using finite elements while capturing the far-field radiation conditions using boundary elements [15].

3. Effects of Surface Topography on Stress Wave Propagation

(1). Influence of Surface Irregularities

Surface topography can significantly influence the propagation of stress waves along free surfaces. Irregular surfaces can cause scattering, focusing, and defocusing of waves, leading to complex wave patterns and energy distributions [17]. When body waves encounter irregular surfaces, they can convert into surface waves through a process known as mode conversion. This process is particularly important for the generation of Rayleigh waves and Love waves in realistic geological settings [13]. The effects of surface irregularities on stress wave propagation can be analyzed using perturbation methods for weak topography variations.

(2). Topographic Amplification and Deamplification

Surface topography significantly influences ground motion through amplification at ridge crests and convex features, and deamplification in valleys and concave features, with amplification factors potentially exceeding 2 at resonant frequencies. The magnitude of this topographic effect is governed by wave frequency content, dimensional ratios between topographic features and wavelength, slope steepness, and impedance contrasts between surface and bedrock layers [12].

5.2 Stress Wave Propagation Under Different Boundary Conditions

(3). Phase Velocity Variations Due to Topography

Surface topography can induce variations in the phase velocity of surface waves. These variations are important for accurately interpreting surface wave dispersion data and for applying travel-time corrections in seismic analyses [17]. The relative phase speed perturbation due to topography for Love waves can be expressed as [17]:

$$\left[\frac{1}{c}\frac{\partial c}{\partial h}\right]^L = -\frac{1}{4cUI_1}\rho^0 l_1^2 (c^2 - \beta^2)(z=0) \tag{5.25}$$

And for Rayleigh waves [17]:

$$\left[\frac{1}{c}\frac{\partial c}{\partial h}\right]^R = -\frac{1}{4cU_1}\rho^0 \left[r_2^2 c^2 + r_1^2 \left(c^2 - 4\left(1 - \frac{\beta^2}{\alpha^2}\right)\beta^2\right)\right](z=0) \tag{5.26}$$

where c is the phase velocity, h is the topography height, U and U_1 are group velocity-related terms, l_1, r_1, and r_2 are displacement eigenfunctions, α and β are P-wave and S-wave velocities, ρ^0 is the surface density. These expressions show that the effect of topography on phase speed is generally largest for periods around 20 s, with relative perturbations on the order of 0.5% for topography of 1 km [17].

4. **Practical Implications of Stress Wave Propagation on Free Surfaces**

(1). Seismic Hazard Assessment

The propagation of stress waves on free surfaces plays a vital role in seismic hazard assessment, particularly in regions characterized by complex topography, where amplification and deamplification effects can substantially impact ground motion distribution during seismic events. Various methodologies have been developed to incorporate these effects into hazard analyses, including topographic amplification factors, site-specific response analysis through detailed numerical simulations, comprehensive microzonation studies for large-scale mapping of topographic effects, and probabilistic approaches that account for uncertainties in topographic influences to provide more robust risk assessments [12].

(2). Rock Engineering Stability Analysis

The propagation of stress waves on free surfaces significantly influences rock engineering stability under dynamic loading conditions such as earthquakes or blasting. The interaction between stress waves and rock surfaces can trigger various failure mechanisms, including tensile fracturing from reflected waves, shear failure along pre-existing discontinuities, toppling through wave-induced dynamic moments, and liquefaction due to pore pressure accumulation. These phenomena can be analyzed through multiple methodologies, ranging from simplified pseudostatic analysis and displacement-based techniques like Newmark sliding block analysis to comprehensive stress-based numerical simulations that capture the complete wave propagation and stress–strain behavior of slope materials [1].

(3). Non-destructive Testing of Rock and Concrete Structures

Stress wave propagation along free surfaces serves as the fundamental principle for non-destructive testing (NDT) in rock and concrete structures, enabling the assessment of material properties and structural integrity [18]. The methodology encompasses several specialized techniques, including Ultrasonic Pulse Velocity (UPV) testing for quality evaluation through P-wave transmission, Impact-Echo testing for flaw detection via wave reflection, Spectral Analysis of Surface Waves (SASW) for elastic property determination in layered systems, and AE testing for damage monitoring through elastic wave release. Effective interpretation of these methods necessitates comprehensive understanding of stress wave propagation phenomena, particularly mode conversion, attenuation, and scattering effects.

Understanding the propagation of stress waves on free surfaces is essential for designing effective survey geometries, processing seismic data, and interpreting the results in terms of subsurface properties and structures.

5. Advanced Topics in Stress Wave Propagation on Free Surfaces

(1). Nonlinear Effects in High-Amplitude Wave Propagation

High-amplitude stress waves on free surfaces exhibit significant nonlinear behavior, particularly evident in strong ground motion during major earthquakes and near-field blast-induced vibrations [7]. These nonlinear phenomena encompass material nonlinearity manifesting through shear modulus reduction and hysteretic damping, geometric nonlinearity due to large deformations, contact nonlinearity from crack/joint dynamics, and nonlinear site response including soil liquefaction. Accurate modeling of these complex behaviors requires sophisticated numerical approaches, including nonlinear finite element methods, discrete element methods for discontinuous media, and hybrid continuum-discontinuum techniques, enabling precise prediction of wave propagation under extreme loading conditions.

(2). Poroelastic Effects in Saturated Rock

Stress wave propagation on free surfaces in saturated rock is governed by poroelasticity theory, which accounts for fluid–solid coupling through equations describing bulk material motion, pore fluid motion, and fluid flow continuity [19]. This theoretical framework identifies three distinct body waves: the fast P-wave (P1), analogous to elastic P-waves; the highly attenuative slow P-wave (P2) associated with fluid flow; and the modified S-wave. At free surfaces, complex boundary conditions involving both stress and fluid pressure enable additional modes such as the Biot wave. The wave propagation characteristics are significantly influenced by rock porosity and permeability, fluid properties, and frequency-dependent fluid–solid coupling, with critical applications in reservoir characterization, hydraulic fracturing, and earthquake-induced liquefaction studies [19].

5.2 Stress Wave Propagation Under Different Boundary Conditions

(3). Thermoelastic Effects on Wave Propagation

Thermoelastic coupling significantly influences stress wave propagation on free surfaces, particularly in applications involving laser-generated ultrasound, thermal shock, and geothermal energy extraction [19]. The governing equations, comprising motion with thermal stresses and heat conduction with mechanical coupling, yield three distinct wave types: quasi-P wave (qP), quasi-SV wave (qSV), and thermal wave (qT). Wave behavior at free surfaces exhibits complex reflection and mode conversion phenomena, governed by both mechanical and thermal boundary conditions. The propagation characteristics are fundamentally determined by material thermal properties (conductivity, specific heat), thermal expansion coefficient, and thermal relaxation times, necessitating specialized numerical techniques to address the disparate mechanical and thermal time scales [19].

(4). Fracture Propagation Induced by Stress Waves

The interaction between stress waves and free surfaces plays a crucial role in fracture initiation and propagation in rock, with significant implications for hydraulic fracturing, explosive excavation, and earthquake rupture propagation [1]. The process is characterized by dynamic stress intensity factors at crack tips, elevated dynamic fracture toughness compared to static conditions, high-velocity crack branching phenomena, and mixed-mode fracture behavior under complex stress states. Advanced numerical modeling techniques, including dynamic finite element methods with cohesive zone models, extended finite element methods (XFEM), and peridynamics, enable simulation of complex fracture patterns and their interaction with propagating stress waves.

(5). Stress Wave Propagation in Layered and Heterogeneous Media

The complexity of stress wave propagation on free surfaces significantly increases in layered and heterogeneous rock masses, demanding sophisticated analysis for seismic data interpretation and wave behavior prediction in complex geological settings [20]. Key phenomena include wave guiding in low-velocity layers, heterogeneity-induced wave scattering generating complex wave fields and coda waves, frequency-dependent dispersion in layered media, and direction-dependent wave velocities due to rock anisotropy. Advanced numerical techniques, encompassing full waveform inversion for velocity model reconstruction, multiscale modeling for cross-scale heterogeneities, and stochastic methods for property uncertainties, enable accurate prediction and interpretation of wave behavior in realistic geological environments.

The propagation of stress waves on free surfaces in rock, encompassing fundamental wave reflection principles and advanced nonlinear effects, is crucial for understanding and predicting wave behavior in geoscience and engineering applications. Continuing research advances enhance methodologies for seismic hazard assessment, non-destructive testing, and geophysical exploration.

5.2.2 Stress Wave Propagation on a Fixed Surface

The propagation of stress waves on fixed surfaces represents a fundamental phenomenon in rock dynamics, encompassing reflection and mode conversion at rigid boundaries or stiff material interfaces. The interaction of compressional, shear, and surface waves at fixed boundaries, governed by wave mechanics and elastodynamics principles, is crucial for understanding wave behavior in underground excavations, rock blasting, and seismic wave propagation applications.

1. **Theoretical Foundations**

The theoretical basis for understanding stress wave propagation on fixed surfaces is rooted in the fundamental equations of elastodynamics. For a linear elastic, isotropic medium, the governing equations for wave propagation can be expressed as [1]:

$$\rho \frac{\partial^2 u}{\partial t^2} = E \frac{\partial^2 u}{\partial x^2} \qquad (5.27)$$

where ρ is the density of the medium, E is Young's modulus. These equations describe the propagation of both P-waves and S-waves within the rock medium.

When a stress wave encounters a fixed surface, it must satisfy specific boundary conditions. For a perfectly rigid, fixed boundary, the displacement at the boundary is zero in all directions:

$$u_i = 0 \text{ at the fixed boundary} \qquad (5.28)$$

This condition leads to the reflection of incident waves and the potential generation of new wave modes. The reflection coefficients for P-waves and S-waves at a fixed boundary can be derived using the method of potentials [2, 21]. For a normally incident P-wave on a fixed surface, the reflection coefficient is:

$$R_P = -1 \qquad (5.29)$$

This means that the reflected P-wave has the same amplitude as the incident wave but with a phase reversal. For a normally incident S-wave, the reflection coefficient is:

$$R_S = 1 \qquad (5.30)$$

indicating that the S-wave is reflected without phase change.

However, for oblique incidence, the situation becomes more complex due to mode conversion. When a P-wave strikes a fixed boundary at an angle, it generates both reflected P-waves and mode-converted S-waves. Similarly, an incident S-wave can produce reflected S-waves and mode-converted P-waves. The amplitudes of these reflected and converted waves depend on the angle of incidence and the material properties of the rock.

5.2 Stress Wave Propagation Under Different Boundary Conditions

The reflection and transmission coefficients for oblique incidence can be derived using the continuity of stress and displacement at the boundary. The behavior of Rayleigh waves, while typically associated with free surfaces, undergoes significant modification near fixed boundaries, leading to complex wave scattering and mode conversion [22]. Additionally, fixed boundaries substantially alter the dispersion relations of both body and surface waves, particularly evident in the complex dispersion characteristics of flexural waves in fixed-boundary plates [23].

2. Experimental Observations

Experimental studies employing photoelasticity, strain gauge measurements, and high-speed photography have provided comprehensive insights into stress wave behavior on fixed surfaces in rock materials [24, 25]. These investigations have verified fundamental theoretical predictions, including phase reversal of P-waves and phase maintenance of S-waves during normal incidence reflection [1], while also revealing complex mode conversion phenomena during oblique incidence [24]. Advanced techniques such as laser interferometry have demonstrated intricate interactions between surface waves and fixed boundaries, particularly in the context of Rayleigh wave propagation [23]. Research has highlighted several key phenomena: increased wave attenuation due to multiple reflections and mode conversions [26], significant stress concentrations near fixed boundaries [27], altered transmission characteristics in layered structures [22], generation of higher-frequency components in reflected waves [28], and modified acoustic emission patterns during rock fracturing [29]. Furthermore, ultrasonic studies have shown that fixed boundaries substantially influence guided wave propagation characteristics, which has important implications for non-destructive testing applications (Fig. 5.10).

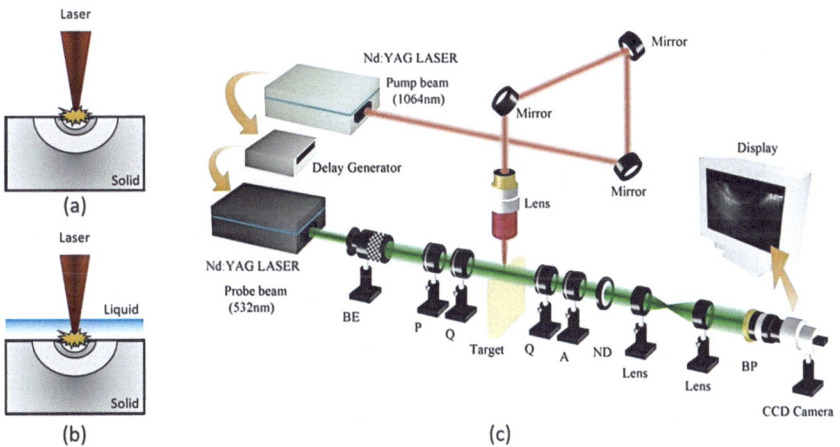

Fig. 5.10 Simplified schematic of the experimental setup. Laser-induced shock excitation of elastic waves at a free surface (**a**) and at a soft solid–liquid interface (**b**) (**c**) Schematic diagram of the photoelasticity imaging system (BE: beam expander, P: polarizer, Q: quarter wave plate, A: analyzer, ND: neutral density filter, and BP: band- pass filter) [28]

3. **Numerical Modeling Approaches**

Numerical modeling techniques have been extensively developed to simulate stress wave propagation on fixed surfaces in rock materials. The FDM offers efficient solutions for wave equations with fixed boundary conditions, though challenges in representing curved boundaries have led to advanced techniques like the immersed boundary method [30]. The finite element method (FEM) provides superior handling of complex geometries and material heterogeneities [27], while the boundary element method (BEM) excels in modeling wave scattering and radiation problems. The spectral element method (SEM) has emerged as a powerful tool for simulating wave propagation in layered structures [22]. Recent developments include the incorporation of viscoelastic and poroelastic models, combined finite-discrete element methods (FDEM) for dynamic fracture modeling [29], and parallel computing techniques for large-scale simulations [26]. The field has further evolved with the integration of uncertainty quantification [31] and machine learning approaches [32], enhancing the accuracy and efficiency of wave propagation predictions in rock masses with fixed boundaries.

4. **Applications and Practical Implications**

The understanding of stress wave propagation on fixed surfaces has significant practical applications across various fields of rock engineering and geophysics. In underground excavation and tunneling, the interaction between stress waves and fixed boundaries critically influences structural stability and support system design [27]. This knowledge is essential for optimizing rock blasting operations and evaluating seismic wave effects in earthquake engineering [1]. The applications extend to rock slope stability assessment, non-destructive testing, and wellbore stability analysis in oil and gas operations [28]. Each application requires careful consideration of wave reflection, mode conversion, and stress concentration phenomena at fixed boundaries to ensure safety and operational efficiency.

5.2.3 Stress Wave Propagation at an Interface Between Two Media

The interaction between stress waves and material interfaces represents a fundamental phenomenon in rock dynamics [33]. This behavior, particularly at the boundaries between distinct rock types or rock-material interfaces, holds critical importance for understanding rock mass responses under dynamic loading conditions. Such knowledge underpins various applications across mining, tunneling, earthquake engineering, and non-destructive testing of geological structures [33].

1. Theoretical Framework of Wave Propagation at Interfaces

To comprehensively analyze stress wave propagation at interfaces, we must first establish a solid theoretical foundation. The behavior of stress waves at interfaces is governed by the principles of continuum mechanics and wave theory, which we will explore in detail.

(1). Governing Equations

The propagation of stress waves in elastic media is described by the equations of motion, which are derived from Newton's second law and the constitutive relationships of the materials involved. For a linear elastic, isotropic medium, the equations of motion in vector form can be expressed as:

$$\nabla \cdot \sigma = \rho \frac{\partial^2 \mathbf{u}}{\partial t^2} \tag{5.31}$$

where σ is the stress tensor, ρ is the density of the medium, and $\mathbf{u}$ is the displacement vector. This equation relates the stress gradients to the acceleration of particles within the medium. In component form, for a two-dimensional problem, these equations can be written as:

$$\frac{\partial \sigma_{xx}}{\partial x} + \frac{\partial \sigma_{xy}}{\partial y} = \rho \frac{\partial^2 u_x}{\partial t^2} \tag{5.32}$$

$$\frac{\partial \sigma_{xy}}{\partial x} + \frac{\partial \sigma_{yy}}{\partial y} = \rho \frac{\partial^2 u_y}{\partial t^2} \tag{5.33}$$

where σ_{xx}, σ_{yy}, and σ_{xy} are components of the stress tensor, and u_x and u_y are components of the displacement vector. The constitutive relationships for a linear elastic material are given by Hooke's law:

$$\sigma = c\varepsilon \tag{5.34}$$

where c is the fourth-order elasticity tensor and ε is the strain tensor. For an isotropic material, this relationship can be simplified using Lamé constants λ and μ:

$$\sigma_{ij} = \lambda \varepsilon_{kk} \delta_{ij} + 2\mu \varepsilon_{ij} \tag{5.35}$$

where ε_{kk} is the volumetric strain and δ_{ij} is the Kronecker delta.

(2). Wave Equations

From the equations of motion and constitutive relationships, we can derive the wave equations for longitudinal (P) and shear (S) waves:

$$\nabla^2 \phi = \frac{1}{c_L^2} \frac{\partial^2 \phi}{\partial t^2} \tag{5.36}$$

$$\nabla^2 \psi = \frac{1}{c_T^2} \frac{\partial^2 \psi}{\partial t^2} \tag{5.37}$$

where φ and ψ are scalar and vector potentials, respectively, and c_L and c_T are the longitudinal and transverse wave velocities given by:

$$c_L^2 = \frac{\lambda + 2\mu}{\rho} \tag{5.38}$$

$$c_T^2 = \frac{\mu}{\rho} \tag{5.39}$$

These wave equations form the basis for analyzing the propagation of stress waves in elastic media and their behavior at interfaces.

(3). Reflection and Refraction at Interfaces

At media interfaces, stress wave behavior is governed by two fundamental principles: displacement continuity and stress continuity (both normal and tangential components) across boundaries [3]. These principles form the basis for determining reflection and transmission coefficients, which quantify the relative amplitudes of reflected and transmitted waves compared to the incident wave. The mathematical expressions for these coefficients are dependent on both the incident wave type (P- or S-wave) and the material properties of the adjacent media.

For a P-wave incident on an interface between two elastic solids, the reflection and transmission coefficients for the P-wave component are given by [34]:

$$R_{pp} = \frac{(\rho_2 c_P 2\cos^2 2\theta_t - \rho_1 c_P 1\cos^2 2\theta_s) - (\rho_2 c_S 2\sin^2 2\theta_t - \rho_1 c_S 1\sin^2 2\theta_s)}{D} \tag{5.40}$$

$$T_{pp} = \frac{2\rho_1 c_{P1} \cos\theta_i (\rho_2 c_{S2} \sin^2 2\theta_t)}{D} \tag{5.41}$$

where

$$D = (\rho_2 c_{P2} \cos^2 2\theta_t + \rho_1 c_{P1} \cos^2 2\theta_s) + (\rho_2 c_{S2} \sin^2 2\theta_t + \rho_1 c_{S1} \sin^2 2\theta_s) \tag{5.42}$$

Here, ρ_1 and ρ_2 are the densities of the first and second medium, respectively, c_{P1} and c_{P2} are the P-wave velocities, c_{S1} and c_{S2} are the S-wave velocities, θ_i is the incident angle, θ_t is the transmitted angle for the P-wave, and θ_s is the reflected angle for the S-wave.

Similar expressions can be derived for S-wave incidence and for the coefficients of mode-converted waves (P to S and S to P). These coefficients are crucial for understanding the partitioning of energy at interfaces and the resulting wave field in layered media.

2. Stress Wave Propagation in Layered Composite Rock Masses

Real geological formations often consist of layered structures with different rock types. Understanding stress wave propagation in such composite rock masses is essential for many practical applications. We will now explore the behavior of stress waves in layered composite rock masses, considering both the effects of interfaces between different rock types and the presence of discontinuities such as joints.

(1). Wave Propagation in Layered Media without Joints

In layered composite rock masses without joints, stress wave propagation involves multiple reflections and transmissions through alternating rock layers, resulting in complex wave fields [35, 36]. The analysis of wave transmission through such layered media can be accomplished using matrix-based approaches, specifically the transfer matrix method or global matrix method, enabling the calculation of overall transmission and reflection coefficients for multi-layered systems [36].

The results show that the transmission characteristics of layered media can be quite different from those of homogeneous media, with frequency-dependent behavior and the potential for pass bands and stop bands in wave transmission [36].

(2). Wave Propagation in Jointed Layered Media

In real rock masses, layers are often separated by joints or fractures, which significantly affect wave propagation. The presence of joints introduces additional complexities in the analysis of wave transmission. To model wave propagation in jointed layered media, researchers have developed various approaches. One common method is to treat the joints as displacement discontinuities with specific stiffness characteristics [37]. In this approach, the stress-displacement relationship across a joint is given by:

$$\sigma = K_n u \tag{5.43}$$

$$\tau = K_s u \tag{5.44}$$

where σ and τ are the normal and shear stresses, u is the displacement discontinuity, and K_n and K_s are the normal and shear stiffnesses of the joint. Incorporating these joint models into the wave propagation analysis leads to modified transmission and reflection coefficients. For a P-wave normally incident on a single joint, the transmission coefficient can be expressed as:

$$T_v = \frac{2}{2 + i\omega z/K_n} \tag{5.45}$$

where ω is the angular frequency of the wave, and z is the acoustic impedance of the rock. For multiple parallel joints in a layered composite rock mass, the wave transmission becomes even more complex due to multiple reflections between joints.

The overall transmission coefficient can be expressed as the sum of direct and indirect transmissions [36]:

$$T = T_d + T_i \qquad (5.46)$$

where T_d is the direct transmission coefficient (considering only the first transmitted wave); and T_i is the indirect transmission coefficient (accounting for multiple reflections between joints).

Research demonstrates that indirect transmission significantly impacts wave propagation, with its influence on particle velocity and stress transmission coefficients varying inversely with incident wave frequency, proportionally with joint numbers, non-linearly with joint stiffness, and differentially with wave impedance ratio–increasing for particle velocity while decreasing for stress [36]. These relationships underscore the complexity of wave propagation in jointed layered media and the necessity of incorporating multiple reflections in analysis.

(3). Effects of Wave Impedance Ratio

The wave impedance ratio between different rock layers plays a crucial role in determining the transmission characteristics of stress waves. The wave impedance Z is defined as the product of density ρ and wave velocity c:

$$Z = \rho c \qquad (5.47)$$

For layered composite rock masses, the wave impedance ratio n is typically defined as [36]:

$$n = \frac{Z_f}{Z_b} \qquad (5.48)$$

where Z_f is the wave impedance of the layer in front of the joint, and Z_b is the wave impedance of the layer behind the joint.

Wave impedance ratio exerts a critical influence on stress wave transmission through layered composite rock masses, with soft-to-hard interfaces (ratio < 1) yielding particle velocity transmission coefficients below unity and potential stress transmission coefficients above unity, while hard-to-soft interfaces (ratio > 1) exhibit opposite characteristics–findings particularly relevant for wave propagation analysis in alternating geological strata [36, 37].

3. **Experimental Observations and Numerical Simulations**

While theoretical models provide valuable insights into stress wave propagation at interfaces, experimental observations and numerical simulations are essential for validating these models and exploring more complex scenarios. In this section, we will discuss key experimental findings and numerical simulation results related to stress wave propagation at interfaces between different media.

5.2 Stress Wave Propagation Under Different Boundary Conditions

(1). Experimental Studies

Experimental investigations on stress wave propagation at material interfaces were conducted using a powder gun system to generate planar impact shock loading on layered composites including Polycarbonate/Aluminum (PC/Al), Polycarbonate/Stainless Steel (PC/SS), and Polycarbonate/Glass (PC/GS) with varying layer thicknesses and impedance mismatches [2]. The study revealed that periodically layered composites can support steady structured shock waves, while interface scattering affects both bulk and deviatoric responses by reducing wave propagation velocity and increasing shock viscosity, respectively. Higher interface density resulted in steeper shock front slopes and increased profile oscillations, whereas larger impedance mismatches between components led to enhanced wave dispersion and more significant effects on shock wave velocity. In the layered composites, the relationship between strain rate at the shock front and shock stress was found to be: $\dot{\varepsilon} \propto \sigma^n$, where $n \approx 1.8 - 2.4$. This indicates much larger shock viscosity in the layered composites compared to homogeneous materials, where typically $n \approx 4$. A laser-induced shock technique was employed to study elastic surface waves at solid–liquid interfaces [38]. The experiment identified two distinct surface waves at free surfaces: a Rayleigh wave propagating at 0.93 times the shear wave speed and a super-shear evanescent wave (SEW) at 1.9 times the shear wave speed. At soft solid–liquid interfaces, four wavefronts were observed: a non-attenuating Scholte wave at 0.75 times the shear wave speed, a weakly attenuating Rayleigh wave below shear wave speed, an intermediate wavefront at 1.35 times the shear wave speed, and a rapidly attenuating SEW traveling at 1.7 times the shear wave speed.

These experimental observations provide valuable insights into the complex behavior of stress waves at interfaces, validating some theoretical predictions and revealing new phenomena that require further investigation.

(2). Numerical Simulations

Numerical simulations play a crucial role in studying stress wave propagation at interfaces, allowing researchers to explore a wide range of scenarios and material combinations that may be difficult or impossible to investigate experimentally. Various numerical methods have been employed, including finite element analysis (FEA), finite difference methods, and discrete element methods (DEM).

A numerical study using LS-DYNA finite element analysis investigated blast stress wave propagation and fracture development in layered composite rock formations [39]. The simulation, employing HJC constitutive model, JWL equation of state, fluid–structure coupling algorithm, and element deletion technique, revealed distinct wave behaviors at interfaces: compressional reflection waves occur in soft-to-hard rock transitions, while tensile reflection waves form in hard-to-soft rock transitions, with total reflection occurring beyond critical incident angles. Characteristic triangular fracture zones developed near the blast hole, with tensile cracks forming at interfaces due to surface waves. In deep-hole blasting, the study demonstrated that detonation initiated in hard rock layers achieved superior fragmentation due to interface-reflected tensile waves, whereas soft rock initiation resulted in under-breaking of hard rock layers (Fig. 5.11).

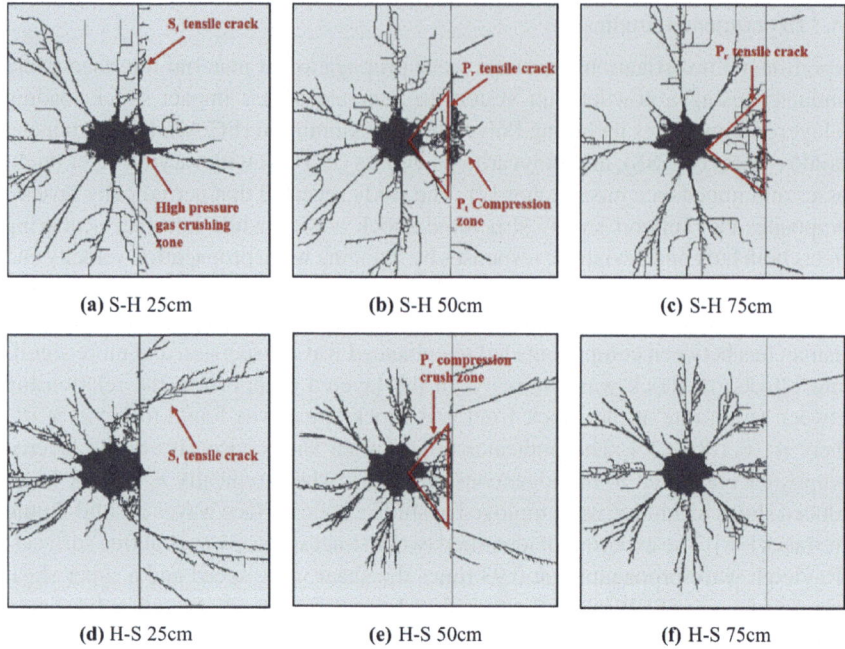

Fig. 5.11 Numerical Simulations of crack propagation under different incidence modes [39]

A numerical investigation of dynamic stress analysis in bimaterial composites was conducted using 2D finite element analysis and ultrasonic wave propagation simulation [33]. The study examined wave propagation patterns and stress distributions in a model containing an elliptical interfacial defect across varying fiber/matrix elastic modulus ratios (E_f/E_m). Results demonstrated that increasing E_f/E_m from 1 to 8 enhanced stress concentration at free edges and interfaces, reversed wave propagation direction at interfaces, and differentially affected stress distribution at defect tips–increasing maximum stress in the fiber layer while decreasing it in the matrix layer. The analysis revealed complex stress behaviors influenced by free edge effects, material anisotropy, and internal defects, with notable normal stress decay from free edges toward defects and singular shear stress behavior at free edges for high E_f/E_m ratios.

These numerical studies demonstrate the power of computational methods in elucidating the complex behavior of stress waves at interfaces between different media, providing insights that complement and extend experimental observations.

4. **Special Cases and Advanced Topics**

Having established the fundamental principles and general behavior of stress wave propagation at interfaces, we now turn our attention to several special cases and advanced topics that are of particular interest in rock dynamics and engineering applications.

5.2 Stress Wave Propagation Under Different Boundary Conditions

(1). Spall Fracture at Interfaces

Spall fracture is a dynamic failure mechanism that occurs when tensile stresses generated by reflected stress waves exceed the material's tensile strength. This phenomenon is particularly relevant at interfaces between different media, where wave reflections can lead to complex stress states. The study by [40] provides a comprehensive overview of spall fracture mechanics in rock and concrete materials.

For a triangular compressive pulse reflecting off a free surface, the spall distance ds and spall time t_s are given by [40]:

$$t_s = \frac{x_r}{c_p} \tag{5.49}$$

$$\sigma_f = f(-d_s - x_r) + Rf(-d_s + x_r) \tag{5.50}$$

where x_r is the distance the wave travels before reflecting, c_p is the wave speed, σ_f is the fracture strength, and R is the reflection coefficient (-1 for a free surface). In layered composite rock masses, spall fracture occurrence at interfaces is governed by three key factors: impedance mismatch between layers, incident stress wave characteristics (amplitude and duration), and material tensile strength. A notable example occurs when compressive waves transition from high to low-impedance materials, generating reflected tensile waves that can induce spalling in the high-impedance material near the interface, a phenomenon particularly significant in blast-induced fracturing of layered rock masses [39].

(2). Nonlinear Effects at High Stress Levels

While linear elastic theory provides a good approximation for stress wave propagation at low to moderate stress levels, nonlinear effects become significant at high stress levels, such as those encountered in blast-induced waves or high-energy impacts. Nonlinear behavior can lead to shock wave formation and more complex wave-interface interactions.

For high-intensity loading, nonlinear material behavior must be considered using shock wave theory. The Rankine-Hugoniot jump equations describe conservation of mass, momentum, and energy across a shock front [40]:

$$\frac{\rho_1}{\rho_0} = \frac{U - v_0}{U - v_1} \tag{5.51}$$

$$P_1 - P_0 = \rho_0(v_1 - v_0)(U - v_0) \tag{5.52}$$

$$e_1 - e_0 = (P_1 v_1 - P_0 v_0)/[\rho_0(U - v_0)](v_1^2 - v_0^2)/2 \tag{5.53}$$

where ρ, P, v, e are density, pressure, particle velocity, and internal energy, respectively; and U is the shock velocity. At media interfaces, nonlinear effects manifest through alterations in critical angles for mode conversion and total reflection, amplitude-dependent transmission and reflection coefficients, and the generation of higher harmonics with wave shape distortion. These nonlinear phenomena significantly influence stress distribution and fracture patterns in layered composite rock masses under high-intensity dynamic loading conditions [39, 40].

(3). Interfacial Waves in Layered Media

Interfaces between different media support various interfacial waves crucial for stress wave propagation and energy transfer. These include Rayleigh waves (surface waves with exponentially decaying amplitude), Stoneley waves (propagating along fluid–solid interfaces), Love waves (horizontally polarized shear waves in layered media), and Scholte waves (fluid–solid interface waves similar to Rayleigh waves). Experimental validation of these waves, including super-shear evanescent waves, was provided [38]. In layered composite rock masses, these interfacial waves significantly influence energy transport and stress redistribution, with their propagation characteristics dependent on adjacent media properties and structural features such as joints or fractures [36].

(4). Wave Propagation in Fractured and Jointed Media

In real rock masses, fracture and joint networks introduce complex wave behaviors including scattering, attenuation, multiple wave generation, frequency-dependent transmission, and wave localization. Studies demonstrated that joint spacing critically influences wave transmission, with the particle velocity transmission coefficient exhibiting non-linear behavior as spacing increases, and wave impedance ratio significantly affecting transmission characteristics [37]. Various modeling approaches have been developed, including treating joints as displacement discontinuities with specific stiffness, using equivalent anisotropic continuum representations, and explicit fracture network modeling, enabling analysis of complex phenomena such as frequency-dependent transmission, multiple scattering, and wave propagation band formations.

(5). Time-Dependent Effects and Viscoelasticity

Viscoelastic behavior in rocks under dynamic loading introduces complexity to interface stress wave propagation, manifesting as frequency-dependent wave velocities and attenuation, wave packet dispersion, and interface creep and relaxation phenomena. The elastic wave interaction boundary conditions at solid interfaces with thin viscoelastic layers, incorporating interface stiffnesses, inertia terms, and normal-tangential stress-displacement coupling were investigated [34]. Their findings revealed the significance of coupling terms for thin layer representation, mass/inertia terms for dense interfaces, and the limited applicability of spring-type boundary conditions to very thin viscous layers.

(6). Dynamic Fracture Propagation at Interfaces

The interaction between stress waves and interfaces plays a critical role in dynamic fracture processes, encompassing stress concentration and singularities at interface corners and crack tips, dynamic stress intensity factors, energy release rates, and mixed-mode fracture phenomena. Numerical investigations revealed that stress concentration at defect tips varies with material property ratios, demonstrating that increasing fiber/matrix elastic modulus ratios elevates maximum stress at defect tips in fiber layers while reducing it in matrix layers [33]. These insights are fundamental for predicting and controlling fracture patterns in dynamically loaded layered composite rock masses, particularly in applications such as blast-induced fracturing [33, 39] engineering fields, from blast design in mining to seismic hazard assessment and the development of advanced protective structures.

5.3 Stress Wave Propagation in Complex Space Conditions

5.3.1 Stress Waves in Rocky Slopes and Landslides

Stress wave propagation through rocky slopes and its interaction with landslides represent a complex phenomenon, which is of crucial significance for the stability and dynamic behavior of such geological structures. Stress wave propagation through rocky slopes and its interaction with landslides represent a complex phenomenon, which is of crucial significance for the stability and dynamic behavior of such geological structures. This section explores the various aspects of stress wave propagation in rocky slopes and landslides, including the effects of topography, discontinuities, and material properties on wave characteristics, as well as the implications for slope stability and landslide triggering mechanisms.

1. **Topographic Effects on Stress Wave Propagation**

Slope topography significantly influences stress wave propagation in rocky slopes through its surface geometry and internal structure, leading to complex patterns of wave energy distribution [41, 42]. Topographic amplification, most pronounced near slope crests, can increase ground motion intensity with amplification factors up to 2.5, particularly when the ratio of slope height to seismic wavelength (H/λ) falls between 0.2 and 0.8 [42]. Experimental shaking table tests on a 60° gradient slope demonstrated horizontal and vertical acceleration amplification coefficients of 3–5 and 2–2.5 respectively at the slope tip [43]. Slope geometry further affects wave propagation through focusing and defocusing effects: concave surfaces concentrate wave energy at specific points, while convex surfaces scatter it [41]. Additionally, slope orientation relative to wave direction influences amplification patterns, with back slopes experiencing greater amplification than facing slopes due to tension wave formation during compression wave reflection [44].

Field observations and numerical studies of earthquake-induced landslides demonstrate the critical role of specific slope geometry in wave amplification and slope stability. Analysis of the 1911 Kemin earthquake-triggered Ananevo rockslide revealed significant topographic amplification at the slope crest, with resonance frequencies ranging from 2.5 to 4 Hz in the lower slope to 1 Hz in the upper slope

Fig. 5.12 Landslide activity indices for possible ranges of slope height over wavelength ratios (H/λ), for different slope aspects [46]

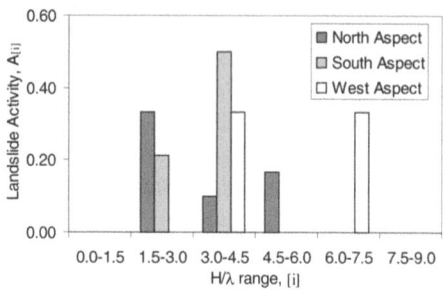

[45]. Similarly, investigation of coseismic rock slides following the 1999 Chi-Chi earthquake in Taiwan's Tachia Valley established a strong correlation between landslide occurrence and specific ranges of slope height to seismic wavelength (H/λ) ratios, with variations dependent on slope aspect [46] (Fig. 5.12).

2. Influence of Discontinuities on Wave Propagation

Discontinuities (joints, fractures, and bedding planes) in rocky slopes significantly influence stress wave propagation through complex wave interactions, attenuation, and reflection patterns [43]. Numerical studies using finite element analysis demonstrate that discontinuities generally decelerate and amplify wave propagation, particularly for horizontal waves. The impact varies with joint orientation: bedding joints (dipping into slope) produce the highest amplification (horizontal coefficient 3.55, vertical coefficient 1.4), compared to toppling joints (2.8, 1.14) and disordered joints (2.82, 1.31) in a 60° gradient slope [43]. Additionally, wave type generation differs based on input wave direction, with horizontal inputs generating dominant shear waves and vertical inputs producing prominent surface waves in jointed slopes, significantly affecting slope stability [43].

Discontinuities in rock masses significantly influence both wave propagation and slope stability during seismic events. When aligned with maximum shear stress directions, these pre-existing weakness planes can concentrate stress and reduce shear strength, potentially initiating landslides [47]. The dynamic opening and closing of joints during seismic activity introduces nonlinear behavior and energy dissipation mechanisms [45], which can lead to progressive damage accumulation and increased vulnerability to future seismic events [42]. These complex interactions underscore the limitations of traditional homogeneous continuum approaches in seismic stability analyses, as they may underestimate local amplification effects and stress concentrations [42].

3. Material Contrasts and Their Effects on Wave Propagation

Material contrasts within rocky slopes, arising from variations in lithology, weathering profiles, or water-saturated zones, significantly influence stress wave propagation and seismic slope stability [42]. Wave trapping phenomena in lower-velocity materials can produce substantial seismic wave amplification, with numerical models

5.3 Stress Wave Propagation in Complex Space Conditions   223

demonstrating amplification factors up to 7. The magnitude of these effects is strongly correlated with the velocity contrast ratio between stable and unstable materials, where a contrast ratio of 3 can result in displacement amplifications of 2–5.5 times compared to homogeneous slopes, depending on input motion characteristics [42].

Material contrasts in slopes generate complex wave refraction and reflection patterns at interfaces of varying seismic velocities. These interactions can focus energy and create constructive interference zones, resulting in localized high stress and strain regions that may initiate slope failure [45]. Low-velocity zones, particularly in weathered or fractured rock layers, can function as waveguides, channeling seismic energy along specific paths. This waveguide effect can produce unexpected ground motion amplification patterns and potentially trigger failures in otherwise seemingly stable regions [45].

Water saturation in rocky slopes creates significant material contrasts by altering bulk density and elastic properties, while water-filled fractures can induce complex pore pressure responses during seismic loading, potentially reducing effective stress and shear strength [41]. The impact of material contrasts is exemplified by the Ananevo rockslide site in Kyrgyzstan, where low-velocity weathered granitic rocks overlying competent bedrock led to significant ground motion amplification, particularly in the low-frequency domain [45]. Post-earthquake landslide distribution patterns frequently concentrate along geological boundaries and lithological contrasts, highlighting these zones' crucial role in focusing seismic energy and triggering slope failures [46].

4. **Wave Attenuation and Dispersion in Rocky Slopes**

Stress wave propagation in rocky slopes is characterized by two primary attenuation mechanisms: geometric spreading and material damping [41]. Geometric spreading causes wave amplitude reduction proportional to $1/r$ (where r is the distance from the source) for body waves and $1/\sqrt{r}$ for surface waves as the wavefront expands. Material damping occurs through energy conversion to heat via friction along grain boundaries and crystal structure dislocations in crystalline rocks, with additional mechanisms in sedimentary rocks including viscous losses in pore fluids [41].

Wave attenuation and dispersion effects are significantly influenced by wave frequency and rock mass characteristics. Higher-frequency waves experience more rapid attenuation due to increased scattering and absorption [41]. Dispersion, particularly pronounced in layered or heterogeneous slopes, causes separation of wave components with different frequencies due to velocity variations. Discontinuities and fractures enhance these effects, with attenuation increasing with higher fracture density and lower fracture stiffness [41, 43].

These wave propagation phenomena have crucial implications for slope stability assessment, particularly in earthquake-induced landslides. While high-frequency components attenuate rapidly, persistent lower-frequency waves may induce resonance effects, potentially increasing failure susceptibility [42]. Accurate modeling of these processes is essential for reliable seismic hazard assessments, as simplified approaches may lead to unreliable stability estimates [42]. This necessitates the use of advanced numerical modeling techniques that incorporate realistic attenuation and dispersion behaviors.

5. Seismic Wave Interaction with Slope Geometry

The interaction between seismic waves and rocky slopes involves various wave phenomena, including reflection, refraction, diffraction, and mode conversion [41, 42]. Wave reflection at material boundaries or free surfaces can generate both P and S waves through mode conversion, particularly at oblique incidence angles [41]. These reflections can create complex interference patterns and standing waves, potentially leading to localized ground motion amplification. Wave refraction at velocity boundaries and diffraction around sharp discontinuities further contribute to complex wave propagation patterns, with diffraction enabling wave propagation into shadow zones [41, 45].

Topographic amplification near slope crests and ridges is a well-documented phenomenon, primarily resulting from constructive interference of incoming and reflected waves [42]. Amplification factors typically range from 2 to 5, with maximum effects occurring when the ratio of slope height to predominant seismic wavelength (H/λ) falls between 0.2 and 0.8 [42, 43, 46]. This relationship between slope dimensions and seismic wavelength significantly influences the spatial distribution of earthquake-induced landslides, as demonstrated by studies of the 1999 Chi-Chi earthquake in Taiwan [46].

The generation of surface waves, particularly Rayleigh waves, along slope faces can significantly impact stability through large displacements [41]. The effectiveness of surface wave generation and propagation depends on slope angle, material properties, and incident wave characteristics. These complex wave-slope interactions highlight the limitations of traditional stability analysis methods and necessitate more advanced numerical modeling approaches [42, 43].

Recent developments in computational capabilities have enabled sophisticated three-dimensional modeling of wave-slope interactions, accounting for realistic geometry, material heterogeneity, and discontinuities [42, 45]. However, model accuracy heavily depends on input parameter quality and underlying assumptions. While these advanced models provide valuable insights into spatial and temporal ground motion variations, practical implementation often requires balancing model complexity with computational feasibility, particularly for regional-scale assessments [46].

6. Landslide Triggering Mechanisms Under Seismic Loading

The initiation of earthquake-induced landslides is primarily driven by inertial forces generated through seismic wave propagation, which produces cyclic accelerations and temporary increases in shear stress along potential failure surfaces. While threshold earthquake magnitudes for triggering landslides range from ML 4.0 for rock falls to MS 6.0–6.5 for large rock avalanches, these values are significantly influenced by local geological and topographic conditions. The spatial distribution of

5.3 Stress Wave Propagation in Complex Space Conditions

failures shows strong correlation with ground motion amplification patterns, particularly evident near slope crests, as demonstrated during the 1999 Chi-Chi earthquake [42, 45].

Multiple mechanisms contribute to landslide triggering under cyclic loading, including shear strength reduction through cementation breakdown, rock fragment reorientation, and pore pressure alterations. The phase relationship between dynamic normal and shear stresses plays a crucial role, with out-of-phase, high-magnitude stresses leading to disproportionate strain responses. Progressive damage accumulation across multiple seismic events can increase slope susceptibility through gradual weakening processes, particularly in the presence of groundwater, which can induce rapid pore pressure changes and potential liquefaction [41, 42, 47].

Earthquake-triggered landslides in rocky slopes manifest in various failure modes, including planar sliding along pre-existing discontinuities, wedge failure along intersecting discontinuity planes, toppling of rock columns, and complex failures involving multiple mechanisms. Research has identified distinct failure patterns in deep-seated rock slope instabilities, characterized by tensile failure in the uppermost 1–2 m and shear failure at depths of 3–10 m, with failure depths correlating to specific seismic parameters [41, 42, 46] (Fig. 5.13).

The understanding of these failure mechanisms is crucial for developing accurate hazard assessments and mitigation strategies, informing slope reinforcement design, monitoring system placement, and early warning system development [46].

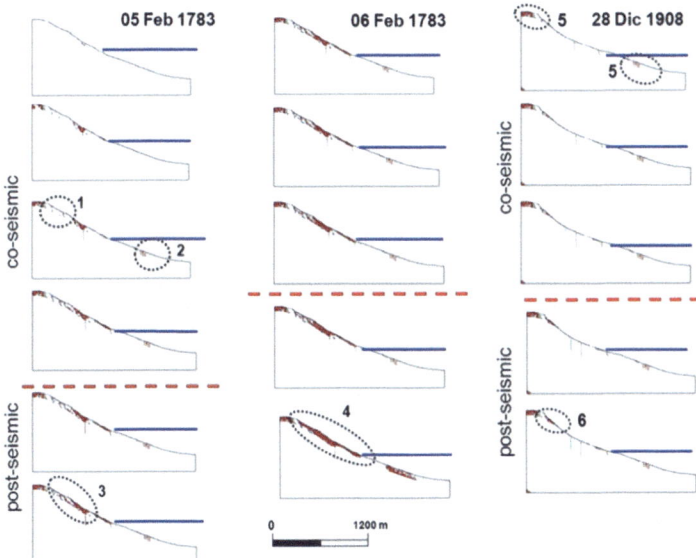

Fig. 5.13 Rupture model derived from the dynamic numerical modelling by FLAC 6.0. (1) Subvertical cracks; (2) crack propagation within the submarine slope; (3) opening of cracks along the crown area; (4) generalised collapse of the slope; (5) previously existing cracks; (6) falls in correspondence with the crown area. The blueline shows the present sea level [48]

Given the complexity of triggering processes and site-specific variability, probabilistic approaches are increasingly employed for regional-scale hazard assessment in seismically active areas [46].

7. **Wave-Induced Pore Pressure Effects**

The interaction between seismic waves and pore fluids in rocky slopes significantly influences slope stability and landslide triggering through wave-induced pore pressure effects that alter the rock mass's effective stress state [41]. In saturated or partially saturated rock masses, seismic waves can induce rapid pore pressure changes through multiple mechanisms: volumetric strain causing pore pressure fluctuations under undrained conditions, shear-induced dilation of fractures leading to localized pressure reductions, fluid flow redistribution particularly in fractured zones, and resonance effects when seismic wave frequencies match natural frequencies of fluid-filled fractures [41].

The magnitude and distribution of these wave-induced pore pressure changes are governed by various factors, including seismic wave characteristics (amplitude, frequency, duration), rock mass hydraulic properties (permeability, storage coefficient), degree of saturation, and initial pore pressure conditions [41]. These interactions can lead to reduced shear strength and increased failure susceptibility in rocky slopes.

5.3.2 Influence of Geological Structures on Wave Propagation

The propagation of stress waves in rock masses is significantly influenced by the presence of various geological structures. These structures, including faults, joints, layered formations, and complex topography, can dramatically alter the characteristics of seismic wave propagation. Understanding these influences is crucial for accurately predicting ground motion, assessing seismic hazards, and designing structures in seismically active regions. This section provides a comprehensive review of the effects of geological structures on stress wave propagation in rock masses, drawing from recent research and advanced modeling techniques.

1. **Fault Zones and Their Effects on Wave Propagation**

Fault zones are among the most prominent geological structures that affect seismic wave propagation. These structures typically consist of a low-velocity core surrounded by higher-velocity country rock, creating a waveguide effect that can trap and channel seismic energy. Recent studies have provided significant insights into the behavior of seismic waves in and around fault zones.

(1). Fault Zone Waves

Fault zone waves, including fault zone head waves (FZHWs), internally reflected waves, and fault zone trapped waves (FZTWs), are seismic waves that generate and

5.3 Stress Wave Propagation in Complex Space Conditions

propagate within fault zone low-velocity layers [49]. These waves can be modeled using analytical solutions for scalar waves in simplified fault zone structures, such as the two vertical fault zone layers between quarter-spaces model described by Ben-Zion [49]. The characteristics of fault zone waves are governed by several key parameters: the velocity contrast ratio between the fault zone and surrounding rock, attenuation coefficients (with low Q values <50 producing strong attenuation effects), fault zone layer widths influencing interference patterns, and source-receiver geometry affecting wave amplitude distribution [49]. Understanding and accurately modeling these waves is crucial for high-resolution fault zone structure imaging, improving the accuracy of velocity structure derivation, earthquake location determination, and fault plane solutions [49].

(2). Wave Attenuation and Reflection in Fault Zones

The presence of fault zones significantly affects the attenuation and reflection of seismic waves. A 3D numerical study investigated stress wave propagation characteristics in interacting fault systems under blast loading conditions [50]. The research revealed complex wave behaviors including stress concentration near fault zones due to refraction and reflection effects, discontinuous propagation between interacting faults with notable intensity variations at fault tips, and significant energy attenuation following a power function decay along fault lines. The study also identified the formation of high-acceleration regions between fault contact points and wave energy amplification through interactive fault areas due to stress wave superposition and reflection, providing crucial insights for blast design and risk assessment in geotechnical applications.

(3) Fault Properties and Their Influence on Wave Propagation

Research into stochastic seismic wave interactions with slippery rock faults revealed distinct wave-type dependencies: P-waves exhibit near-complete transmission with minimal reflection, while S-waves demonstrate significant attenuation and reflection patterns. S-wave behavior shows complex angle-dependent characteristics, with normal incidence promoting relative slip more than oblique incidence, and transmitted velocities capped at 0.037 m/s for normal incidence versus 0.05 m/s at 10° incidence. The study established that relative slip coefficients increase with impinging angle, while reflection coefficients display non-linear angle dependence, peaking at 24°, with loading intensity positively correlating to reflection coefficients within the 0–24° range [51]. These findings demonstrate the complex relationship between fault properties, wave characteristics, and the resulting wave propagation behavior. Understanding these relationships is crucial for accurately predicting seismic wave behavior in faulted rock masses.

(4) 3D Variations in Fault Zone Structure

Three-dimensional finite-difference investigations of irregular fault zone structures revealed hierarchical effects on guided waves: negligible impact from vertical property gradients and sub-wavelength variations, moderate influence from gradual shape variations and surface-ward widening, and significant effects from source positioning and lateral discontinuities exceeding fault zone width. The study demonstrated that split-fault configurations can support wave propagation in multiple segments when sourced below bifurcation points, indicating potential structural connectivity assessment applications [52]. These findings suggest that while 3D variations in fault zone structure can have significant effects on wave propagation, they may not be as resolvable from guided-wave data as previously thought. This highlights the need for careful additional observational and theoretical studies to obtain detailed knowledge of fault zone structure at depth [52].

2. **Influence of Joints and Fractures**

In addition to major fault structures, smaller-scale discontinuities such as joints and fractures can significantly affect stress wave propagation in rock masses. These features are common in most rock formations and can have substantial impacts on wave attenuation, reflection, and transmission.

(1). Wave Attenuation in Jointed Rock Masses

Investigation of blasting stress waves in layered and jointed rock caverns demonstrated significant wave attenuation by joints, with an average attenuation rate of 76.69% per joint interval. The study revealed inverse relationships between joint inclination and attenuation, while joint spacing exhibited positive correlation with attenuation rate but negative correlation with peak velocity [53]. These findings highlight the significant impact that joints and fractures can have on stress wave propagation in rock masses, particularly in the context of blasting operations.

(2). Wave Transmission and Reflection at Joints

The transmission and reflection of stress waves at joints are complex phenomena that depend on various factors, including joint properties and wave characteristics. 2D finite-difference elastic wave modeling was used to investigate these effects in the presence of complex surface topography [54]. A numerical scheme utilizing a fictitious surface layer successfully simulated various seismic wave types and their interface behaviors, with particular emphasis on Rayleigh waves. The study revealed significant influence of irregular bedrock topography (modeled sinusoidally) on surface wave propagation, demonstrating the critical importance of incorporating subsurface structural complexity in wave propagation models [54].

3. **Effects of Layered Formations**

Many geological settings consist of layered formations with varying physical properties. These layered structures can have significant effects on stress wave propagation, leading to complex wave behavior and ground motion patterns.

(1). Wave Propagation in Layered Media

Two-dimensional finite-difference elastic wave modeling in layered media demonstrated the complex interaction of P-waves, S-waves, converted waves, and Rayleigh waves at layer interfaces, including reflection, refraction, and mode conversion phenomena, with surface waves exhibiting frequency-dependent dispersive behavior [54].

5.3 Stress Wave Propagation in Complex Space Conditions

(2). Influence of Layer Properties on Wave Attenuation

A study examining seismic wave propagation characteristics in the Western Caucasus region demonstrated the significant influence of layer properties on wave attenuation [55]. The research revealed varying Q-factor estimates across different regions and identified distinct local variations in seismic radiation and propagation patterns associated with geological features including fault structures, thrusts, shatter zones, and sedimentary cover characteristics. Notable differences in attenuation patterns were observed between continental and subcontinental crustal sources, emphasizing the impact of large-scale geological structures on wave attenuation behavior.

(3). Guided Waves in Low-Velocity Layers

A study investigating guided wave behavior in low-velocity layers demonstrated their function as seismic waveguides [52]. The research revealed that these layers trap seismic energy, producing dispersive wave trains characteristic of the layer structure, with wave properties primarily controlled by the ratio of propagation distance to layer width. The study also showed that gradual variations in layer shape moderately affect trapped waves, while layer widening and merging with horizontal low-velocity layers results in spatially distributed regions of enhanced guided wave amplitudes, emphasizing the significant influence of low-velocity layers on seismic wave propagation in layered formations.

4. Influence of Complex Topography

Surface topography can have a significant impact on seismic wave propagation, particularly for surface waves and shallow seismic events. Recent studies have provided valuable insights into how complex topography affects wave propagation and ground motion.

(1). Effects of Surface Irregularities on Wave Propagation

2D finite-difference elastic wave modeling was employed to investigate the effects of irregular surface topography on seismic wave propagation [54]. Their study implemented an improved vacuum formulation for complex topography free-surface boundary conditions, ensuring zero stress components perpendicular to the surface; modeled sinusoidal surface topography with amplitude $0.5\lambda R$ and period $10\lambda R$ (where λR is the Rayleigh wavelength), demonstrating significant topographic effects on wave propagation; and developed a parameter averaging scheme introducing a fictitious layer above the surface to enable accurate simulation of Rayleigh waves with surface topography [54]. These findings highlight the importance of considering surface topography in seismic wave propagation models, particularly for near-surface phenomena and surface wave behavior (Fig. 5.14).

(2). Scattering of Seismic Waves by Topographic Features

Seismic wave scattering by topographic features can result in complex wave patterns and localized amplification or deamplification of ground motion. A study [56] investigated the scattering of anti-plane (SH) waves by a semi-elliptical hill on an elastic

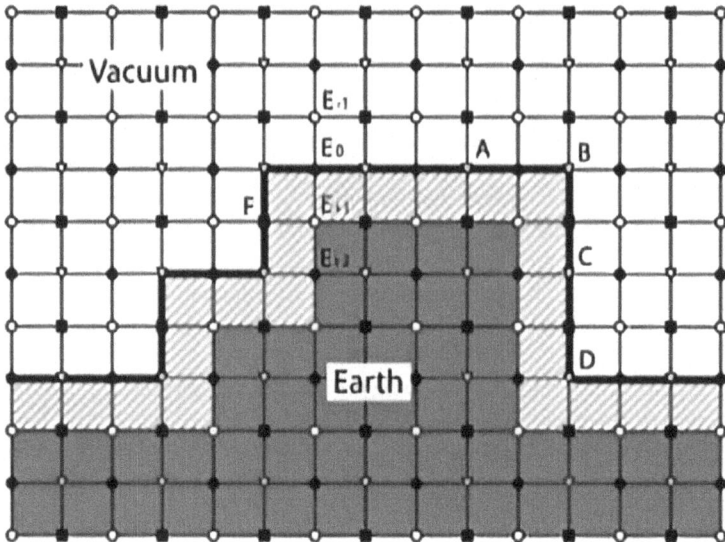

Fig. 5.14 Definition of the conditions in the grid using the improved vacuum method [54]

half-space, revealing that the hill's influence intensifies with increasing aspect ratio ($R = B/A$, where B is hill height and A is half-width); at grazing incidence angles, the hill acts as a barrier forming standing wave patterns; through the introduction of a dimensionless frequency parameter $\eta = 2A/\lambda$ (where λ is wavelength), they demonstrated that wave patterns become more complex at higher frequencies; and due to wave reflection and local concentrations, abrupt displacement amplitude jumps occur at the right edge of steeper hills]. These findings demonstrate the significant impact that topographic features can have on seismic wave propagation and highlight the need to consider such effects in seismic hazard assessments.

(3). Topographic Amplification of Ground Motion

Topographic features significantly influence ground motion amplification through multiple mechanisms, presenting crucial implications for seismic hazard assessment and engineering design. These mechanisms include the focusing of seismic waves by convex topographic features, constructive interference between reflected and diffracted waves, and resonance effects occurring when incident wave wavelengths match topographic feature dimensions. The amplification magnitude is controlled by various parameters, including topographic feature geometry, incident wave characteristics (frequency content and angle), and underlying rock mechanical properties. While existing studies [54, 56] have enhanced our understanding of topographic effects on wave propagation, comprehensive research remains necessary to fully quantify and predict topographic amplification in complex geological settings.

5.3 Stress Wave Propagation in Complex Space Conditions 231

5. **Combined Effects of Geological Structures**

In real geological settings, multiple types of structures often coexist, leading to complex interactions in their effects on seismic wave propagation. Understanding these combined effects is crucial for accurate prediction of ground motion and seismic hazard assessment.

(1). Interaction Between Faults and Layered Structures

Wave propagation analysis in crossing-fault tunnels revealed complex interactions between fault zones and layered geological structures, with faults serving as primary seismic energy transmission channels. The layered structure of surrounding rock significantly influences wave propagation and attenuation patterns, with the crossing-fault section experiencing maximum dynamic amplification and heightened vulnerability to earthquake damage. Modal and Fourier spectrum analyses demonstrated frequency-dependent deformation patterns, where low-order frequencies induced overall structural deformation, while higher-order frequencies caused localized deformation in the crossing-fault section. Additionally, while tunnels minimally impacted spectral characteristics in intact rock, they significantly amplified spectral peaks within fault zones, with peak amplitudes maximizing at the crossing-fault section and decreasing with distance [57].

(2). Effects of Joints and Topography on Wave Propagation

Research examining the combined effects of joints and surface topography on seismic wave propagation reveals more complex wave behavior patterns than when these factors are considered independently [53, 54, 56]. The interaction between joint-induced wave scattering and attenuation, coupled with topographic focusing and amplification, creates intricate ground motion patterns, while the combination of joint-induced anisotropic wave propagation and topographic effects produces directional variations in ground motion that exceed the complexity of either factor alone. The frequency-dependent nature of both joint and topographic effects on wave propagation results in complex amplification and attenuation patterns, particularly pronounced in the near-surface region, necessitating further research to fully quantify and model these combined effects in realistic geological settings.

(3). Complex Wave Propagation in Fault-Bounded Sedimentary Basins

Fault-bounded sedimentary basins represent complex geological environments where the interaction of multiple structural features, including fault zones, layered media, and varying topography, creates sophisticated wave propagation patterns. In these settings, fault zones along basin margins can both generate and trap seismic waves, potentially amplifying ground motion near the edges, while the layered sedimentary structure can induce resonance effects, resulting in seismic energy trapping and amplification within the basin. The interaction between basin sediments and surrounding bedrock generates significant surface waves, particularly Love waves, while the three-dimensional basin geometry creates complex wave propagation patterns through focusing and defocusing of seismic energy. This comprehensive understanding of wave behavior in fault-bounded basins is crucial for accurate seismic hazard assessment in seismically active urban regions.

6. **Implications for Seismic Hazard Assessment and Engineering**

The influence of geological structures on seismic wave propagation significantly impacts seismic hazard assessment, as demonstrated by multiple studies [50–57]. These structures, including fault zones, joint patterns, layered formations, and surface topography, create localized ground motion amplification zones that necessitate modification of standard ground motion prediction equations. This variability needs to be incorporated into probabilistic seismic hazard analyses to ensure accurate assessment of potential ground motions in complex geological settings.

The presence of geological structures has profound implications for earthquake-resistant design and infrastructure development. Complex geological settings often require site-specific design spectra for critical structures, particularly affecting foundation and underground structure design. Detailed microzonation studies incorporating local geological structural effects provide essential input for urban planning and building code development in seismically active regions, ensuring more effective seismic risk mitigation.

The understanding of wave propagation effects in geological structures significantly enhances the effectiveness of seismic monitoring and early warning systems through optimal sensor placement and improved real-time ground motion predictions. However, despite significant advances, current studies [50–57] primarily utilize 2D models or simplified 3D representations, highlighting the need for developing comprehensive 3D models that incorporate complex geological structures to improve our ability to predict and mitigate seismic hazards in real-world settings.

5.3.3 Stress Waves in Geothermal and Mining Applications

Stress waves in geothermal and mining applications play a crucial role in understanding and managing the complex subsurface environments associated with these industries. The propagation of stress waves in these settings is influenced by a myriad of factors, including rock properties, in-situ stress conditions, fluid presence, and anthropogenic activities. This section explores the various aspects of stress wave propagation in geothermal and mining applications, highlighting recent research findings and their implications for resource extraction, safety, and environmental management.

1. **Geothermal Applications**

In geothermal systems, stress wave propagation is of particular interest for several reasons. First, it provides valuable insights into the subsurface structure and properties of geothermal reservoirs. Second, it allows for the monitoring of reservoir

5.3 Stress Wave Propagation in Complex Space Conditions

changes during exploitation. Third, it helps in assessing the potential risks associated with geothermal activities, such as induced seismicity.

(1). Characterization of Geothermal Reservoirs

One of the primary applications of stress wave analysis in geothermal systems is the characterization of reservoir properties. This includes understanding the fracture network, fluid content, and stress conditions within the reservoir. Several techniques have been developed to extract this information from stress wave data.

(1) Shear-Wave Splitting Analysis

Shear-wave splitting analysis has emerged as an effective non-invasive technique for investigating fracture systems in geothermal reservoirs, based on the principle that shear waves propagating through anisotropic fractured rock split into two components with distinct velocities and polarizations, where the faster shear wave's polarization direction typically aligns with the dominant fracture set's strike, and the time delay between components correlates with fracture density. This method offers significant advantages for geothermal applications, including the ability to assess fracture orientation and density around wells, monitor temporal changes in subsurface conditions related to production activities or natural processes, and support geothermal resource assessment and waste disposal planning. However, interpretation challenges arise from complex fracture networks, stress-induced anisotropy, and multiple anisotropic layers, necessitating integration with other geophysical and geological data for comprehensive reservoir characterization [58].

(2) P-Wave Anisotropy Analysis

P-wave anisotropy analysis was employed as a complementary method to shear-wave splitting for investigating stress states and crack distributions in geothermal reservoirs, utilizing a novel inversion method developed at the Coso geothermal field that represents direction-dependent P-wave velocity through a symmetric positive definite matrix A. The comprehensive study analyzed traveltime data from 2,104 microseismic events encompassing 17,758 P-wave picks across 29 stations, revealing a circular dome-like structure in the southwestern geothermal region with deviatoric stress of 3–6 MPa at production depths and distinct compressional NNE-SSW and dilational WNW-ESE stress fields. The spatial distribution analysis showed significant variations in anisotropy, with maximum values exceeding 8.0% at specific locations, while fast directions exhibited depth-dependent orientation patterns, transitioning from N-S at shallow depths to easterly orientations in deeper regions, enabling the estimation of differential stress and permeability distribution that correlated with areas of dense seismicity [59].

(3) Microseismic Monitoring

Microseismic monitoring serves as a vital tool for geothermal reservoir characterization, enabling analysis of spatial and temporal distributions of microseismic events and their source characteristics to understand stress states, fluid flow patterns, and induced seismicity potential. A comprehensive study of the Basel enhanced geothermal system (EGS) demonstrated that microseismic cloud expansion occurred perpendicular to the minimum principal stress direction, with cloud shape influenced by intermediate principal stress while maintaining stable growth orientation aligned with maximum horizontal stress. The research revealed significant correlations between microseismic cloud aspect ratios and in-situ effective stress ratios, with the estimated apparent permeability tensor showing alignment with observed cloud growth patterns [60].

Another significant application of microseismic monitoring was demonstrated at the Kakkonda geothermal field in Northeast Japan, where researchers investigated potential supercritical geothermal resources using multiple seismic analysis techniques [61]. The study employed precise hypocenter determination, seismic wave tomography, non-double couple component estimation, and pre-stack depth migration imaging. Key findings revealed seismic events predominantly above -3000 m a.s.l., with a low V_p/V_s structure ($V_p/V_s \sim 1.6$) around the shallower seismic cloud and a transition zone to medium–high V_p/V_s s (>1.75) near the deeper cloud. Strong reflectors were observed in the low-resistivity anomaly above -2000 m a.s.l., potentially indicating precipitated silica or Kakkonda granite presence [61].

(2). Monitoring Reservoir Changes

Stress wave analysis plays a crucial role in monitoring changes in geothermal reservoirs over time. These changes can be related to natural processes, such as fluid migration and tectonic activity, or anthropogenic factors like fluid injection and extraction. Monitoring these changes is essential for optimizing reservoir management and ensuring the long-term sustainability of geothermal resources.

(1) Temporal Variations in Seismic Properties

At the Coso geothermal field, analysis of temporal variations in shear-wave splitting parameters over a five-year period (1996–2000) revealed stable fast shear wave polarization directions, indicating consistent fracture orientation, while significant variations in time delays (δt) were observed in 1999 across multiple stations. Notable changes included increased prominence of N10E polarization at stations CE1 and CE4 in 1999, with recovery of longer time delays in 2000 limited to a N10E-striking zone. These temporal variations were attributed to either episodic hydrothermal venting, involving thermo-elastic sealing and reopening of secondary cracks during fluid upwelling through a major NNE-trending fracture conduit, or stress-induced changes related to a M5.2 earthquake sequence in March 1998, where stress loading and subsequent relaxation affected crack properties, demonstrating the effectiveness of shear-wave splitting analysis for monitoring both strain changes preceding earthquakes and geothermal production effects [58].

(2) Microseismic Monitoring During Stimulation

At the Basel EGS, researchers implemented an innovative multiple sensor multiplet analysis method during microseismic monitoring to enhance event location resolution, utilizing data from multiple receivers and cross-correlation techniques for refined shear-wave traveltime picks, which significantly improved relative location accuracy within multiplets from 70 to 26 m. The enhanced resolution analysis revealed distinct fracture planes with diverse orientations and reduced the initial seismically active volume estimation by 15%, while also identifying temporal variations in multiplet activity patterns. This high-resolution analysis of spatial multiplet distribution demonstrated complex fracture network evolution during stimulation, highlighting the importance of advanced microseismic analysis techniques for optimizing stimulation procedures and understanding the long-term behavior of enhanced geothermal systems [4].

(3) Stress Wave Propagation Modeling

A numerical investigation employing finite-difference method was conducted to study elastic wave propagation in two-phase vapor-dominated geothermal reservoirs [62]. The model utilized an explicit finite difference scheme with staggered grid method (second-order accuracy in space and time), implementing a 1 km^3 computational domain with 5 m grid spacing, 0.8 ms time step, and absorbing boundaries except at the top surface. For the reservoir configuration, dimensions of 250 m width and depth with 125 m thickness were set at 250 m below the model top. The results demonstrated that increased water grid points in the reservoir led to enhanced amplitude and phase delays in waveforms, with significant variations observed after 0.5 s between models with and without water content. Notably, a model incorporating 3.4% connected water grid points exhibited distinct wave propagation characteristics, while snapshots at 0.432 s revealed substantial differences near the reservoir region, indicating the method's potential for detecting reservoir property changes and its value in field data interpretation and monitoring strategy design.

(3). Induced Seismicity in Geothermal Systems

The assessment and management of induced seismicity is a critical aspect of geothermal energy development, particularly for EGS that involve hydraulic stimulation of low-permeability reservoirs. Stress wave analysis plays a crucial role in understanding the mechanisms of induced seismicity and developing strategies to mitigate associated risks.

(1) Characteristics of Induced Seismicity

A seismic monitoring study during the drilling of the Venelle 2 geothermal well in the Larderello-Travale geothermal field (LTGF), Italy, identified two distinct types of seismic events [63]. Type-1 events exhibited clear P and S wave arrivals with broad frequency content, magnitudes from −1 to 2.3, and formed clusters aligned with fault systems at depths of 1–4.5 km, indicating natural tectonic origins. In contrast, Type-2 events displayed emergent waveforms within a 7–16 Hz frequency band, occurred at shallower depths (<2 km), and showed high waveform similarity within swarms, suggesting correlation with drilling activities. The study revealed that while Type-2 events preceded circulation loss and were likely induced by drilling-related fluid movements, the absence of large magnitude induced earthquakes demonstrated the feasibility of drilling into supercritical conditions without triggering significant induced seismicity.

(2) Mechanisms of Induced Seismicity

Understanding the mechanisms of induced seismicity is crucial for developing effective mitigation strategies. A microseismic monitoring study of the Basel EGS project in Switzerland analyzed the correlation between hydraulic stimulation and induced seismicity during the injection of 11,500 m^3 water [64]. The study documented over 13,000 microseismic events, revealing strong correlations between seismic activity and injection parameters (reaching maximum pressure of 29.6 MPa and flow rate of 3300 L/min). Spatial and temporal analysis showed most events occurred above −3000 m sea level, with deeper events concentrated in western areas, while the seismically active volume expanded with continued injection. The research demonstrated that induced seismicity was primarily controlled by pore pressure increases and stress perturbations from fluid injection, suggesting the potential for seismicity management through controlled injection operations.

(3) Risk Assessment and Mitigation

Assessing and mitigating the risks associated with induced seismicity is crucial for the sustainable development of geothermal resources. Research from the Basel EGS project provided significant insights into induced seismicity risk assessment and mitigation [64]. Statistical analysis revealed a typical b-value of 1.3 for fluid-induced seismicity, with temporal variations observed and maximum predicted magnitudes ranging from 1.98 to 3.0. Complementary findings from the Larderello-Travale geothermal field study [63] demonstrated the feasibility of drilling into near-supercritical conditions without triggering significant seismic events. These studies emphasize the effectiveness of comprehensive risk mitigation strategies, including careful site selection, implementation of traffic light systems, gradual injection rate adjustment, continuous microseismic monitoring, and public engagement, while highlighting the potential for managing induced seismicity through adaptive injection strategies based on real-time seismic monitoring.

2. **Mining Applications**

In the mining industry, stress wave analysis plays a crucial role in understanding rock mass behavior, optimizing extraction processes, and ensuring mine safety. The propagation of stress waves in mining environments is influenced by factors such as in-situ stress conditions, rock properties, mining-induced stress changes, and the presence of discontinuities like fractures and faults.

(1). Characterization of Rock Mass Properties

Stress wave analysis provides valuable tools for characterizing rock mass properties in mining environments. These properties are essential for designing safe and efficient mining operations, as well as for predicting and managing potential geotechnical hazards.

(1) Seismic Velocity Structure

5.3 Stress Wave Propagation in Complex Space Conditions

A comprehensive microseismic monitoring study [65] was conducted at a limestone mine to investigate seismic velocity structure and its implications for rock mass characterization. Through analysis of 157 microseismic events coinciding with a major roof fall, researchers implemented optimized monitoring strategies, including enhanced sensor array geometry and advanced data processing techniques with a Simplex algorithm-based source location method. The study achieved notable location accuracy of 7.6 m, with event clusters concentrated within a 61 m × 244 m zone that strongly correlated with observed roof falls. These findings demonstrated the effectiveness of microseismic monitoring in mapping spatial distributions of seismic events and evaluating rock mass properties in mining environments.

(2) Anisotropy Analysis

The anisotropy analysis techniques developed for geothermal applications demonstrate significant potential for adaptation to mining environments. The P-wave anisotropy analysis method, originally developed for the Coso geothermal field [59], employs a symmetric positive definite matrix to represent direction-dependent P-wave velocity, offering valuable insights into stress distributions and fracture patterns in rock masses. Complementarily, shear-wave splitting analysis techniques [58] can be applied to mining microseismic data to characterize fracture orientations and densities, providing crucial information for assessing mine opening stability and predicting potential failure zones.

(3) Stress State Estimation

Stress wave analysis offers multiple approaches for estimating in-situ stress states, which is essential for mine design and hazard assessment. These methods include focal mechanism analysis of microseismic events for determining local stress field orientation, analysis of mining-induced seismicity distribution patterns for identifying high stress concentration zones, and anisotropy analysis for understanding stress-induced variations in rock properties. Together, these complementary techniques provide comprehensive insights into the stress conditions within rock masses.

A microseismic monitoring study at China's Hongtoushan copper mine [66] investigated stress distributions in deep mining conditions through analysis of key microseismic parameters, including energy index (EI), apparent volume (VA), and b-value. The research revealed progressive stress concentration at the goaf roof during mining operations, with spatial variations in stress-deformation characteristics indicating diverse potential hazards. The findings demonstrated that microseismic monitoring parameters effectively provide real-time insights into stress distributions, contributing to mining safety and operational efficiency.

(2). Blast-Induced Stress Wave Propagation

Understanding blast-induced stress wave propagation is essential for optimizing blasting operations in mining and minimizing unwanted damage to the surrounding rock mass. Several studies have investigated various aspects of blast-induced stress waves in mining contexts.

(1) Stress Wave Interaction Between Adjacent Blast Holes

Stress wave interactions between adjacent blast holes was examined through analytical modeling and numerical simulation approaches [67]. The study revealed that significant stress wave interaction occurs within a narrow delay time window (0–4 ms), with tensile stress enhancement confined to a limited interaction zone (less than 4 times the hole radius). Their numerical findings challenged the hypothesis that very short delay times improve rock fragmentation through stress wave interaction effects. These findings have important implications for blast design in mining operations, suggesting that attempting to achieve stress wave superposition through very short delay times may not be an effective strategy for improving fragmentation.

(2) Effect of Geological Structures on Blast-Induced Stress Waves

The effects of geological discontinuities on delayed blasting performance were investigated through numerical modeling and field experiments [66]. Their research revealed that joints and fractures create asymmetric stress fields, with peak effective stress reaching 10–12 MPa before decay. The study established optimal delay timing of 3–5 ms between upper and lower charges for highwall bench blasting, while highlighting the significance of stress wave attenuation and sub-layer shock wave supplementation in rock fragmentation mechanisms. These findings provide valuable insights for optimizing blast design in mining operations with complex geological structures, highlighting the importance of considering joint and fracture patterns in determining optimal delay times and charge configurations.

(3) Stress Wave Attenuation in Mining Environments

Explosion wave interactions and rock breakage were examined during dual initiation blasting [68]. Their research demonstrated that shock wave velocity in the collision zone exceeds radial shock wave velocity, with waves propagating at an average angle of 60°. The study revealed that superimposed stress on the blast hole wall reached 1.73 times the single explosion wave incident stress under experimental conditions.

5.4 Stress Wave Propagation in Different Medium

5.4.1 Stress Wave Behavior in Saturated and Unsaturated Rocks

The propagation of stress waves in saturated and unsaturated rocks is a complex phenomenon that has significant implications for various fields, including seismology, petroleum engineering, and civil engineering. Understanding the behavior of stress waves in these media is crucial for accurately interpreting seismic data, characterizing reservoirs, and assessing the stability of underground structures. This section delves into the intricate dynamics of stress wave propagation in saturated and unsaturated rocks, exploring the various factors that influence wave behavior and the theoretical models developed to describe these processes.

1. **Theoretical Foundations**

The theoretical foundation for understanding stress wave propagation in saturated and unsaturated rocks is largely based on Biot's theory of poroelasticity, which provides a comprehensive framework for analyzing wave behavior in fluid-saturated porous media. Initially developed for low-frequency waves and later extended to higher frequencies through the introduction of a complex viscosity correction factor, Biot's theory predicts three distinct wave types in fluid-saturated porous media: one rotational wave and two dilatational waves, enabling more accurate representation of wave behavior across broad frequency ranges [69].

For rotational waves, the phase velocity (v_r), group velocity (U_r), and attenuation (α_r) are given by:

$$v_r = \sqrt{\frac{N}{\rho}} \times \sqrt{1 - \frac{i\omega_f}{\omega}} \tag{5.54}$$

$$U_r = v_r \times \frac{1 + \frac{\omega_f}{2\omega}}{1 + \left(\frac{\omega_f}{2\omega}\right)^2} \tag{5.55}$$

$$\alpha_r = \frac{\omega}{v_r} \times \frac{\frac{\omega_f}{2\omega}}{1 + \left(\frac{\omega_f}{2\omega}\right)^2} \tag{5.56}$$

where N is the shear modulus of the dry frame. ρ is the bulk density, ω is the angular frequency, and ω_f is the characteristic frequency related to fluid flow [70].

For dilatational waves, Biot's theory predicts two types: waves of the first kind (fast P-waves) and waves of the second kind (slow P-waves). The behavior of these waves is more complex and depends on various parameters of the porous medium and the saturating fluid [69].

Building upon Biot's theory, several models have been developed to address specific aspects of wave propagation in partially saturated rocks. One such model is White's model, which focuses on the effects of patchy saturation on wave propagation [70]. White's model assumes spherical gas pockets much larger than grains but smaller than the wavelength, and predicts a relaxation mechanism due to pressure equilibration between gas pockets and surrounding water-saturated rock. The model provides expressions for complex P-wave velocity as a function of frequency, permeability, porosity, and fluid properties [70].

Another important theoretical framework is the Gassmann-Biot model, which relates the elastic properties of fluid-saturated rocks to those of dry porous rocks [71]. This model is particularly useful for understanding the low-frequency behavior of saturated rocks. The Gassmann equation expresses the undrained static bulk modulus of the porous medium as a function of the dry frame and saturating fluid properties:

$$K_{\text{sat}} = K_d + \alpha^2 M \tag{5.57}$$

where K_{sat} is the saturated bulk modulus, K_d is the dry frame bulk modulus, α is the Biot-Willis coefficient, and M is a poroelastic modulus related to fluid and solid properties [71].

2. **Velocity Dispersion and Attenuation**

Stress wave propagation through saturated and unsaturated rocks exhibits velocity dispersion, where wave velocity varies with frequency, closely related to wave attenuation, both indicating energy dissipation within the rock-fluid system. In fully saturated rocks, Biot's theory [69] explains that the interplay between inertial and viscous forces governs velocity dispersion and attenuation. At low frequencies, viscous forces dominate, causing the wave velocity to approach the Gassmann-Biot low-frequency limit, while at higher frequencies, inertial forces become significant, increasing wave velocity and resulting in an attenuation peak [71].

The situation becomes more complex in partially saturated rocks due to multiple fluid phases. White's model predicts a relaxation mechanism caused by pressure equilibration between gas pockets and surrounding water-saturated rock, leading to significant velocity dispersion and attenuation at a characteristic frequency related to gas pocket size and rock properties [70]. Experimental research by Cadoret et al. [72] on limestone samples revealed that P-wave velocity initially decreases with increasing water saturation at seismic frequencies but increases at ultrasonic frequencies, reflecting the transition from Gassmann-Wood (low-frequency) to Gassmann-Hill (high-frequency) predictions.

Fluid mobility, influenced by rock permeability and fluid viscosity, strongly impacts velocity dispersion and attenuation in partially saturated rocks [73]. High permeability samples saturated with low-viscosity fluids like water show minimal dispersion at seismic frequencies, whereas low permeability samples saturated with high-viscosity fluids like glycerin exhibit strong dispersion. This behavior is associated with the characteristic frequency for fluid pressure equilibration, which shifts to lower frequencies as fluid mobility decreases.

Furthermore, the relationship between attenuation and saturation is complex, with attenuation peaks often occurring at intermediate saturation levels around 50–80% water saturation [70, 74]. This is attributed to optimal conditions for wave-induced fluid flow (WIFF) between regions of different compressibility. The exact saturation level at which peak attenuation occurs depends on factors such as frequency, rock properties, and fluid distribution [74].

3. **Fluid Distribution Effects**

The spatial distribution of fluids within pore spaces significantly influences stress wave behavior in partially saturated rocks. Two primary fluid distribution types are considered: homogeneous saturation and patchy saturation. In homogeneous saturation, fluids are uniformly mixed at the pore scale, typically modeled using the Gassmann-Wood approach, which employs an effective fluid bulk modulus based on the volume-weighted harmonic average of individual fluid bulk moduli [71]. This model predicts low velocities and minimal dispersion at low frequencies.

In contrast, patchy saturation involves discrete regions of different fluids larger than pore size but smaller than wavelength, leading to significant velocity dispersion and attenuation due to wave-induced fluid flow between patches [70]. The Gassmann-Hill model describes the high-frequency limit for patchy saturation, where pressure equilibration between patches during wave passage is insufficient [71]. Experimental studies, such as those by Knight and Nolen-Hoeksema [75] on sandstone samples, have shown that different saturation methods result in distinct velocity-saturation relationships attributed to variations in fluid distribution patterns.

Field-scale observations, like those from time-lapse seismic surveys in hydrocarbon reservoirs, also provide evidence of patchy saturation effects [76]. These observations emphasize the importance of considering fluid distribution when interpreting seismic data for reservoir characterization and monitoring.

Recent theoretical advancements aim to offer more realistic models of fluid distribution and its impact on wave propagation. Fractal models, for instance, describe the spatial distribution of fluid patches [77, 78], recognizing self-similarity across scales with significant implications for wave behavior. A model for P-wave attenuation and velocity dispersion in partially saturated porous media with fluid patches forming random fractals was introduced [77]. This model uses the von Karman function to characterize spatial correlation and the Hurst exponent to control anisotropic scaling. The study found that the velocity-saturation relation can vary between Gassmann-Wood and Gassmann-Hill bounds depending on the Hurst exponent [77].

This fractal approach provides a nuanced understanding of wave behavior in partially saturated rocks, explaining why field experiments often show velocity-saturation relationships closer to the Gassmann-Wood bound despite fluid patch sizes suggesting proximity to the Gassmann-Hill bound. This behavior is attributed to the fractal nature of fluid distribution, where the high-frequency regime is never fully realized, and pore pressure equilibration occurs at all length scales [77].

4. **Influence of Rock Properties**

The behavior of stress waves in saturated and unsaturated rocks is strongly influenced by the intrinsic properties of the rock matrix. Factors such as porosity, permeability, grain size, and elastic moduli of the solid frame all play significant roles in determining wave velocities, attenuation, and dispersion characteristics.

Porosity is a fundamental property that affects wave propagation in porous media. Higher porosity generally leads to lower wave velocities due to the increased influence of the more compliant pore fluid on the overall rock stiffness [71].

Permeability is another critical factor, particularly in the context of wave-induced fluid flow mechanisms. Higher permeability allows for easier fluid movement within the pore space, which can enhance attenuation at certain frequencies [73]. The study on seismic attenuation due to heterogeneities of rock fabric and fluid distribution demonstrated that the characteristic frequency for maximum attenuation scales with permeability. This relationship highlights the importance of permeability in determining the frequency range over which WIFF mechanisms are most effective.

In addition to these fundamental properties, the presence of fractures or microcracks in the rock can significantly affect wave propagation. Fractures introduce additional compliance to the rock frame and can create preferential pathways for fluid flow. This can lead to anisotropic wave behavior and frequency-dependent effects due to squirt flow mechanisms [79].

The influence of rock properties on wave behavior is not static but can evolve over time due to various processes. For instance, stress-induced changes in pore structure or chemical interactions between fluids and the rock matrix can alter the rock's elastic properties and fluid flow characteristics. These dynamic effects are particularly important in the context of time-lapse seismic monitoring of reservoirs undergoing production or injection [76].

5. Frequency-Dependent Behavior

The frequency dependence of stress wave behavior in saturated and unsaturated rocks is a key aspect that distinguishes these media from simple elastic solids. This frequency dependence manifests in various ways, including velocity dispersion, attenuation spectra, and changes in the dominant attenuation mechanisms across different frequency ranges.

At very low frequencies, corresponding to the seismic exploration band (typically <100 Hz), the behavior of saturated rocks is often well-described by the Gassmann-Biot theory [71]. In this regime, there is sufficient time for pore pressure equilibration to occur during wave passage, leading to a relaxed state where the rock behaves as if it were saturated with a homogeneous fluid. For partially saturated rocks, this low-frequency limit is often approximated by the Gassmann-Wood model, which assumes a uniform mixture of fluids at the pore scale [71].

As frequency increases, various relaxation mechanisms come into play, leading to velocity dispersion and attenuation. One of the primary mechanisms in this intermediate frequency range is wave-induced fluid flow (WIFF) between regions of different compressibility [63]. This can occur at various scales, from the pore scale (squirt flow) to the mesoscopic scale (patchy saturation effects).

The frequency-dependent behavior of stress waves in rocks exhibits distinct characteristics across different frequency ranges through several key mechanisms. At ultrasonic frequencies (>100 kHz), wave propagation is primarily governed by two phenomena: scattering from heterogeneities and inertial effects described by Biot's theory [69]. In this range, saturated rocks approach either the Gassmann-Hill limit or Biot's high-frequency limit [71]. The attenuation patterns follow a characteristic curve, showing an increase at low frequencies, reaching a peak at intermediate frequencies, before decreasing at higher frequencies [70, 74]. For high-frequency behavior, the continuous random media (CRM) model predicts attenuation scaling as $Q^{-1} \propto \omega^{-\nu}$, where the exponent is related to the fluid distribution's fractal dimension [77]. These frequency-dependent characteristics are crucial for bridging the gap between laboratory measurements at ultrasonic frequencies and lower-frequency field seismic data [73].

6. Experimental Observations and Challenges

Experimental studies have advanced our understanding of stress wave behavior in porous media through four key investigations. Research [75] revealed that sandstone exhibits distinct velocity-saturation relationships between imbibition and drainage processes, attributed to variations in fluid distribution patterns. Studies [72] documented frequency-dependent P-wave velocity behavior in limestone, demonstrating a transition from Gassmann-Wood to Gassmann-Hill predictions as frequency increases. Investigation [80] discovered that carbonate rocks show dominant bulk compressibility losses over shear wave losses, contrasting with typical sandstone behavior. Research [73] established that fluid mobility significantly influences velocity dispersion and attenuation in partially saturated sandstones, identifying permeability and fluid viscosity as critical controlling factors.

Experimental investigations of wave propagation in saturated and unsaturated rocks encounter several significant challenges, including scale effects in laboratory samples [76], frequency range limitations [74], difficulties in controlling fluid distribution [72], and complexities in isolating individual attenuation mechanisms [74]. Additional challenges involve maintaining experimental repeatability due to sample alterations during saturation cycles [80] and controlling environmental conditions [80]. Nevertheless, recent advances in X-ray computed tomography for fluid distribution characterization [76] and broader frequency range measurement capabilities [73] are advancing our understanding of stress wave behavior in these materials.

7. Numerical Modeling Approaches

Numerical modeling has become an increasingly important tool for studying stress wave propagation in saturated and unsaturated rocks. These approaches allow for the simulation of complex scenarios that may be difficult or impossible to replicate in laboratory experiments. They also provide a means of testing theoretical models and exploring the interplay between various physical mechanisms affecting wave behavior.

One of the primary numerical approaches used in this field is based on solving the poroelastic equations derived from Biot's theory [69]. These simulations typically employ finite difference or finite element methods to discretize the governing equations in both space and time [70]. Numerical modeling has advanced to address wave propagation in fractured porous media, a double double-porosity model that integrates fabric heterogeneity and patchy fluid saturation was developed. Recent computational advances have enabled sophisticated multi-physics simulations that simultaneously model fluid flow, geomechanical deformation, and chemical reactions, enhancing understanding of wave behavior in complex geological environments [76].

Numerical modeling approaches offer significant advantages in studying wave propagation, including flexibility in simulating complex scenarios, precise control over physical mechanisms, capability to bridge multiple scales, and efficient parameter space exploration. However, these methods face notable challenges such as computational costs for high-resolution simulations, validation requirements, upscaling difficulties in heterogeneous media, and intensive uncertainty quantification. Despite these limitations, numerical modeling remains a vital tool in advancing our understanding of stress wave propagation in saturated and unsaturated rocks, with continuing improvements in computational capabilities and algorithmic sophistication enhancing its utility for studying complex wave phenomena in porous media.

Understanding stress wave behavior in saturated and unsaturated rocks is crucial for geophysical exploration, reservoir characterization, and geotechnical engineering. The integration of theoretical models, experimental studies, and numerical simulations enhances our understanding of these complex phenomena. In seismic exploration and reservoir characterization, this knowledge is essential for fluid detection and characterization, time-lapse monitoring [76], velocity model improvement [73], and AVO (Amplitude Versus Offset) analysis [71]. For geotechnical engineering applications, it facilitates site characterization, earthquake response assessment [69], and underground construction design [81].

5.4.2 Stress Wave Behavior in Anisotropic Rocks

Anisotropy in rocks, characterized by directional dependence of physical properties, significantly influences the propagation of stress waves. This section delves into the complex behavior of stress waves in anisotropic rocks, exploring the theoretical foundations, experimental observations, and implications for various geophysical and engineering applications.

1. **Fundamentals of Rock Anisotropy**

Rock anisotropy arises from various sources, including preferred orientation of minerals, layered structures, aligned cracks, and stress-induced effects. Understanding these sources is crucial for interpreting stress wave behavior in anisotropic media.

(1). Sources of Anisotropy

Rock anisotropy can originate from four primary mechanisms: crystallographic preferred orientation in metamorphic rocks due to mineral alignment during deformation; sedimentary layering caused by alternating materials or grain sizes; aligned microcracks and pores, particularly significant under low confining pressures; and stress-induced preferential crack closure, which can transform initially isotropic rocks into anisotropic ones [82].

(2). Classification of Anisotropy

Anisotropic media are categorized into distinct symmetry systems, ranging from isotropic materials (exemplified by unstressed granite) with uniform properties in all directions, to transversely isotropic media (such as layered sedimentary rocks) exhibiting axial symmetry, orthorhombic systems (common in metamorphic rocks) characterized by three orthogonal symmetry planes, monoclinic systems (typical of sheared rocks) with a single symmetry plane, and triclinic systems representing the

5.4 Stress Wave Propagation in Different Medium

most complex case with no symmetry planes, though rarely encountered in practical applications.

2. **Experimental Techniques for Measuring Anisotropy**

Various experimental methods have been developed to measure and characterize anisotropy in rock samples. These techniques provide crucial data for understanding stress wave behavior in anisotropic rocks.

(1). Ultrasonic Velocity Measurements

A sophisticated laboratory methodology [83] enables comprehensive anisotropic characterization through P-wave velocity measurements in spherical rock samples. The system utilizes 50 mm diameter specimens to measure velocities across 132 independent directions at 15° spherical coordinate intervals under hydrostatic pressures up to 400 MPa. This technique, through complete ultrasonic wavelet capture, provides detailed insights into velocity anisotropy patterns and their pressure-dependent evolution (Fig. 5.15).

(2). Anisotropy of Magnetic Susceptibility (AMS)

Research [84] establishes the complementary nature of AMS and ultrasonic velocity measurements in rock fabric anisotropy characterization. Studies on sandstone samples demonstrate AMS's effectiveness in identifying pore space anisotropy without directional bias, revealing strong correlations between AMS and velocity anisotropy principal directions. The findings validate AMS as a reliable proxy for permeability anisotropy assessment in rocks characterized by minimal matrix anisotropy.

(3) Stress-Induced Anisotropy Measurements

Experimental studies [85] on stress-induced anisotropy utilized cubic Springwell sandstone samples under uniaxial stress conditions up to 30–40 MPa, measuring ultrasonic wave velocities in both axial and lateral directions. The results revealed distinctive velocity behaviors: the axial P-wave velocity (P1) showed rapid increases, particularly at low stress levels, while the lateral P-wave velocity (P3) exhibited only slight increases up to ~ 10 MPa before plateauing. The observed 27.4% difference between P1 and P3 at 30 MPa axial stress provides quantitative evidence of significant stress-induced anisotropy (Fig. 5.16).

3. **Anisotropic Behavior of Different Rock Types**

Different rock types exhibit varying degrees and types of anisotropy, which significantly influence stress wave propagation.

(1). Sedimentary Rocks

Sedimentary rocks exhibit varying degrees of anisotropy influenced by their microstructural characteristics. Comparative analysis of sandstones [19] revealed that Bentheim sandstone shows minimal anisotropy (P-waves: 4.7%, S-waves: 3.0%)

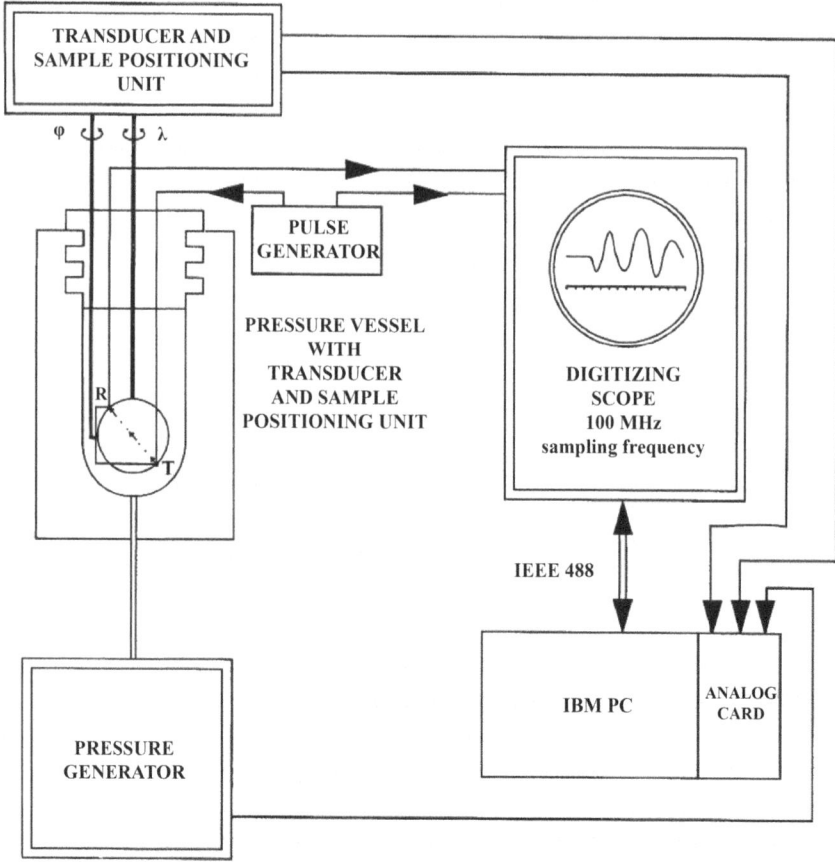

Fig. 5.15 Equipment for the laboratory study of P-wave anisotropy on spherical samples under hydrostatic pressure of up to 400 MPa. T—transmitter of ultrasonic signal, R—receiver of ultrasonic signal. Transducers are located on opposite sides of the sample [83]

with low pressure sensitivity due to its equant pore-dominated structure, while Crab Orchard sandstone displays significant anisotropy (P-waves: 19.1%, S-waves: 7.6%) with pressure dependence up to 30 MPa. Studies of Bakken formation black shales [24] demonstrated extreme velocity anisotropy (40-50%) with pronounced differences between perpendicular and parallel bedding orientations, persisting under high confining pressures and influenced by kerogen properties, providing valuable insights for seismic exploration.

(2). Igneous Rocks

Igneous rocks, while often considered more isotropic than sedimentary rocks, can still exhibit significant anisotropy, particularly due to aligned microcracks. Studies on granodiorite velocity anisotropy [86] demonstrated significant pressure-dependent behavior, exhibiting pronounced velocity differences of 1.6 km/s at atmospheric

5.4 Stress Wave Propagation in Different Medium

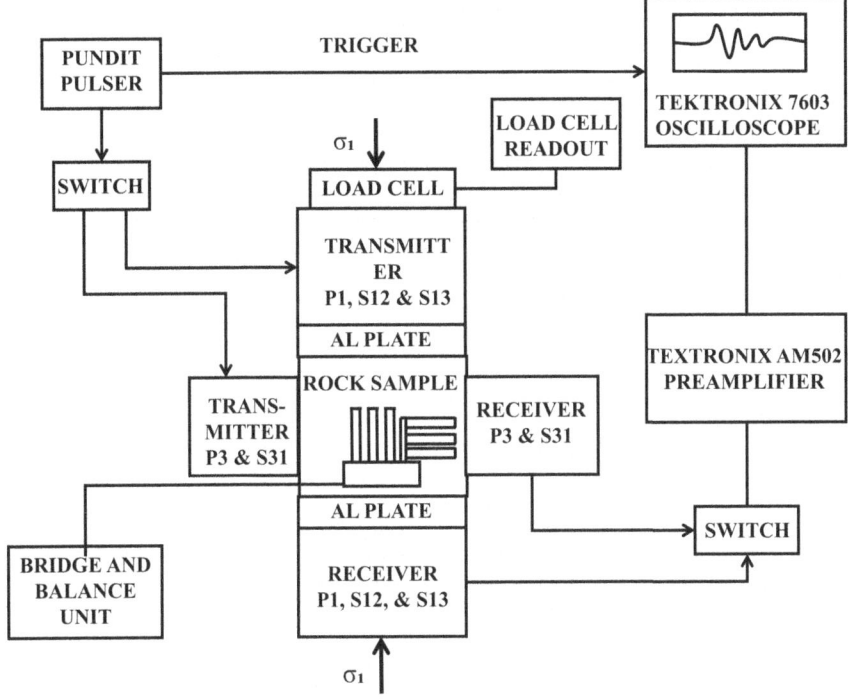

Fig. 5.16 Block diagram of apparatus [85]

pressure that diminished to 0.27 km/s at 300 MPa. The research revealed a strong correlation between velocity anisotropy and the orientation of cleavage cracks in biotite and hornblende minerals, establishing microcrack orientation as a primary control mechanism for anisotropic behavior in granitic rocks.

(3). Metamorphic Rocks

Metamorphic rocks often display strong anisotropy due to preferred mineral orientation and foliation. Research on quartzite anisotropy [86] revealed distinct characteristics compared to granodiorite, displaying more moderate anisotropy levels and reduced pressure sensitivity while maintaining consistent velocity patterns across the pressure range. The maximum velocity increased from 5.62 to 6.30 km/s and minimum velocity from 5.10 to 5.89 km/s with increasing pressure, indicating that quartzite anisotropy is predominantly controlled by intrinsic mineral orientation rather than microcrack distribution.

4. Mathematical Models for Wave Propagation in Anisotropic Rocks

Various mathematical models have been developed to describe wave propagation in anisotropic rocks, ranging from simple analytical solutions to complex numerical models.

(1). Analytical Solutions for Simple Anisotropic Systems

Research [87] established analytical solutions for seismic wave propagation eigenvalue problems in stratified anisotropic media, encompassing monoclinic symmetry and its higher symmetry subclasses. The methodology provides analytical expressions for eigenvalues and eigenvectors, offering enhanced computational efficiency compared to numerical approaches. While applicable primarily to simplified anisotropic systems, these solutions serve dual purposes: providing fundamental insights into wave propagation behavior and establishing benchmark standards for complex numerical modeling.

(2). Numerical Models for Complex Anisotropic Systems

A comprehensive numerical model [88] for wave propagation in anisotropic, viscoelastic, porous media with viscous fluid saturation integrates arbitrary frame anisotropy, viscoelasticity, anisotropic permeability, and fluid viscosity effects. The model characterizes four attenuating quasi-waves (qP1, qP2, qS1, and qS2), providing complex velocities and attenuation quality factors. Numerical simulations revealed that fluid viscosity significantly impacts wave velocities, particularly for slower waves, while hydraulic anisotropy primarily influences attenuation in the high-frequency regime. Wave velocities increase with frequency, approaching non-dissipative conditions, with maximum attenuation observed in the slowest wave mode. This sophisticated modeling approach enables detailed analysis of wave behavior in complex anisotropic rock systems.

(3). Ray Tracing in Anisotropic Media

Ray tracing techniques enable effective modeling of wave propagation in anisotropic media, particularly for high-frequency approximations. The methodology [89] provides a comprehensive framework for calculating group velocity along arbitrary ray directions in general anisotropic media through three key components: three-dimensional phase-to-group velocity transformation, Newton's algorithm implementation for phase direction determination, and singularity management at qS1-qS2 velocity convergence points. This approach facilitates efficient computation of ray paths and travel times in complex anisotropic structures, serving as a fundamental tool for seismic imaging and anisotropic media inversion.

5. Effects of Anisotropy on Stress Wave Propagation

Anisotropy significantly influences various aspects of stress wave propagation, including velocity variations, wave splitting, and attenuation.

(1). Velocity Variations

One of the most apparent effects of anisotropy is the variation of wave velocities with propagation direction. A set of parameters to quantify anisotropy in transversely isotropic media was introduced [90]:

5.4 Stress Wave Propagation in Different Medium

$$\varepsilon = \frac{C_{11} - C_{33}}{2C_{33}}, \gamma = \frac{C_{66} - C_{44}}{2C_{44}} \quad (5.58)$$

$$\delta = \frac{(C_{13} + C_{44})^2 - (C_{33} - C_{44})^2}{2C_{33}(C_{33} - C_{44})} \quad (5.59)$$

where ε describes P-wave anisotropy, γ describes S-wave anisotropy, and δ controls the behavior of waves propagating at angles oblique to the symmetry axis.

(2). Shear Wave Splitting

In anisotropic media, shear waves split into two orthogonally polarized waves traveling at different velocities. This phenomenon, known as shear wave splitting or birefringence, is a powerful indicator of anisotropy in rocks (Fig. 5.17).

A method for analyzing shear wave splitting parameters in the presence of two anisotropic layers was developed [91]. Their approach relates the apparent splitting

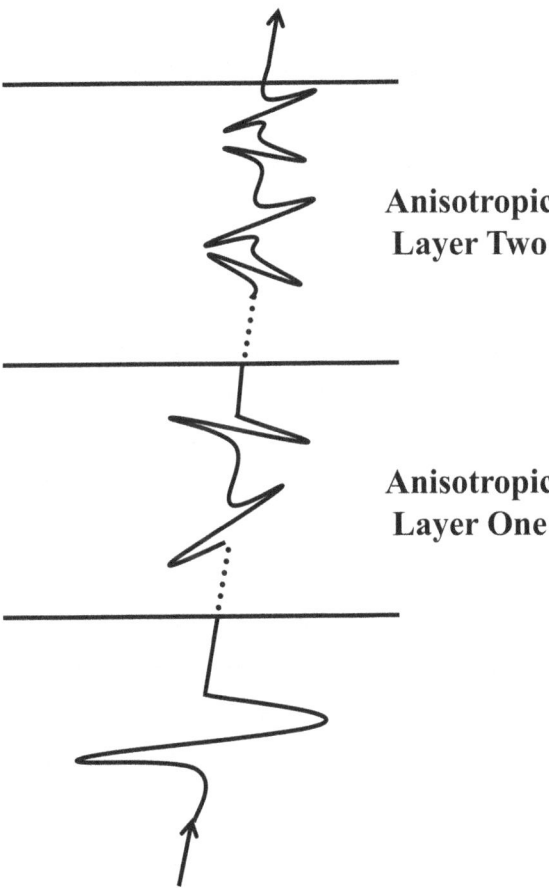

Fig. 5.17 Schematic of shear-wave splitting in the case of two Anisotropic layers. The incoming shear wave is split twice, leading to four individual waves at the receiver (from Yardley & Crampin 1991). In most cases the individual arrivals are unresolved [91]

parameters to those of the individual layers:

$$\tan \alpha_a = \left(a_{p\perp}^2 + C_s^2\right) / \left(a_{p\perp} a_p + C_s C_c\right) \quad (5.60)$$

$$\tan \theta_a = a_{p\perp} / (C_s \cos \alpha_a - C_c \sin \alpha_a) = C_s / \left(a_p \sin \alpha_a - a_{p\perp} \cos \alpha_a\right) \quad (5.61)$$

where $\alpha_a = 2\varphi_a$, $\theta_a = \omega \delta t_a / 2$, and a_p, $a_{p\perp}$, C_c, C_s are coefficients defined in terms of the individual layer parameters.

The apparent splitting parameters exhibit distinctive characteristics: π/2 periodic behavior with respect to incoming polarization direction, presence of minimal splitting along "null" directions, approximate weighted averaging of layer fast directions in the low-frequency domain, and cumulative delay times that sum or subtract based on the relative orientation of fast directions. These fundamental properties provide essential insights for interpreting shear wave splitting measurements in complex anisotropic media.

(3). Attenuation Anisotropy

Beyond velocity anisotropy, rock media demonstrate anisotropic attenuation characteristics. Research on poroelastic media [88] reveals distinct frequency-dependent behaviors: predominant attenuation in the slowest wave due to viscosity effects, modest attenuation reduction in faster waves under hydraulic anisotropy in low-frequency conditions, and a complex high-frequency response featuring enhanced attenuation across three waves while diminishing in the slowest wave. These findings underscore the intricate relationship between frequency regimes and hydraulic anisotropy in determining rock attenuation patterns.

6. Stress-Induced Anisotropy

Applied stress can induce anisotropy in initially isotropic rocks or modify existing anisotropy. This stress-induced anisotropy has important implications for wave propagation in the Earth's crust.

(1). Mechanisms of Stress-Induced Anisotropy

Stress-induced anisotropy primarily develops through selective closure of cracks perpendicular to maximum compressive stress. A predictive methodology [92] for stress-induced velocity anisotropy utilizes hydrostatic pressure-based measurements of isotropic wave velocities, incorporating three key components: characterization of crack-like pore spaces through generalized compliances, pressure-dependence mapping between hydrostatic and non-hydrostatic stress states, and determination of stress-induced anisotropic compliance tensors. This framework enables quantitative assessment of stress-induced velocity variations in rock media.

(2). Experimental Observations of Stress-Induced Anisotropy

Experimental studies on Springwell sandstone under uniaxial stress [85] revealed distinctive wave velocity behaviors: significant enhancement of P-wave velocity parallel to the stress axis particularly at low stress levels, minimal velocity increase

5.4 Stress Wave Propagation in Different Medium

perpendicular to the stress axis up to ~ 10 MPa followed by stabilization, and substantial velocity anisotropy development reaching 27.4% differential between axial and lateral velocities at 30 MPa. These findings demonstrate the profound impact of applied stress on wave propagation characteristics in rock media.

(3). Implications for Crustal Seismology

Stress-induced anisotropy plays a fundamental role in crustal seismology through three primary aspects: the correlation between seismic velocities and stress states, the interpretation of crustal shear wave splitting patterns, and the analysis of temporal velocity variations associated with stress perturbations, particularly in pre-seismic contexts. These relationships provide crucial insights for understanding crustal dynamics and seismic processes.

7. Anisotropy in Fluid-Saturated Rocks

The presence of fluids in rock pores can significantly influence anisotropic behavior, particularly in sedimentary rocks. Understanding wave propagation in fluid-saturated anisotropic rocks is crucial for reservoir characterization and hydrocarbon exploration.

(1). Biot Theory for Anisotropic Poroelastic Media

A comprehensive wave propagation model for general anisotropic poroelastic media with anisotropic permeability [90], founded on Biot's theory, characterizes four attenuating quasi-waves (qP1, qP2, qS1, and qS2). The wave behavior is mathematically described through complex velocities obtained from biquadratic equation solutions, providing a theoretical framework for understanding wave attenuation in complex poroelastic systems.

(2). Effects of Fluid Properties on Anisotropy

Fluid presence fundamentally alters rock anisotropic behavior through velocity reduction, frequency-dependent responses, and modified attenuation patterns. Research findings [88] demonstrate that pore fluid viscosity significantly diminishes wave velocities, particularly affecting slower waves, while higher frequencies drive velocities toward non-dissipative conditions. Additionally, increased pore width or reduced viscosity enhances slower wave velocities, and viscosity effects manifest most prominently in slowest wave attenuation. These relationships highlight the complex interplay between fluid properties and wave propagation characteristics in fluid-saturated anisotropic media.

(3). Hydraulic Anisotropy

Research on rock media reveals the coexistence of elastic and hydraulic anisotropy, the latter arising from preferential orientation of pores or fractures. Modeling studies [90] demonstrate frequency-dependent effects of hydraulic anisotropy: minimal velocity impact and slight attenuation reduction for faster waves in low-frequency regimes, contrasting with substantial attenuation enhancement at high frequencies. These findings emphasize the critical role of considering both elastic and hydraulic anisotropic properties in analyzing wave propagation through fluid-saturated rock systems.

8. Applications of Anisotropic Wave Propagation in Rock Mechanics and Engineering

Understanding stress wave behavior in anisotropic rocks has numerous applications in rock mechanics and engineering.

(1). Seismic Exploration and Reservoir Characterization

Anisotropy plays a crucial role in seismic exploration and reservoir characterization through three primary mechanisms: enhanced seismic imaging via refined velocity modeling, fracture system characterization through shear wave splitting analysis, and lithological differentiation using anisotropic parameters. Research [93] has demonstrated that hydrocarbon source rock seismic properties exhibit strong dependence on kerogen content and maturity, establishing a framework for seismic-based source rock property mapping. These relationships provide fundamental tools for reservoir assessment and hydrocarbon exploration.

(2). Mining Engineering

In mining engineering, stress wave behavior in anisotropic rocks critically influences three key operational aspects: blast pattern optimization for improved fragmentation control, refined stability analysis for rock mass behavior prediction, and ground support system design based on excavation stress distributions. Research [94] reveals that anisotropy significantly impacts stress wave propagation near mine workings, resulting in heterogeneous stress distributions and localized high-stress concentration zones.

(3). Earthquake Science

Anisotropy serves as a fundamental parameter in earthquake science across multiple domains: crustal stress orientation mapping through shear wave splitting analysis, potential earthquake prediction through temporal anisotropic variations, and enhanced ground motion prediction modeling. Research [91] has established that multi-layer anisotropic shear wave splitting analysis provides crucial insights into crustal structure complexity and deformation patterns, advancing our understanding of seismogenic processes and hazard assessment capabilities.

(4). Non-Destructive Testing

Anisotropic wave propagation principles extend to non-destructive testing applications across diverse materials, encompassing fiber-reinforced composite characterization, oriented crack detection in concrete structures, and wood quality assessment based on grain orientation. Research [83] has established measurement techniques for elastic anisotropy in rock samples that provide adaptable methodologies for broader non-destructive material testing applications.

5.4.3 Stress Wave Propagation in Fractured Rock

Stress wave propagation in fractured rock is a complex phenomenon that plays a crucial role in various geotechnical and geological applications, including seismic exploration, rock blasting, and earthquake engineering. The presence of fractures in rock masses significantly alters the propagation characteristics of stress waves, leading to attenuation, dispersion, and mode conversion. This section provides a comprehensive review of the current understanding of stress wave propagation in fractured rock, encompassing theoretical models, experimental studies, and practical implications.

1. **Theoretical Models for Wave Propagation in Fractured Rock**

(1). Displacement Discontinuity Method (DDM)

The Displacement Discontinuity Method is one of the fundamental approaches used to model wave propagation across rock fractures. This method assumes that the fracture behaves as a displacement discontinuity while maintaining stress continuity across the interface [95]. The DDM has been widely applied in various studies due to its ability to capture the essential physics of wave-fracture interaction.

In the DDM, the relationship between stress and displacement discontinuity is typically expressed as:

$$\sigma = K \cdot \Delta u \tag{5.62}$$

where σ is the stress, K is the fracture stiffness, and Δu is the displacement discontinuity across the fracture.

The DDM encompasses three main implementation approaches [20]: the Method of Characteristics (MC), operating in the time domain and suitable for nonlinear joint behavior and Coulomb slip; the Scattering Matrix Method (SMM), functioning in the frequency domain for deriving analytical transmission and reflection coefficients; and the Virtual Wave Source Method (VWS), representing fractures as virtual wave sources, particularly effective for analyzing transmitted waves and comparing with experimental results.

(2). Equivalent Medium Model (EMM)

The Equivalent Medium Model treats the fractured rock mass as a continuous medium with equivalent properties [95]. This approach is particularly useful for analyzing wave propagation in rock masses with multiple fractures or complex fracture networks. The EMM often employs a viscoelastic medium model combined with the VWS concept to obtain effective moduli of the jointed rock mass.

The EMM is particularly advantageous for quantitative wave attenuation calculations and can provide insights into the overall behavior of fractured rock masses. However, it requires additional parameter computations and may not capture some of the fine-scale effects of individual fractures.

(3). Three-Phase Medium Model

Recent advancements in modeling wave propagation through filled fractures have led to the development of more sophisticated approaches, such as the three-phase medium model [96]. This model considers the filled fracture as a composite medium consisting of solid particles, water, and air. The equations of state for each phase are given as:

For air:

$$p_g = p_0 \left(\frac{\rho_g}{\rho_{g0}}\right)^{k_g} \tag{5.63}$$

where p_0 is the atmospheric pressure, ρ_{g0} and ρ_g are the densities of the air phase at the pressures p_0 and p_g, respectively, $\rho_{g0} = 1.2$ kg/m³, $c_{g0} = 340$ m/s is the wave velocity in air and $k_g = 1.4$.

For water:

$$p_w = p_0 + \frac{\rho_{w0} c_{w0}^2}{k_w}\left[\left(\frac{\rho_w}{\rho_{w0}}\right)^{k_w} - 1\right] \tag{5.64}$$

where $c_{w0} = 1500$ m/s, ρ_{w0} and ρ_w are the densities of the water phase at the pressures p_0 and p_w, respectively, and $\rho_{w0} = 1000$ kg/m³, k_w is recommended to be equal to 3 by Henrych [97].

For solid particles:

$$p_s = p_0 + \frac{\rho_{s0} c_{s0}^2}{G_s k_s}\left[\left(\frac{\rho_s}{\rho_{s0}}\right)^{k_s} - 1\right] \tag{5.65}$$

where c_{s0} is the wave velocity in solid, ρ_{s0} and ρ_s are the densities of the solid phase at the pressures p_0 and p_s, respectively, $k_s = 3$, and G_s is a correction parameter related to the solid phase.

The average pressure of the three-phase medium is expressed as:

$$p - p_0 = \alpha_{s0}(p_s - p_0) + \alpha_{w0}(p_w - p_0) + \alpha_{g0}(p_g - p_0) \tag{5.66}$$

where α_{s0}, α_{w0}, and α_{g0} denote the volume ratios of solid particle, water, and air, respectively.

This model provides a more comprehensive description of wave propagation through filled fractures, accounting for the complex interactions between different phases within the fracture [96].

5.4 Stress Wave Propagation in Different Medium

2. Wave Attenuation and Transmission in Fractured Rock

(1). Single Fracture

The interaction of stress waves with a single fracture is fundamental to understanding wave propagation in fractured rock masses. When a stress wave encounters a fracture, part of the wave energy is reflected, part is transmitted, and part is dissipated due to the fracture's properties.

The transmission coefficient (T) is a key parameter used to quantify the proportion of wave energy that passes through the fracture. For a single fracture, the transmission coefficient can be expressed as:

$$T = \frac{2k_n/Z}{-i\omega + 2k_n/Z} \quad (5.67)$$

where i is the sign of imaginary part of a complex number, ω is the angular frequency, C is the velocity of stress wave in the intact rock, k_n denotes the normal specific stiffness of joint, and Z denotes the wave impedance [98].

Research has identified several key factors affecting wave transmission coefficients in fractured media: fracture stiffness positively correlates with wave transmission [95], frequency demonstrates an inverse relationship exhibiting low-pass filter characteristics [98], seismic impedance contrast between intact rock and fracture filling material influences transmission [99], and fracture thickness generally shows an inverse relationship with transmission coefficients [100].

(2). Multiple Parallel Fractures

In rock masses with multiple parallel fractures, wave propagation becomes more complex due to multiple reflections and interactions between fractures. The overall transmission coefficient for N parallel fractures is not simply the product of individual transmission coefficients, as multiple reflections between fractures can lead to constructive or destructive interference [101].

The effects of the multiple parallel fractures on apparent wave attenuation are examined especially in terms of spacing and the number of fractures. The parametric study shows that the dependence of the magnitude of transmission coefficient ($|T_N|$) on fracture spacing and fracture number is governed by the ratio (ξ) of fracture spacing to wavelength. Two important indices of ξ, the threshold value (ξ_{thr}) and the critical value (ξ_{cri}), are identified. They determine whether and how $|T_N|$ is affected by multiple reflections, for any spacing or number of fractures [6].

3. Effects of Fracture Properties on Wave Propagation

(1) Fracture Stiffness

Fracture stiffness is a crucial parameter that significantly influences wave propagation in fractured rock. It is defined as the ratio of stress change to displacement across the fracture. The relationship between fracture stiffness and wave transmission is generally positive, with higher stiffness leading to increased transmission [95]. For a single fracture, the relationship between transmission coefficient and normalized fracture stiffness ($K_n = k_n/\omega Z$) can be approximated as:

$$T \approx \frac{2K_n}{-i + 2K_n} \tag{5.68}$$

This relationship shows that $|T|$ approaches 1 as K_n approaches infinity (welded interface) and approaches 0 as K_n approaches 0 (free interface) [95]. In the case of multiple fractures, the effect of fracture stiffness becomes more complex due to wave interactions between fractures. However, the general trend of increased transmission with higher stiffness remains valid [101].

(2). Fracture Filling

The material filling rock fractures plays a significant role in determining wave propagation characteristics. Filled fractures are common in natural rock masses and can significantly alter wave attenuation and transmission properties compared to unfilled fractures [99]. Research has revealed significant effects of fracture filling on wave propagation characteristics. Studies have shown that filling material composition influences wave transmission, with clay-filled fractures exhibiting higher attenuation than sand-filled ones [96]. Water content demonstrates an inverse relationship with wave transmission, particularly in clay-filled fractures [96], while optimal packing density for wave transmission in sand-clay mixtures occurs at approximately 30% clay weight fraction [99]. The viscoelastic behavior of filled fractures, modelable using Kelvin or Maxwell approaches, leads to frequency-dependent wave attenuation and dispersion [102].

(3). Fracture Roughness

The roughness of fracture surfaces is another important factor affecting wave propagation. Rough fractures can lead to complex wave scattering and increased attenuation compared to smooth fractures [103]. Studies have established significant correlations between fracture roughness and wave propagation characteristics. Research demonstrates that higher Joint Roughness Coefficient (JRC) values correspond to increased wave attenuation [103], while rougher fractures exhibit reduced effective contact areas, leading to lower fracture stiffness and enhanced wave attenuation [104]. The influence of fracture roughness on wave attenuation becomes more pronounced at higher frequencies due to increased scattering effects [103].

4. **Complex Fracture Systems and Heterogeneous Rock Masses**

(1) Non-parallel Fracture Sets

Recent research has expanded beyond conventional parallel fracture studies to investigate wave propagation in natural rock masses containing complex fracture networks with diverse orientations, employing numerical methods such as DEM and FDM for system simulation [95]. The key findings reveal that non-parallel fracture sets induce

5.4 Stress Wave Propagation in Different Medium

enhanced wave attenuation through complex wave paths and increased scattering, exhibit pronounced anisotropic behavior manifested in directionally dependent wave velocities and attenuation, and demonstrate significant mode conversion between P-waves and S-waves, collectively contributing to more intricate wave propagation patterns compared to parallel fracture systems [95].

(2). Heterogeneous Rock Properties

Natural rock masses exhibit material property heterogeneity that interacts with fractures to create complex wave propagation patterns, which have been investigated through advanced numerical modeling techniques [105]. The interaction between heterogeneity and fractures produces significant impedance contrast effects, particularly evident in "soft-to-hard" rock masses where particle velocity transmission ratios remain below 1.0, while stress transmission ratios may exceed 1.0 under specific conditions [105]. The influence of rock heterogeneity on wave propagation demonstrates notable frequency-dependent characteristics, with increased complexity manifesting at higher frequencies [105]. In cases of highly heterogeneous rock masses containing dense fracture networks, effective medium approximations may provide more suitable modeling approaches compared to discrete fracture models [95].

5. **Experimental Studies and Validation**

Experimental studies play a crucial role in validating theoretical models and providing insights into wave propagation phenomena in fractured rock. Various experimental techniques have been developed to study wave propagation in fractured rock, including:

(1). Split Hopkinson Pressure Bar (SHPB) Tests

The SHPB technique, modified for rock testing, has been extensively applied to study wave propagation across single and multiple fractures [100, 104, 106, 107]. This technique offers distinct advantages in generating controlled stress waves and precisely measuring incident, reflected, and transmitted waves, while accommodating tests with various filling materials and fracture geometries. Recent studies have significantly advanced our understanding of wave propagation through filled fractures [99], the influence of fracture roughness [103], and wave transmission across multiple parallel fractures [100] (Fig. 5.18).

(2) Ultrasonic Testing

Ultrasonic testing techniques have been employed to investigate wave propagation in fractured rock at smaller scales and higher frequencies. The methodology primarily comprises pulse transmission tests for measuring wave velocities and attenuation, and acoustic emission monitoring for studying microcrack development and coalescence. These studies have proved particularly valuable in exploring the frequency-dependent characteristics of wave propagation in fractured rock and validating theoretical models at high frequencies [108].

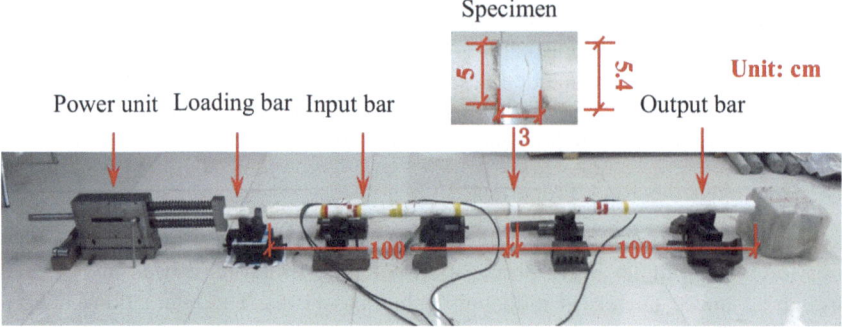

(a) Modified Split Hopkinson Pressure Bar (SHPB) test equipment

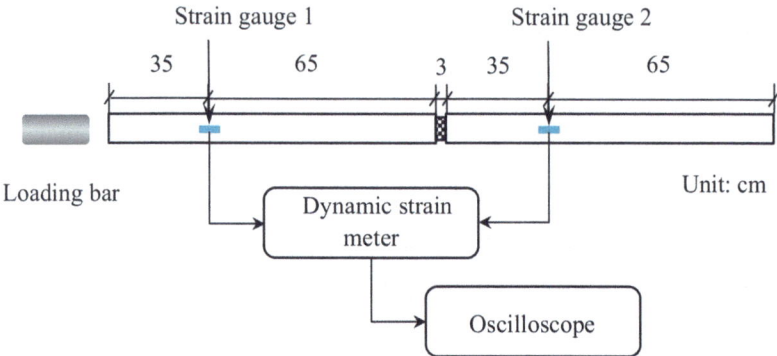

(b) Schematic view of modified SHPB test equipment

Fig. 5.18 Modified SHPB test equipment [103]

(3). Large-scale Field Tests

Field-scale experiments complement laboratory testing in understanding wave propagation through rock masses, encompassing seismic surveys, blast-induced vibration monitoring, and cross-hole seismic testing in fractured formations. These large-scale investigations have proven instrumental in validating theoretical models for practical applications across mining, tunneling, and earthquake engineering disciplines [95].

5.5 Conclusion

This chapter provides a comprehensive and nuanced exploration of stress wave propagation in rock mechanics, illuminating the intricate scientific and engineering challenges at the intersection of theoretical physics, materials science, and practical applications. Research demonstrates the profound complexity of stress wave behavior across diverse geological contexts, where multiple interconnected factors - including surface topography, material properties, geological structures, rock composition, and environmental conditions - dynamically influence wave propagation characteristics. From rocky slopes to geothermal and mining engineering, stress wave analysis has emerged as a sophisticated analytical approach for understanding rock mass characteristics, predicting potential hazards, and optimizing engineering interventions.

While computational advances and numerical modeling techniques have significantly expanded research capabilities, substantial challenges persist in bridging idealized theoretical models with the complex, often unpredictable behavior of natural rock masses. The chapter underscores the critical importance of continued interdisciplinary research, suggesting that future developments in emerging technologies, advanced constitutive modeling, and sophisticated multi-physics process understanding will be pivotal in enhancing predictive capabilities. These advances promise to provide more reliable theoretical guidance and technical support for solving practical engineering problems.

Through multidisciplinary integration, these developments are poised to drive transformative improvements across geophysics, petroleum engineering, and civil engineering disciplines. Their impact will be particularly significant in enhancing rock engineering safety, promoting sustainable resource development, and advancing geological hazard mitigation strategies. The synthesis of these various fields and approaches represents a crucial step forward in addressing the complex challenges faced in modern rock mechanics and engineering applications.

References

1. Zhang Z. Rock Fracture and blasting: theory and applications [M]. Elsevier Science;2016.
2. Kolsky H. Stress waves in solids. J Sound Vib. 1964;1(1):88–110.
3. Wang L. Foundations of stress waves. Elsevier;2011.
4. Julius MA, Reviewer R. The theory of elastic waves and waveguides. North Holland Pub. Co.;1978.
5. Rayleigh L. On waves propagated along the plane surface of an elastic solid. Proc Lond Math Soc. 1885;1(1):4–11.
6. Abramson H, Plass H, Ripperger E. Stress wave propagation in rods and beams. Adv Appl Mech. 1958;5:111–94.
7. Johansson D, Ouchterlony F. Shock wave interactions in rock blasting: the use of short delays to improve fragmentation in model-scale. Rock Mech Rock Eng. 2013;46:1–18.
8 He C, Gao J, Chen D, Xiao J. Investigation of stress wave interaction and fragmentation in granite during multihole blastings. IEEE Access. 2020;8:185187–97.
9. Toffoli A, Bitner-Gregersen EM. Types of ocean surface waves, wave classification. In: Encyclopedia of maritime and offshore engineering;2017. p. 1–8.
10. Yang XY, Zhang B, Zhang DF. Reflected amplitude of elastic wave at free surface. Adv Mater Res. 2011;287:1888–91.
11. Wong H. Effect of surface topography on the diffraction of P, SV, and Rayleigh waves. Bull Seismol Soc Am. 1982;72(4):1167–83.
12. Wang L, Xu Y, Xia J, Luo Y. Effect of near-surface topography on high-frequency Rayleigh-wave propagation. J Appl Geophys. 2015;116:93–103.

13. Edelman I. Surface waves at vacuum/porous medium interface: low frequency range. Wave Motion. 2004;39(2):111–27.
14. Takekawa J, Mikada H. Free-surface implementation in a mesh-free finite-difference method for elastic wave propagation in the frequency domain. Geophys Prospect. 2019;67(8):2104–14.
15. Wang X, Zhang H. Modeling of elastic wave propagation on a curved free surface using an improved finite-difference algorithm. Sci China Ser G. 2004;47:633–48.
16. Xu H, Cantwell CD, Monteserin C, Eskilsson C, Engsig-Karup AP, Sherwin SJ. Spectral/hp element methods: Recent developments, applications, and perspectives. J Hydrodyn. 2018;30:1–22.
17. Snieder R. The influence of topography on the propagation and scattering of surface waves. Phys Earth Planet Inter. 1986;44(3):226–41.
18. Chen H, Zhou M, Gan S, et al. Review of wave method-based non-destructive testing for steel-concrete composite structures: Multiscale simulation and multi-physics coupling analysis. Constr Build Mater. 2021;302(3): 123832.
19. Othman MI, Song Y. Reflection of plane waves from an elastic solid half-space under hydrostatic initial stress without energy dissipation. Int J Solids Struct. 2007;44(17):5651–64.
20. Liu X, Chen J, Zhao Z, Lan H, Liu F. Simulating seismic wave propagation in viscoelastic media with an irregular free surface. Pure Appl Geophys. 2018;175:3419–39.
21. Ostad H, Mohammadi S. Analysis of shock wave reflection from fixed and moving boundaries using a stabilized particle method. Particuology. 2009;7(5):373–83.
22. Mukhomodyarov R, Rogerson G. Long-wave dispersion phenomena in a layer subject to elastically restrained boundary conditions. Z Angew Math Phys. 2012;63:171–88.
23. Lutianov M, Rogerson GA. The influence of boundary conditions on dispersion in an elastic plate. Mech Res Commun. 2010;37(2):219–24.
24. Carr M, Davies P. Boundary layer flow beneath an internal solitary wave of elevation. Phys Fluids 2010; 22(2).
25. Wu J, Gao Y-T, Tang S-H, et al. Experimental study on the boundary reflection effect of stress wave propagation based on the newly developed test apparatus. Adv Civil Eng. 2024; 2024(1):7170963.
26. Jiang M, Zhang Y. Existence and asymptotic behavior of global smooth solution for p-system with nonlinear damping and fixed boundary effect. Math Methods Appl Sci. 2014;37(17):2585–96.
27. Tao C, Zhenpeng L. Effects of the boundary conditions at fixed end on the flexural wave propagation in the periodic beam. Arch Appl Mech. 2015;85:191–203.
28. Nguyen TTP, Tanabe-Yamagishi R, et al. Rayleigh wave and super-shear evanescent wave excited by laser-induced shock at a soft solid–liquid interface observed by photoelasticity imaging technique. J Appl Phys 2022; 131(12).
29. Ashirbayev N, Ashirbayeva Z, Shomanbayeva M. Influence of heterogeneity of nature of border fixing on the propagation of two-dimensional waves. Adv Math Sci. 2015;1676(1): 020067.
30. Mitrea I, Ott K. Electromagnetic scattering from perturbed surfaces. Math Methods Appl Sci. 2007;30(7):861–76.
31. Godoy E, Durán M, Nédélec J-C. On the existence of surface waves in an elastic half-space with impedance boundary conditions. Wave Motion. 2012;49(6):585–94.
32. Bérenger J-P. Theoretical investigation of the reflection from impedance absorbing boundary conditions. IEEE Microwave Wirel Compon Lett. 2018;28(7):543–5.
33. Li R, Natsuki T, Ni Q-Q. A novel dynamic stress analysis in bimaterial composite with defect using ultrasonic wave propagation. Compos Struct. 2015;132:255–64.
34. Rokhlin S, Wang Y. Analysis of boundary conditions for elastic wave interaction with an interface between two solids. J Acoust Soc Am. 1991;89(2):503–15.
35. Lyu X-F, Lin X, Ouyang W, et al. Stress wave propagation attenuation law and energy dissipation characteristics of rock-barrier-coal composite specimen under dynamic load. J Central South Univ 2023; 30(1):276–88

36. Fan L, Shi X, Wang M, Jiang F. Wave transmission in layered composite rock mass comprising parallel joints and different rock materials. Int J Rock Mech Min Sci. 2024;181: 105858.
37. Fan L, Shi X, Wang M, et al. The effects of joint spacing on transmission characteristics of stress waves through layered composite rock masses. Int J Numer Anal Meth Geomech. 2024;48(3):822–36.
38. Lv X, Chen C, Ma C. Interaction between a semi-cylindrical hill and a shallow buried fixed rigid cylindrical inclusion under SH wave. In: Proceedings of the IOP conference series: earth and environmental science, F. IOP Publishing;2019
39. Ding X, Yang Y, Zhou W, et al. The law of blast stress wave propagation and fracture development in soft and hard composite rock. Sci Rep. 2022;12(1):17120.
40. Rossmanith H, Uenishi K. The mechanics of spall fracture in rock and concrete. Fragblast. 2006;10(3–4):111–62.
41. Xu C, Liu Q, Tang X, et al. Dynamic stability analysis of jointed rock slopes using the combined finite-discrete element method (FDEM). Comput Geotech. 2023;160: 105556.
42. Gipprich TL, Snieder R, Jibson R, et al. The role of shear and tensile failure in dynamically triggered landslides. Geophys J Int. 2008;172(2):770–8.
43. Che A, Yang H, Wang B, et al. Wave propagations through jointed rock masses and their effects on the stability of slopes. Eng Geol. 2016;201:45–56.
44. Zhou H, Ye F, Fu W, et al. Dynamic effect of landslides triggered by earthquake: a case study in Moxi Town of Luding County, China. J Earth Sci. 2024;35(1):221–34.
45. Havenith H-B, Vanini M, Jongmans D, Faccioli E. Initiation of earthquake-induced slope failure: influence of topographical and other site specific amplification effects. J Seismolog. 2003;7:397–412.
46. Sepúlveda S, Murphy W, Petley D. Topographic controls on coseismic rock slides during the 1999 Chi-Chi earthquake, Taiwan. Quart J Eng Geol Hydrogeol. 2005;38(2):189–96.
47. Brain MJ, Rosser NJ, Sutton J, et al. The effects of normal and shear stress wave phasing on coseismic landslide displacement. J Geophys Res Earth Surf. 2015;120(6):1009–22.
48. Bozzano F, Lenti L, Martino S, et al. Earthquake triggering of landslides in highly jointed rock masses: reconstruction of the 1783 Scilla rock avalanche (Italy). Geomorphology. 2011;129(3–4):294–308.
49. Ben-Zion Y. Properties of seismic fault zone waves and their utility for imaging low-velocity structures. J Geophys Res Solid Earth. 1998;103(B6):12567–85.
50. Feng X, Zhang Q, Wang E, et al. 3D modeling of the influence of a splay fault on controlling the propagation of nonlinear stress waves induced by blast loading. Soil Dyn Earthq Eng. 2020;138: 106335.
51. Li J, Ma G, Zhao J. Analysis of stochastic seismic wave interaction with a slippery rock fault. Rock Mech Rock Eng. 2011;44:85–92.
52. Jahnke G, Igel H, Ben-Zion Y. Three-dimensional calculations of fault-zone-guided waves in various irregular structures. Geophys J Int. 2002;151(2):416–26.
53. Zeng S, Wang S, Sun B, et al. Propagation characteristics of blasting stress waves in layered and jointed rock caverns. Geotech Geol Eng. 2018;36:1559–73.
54. Sánchez I, Sierra D, Duarte C, et al. Simulation of surface seismic waves propagation by 2D finite-difference elastic wave modeling in the presence of complex surface topography. In: Proceedings of the XI Congreso Colombiano de Métodos Numéricos, F; 2011.
55. Kharazova YV, Pavlenko O, Dudinskii K. The correlation between the characteristics of seismic wave propagation in Western Caucasus and the geological–tectonic features of the region, Izvestiya. Phys Solid Earth. 2016;52:399–412.
56. Lee VW, Amornwongpaibun A. Scattering of anti-plane (SH) waves by a semi-elliptical hill: I—shallow hill. Soil Dyn Earthq Eng. 2013;52:116–25.
57. Song D, Shi W, Liu M, et al. Wave propagations in crossing-fault tunnels and their effects on the dynamic response characteristics of tunnel surrounding rock. Bull Eng Geol Env. 2024;83(6):1–19.
58. Vlahovic G, Elkibbi M, Rial J. Shear-wave splitting and reservoir crack characterization: the Coso geothermal field. J Volcanol Geoth Res. 2003;120(1–2):123–40.

59. Lees JM, Wu H. P wave anisotropy, stress, and crack distribution at Coso geothermal field, California. J Geophys Res Solid Earth. 1999;104(B8):17955–73.
60. Mukuhira Y, Ito T, Asanuma H, et al. Evaluation of flow paths during stimulation in an EGS reservoir using microseismic information. Geothermics. 2020;87: 101843.
61. Okamoto K, Imanishi K, Asanuma H. Structures and fluid flows inferred from the microseismic events around a low-resistivity anomaly in the Kakkonda geothermal field, Northeast Japan. Geothermics. 2022;100: 102320.
62. Kikuchi T, Nakao S. Wave propagation in geothermal reservoirs: a finite-difference method. 2005.
63. Minetto R, Montanari D, Planès T, et al. Tectonic and anthropogenic microseismic activity while drilling toward supercritical conditions in the Larderello-Travale geothermal field, Italy. J Geophys Res Solid Earth 2020; 125(2):e2019JB018618.
64. Liu J, Liu Z, Wang S, et al. Analysis of microseismic activity in rock mass controlled by fault in deep metal mine. Int J Min Sci Technol. 2016;26(2):235–9.
65. Ge M, Mrugala M, Iannacchione A. Microseismic monitoring at a limestone mine. Geotech Geol Eng. 2009;27:325–39.
66. Zhang P, Bai R, Sun X, et al. Investigation of rock joint and fracture influence on delayed blasting performance. Appl Sci. 2023;13(18):10275.
67. Yi C, Johansson D, Nyberg U, Beyglou A. Stress wave interaction between two adjacent blast holes. Rock Mech Rock Eng. 2016;49:1803–12.
68. Yang R, Zuo J, Ma L, et al. Analysis of explosion wave interactions and rock breaking effects during dual initiation. Int J Miner Metall Mater. 2024;31(8):1788–98.
69. Range IL-F, Biot M. Theory of propagation of elastic waves in a fluid-saturated porous solid. J Acoust Soc Am. 1956;28(2):179–91.
70. Carcione JM, Helle HB, Pham NH. White's model for wave propagation in partially saturated rocks: comparison with poroelastic numerical experiments. Geophysics. 2003;68(4):1389–98.
71. King M, Marsden J, Dennis J. Biot dispersion for P-and S-wave velocities in partially and fully saturated sandstones. Geophys Prospect. 2000;48(6):1075–89.
72. Cadoret T, Mavko G, Zinszner B. Fluid distribution effect on sonic attenuation in partially saturated limestones. Geophysics. 1998;63(1):154–60.
73. Ma X-Y, Wang S-X, Zhao J-G, et al. Velocity dispersion and fluid substitution in sandstone under partially saturated conditions. Appl Geophys. 2018;15(2):188–96.
74. Ba J, Carcione JM, Sun W. Seismic attenuation due to heterogeneities of rock fabric and fluid distribution. Geophys J Int. 2015;202(3):1843–7.
75. Toms J, Müller T, Ciz R, Gurevich B. Comparative review of theoretical models for elastic wave attenuation and dispersion in partially saturated rocks. Soil Dyn Earthq Eng. 2006;26(6–7):548–65.
76. Caspari E, Qi Q, Lopes S, et al. Wave attenuation in partially saturated porous rocks—new observations and interpretations across scales. Lead Edge. 2014;33(6):606–14.
77. Zhang L, Ba J, Carcione JM, Wu C. Seismic wave propagation in partially saturated rocks with a fractal distribution of fluid-patch size. J Geophys Res Solid Earth 2022; 127(2):e2021JB023809.
78. Müller T, Toms-Stewart J, Wenzlau F. Velocity-saturation relation for partially saturated rocks with fractal pore fluid distribution. Geophys Res Lett 2008; 35(9).
79. Cadoret T, Marion D, Zinszner B. Influence of frequency and fluid distribution on elastic wave velocities in partially saturated limestones. J Geophys Res Solid Earth. 1995;100(B6):9789–803.
80. Adam L, Batzle M, Lewallen K, et al. Seismic wave attenuation in carbonates. J Geophys Res Solid Earth 2009; 114(B6).
81. Ricketts TE, Goldsmith W. Dynamic properties of rocks and composite structural materials. Int J Rock Mech. 1970;7(3):315–35.
82. Pros Z, Lokajíček T, Přikryl R, et al. Direct measurement of 3D elastic anisotropy on rocks from the Ivrea zone (Southern Alps, NW Italy). Tectonophysics 2003; 370(1–4):31–47.

References

83. Pros Z, Lokajíček T, Klíma K. Laboratory approach to the study of elastic anisotropy on rock samples. Pure Appl Geophys. 1998;151(2–4):619.
84. Benson PM, Meredith PG, Platzman ES, White RE. Pore fabric shape anisotropy in porous sandstones and its relation to elastic wave velocity and permeability anisotropy under hydrostatic pressure. Int J Rock Mech Min Sci. 2005;42(7–8):890–9.
85. Wu B, King M, Hudson J. Stress-induced ultrasonic wave velocity anisotropy in a sandstone. Int J Rock Mech Min Sci Geomech Abs F 1991; Elsevier.
86. Babuška V, Pros Z. Velocity anisotropy in granodiorite and quartzite due to the distribution of microcracks. Geophys J Int. 1984;76(1):121–7.
87. Fryer GJ, Frazer LN. Seismic waves in stratified anisotropic media—II. Elastodynamic eigensolutions for some anisotropic systems. Geophys J Int 1987; 91(1):73–101.
88. Vashishth A, Sharma M. Propagation of plane waves in poroviscoelastic anisotropic media. Appl Math Mech. 2008;29:1141–53.
89. Sharma M. Group velocity along general direction in a general anisotropic medium. Int J Solids Struct. 2002;39(12):3277–88.
90. Sharma M. Effect of initial stress on the propagation of plane waves in a general anisotropic poroelastic medium. J Geophys Res Solid Earth 2005; 110(B11).
91. Silver PG, Savage MK. The interpretation of shear-wave splitting parameters in the presence of two anisotropic layers. Geophys J Int. 1994;119(3):949–63.
92. Mavko G, Mukerji T, Godfrey N. Predicting stress-induced velocity anisotropy in rocks. Geophysics. 1995;60(4):1081–7.
93. Vernik L, Nur A. Ultrasonic velocity and anisotropy of hydrocarbon source rocks. Geophysics. 1992;57(5):727–35.
94. Baranowski Z, Lugovoi P. Stress-strain state near mine workings in anisotropic rock masses under the action of discontinuous waves. Int Appl Mech. 2008;44:406–12.
95. Perino A, Zhu J, Li J, et al. Theoretical methods for wave propagation across jointed rock masses. Rock Mech Rock Eng. 2010;43:799–809.
96. Ma G, Li J, Zhao J. Three-phase medium model for filled rock joint and interaction with stress waves. Int J Numer Anal Meth Geomech. 2011;35(1):97–110.
97. Henrych J, Abrahamson GR. The dynamics of explosion and its use. J Appl Mech. 1980;47(1):218.
98. Fan L, Wu Z. Evaluation of stress wave propagation through rock mass using a modified dominate frequency method. J Appl Geophys. 2016;132:53–9.
99. Wu W, Li J, Zhao J. Role of filling materials in a P-wave interaction with a rock fracture. Eng Geol. 2014;172:77–84.
100. Chen X, Li J, Cai M, et al. Experimental study on wave propagation across a rock joint with rough surface. Rock Mech Rock Eng. 2015;48:2225–34.
101. Zhao BX, Zhao J, Cai JG. P-wave transmission across fractures with nonlinear deformational behaviour. Int J Numer Anal Methods Geomech 2006; 30(11):1097–112.
102. Zhu J, Perino A, Zhao G, Barla G, Li J, Ma G, Zhao J. Seismic response of a single and a set of filled joints of viscoelastic deformational behaviour. Geophys J Int. 2011;186(3):1315–30.
103. Li J, Rong L, Li H, Hong S. An SHPB test study on stress wave energy attenuation in jointed rock masses. Rock Mech Rock Eng. 2019;52:403–20.
104. Li J, Li N, Li H, Zhao J. An SHPB test study on wave propagation across rock masses with different contact area ratios of joint. Int J Impact Eng. 2017;105:109–16.
105. Fan L, Wang L, Wu Z. Wave transmission across linearly jointed complex rock masses. Int J Rock Mech Min Sci. 2018;112:193–200.
106. Wu W, Zhu J, Zhao J. A further study on seismic response of a set of parallel rock fractures filled with viscoelastic materials. Geophys J Int. 2013;192(2):671–5.
107. Li JC, Wu W, Li H, et al. A thin-layer interface model for wave propagation through filled rock joints. J Appl Geophys. 2013;91:31–8.
108. Cook N. Natural joints in rock: Mechanical, hydraulic and seismic behaviour and properties under natural stress, Jaeger Memorial Lecture. Int Journal Rock Mech Min Sci & Geomech Abstr. 1992;29(3):198–223.

Open Access This chapter is licensed under the terms of the Creative Commons Attribution-NonCommercial-NoDerivatives 4.0 International License (http://creativecommons.org/licenses/by-nc-nd/4.0/), which permits any noncommercial use, sharing, distribution and reproduction in any medium or format, as long as you give appropriate credit to the original author(s) and the source, provide a link to the Creative Commons license and indicate if you modified the licensed material. You do not have permission under this license to share adapted material derived from this chapter or parts of it.

The images or other third party material in this chapter are included in the chapter's Creative Commons license, unless indicated otherwise in a credit line to the material. If material is not included in the chapter's Creative Commons license and your intended use is not permitted by statutory regulation or exceeds the permitted use, you will need to obtain permission directly from the copyright holder.

Chapter 6
Rockburst Dynamics and Engineering Protection

6.1 Phenomenon and Cause of Rockburst

6.1.1 Definition and Phenomena of Rockburst

Rockburst is a complex and dangerous phenomenon that occurs in deep underground excavations, particularly in hard rock environments. It is characterized by the sudden and violent failure of rock mass around an excavation, often accompanied by the ejection of rock fragments at high velocities. This section will provide a comprehensive overview of the definition and phenomena associated with rockbursts, drawing from multiple sources to present a thorough understanding of this critical issue in rock mechanics and underground engineering.

1. **Definition of Rockburst**

There is no single, universally accepted definition of rockburst, as the understanding of this phenomenon has evolved over time [1]. A comprehensive definition that captures these elements is: "Rockburst is a sudden and violent failure of rock associated with a rapid release of accumulated energy around underground openings, resulting in the instantaneous ejection of rock fragments at varying velocities" [2]. It's important to note that rockbursts are distinct from other forms of rock mass instability, such as progressive failure or rock falls, due to their sudden and energetic nature [3].

2. **Phenomena Associated with Rockbursts**

Rockburst events manifest through multiple complex phenomena, characterized by varying intensities and types of burst manifestations that significantly impact underground excavations and their surroundings [1–6]. The primary physical phenomena include the ejection of rock fragments at velocities ranging from a few meters per second to over 50 m/s, accompanied by seismic events with magnitudes up to 5 on the Richter scale, audible effects ranging from popping sounds to explosive reports, and powerful air blasts in confined spaces [1, 2, 4, 5]. Secondary effects encompass ground vibration, extensive dust generation, damage to support systems, and deformation of excavation boundaries through mechanisms such as wall bulging, floor heave, and roof sagging [1, 2, 4, 6]. Furthermore, rockbursts induce significant stress redistribution that may trigger subsequent events, while also potentially generating thermal effects through elastic energy conversion, sudden gas emissions in specific geological settings like coal mines, and increased water inflow through newly created or existing fracture networks [1, 3–5]. The manifestation and intensity of these phenomena vary considerably based on factors including excavation depth, in-situ stress conditions, rock mass properties, and excavation geometry, making their understanding crucial for developing effective prediction, prevention, and mitigation strategies.

It's important to note that the manifestation of rockburst phenomena can differ between hard rock mines and coal mines. In hard rock environments, rockbursts are typically characterized by the brittle failure of intact rock and ejection of rock fragments. In coal mines, the term "coal burst" is often used to describe similar phenomena, which may involve both coal and surrounding rock strata [7].

The phenomena associated with rockbursts highlight the complex and multifaceted nature of this geomechanical hazard. They underscore the need for a comprehensive approach to rockburst management that considers not only the immediate failure mechanism but also the broader geomechanical and operational context in which these events occur [1–3].

In conclusion, the definition and phenomena of rockbursts encompass a wide range of geomechanical processes and observable effects. The sudden and violent nature of these events, combined with their potential for causing significant damage and creating multiple hazards, makes them a critical concern in deep underground excavations. As mining and tunneling activities continue to extend to greater depths and more challenging geological environments, the importance of understanding and managing rockburst risks will only increase [1–7].

6.1.2 Occurence Conditions and Influcing Factors for Rockburst

Rockburst is a complex phenomenon in rock mechanics that poses significant challenges to deep underground engineering projects. The occurrence of rockburst is influenced by various factors, including geological conditions, in-situ stress state, excavation methods, and dynamic disturbances. Understanding these conditions and influencing factors is crucial for predicting, preventing, and mitigating rockburst hazards. This section provides a comprehensive review of the key conditions necessary for rockburst occurrence and the factors that influence its intensity and characteristics.

1. Geological Conditions

The geological setting plays a fundamental role in determining the susceptibility of rock masses to rockburst. Several geological factors contribute to the likelihood and intensity of rockburst events:

(1) Rock Type and Properties

The type of rock and its inherent properties significantly influence rockburst potential. Rocks with high brittleness, high elastic modulus, and high strength are generally more prone to rockburst [8]. Granite, for example, is known to be particularly susceptible to rockburst due to its high brittleness and ability to store large amounts of elastic strain energy [9, 10].

The rock brittleness index (B) is often used to quantify the rockburst potential of different rock types:

$$B = \frac{\sigma_c}{\sigma_T} \tag{6.1}$$

where σ_c is the uniaxial compressive strength (MPa) and σ_T is the tensile strength (MPa) [8]. Rocks with higher brittleness indices are more likely to experience violent failure under high stress conditions. Additionally, the elastic modulus and Poisson's ratio of the rock affect its ability to store and release strain energy, which is crucial in rockburst mechanisms [9].

(2) Structural Features

Structural features within rock masses play a crucial role in determining rockburst behavior. These geological characteristics encompass joints and fractures, whose orientation and distribution patterns influence rock mass stability and stress response mechanisms [11], fault systems that can concentrate stress and facilitate its transfer between regions [2], and foliation or bedding planes in metamorphic and sedimentary rocks that may create anisotropic stress conditions leading to localized failure [8].

(3) Structural Planes

Small-scale structural planes significantly influence rockburst behavior in deep tunnels, particularly through their exposure conditions and geometric orientations [11]. Research indicates that unexposed structural planes typically accelerate and intensify rockburst occurrence, while exposed planes tend to mitigate and delay such events. The inclination angle of unexposed planes critically affects failure mechanisms, with angles between 30 and 60° promoting more violent shear failures. These planes also modify stress distributions and energy release patterns, where unexposed planes concentrate stress between the plane and tunnel wall, resulting in higher localized energy release, while exposed planes redistribute stress deeper into the rock mass with reduced energy release [11]. These findings emphasize the critical role of small-scale structural features in rockburst risk assessment for deep tunneling projects.

(4) In-situ Stress State

The in-situ stress state fundamentally controls rockburst occurrence, with high-stress environments commonly associated with deep or tectonically active regions. Critical stress parameters include the absolute magnitude of principal stresses, which determines stored strain energy potential [9, 12], the maximum to minimum principal stress ratio (σ_1/σ_3), which influences failure mode stability [8], and the orientation of principal stresses relative to excavation geometry, which governs stress concentration patterns and subsequent rockburst risk [12].

2. **Excavation-Induced Factors**

The process of excavation introduces significant changes to the stress state and mechanical properties of the surrounding rock mass, which can trigger or exacerbate rockburst conditions. Several excavation-related factors influence rockburst behavior.

(1) Excavation Method

Excavation methodology significantly influences rockburst susceptibility and characteristics. Drill and blast techniques introduce dynamic stresses and create damage zones that increase susceptibility to remotely triggered rockbursts, while TBM excavation, despite causing less initial rock damage, may lead to higher rockburst intensity due to enhanced energy accumulation potential in the intact rock mass [9]. Controlled excavation techniques, including smooth blasting and low-vibration mechanical methods, can effectively minimize surrounding rock damage and mitigate rockburst risk [12] (Fig. 6.1).

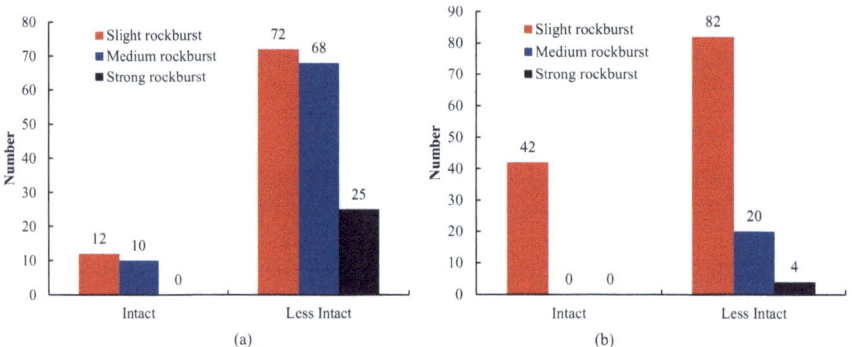

Fig. 6.1 Numbers of rockbursts that occurred during the construction of the Jinping II Hydropower Station during excavation using **a** drilling and blasting and **b** Tunnel Boring Machine (TBM) [9]

6.1 Phenomenon and Cause of Rockburst

(2) Excavation Geometry

The geometrical characteristics of excavations significantly affect stress redistribution patterns and rockburst susceptibility. Opening size exhibits a non-linear correlation with rockburst potential, with larger excavations generally inducing more extensive stress redistribution [12]. The tunnel shape plays a crucial role, as demonstrated by trapezoidal tunnels which show more extensive failure zones along sidewalls compared to circular profiles, necessitating broader support coverage [8]. Additionally, sharp corners in non-circular openings can generate concentrated stress fields, thereby elevating the risk of localized failure and rockburst initiation [8].

(3) Excavation Rate

Experimental investigations have revealed that excavation rate significantly influences rockburst behavior through its effect on rock mass loading rates. Studies demonstrate that elevated loading rates correlate with more severe rockburst phenomena, with granite samples transitioning from static brittle failure to dynamic rockburst failure as loading rates increase from 0.05 to 5.0 MPa/s [10, 13]. The failure mode evolves from tensile-dominated slabbing at lower rates (0.05 MPa/s) to shear-dominated rockburst at higher rates (1.0–5.0 MPa/s) [10, 13]. Higher loading rates generate larger fragments due to increased fracturing energy, with Weibull distribution analysis indicating decreased size parameter k and increased fractal dimensions, suggesting more severe fragmentation [10]. Although total energy dissipation decreases with increasing loading rate, shear failure energy consumption consistently dominates (97–99%) across all rates [10]. These findings suggest that controlling excavation speed could potentially mitigate rockburst risks, though the relationship is complex and depends on various factors including rock type and in-situ stress conditions.

(4) Stress Path

The stress path during excavation plays a critical role in determining rockburst susceptibility. Rapid unloading of confining stress can trigger unstable crack propagation and sudden failure, with higher stress release rates corresponding to increased rockburst risk [8, 12]. The excavation process induces stress concentrations around the opening, whose magnitude and distribution are governed by the excavation geometry, in-situ stress conditions, and rock properties [8]. In certain instances, rockburst may manifest through a progressive failure mechanism, characterized by gradual damage accumulation culminating in an abrupt, violent energy release [12].

3. **Dynamic Disturbances**

Dynamic disturbances, whether natural or induced by human activities, can trigger or exacerbate rockburst conditions. These disturbances introduce additional stress waves into the rock mass, potentially leading to failure in pre-stressed regions.

(1) Seismic Events

Seismic events, both natural and mining-induced, can act as rockburst triggers in vulnerable zones. Earthquakes generate strong ground motions that induce dynamic stress variations in underground openings, potentially initiating rockbursts in high-stress region. Similarly, extensive ore extraction in mining operations can cause substantial stress redistribution and induced seismicity, which may trigger rockbursts in neighboring areas [12].

(2) Blasting Vibrations

Blasting operations, whether conducted for excavation purposes or adjacent construction activities, can induce dynamic loading conditions that trigger rockbursts. In the near-field domain, blasting operations can cause direct damage and generate stress waves that may initiate rockbursts [8]. Even in far-field regions, the propagating stress waves, despite attenuation, retain sufficient energy to potentially trigger rockbursts in rock masses under critical stress conditions [12].

(3) Mechanical Vibrations

Mechanical vibrations from underground construction and mining equipment can contribute to rockburst development through distinct mechanisms. Prolonged exposure to low-amplitude vibrations may deteriorate rock structures and facilitate progressive failure processes [12], while sudden impacts or high-amplitude vibrations from equipment operation can trigger immediate rockburst events in vulnerable zones [8].

(4) Influence of Dynamic Loading Frequency

Dynamic loading frequency significantly influences rockburst characteristics in sandstone, as evidenced by comprehensive experimental investigations [14]. Higher frequencies induce more concentrated damage patterns, manifesting as deeper and narrower shaped blast pits, accompanied by enhanced acoustic emission activity and accelerated energy release. This phenomenon leads to intensified fragmentation, characterized by smaller fragment sizes and higher fractal dimensions, along with expedited particle ejection and reduced time to failure. The mechanism, rooted in elastic mechanics, suggests that elevated frequencies generate increased disturbance and tangential stresses, resulting in rapid energy accumulation beyond uniaxial failure thresholds, ultimately intensifying rockburst phenomena [14].

4. **Environmental Factors**

Environmental conditions, particularly the presence of water, can significantly influence rockburst behavior. Recent experimental studies have provided valuable insights into how moisture content affects rockburst mechanisms and monitoring techniques.

(1) Water Content Effects

Water saturation significantly influences rockburst behavior in tunnels, as demonstrated through acoustic emission (AE) and infrared (IR) monitoring studies [15]. The presence of water reduces rock mechanical properties, evidenced by decreases in elastic modulus (18.3%) and peak strength (10.8%), leading to wider and deeper V-shaped rockburst pits and shortened AE quiet periods. IR monitoring reveals elevated average temperatures (24.78 °C versus 20.21 °C in dry samples) and more abrupt

6.1 Phenomenon and Cause of Rockburst

temperature changes before failure in saturated conditions, particularly along tunnel sidewalls. The damage evolution under saturation is characterized by accelerated early-stage deterioration but diminished dynamic fracturing during final failure [15].

(2) Temperature Effects

Temperature exerts significant influence on rockburst behavior through multiple mechanisms. In deep underground settings or near geothermal sources, elevated temperatures generate thermal stresses that contribute to the overall stress state, potentially enhancing rockburst risk. Cyclic thermal loading can induce thermal fatigue, progressively weakening rock structures and increasing rockburst susceptibilit. Furthermore, prolonged exposure to extreme temperatures may trigger mineral alterations in certain rock types, modifying their mechanical properties and consequently affecting their rockburst susceptibility [4].

5. Stress Field Complexity

The complexity of the stress field surrounding underground excavations plays a crucial role in rockburst occurrence and characteristics. Recent research has highlighted the importance of considering multi-directional stress components, particularly the often-overlooked tunnel axis stress.

(1) Influence of Tunnel Axis Stress

The impact of tunnel axis stress (σ_x) on strain burst behavior in granite was investigated and several key relationships were identified [16]. Rock strength increases with σ_x up to approximately 80 MPa and decreases at higher stress levels. Strain burst intensity, reflected by the kinetic energy of ejected fragments, increases up to 60 MPa, decreases between 60 and 90 MPa, and rises sharply beyond 90 MPa, peaking at σ_x = 110 MPa. Failure modes also change with σ_x: zoning failure dominates at lower stresses (<90 MPa), while higher stresses (>90 MPa) lead to fragmentation into large pieces. Additionally, rockburst pit volume exhibits a similar trend to strain burst intensity, increasing, then decreasing, and rising again with higher σ_x [16].

(2) Mechanisms of Tunnel Axis Stress Effects

The complex relationship between σ_x and strain burst behavior was attributed to the interplay of two effects by a study [7]. The constraint effect, driven by lateral friction from σ_x, restricts outward bending and rock plate ejection while dissipating energy, becoming stronger as σ_x increases. In contrast, the Poisson effect, resulting from expansion perpendicular to the compression direction, promotes outward bending and ejection, becoming significant at high σ_x values. These competing effects explain the observed trends: strength increase dominates at low σ_x (10–60 MPa), the constraint effect hinders ejection in the mid-range (60–90 MPa), and the Poisson effect facilitates rapid energy release and ejection at high σ_x (90–110 MPa) [16].

(3) Implications for Rockburst Assessment

These findings have important implications for rockburst prediction and prevention in deep underground projects. First, comprehensive stress analysis is crucial, as tunnel axis stress significantly influences strain burst intensity and may be underestimated in traditional models focusing on radial and tangential stresses [16]. Second, high tunnel axis stress conditions, common in deep tunnels or tectonically stressed regions, pose a greater risk of intense strain bursts [16]. Third, insights into the effects of tunnel axis stress can guide the design of support systems to mitigate large-scale fragmentation at high stress levels [16]. Lastly, the results may inform excavation strategies, optimizing sequencing and direction to manage the stress state and reduce rockburst risk [16].

6. Geometric Factors

The geometry of underground excavations, including their shape, size, and orientation relative to the in-situ stress field, significantly influences rockburst behavior. Recent research has provided insights into how specific geometric features affect rockburst characteristics.

(1) Influence of Tunnel Shape

A previous study [8] demonstrated that tunnel shape significantly influences stress distribution and rockburst potential, with trapezoidal tunnels showing distinct characteristics. Tangential compressive stress is higher in the sidewalls than in the roof and floor, increasing the likelihood of rockburst initiation in these regions. Tensile stress zones are present near the roof and floor, with their depth expanding as lateral stress increases. Additionally, the lower corners of trapezoidal tunnels experience higher principal compressive stress than the upper corners, making them potential rockburst initiation points. Compared to circular openings, trapezoidal tunnels exhibit larger failure zones on sidewalls, necessitating more extensive support measures to mitigate rockburst risks [8].

(2) Elliptical Caverns

The effects of elliptical cavern geometry on rockburst behavior were investigated by a study [17], with the role of the long axis orientation being highlighted. When the long axis is parallel to the disturbance direction (HD configuration), the structure exhibits higher initial failure stress, greater rockburst resistance, and better load-bearing capacity compared to the perpendicular configuration (VD). However, HD samples showed more severe rockburst phenomena, with larger pit areas, angles, and depths when failure occurred. Fragmentation also varied with orientation: VD samples had finer and more elongated debris, while HD samples produced thicker, plate-like fragments. Additionally, HD samples exhibited higher average ejection velocities (7.98 m/s versus 6.94 m/s for VD), indicating more energetic failures [18] (Fig. 6.2).

6.1 Phenomenon and Cause of Rockburst

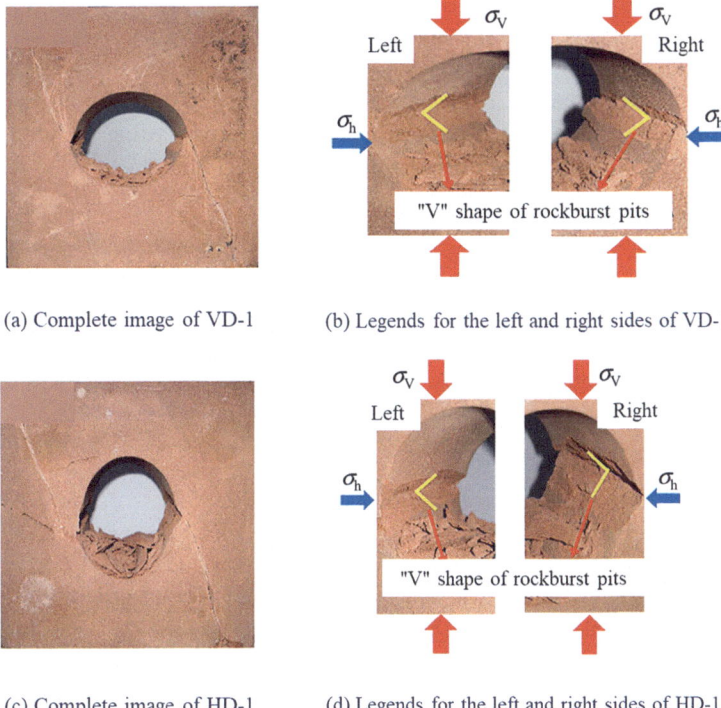

Fig. 6.2 Failure photos of sandstone after rockburst [18]

7. **Energy Accumulation and Release Mechanisms**

The accumulation and release of energy within the rock mass is fundamental to rockburst occurrence. Understanding these mechanisms is crucial for predicting and mitigating rockburst hazards.

(1) Energy Accumulation

Energy accumulation in rock masses is influenced by several factors. High in-situ stress conditions, common in deep underground environments, result in significant elastic strain energy storage [9, 12]. Excavation-induced stress redistribution creates stress concentrations around openings, forming high-stress zones prone to energy accumulation [8, 16]. Additionally, rocks with higher elastic modulus and strength have greater capacity for elastic strain energy storage, increasing rockburst potential [9, 10]. Structural features, such as discontinuities and planes, can further localize stress concentrations, promoting energy accumulation in specific areas [11].

(2) Energy Release Mechanisms

Rockburst events are characterized by the sudden release of accumulated energy, driven by several mechanisms. Stress-induced fracturing occurs when local stresses exceed rock strength, causing fracture initiation and propagation [8, 12]. Slip along discontinuities, where shear stresses surpass frictional resistance on structural planes or joints, also releases energy [11]. Buckling failure in highly stressed thin rock slabs leads to abrupt failure and energy dissipation [8, 15]. Additionally, dynamic triggering from external loads, such as seismic waves or blasting vibrations, can initiate energy release in critically stressed rock masses [12, 14].

(3) Energy Partitioning

During rockburst, the released energy is distributed into various forms. Kinetic energy drives the ejection of rock fragments and is commonly used to quantify rockburst intensity [10, 16]. Fracture energy is consumed in generating new fracture surfaces during the failure process [10]. Thermal energy is dissipated as heat, detectable through infrared monitoring [15]. Additionally, seismic energy is emitted as seismic waves, which can be captured by microseismic monitoring systems [12].

(4) Quantifying Energy Release

Various methods have been proposed to quantify energy release during rockburst. These include estimating the kinetic energy of ejected rock fragments through high-speed video analysis [16], measuring the Strain Energy Release Rate (SERR) to quantify energy released per unit area of new fracture surface [12], and utilizing the energy of acoustic emission events as a proxy for energy release during microfracturing processes [15].

(5) Influence of Loading Conditions on Energy Release

Recent experimental investigations have revealed significant correlations between loading conditions and energy release patterns during rockburst events. Higher loading rates lead to more intense energy release but reduced total fragment energy dissipation, with shear failure energy consumption consistently dominating at 97–99% [10]. The relationship between tunnel axis stress and energy release exhibits non-monotonic behavior in terms of ejected fragment kinetic energy [16]. Water saturation induces earlier damage accumulation while suppressing dynamic fracturing at failure [15]. Higher dynamic loading frequencies intensify energy release, evidenced by enhanced acoustic emission and fragmentation [14]. These findings are crucial for developing effective rockburst prediction and control strategies.

6.2 Precursors and Prediction Methods of Rock Burst

6.2.1 Characteristics and Monitoring of Rockburst Precursors

Rockburst precursors are crucial indicators that can provide early warning of impending rockburst events in underground excavations. Understanding and effectively monitoring these precursors is essential for improving safety and mitigating

6.2 Precursors and Prediction Methods of Rock Burst

risks in deep mining and tunneling operations. This section provides a comprehensive review of the characteristics and monitoring methods for rockburst precursors, drawing from recent research and case studies in the field.

1. **Microseismic Precursors**

Microseismic (MS) activity is one of the most widely studied and reliable precursors for rockburst events. The characteristics of MS activity leading up to a rockburst can provide valuable insights into the impending event's location, intensity, and timing.

(1) Energy and Event Count Variations

MS energy release patterns and event counts serve as crucial precursors to rockburst events. A distinctive pattern emerges, characterized by dramatic fluctuations in MS energy release, typically manifesting as a sudden increase that indicates strain energy accumulation in the rock mass [19]. Notably, this energy surge often coincides with an anomalously low event count, a phenomenon known as the "quiet period" [19]. This inverse relationship between energy release and event frequency, indicating fewer but more energetic events, suggests a critical state in the rock mass. This pattern was validated through a case study at the Qianqiu coal mine in China, where both MS energy and event count decreased 1–3 days prior to a rockburst occurrence [20] (Figs. 6.3 and 6.4).

(2) Fault Total Area

Fault total area, representing the cumulative fracture area within the rock mass, serves as a significant MS precursor to rockburst events. Research demonstrates a characteristic pattern: a steady increase in fault total area during the days preceding a rockburst, followed by a sharp rise 1–2 days before the event, reaching its peak at the time of rockburst occurrence [19]. This pattern reflects the progressive development and coalescence of fractures within the rock mass leading to the final failure event (Fig. 6.5).

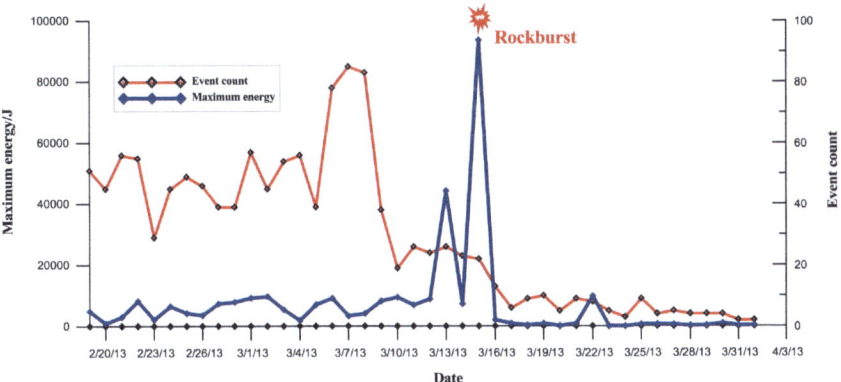

Fig. 6.3 Variation curves of the daily MS maximum energy and event count [19]

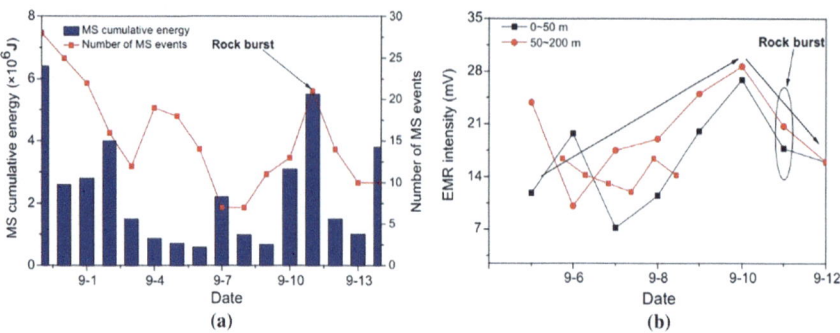

Fig. 6.4 Monitoring results for the "9.11" rock burst. **a** MS results; **b** EMR results [20]

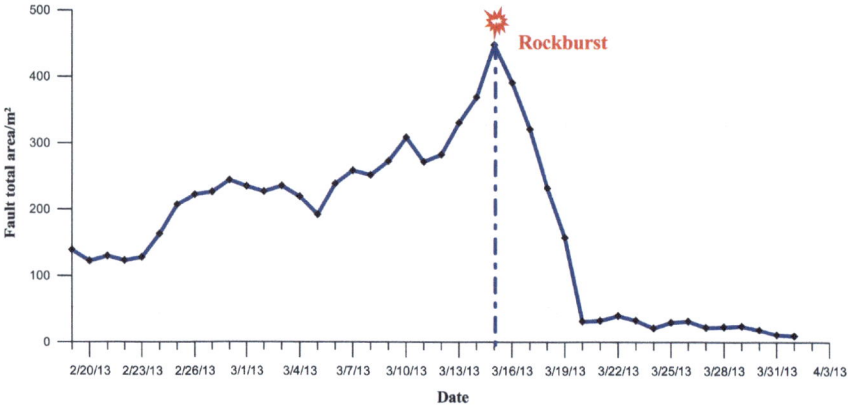

Fig. 6.5 Variation curve of fault total area for MS events (the energy is $10^2 - 10^7$ J) [19]

(3) b-value Analysis

The b-value, a statistical parameter derived from the Gutenberg-Richter relationship, quantifies the relative distribution of large and small seismic events. In rockburst prediction, abnormally low b-values are observed prior to rockburst occurrence, indicating an increased probability of high-energy events and suggesting the rock mass is approaching a critical state [19] (Fig. 6.6).

The b-value can be calculated using the following formula:

$$\log N = a - bM \qquad (6.2)$$

where N is the number of events with magnitude greater than M, and a and b are constants [21].

6.2 Precursors and Prediction Methods of Rock Burst

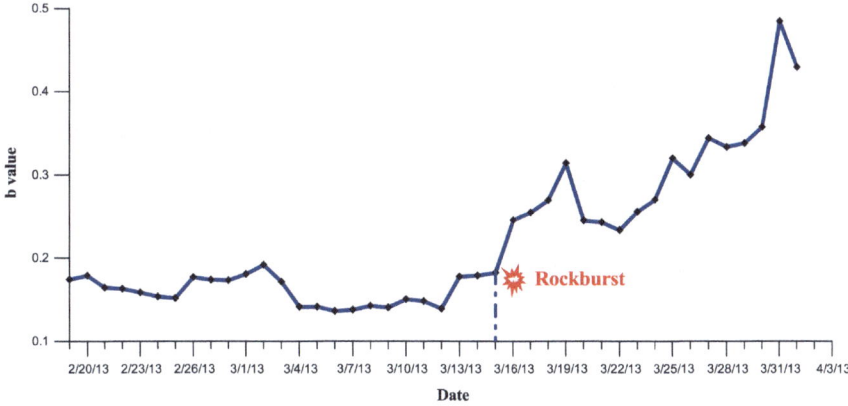

Fig. 6.6 Variation curve of lack of shock b value [19]

(4) Z-value Analysis

The Z-value is another statistical parameter used in MS analysis for rockburst prediction. It is defined as:

$$Z = \frac{R_a - R_m}{S_d} \qquad (6.3)$$

where R_a is the average event rate in the analysis window, R_m is the mean event rate in the background window, and S_d is the standard deviation of the event rate in the background window [21]. The Z-value serves as a critical indicator in rockburst prediction, with two key characteristics: an absolute value exceeding 2 (|Z|>2) indicates high rockburst risk, and a sudden sharp decrease typically occurs in the days immediately before a rockburst event [19] (Fig. 6.7).

(5) Frequency Analysis

MS event frequency characteristics serve as valuable precursors to rockburst events. Two key indicators have been identified: a shift in the dominant frequency to the lowest frequency band, and the occurrence of maximum energy ratio in the low-frequency band prior to failure [19]. These frequency-related changes reflect alterations in the rock mass's mechanical properties and stress state as it approaches critical conditions (Fig. 6.8).

(6) Source Parameter Space Analysis

The "EMS" method, analyzing seismic energy (E), seismic moment (M), and apparent stress (S), enables classification of MS events into six types: stress-adjusting (Types I and II), deformation-driven (Types III and IV), and energy-releasing events

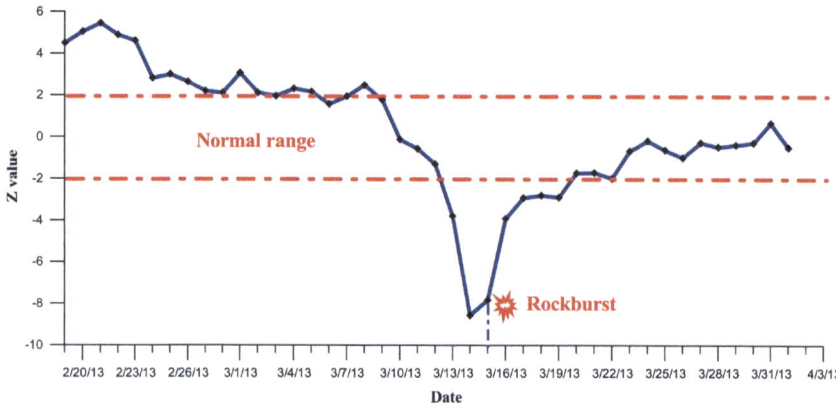

Fig. 6.7 Variation curve of Z value [19]

(Types V and VI) [22]. Analysis reveals distinct stages in rockburst development: energy accumulation, deformation/damage, energy transfer, and energy release. Notably, deformation-driven events increase by approximately 15% during the late development stage, while energy-releasing events predominate during the occurrence stage with high energy release rates [22]. This comprehensive parametric analysis enhances understanding of rock mass behavior evolution and improves rockburst prediction accuracy (Fig. 6.9).

2. **Electromagnetic Radiation Precursors**

Electromagnetic radiation (EMR) has emerged as a complementary precursor to MS activity for rockburst prediction. EMR monitoring can provide additional insights into the stress state and deformation processes in the rock mass.

(1) EMR Intensity Variations

EMR emissions demonstrate characteristic patterns during rockburst evolution. Research shows that EMR intensity typically peaks approximately one day before a rockburst event, followed by a decrease to low levels during the event [20]. Notably, the EMR peak often coincides with the MS quiet period [20], suggesting that integrated EMR and MS monitoring can provide complementary information for enhanced rockburst prediction capabilities.

(2) Theoretical Basis for EMR Precursors

The relationship between EMR intensity and stress in coal and rock masses has been theoretically analyzed using the Weibull distribution and minimum potential energy principle [20]. EMR generation is modeled using the Weibull distribution and related to stress through the equation:

$$E_m = f(\sigma) = b_n \Delta \sigma^n + b_{n-1} \Delta \sigma^{n-1} + \cdots + b_1 \Delta \sigma + b_0 \quad (6.4)$$

Fig. 6.8
Frequency-spectrum distribution of rockburst recorded by 1, 4, 5, and 15 sensors. **a** 1 sensor, **b** 4 sensor, **c** 5 sensor, and **d** 15 sensor [19]

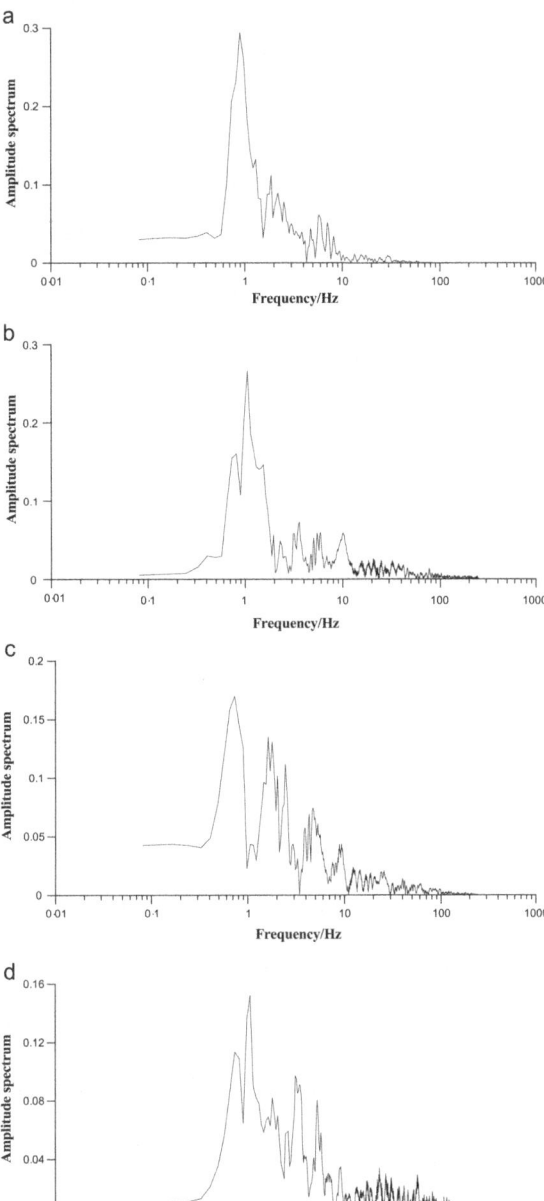

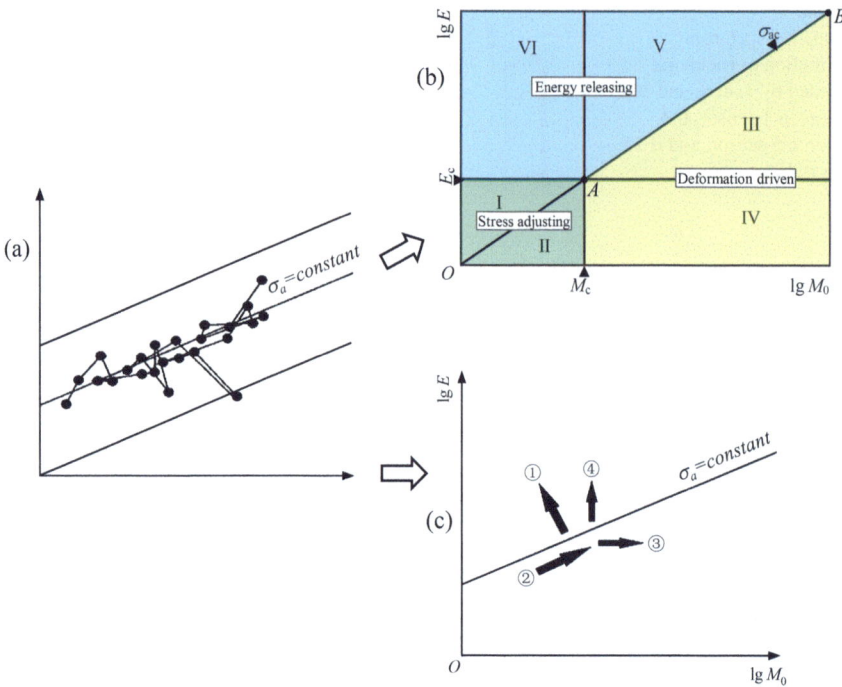

Fig. 6.9 Theory of source-parameter space. **a** Distribution of microseismic events in the space; **b** Classification of microseismic events; **c** Microseismic paths [22]

where E_m is EMR intensity and $\Delta\sigma$ is change in stress [20]. The number of EMR pulses (N) is related to stress change ($\Delta\sigma$) by:

$$N = a_n\Delta\sigma^n + a_{n-1}\Delta\sigma^{n-1} + \cdots + a_1\Delta\sigma + a_0 \qquad (6.5)$$

EMR energy (W_e) is given by:

$$W_e = \frac{1}{2}\varepsilon E_m^2 V \qquad (6.6)$$

where ε is the dielectric constant and V is the coal/rock volume [20]. This theoretical framework provides a basis for understanding the nonlinear positive correlation between EMR intensity and stress in coal and rock masses, explaining the observed EMR patterns before and during rockbursts.

3. **Stress and Deformation Precursors**

Direct monitoring of stress changes and deformation in the rock mass can provide additional precursor information for rockburst prediction.

(1) Stress Changes

Stress monitoring utilizing CCBO (Compact Conical-ended Borehole Overcoring) and CCBM (Compact Conical-ended Borehole Monitoring) techniques has revealed significant precursory stress patterns. Research indicates substantial stress increases 40–200 m ahead of excavation faces in tunneling and longwall mining operations, with maximum stress changes of 8–29 MPa recorded near the excavation face [23]. These stress variations correlate with seismic events and can be modulated by destress blasting activities [23]. The findings emphasize the critical role of continuous stress monitoring in high-risk zones for early rockburst detection.

(2) Deformation Monitoring

Deformation monitoring through multipoint extensometers reveals critical precursory patterns, including accelerating deformation rates in the days or hours preceding rockburst events [24]. Notably, MS activity typically precedes visible deformation by at least one week [24], suggesting its value as an early warning indicator. The integration of deformation monitoring with MS and stress monitoring systems enables a more comprehensive understanding of rock mass behavior evolution prior to rockburst occurrence.

4. **Geological and Structural Precursors**

The presence and characteristics of geological structures can significantly influence rockburst occurrence and should be considered in precursor analysis.

(1) Influence of Structural Planes

Research has shown that structural planes, such as faults and joints, can modify the characteristics of MS precursors. Rockbursts affected by structural planes exhibit a lower proportion of high-energy MS events ($>10^{3.83}$ J) and a higher proportion of low-energy events ($<10^{1.09}$ J) [25]. The presence of structural planes can lead to lower threshold values for the number of MS events and total energy required to trigger a rockburst [25]. Structural planes can influence the evolution patterns of MS events, with large and small events alternating throughout the development process [25]. These findings suggest that the presence of structural planes should be considered when interpreting MS precursors and assessing rockburst risk (Fig. 6.10).

(2) Fault and Joint Mapping

Detailed mapping of faults, joints, and other geological structures is crucial for understanding potential rockburst sources and interpreting precursory signals. Strong rockbursts are often associated with geological structures like faults and interbedded zones [26]. MS event clusters have been found to correlate strongly with geological structures, particularly faults [24]. Integrating geological mapping with MS monitoring can improve the accuracy of rockburst location prediction and risk assessment.

5. **Integration of Multiple Precursor Types**

The most effective approach to rockburst prediction involves the integration of multiple precursor types and monitoring methods. This integrated approach can

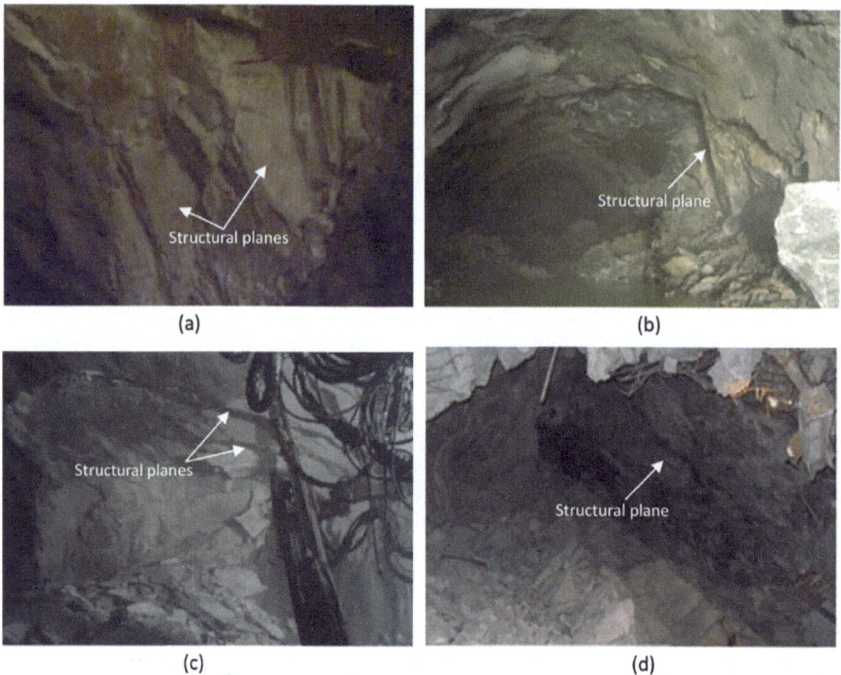

Fig. 6.10 Some photos of the structural planes in rockburst area [25]

provide a more comprehensive understanding of the rock mass behavior and improve prediction accuracy.

(1) Combining MS and EMR Monitoring

The integration of MS and EMR monitoring has shown promise in improving rockburst prediction accuracy. MS and EMR signals show complementary patterns, with EMR peaks often corresponding to MS quiet periods [20]. The combination of MS and EMR monitoring can provide a more complete picture of the stress state and deformation processes in the rock mass [20].

(2) Multi-parameter MS Analysis

Combining multiple MS parameters can enhance prediction accuracy. The "EMS" method, analyzing energy, moment, and apparent stress, provides a more nuanced understanding of rock mass behavior [22]. Integrating parameters such as energy, event count, b-value, Z-value, and frequency characteristics can improve the reliability of predictions [19].

(3) Integration with Numerical Modeling

Numerical modeling of stress distribution and rock mass behavior can complement monitoring data. 2D and 3D numerical models can help interpret observed stress

6.2 Precursors and Prediction Methods of Rock Burst

changes and predict areas of high rockburst risk [23]. Integrating monitoring data with numerical models can improve the understanding of rockburst mechanisms and enhance prediction capabilities [24].

6. **Monitoring System Design and Implementation**

Effective monitoring of rockburst precursors requires careful design and implementation of monitoring systems.

(1) MS Monitoring System Design

The design of MS monitoring systems requires careful consideration of several key technical parameters [21]. The sensor array configuration should encompass the monitoring area with appropriate spacing and increased density in critical zones. System specifications should be tailored to the monitoring scale: geophones (5–200 Hz) for larger areas and accelerometers (500–3000 + Hz) for smaller regions. Data acquisition requires sampling rates 5–10 times the maximum expected frequency using 3–48 channel portable units. Data transmission methods should be selected based on distance requirements: twisted pair cables for short ranges (<300 m) and fiber optic or wireless solutions for extended distances [21] (Fig. 6.11).

(2) EMR Monitoring Implementation

EMR monitoring systems can be deployed in two configurations: portable devices such as KBD5 for periodic tunnel wall measurements [20], and fixed installations that enable continuous data collection in high-risk areas (Fig. 6.12).

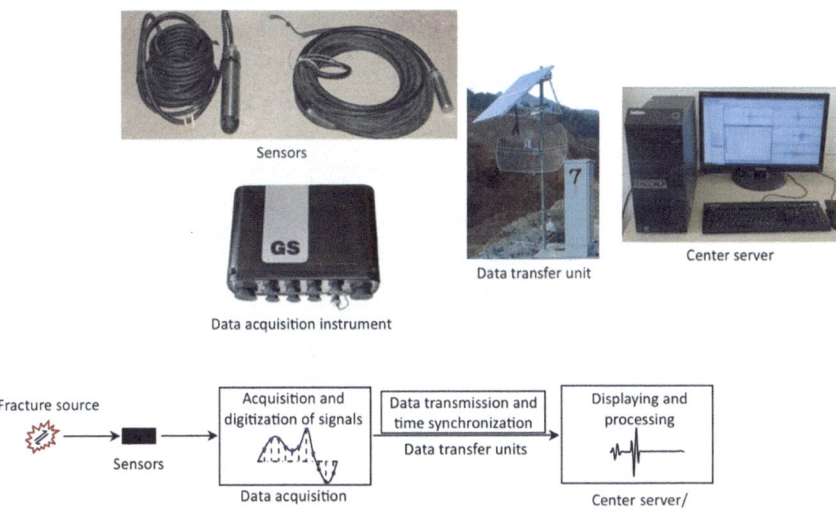

Fig. 6.11 The basic principle of microseismic monitoring [21]

Fig. 6.12 Photograph of a KBD5 EMR monitoring [20]

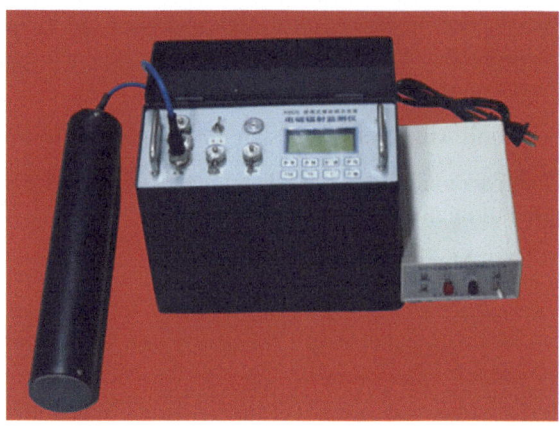

(3) Stress and Deformation Monitoring

The design of stress and deformation monitoring systems requires strategic implementation of specialized techniques in critical areas, incorporating Compact Conical-ended Borehole Overcoring (CCBO) for initial stress measurements and compact conical ended borehole monitoring method (CCBM) for continuous stress monitoring [23], along with multipoint extensometers and complementary deformation monitoring tools in zones of anticipated high strain [24] (Figs. 6.13 and 6.14).

(4) Data Integration and Analysis

The integration and analysis of multi-system monitoring data require a comprehensive approach, encompassing real-time data collection and processing systems

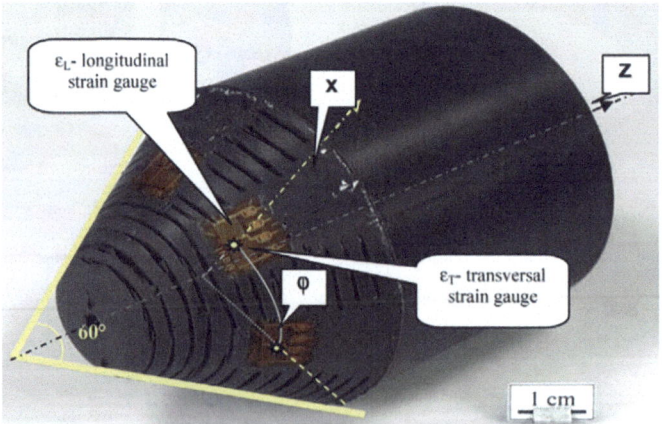

Fig. 6.13 Geometry of CCBO probe [23]

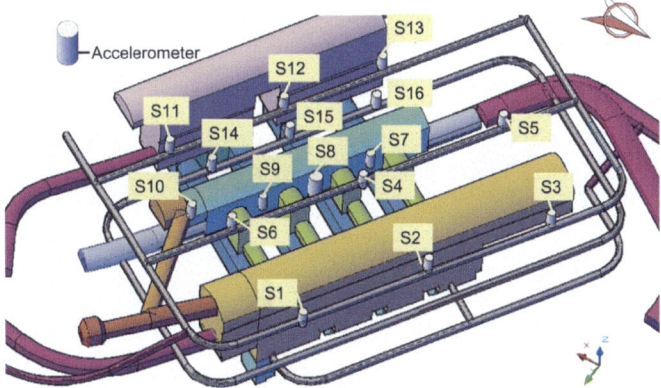

Fig. 6.14 Sketch map of accelerometer array at the underground group caverns of Houziyan hydropower station. The gray cylinders represent accelerometers. The sensor array includes 18 channel swith at riaxial accelerometer S8 [24]

for swift response to precursory signals, coupled with data visualization tools for interpreting multi-parameter precursor data [26]. This is enhanced by the application of machine learning and artificial intelligence techniques to optimize pattern recognition and prediction accuracy [27] (Fig. 6.15).

7. **Case Studies and Practical Applications**

Several case studies demonstrate the successful application of precursor monitoring for rockburst prediction and mitigation.

(1) Jinping II Hydropower Station Tunnels

The implementation of a comprehensive MS monitoring system in the deep tunnels of the Jinping II Hydropower Station demonstrated exceptional predictive capabilities [26]. During the period from August 2010 to November 2011, the system recorded 237 rockbursts, achieving a 94% prediction success rate with 80.6% accuracy in both range and grade predictions. The system's effectiveness was notably exemplified by its accurate prediction of an extremely strong rockburst in the drainage tunnel on November 28, 2009, which was precisely located 14 days prior to its occurrence [26], underscoring the significant potential of MS monitoring in enhancing deep tunneling safety (Figs. 6.16 and 6.17).

(2) Qianqiu Coal Mine

The integrated MS and EMR monitoring system implemented at the Qianqiu coal mine in Henan Province, China, revealed distinctive signal patterns associated with rockburst events [20]. The combined monitoring approach enhanced the understanding of rockburst development processes [7], demonstrating the effectiveness of multi-technique monitoring integration for improved rockburst prediction capabilities (Figs. 6.18 and 6.19).

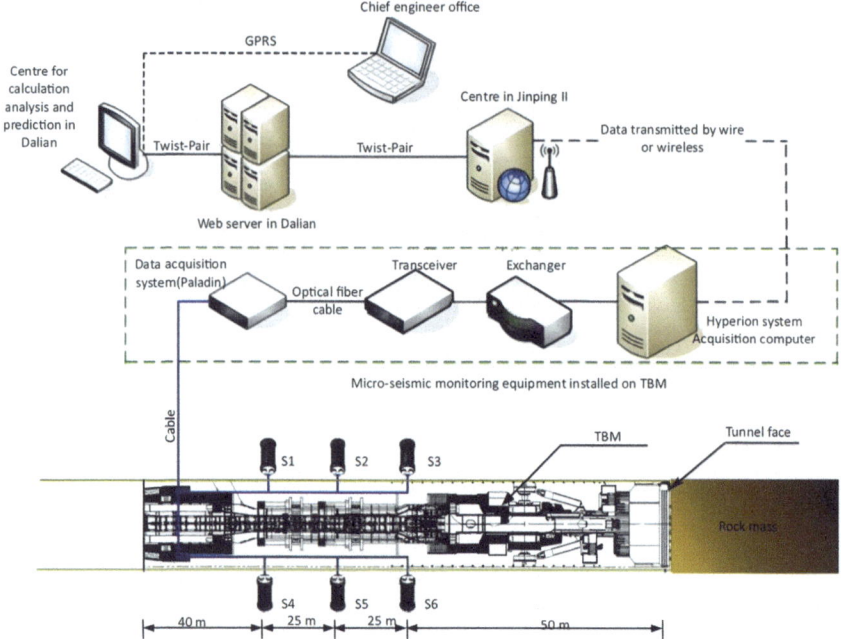

Fig. 6.15 The monitoring and analysis system for rockbursts during TBM tunneling for Jinping II Hydropower Station [26]

Fig. 6.16 Site condition after an extremely strong rockburst in the drainage tunnel on 28 November 2009 [26]

(3) Upper Silesian Coal Basin Longwall Mining

A comprehensive monitoring program of stress and seismicity in a high rockburst risk longwall mining area revealed significant stress increases 40–200 m ahead of the longwall face [23]. The monitoring data effectively guided destress blasting operations, which successfully reduced both seismicity and stress concentrations [23], demonstrating the practical application of monitoring systems for both prediction and active control of rockburst hazards (Fig. 6.20).

6.2 Precursors and Prediction Methods of Rock Burst 287

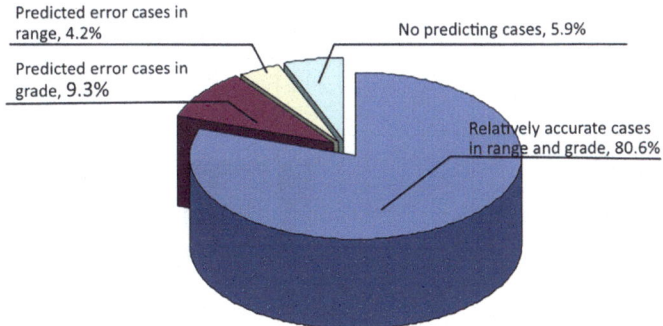

Fig. 6.17 Accuracy of rockburst prediction based on microseismic monitoring [26]

6.2.2 Common Rockburst Prediction Methods

Rockburst prediction is a critical aspect of managing safety in deep underground excavations. Various methods have been developed to forecast the occurrence and intensity of rockbursts, drawing on theoretical principles, empirical data, and monitoring techniques. This section provides a comprehensive overview of common rockburst prediction methods, their underlying principles, and their applications in engineering practice.

1. **Theoretical Criteria-Based Methods**

Theoretical criteria-based methods rely on fundamental rock mechanics principles and measurable rock properties to assess the likelihood and intensity of rockbursts. These methods typically involve calculating specific indices or ratios that correlate with rockburst potential. Some of the most widely used theoretical criteria include:

(1) Stress-Based Criteria

Stress-based criteria focus on the relationship between the induced stresses in the rock mass and the rock's strength properties. One commonly used stress-based criterion is the intensity stress ratio, which compares the maximum tangential stress around an excavation to the uniaxial compressive strength of the rock [28]. The general form of this criterion can be expressed as: σ_θ/σ_c, where σ_θ is the maximum tangential stress, σ_c is the uniaxial compressive strength of the rock. Higher values of this ratio indicate a greater likelihood of rockburst occurrence. Different threshold values have been proposed by researchers to classify rockburst risk levels based on this ratio [27].

Another important stress-based parameter is the stress concentration factor (SCF), which is defined as: $SCF = \sigma_\theta/\sigma_c$. This factor provides insight into the degree of stress concentration around an excavation and its potential to trigger rockbursts [29].

(2) Brittleness Coefficient

The brittleness coefficient is a measure of the rock's tendency to fail in a brittle manner, which is closely related to rockburst potential. Two common formulations of the brittleness coefficient are:

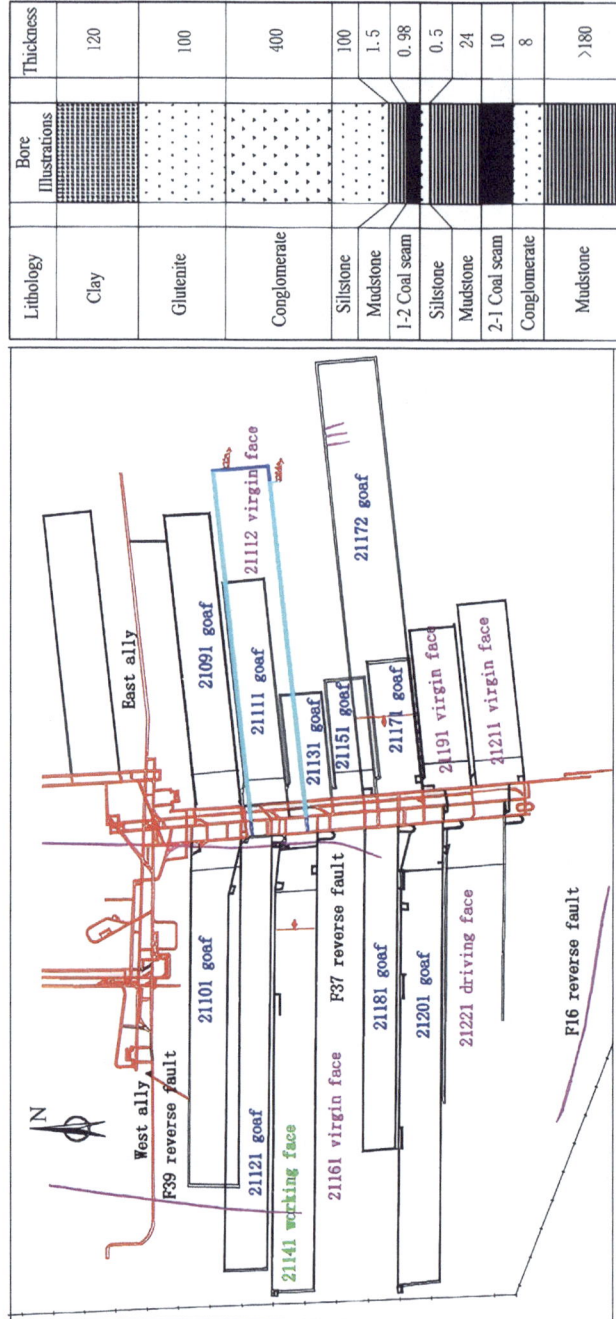

Fig. 6.18 Layout of the working face in the Qianqiu coal mine and the geological map histogram [20]

6.2 Precursors and Prediction Methods of Rock Burst

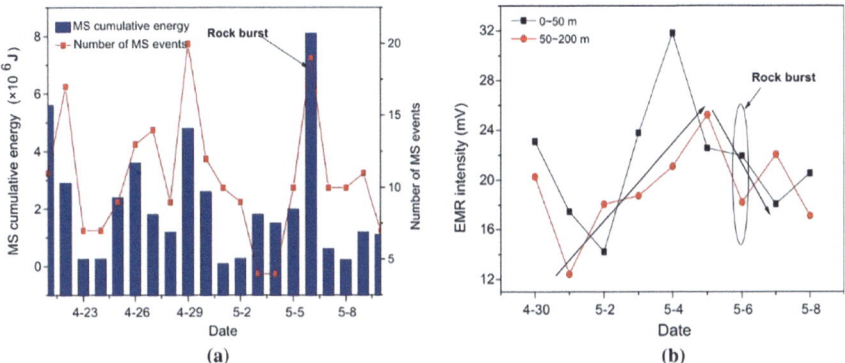

Fig. 6.19 Monitoring results for the rock burst. **a** MS results; **b** EMR results [20]

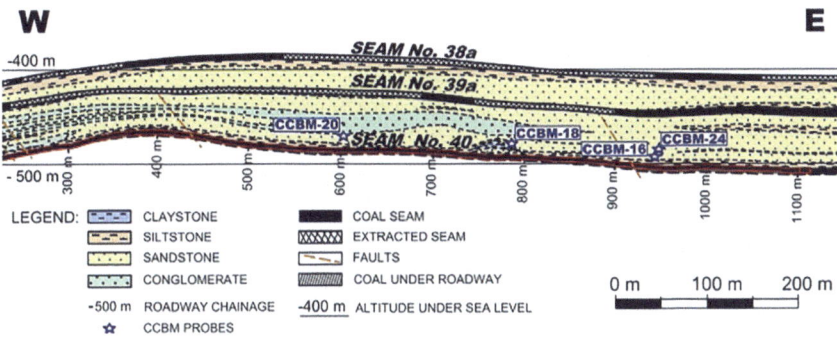

Fig. 6.20 Lithological cross-section across main gate no.40 [23]

$$B_1 = \frac{\sigma_c}{\sigma_t} \tag{6.7}$$

$$B_2 = \frac{\sigma_c - \sigma_t}{\sigma_c + \sigma_t} \tag{6.8}$$

(3) Energy Criteria

Where σ_c is the uniaxial compressive strength, σ_t is the tensile strength of the rock. Higher values of these brittleness coefficients generally indicate a greater susceptibility to rockbursts [29]. The brittleness coefficient is often used in combination with other criteria to provide a more comprehensive assessment of rockburst risk.

Energy-based criteria focus on the potential for sudden release of stored elastic strain energy, which is a fundamental characteristic of rockbursts. Some key energy-based parameters include:

(a) Elastic Energy Index (W_{et}):

$$W_{et} = \frac{\Phi_{sp}}{\Phi_{st}} \quad (6.9)$$

where Φ_{sp} is the stored elastic strain energy, Φ_{st} is the dissipated elastic strain energy. This index provides a measure of the rock's ability to store and release elastic energy [28]. Higher values of W_{et} indicate a greater potential for violent energy release during failure.

(b) Improved Brittleness Index (IBI):

$$IBI = \frac{A_1}{A_3} \quad (6.10)$$

where A_1 is the area under the stress–strain curve up to the peak strength, A_3 is the area under the stress–strain curve from the peak strength to the residual strength. The IBI provides information about the rock's post-peak behavior and its potential for sudden failure [27].

(c) Impact Energy Index (IEI):

$$IEI = \frac{A_1}{A_2} \quad (6.11)$$

where A_1 is the area under the stress–strain curve up to the peak strength, A_2 is the area under the stress–strain curve from the origin to the peak strength. This index reflects the rock's capacity to absorb energy before failure and is related to its rockburst proneness [27].

These energy-based criteria often provide valuable insights into the rock's potential for violent failure, complementing stress-based and brittleness-based approaches.

(4) Limitations of Theoretical Criteria

While theoretical criteria offer a straightforward approach to rockburst assessment, they have several limitations: (a) They often rely on simplified assumptions about rock behavior and stress conditions; (b) They may not adequately account for complex geological structures and heterogeneities in the rock mass; (c) The threshold values for different risk levels can vary depending on the specific geological context and may not be universally applicable; (d) They typically do not consider the dynamic nature of rockbursts and the influence of time-dependent factors. Despite these limitations, theoretical criteria remain valuable tools for initial assessments and are often used in combination with other prediction methods to provide a more comprehensive evaluation of rockburst risk.

2. **Case Analysis-Based Intelligent Methods**

As the complexity of rockburst phenomena often exceeds the capabilities of simple theoretical criteria, researchers have turned to more sophisticated data-driven

6.2 Precursors and Prediction Methods of Rock Burst

approaches. Case analysis-based intelligent methods leverage historical data from past rockburst events to develop predictive models. These methods often employ machine learning techniques to identify patterns and relationships that may not be apparent through conventional analysis. Some key approaches in this category include:

(1) Machine Learning Approaches

Machine learning approaches have been successfully applied to rockburst prediction. Support Vector Machines (SVM) have demonstrated effectiveness in classifying rockburst risk levels. For example, a Support Vector Classification model achieved high accuracy in predicting rockburst intensity in kimberlite pipes using a 246-case dataset. Artificial Neural Networks (ANN) have been employed to capture non-linear relationships between rockburst factors and occurrence [27]. Additionally, Random Forests have proven valuable for rockburst prediction due to their capability to process high-dimensional data and model complex variable interactions [27].

(2) Data Mining Techniques

Data mining approaches aim to extract meaningful patterns and rules from large datasets of rockburst cases. These techniques can help identify critical combinations of factors that lead to rockbursts and provide insights into the relative importance of different parameters [27].

(3) Fuzzy Inference Systems

Fuzzy logic-based methods have been applied to rockburst prediction to handle the inherent uncertainty and imprecision in rockburst assessment. These systems can incorporate expert knowledge and linguistic variables to provide a more nuanced evaluation of rockburst risk [30].

For example, Gong et al. [30] proposed an extended Multi-Attributive Border Approximation Area Comparison (MABAC) method under fuzzy conditions for rockburst risk assessment. This approach uses triangular fuzzy numbers to express uncertain assessment information and incorporates multiple evaluation indexes. The method was successfully applied to assess rockburst risk in a phosphate mine, demonstrating its ability to handle both qualitative and quantitative factors (Fig. 6.21).

Intelligent methods for rockburst prediction offer significant advantages, including the ability to model complex non-linear relationships, process large datasets, improve accuracy through continuous learning, and potentially reveal new insights into rockburst mechanisms. However, these methods also face notable limitations: they require extensive high-quality training data, risk overfitting, present interpretability challenges due to their "black box" nature, and may show reduced performance when applied to geological conditions differing from training data. Despite these constraints, the integration of intelligent methods with domain expertise and traditional prediction approaches shows promise for enhancing rockburst prediction accuracy.

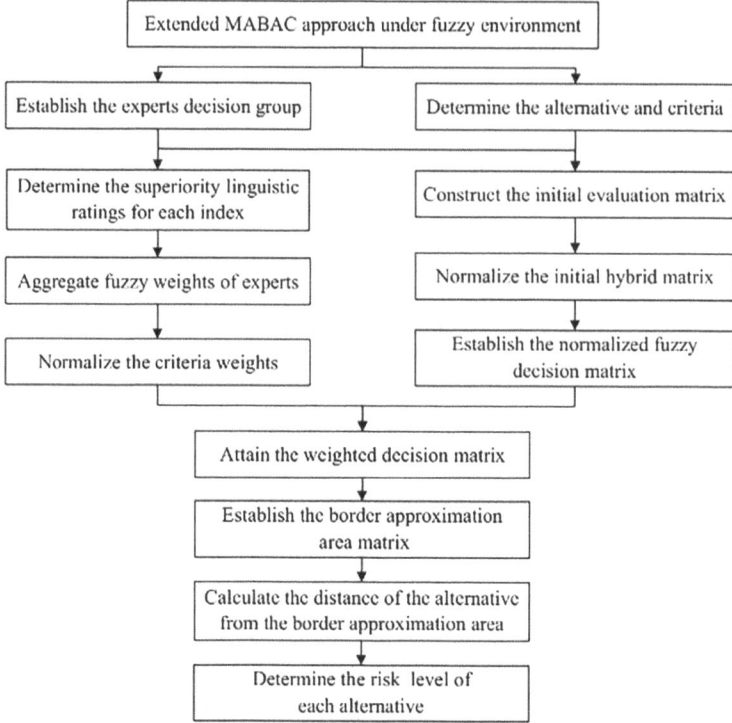

Fig. 6.21 Framework of the extended MABAC approach [30]

3. Field Monitoring-Based Methods

Field monitoring-based methods rely on real-time or near-real-time measurements of various physical parameters to detect precursors to rockbursts. These methods provide valuable insights into the evolving state of the rock mass and can potentially offer more timely warnings of impending rockbursts. The most prominent field monitoring techniques include:

(1) Microseismic Monitoring

Microseismic monitoring has emerged as one of the most powerful tools for rockburst prediction and understanding rockburst mechanisms. This technique involves installing a network of sensors to detect and locate small-scale seismic events (microseisms) within the rock mass [27, 31–34].

Microseismic monitoring provides multiple indicators for rockburst prediction through distinct characteristics. Precursory microseismic events, observed days before major rockbursts, indicate progressive rock mass damage accumulation [33]. The tempo-spatial evolution of microseismic events, conceptualized as the "3S principle" by Feng et al. [27], reflects stress redistribution patterns and potential rockburst locations. Rapid increases in energy release rates and cumulative energy often

6.2 Precursors and Prediction Methods of Rock Burst

precede rockburst occurrences [33]. The magnitude and spatial clustering of events serve as crucial indicators of rockburst potential [27], while sudden increases in cumulative apparent volume calculated from microseismic parameters can also signal impending rockbursts [34].

(2) Acoustic Emission Monitoring

AE monitoring is similar to microseismic monitoring but typically focuses on higher-frequency signals associated with smaller-scale fracturing processes. AE monitoringcan provide early indications of rock damage and stress changes that may lead to rockbursts [27] (Fig. 6.22).

(3) Electromagnetic Radiation Monitoring

Some researchers have explored the use of electromagnetic radiation (EMR) monitoring for rockburst prediction. EMR emissions from rocks under stress can potentially indicate impending failure. However, this technique is less widely used than microseismic monitoring and requires further validation [27].

(4) Integrated Monitoring Approaches

To improve prediction accuracy, researchers often combine multiple monitoring techniques. For example, Wang et al. [33] integrated microseismic monitoring with land-sonar technique for fracture detection. They found that rockburst locations were consistent with intersection points of fractures detected by landsonar, highlighting the role of geological structures in energy concentration and rockburst initiation.

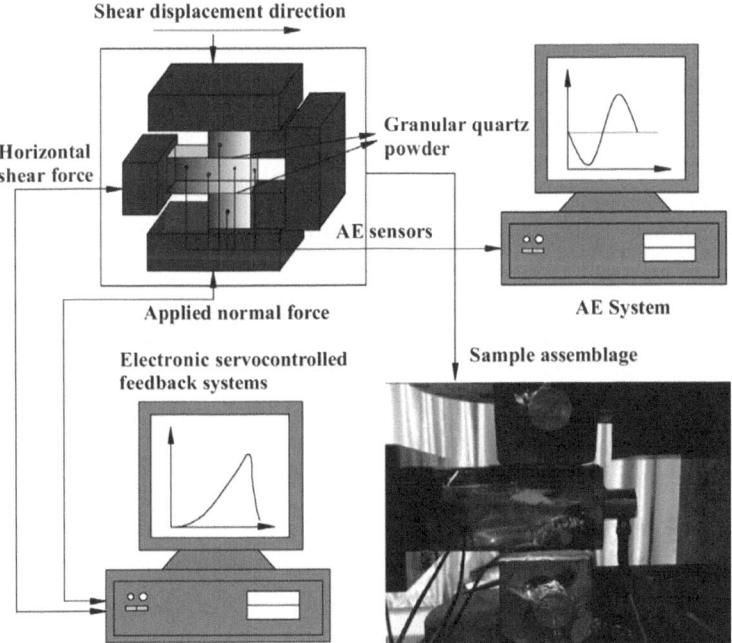

Fig. 6.22 Laboratory configuration sketch and sample assemblage [27]

Field monitoring methods for rockburst prediction offer key advantages: real-time or near-real-time data acquisition, detection of unique precursory signals, spatial–temporal tracking of rock mass response to excavation, and potential integration with numerical modeling. However, these methods face several limitations: substantial equipment and data processing investments, complex data interpretation requiring expert judgment, potentially insufficient warning time for certain rockburst types, and performance dependency on sensor placement and coverage. Despite these constraints, field monitoring methods, particularly microseismic monitoring, have emerged as essential tools in rockburst prediction and management for deep mining and tunneling operations.

4. **Integrated and Hybrid Approaches**

The complexity of rockburst prediction has led researchers to develop sophisticated integrated approaches that synthesize multiple predictive techniques. These hybrid methodologies aim to overcome the limitations of individual prediction methods by combining diverse analytical strategies [27, 29, 33].

One prominent approach involves integrating theoretical criteria with real-time monitoring data. Researchers utilize stress-based theoretical frameworks for initial risk assessment, subsequently refining predictions through continuous microseismic monitoring. Advanced techniques include developing hybrid models that merge data-driven algorithms with fundamental physical principles of rock mechanics, enhancing prediction accuracy while maintaining theoretical interpretability [27–29].

Comprehensive evaluation systems have emerged that leverage multi-parameter analysis. Notably, researchers are increasingly integrating advanced numerical modeling techniques with empirical monitoring data. Coupled static-dynamic simulations offer more holistic rockburst risk assessments by combining sophisticated computational models with real-world observational insights [27, 35]. These integrated methodologies represent a significant advancement in understanding and mitigating rock mass failure risks.

6.2.3 *Major Criteria for Rockburst Prediction*

Rockburst prediction is a critical aspect of ensuring safety in deep underground excavations. Over the years, researchers have developed various criteria and methods to assess rockburst potential. This section provides a comprehensive overview of the major criteria used for rockburst prediction, their underlying principles, advantages, limitations, and recent advancements.

1. **Stress-Based Criteria**

Stress-based criteria are among the earliest and most widely used methods for rockburst prediction. These criteria typically compare the in-situ stress state to the rock strength to assess the likelihood of violent failure.

6.2 Precursors and Prediction Methods of Rock Burst

(1) Tangential Stress Criterion

One of the fundamental stress-based criteria is the tangential stress criterion, which compares the maximum tangential stress around an excavation to the uniaxial compressive strength of the rock. The criterion is often expressed as:

$$R_b = \frac{\sigma_\theta}{\sigma_c} \qquad (6.12)$$

where R_b is the rockburst proneness index, σ_θ is the maximum tangential stress, σ_c is the uniaxial compressive strength of the rock. This criterion has been widely used due to its simplicity and ease of application. However, it has limitations in capturing the complex stress states in deep underground environments [36, 37].

(2) Stress Ratio Critteria

Various stress ratio criteria have been proposed to account for different stress components. For example, Hoek and Brown [38] suggested using the ratio of major principal stress to uniaxial compressive strength:

$$\frac{\sigma_1}{\sigma_c} \qquad (6.13)$$

where σ_1 is the major principal stress. This criterion has been modified and expanded by various researchers to include additional stress components or rock properties [37, 39].

(3) Strength/Stress Ratio Criteria

These criteria consider both the rock strength and the in-situ stress state. For example, the criterion proposed by Barton et al. [40] uses the ratio:

$$\frac{\sigma_c}{\sigma_1} \qquad (6.14)$$

where σ_c is the uniaxial compressive strength and σ_1 is the major principal stress. This approach aims to account for both the rock's ability to resist failure and the stress conditions it experiences [41].

(4) Limitations of Stress-Based Criteria

While stress-based criteria are widely used due to their simplicity, they have several limitations: (a) They often do not account for rock mass properties or discontinuities; (b) They may not capture the dynamic nature of rockbursts; (c) They can be overly conservative in some cases, leading to overdesign of support systems [37, 39].

2. **Energy-Based Criteria**

Energy-based criteria focus on the energy accumulation and release processes associated with rockbursts. These criteria often provide a more comprehensive assessment of rockburst potential by considering the rock's ability to store and release elastic strain energy.

(1) Elastic Strain Energy Index (W_{et})

The elastic strain energy index is defined as:

$$W_{et} = \frac{\sigma_c^2}{2E} \tag{6.15}$$

where σ_c is the uniaxial compressive strength, E is the elastic modulus. This index represents the capacity of the rock to store elastic strain energy. Higher values of Wet indicate a greater potential for violent failure [37, 39].

(2) Bursting Energy Index (K_{et})

The bursting energy index, introduced by Kidybinski [42], is calculated as:

$$K_{et} = \frac{\sigma_c^2}{2E_{avg}} \tag{6.16}$$

where E_{avg} is the average modulus of the complete stress–strain curve. This index aims to capture both the energy storage and release characteristics of the rock [39].

(3) Rockburst Energy Release Rate (*ERR*)

The energy release rate concept, originally developed for deep gold mines in South Africa, quantifies the rate of energy release during excavation:

$$ERR = \frac{W_1 - W_2}{\Delta V} \tag{6.17}$$

where W_1 is the work done by pre-mining stresses, W_2 is the work done by post-mining stresses; ΔV is the volume of excavated rock. Higher ERR values indicate a greater potential for rockbursts [37, 40].

3. **Brittleness-Based Criteria**

Brittleness-based criteria focus on the rock's failure characteristics, particularly its tendency for sudden, violent failure under stress.

(1) *B*

The brittleness coefficient, proposed by Hucka and Das [43], is defined as:

$$B = \frac{\sigma_c - \sigma_t}{\sigma_c + \sigma_t} \tag{6.18}$$

where σ_c is the uniaxial compressive strength, σ_t is the tensile strength. Higher values of B indicate more brittle behavior and a greater propensity for rockbursts [44].

Fig. 6.23 Contrast diagram of elastic energy of coal seam and rock formation accumulation under uniform stress condition [39]

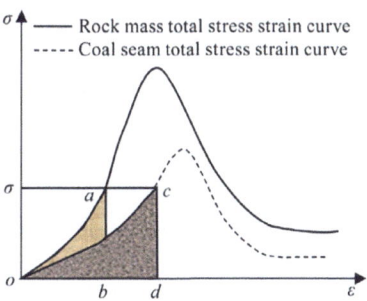

(2) Stress–Strain Curve Shape

Some researchers have proposed using the shape of the complete stress–strain curve to assess rockburst potential. For example, the ratio of post-peak modulus to pre-peak modulus has been used as an indicator of brittle behavior [37, 39] (Fig. 6.23).

4. **Rock Mass Classification-Based Criteria**

These criteria incorporate rock mass classification systems to assess rockburst potential, aiming to account for the overall quality and behavior of the rock mass.

(1) Q-System Based Criterion

Barton et al. (1974) proposed using the Q-system rock mass classification to assess rockburst potential. The criterion considers the ratio of rock mass quality (Q) to excavation dimension and in-situ stress [37, 40].

(2) RMR-Based Criteria

Some researchers have developed rockburst prediction methods based on the Rock Mass Rating (RMR) system, often in combination with other factors such as stress conditions or energy parameters [37].

5. **Multi-factor Criteria**

Recognizing the complex nature of rockbursts, many researchers have proposed multi-factor criteria that combine various aspects of stress conditions, rock properties, and excavation characteristics.

(1) *EVP*

The *EVP* criterion, proposed by Heal et al. [45], combines three factors:

$$EVP = \left(\frac{\sigma_1}{\sigma_c}\right) \cdot (\sigma_1 - \sigma_3) \cdot E_{un} \quad (6.19)$$

where σ_1 and σ_3 are the major and minor principal stresses, σ_c is the uniaxial compressive strength; Eun is the modulus of the unloading portion of the stress–strain curve. This criterion aims to capture stress conditions, rock strength, and energy release characteristics [37].

(2) *RVI*

The *RVI*, developed by Feng et al. [46], incorporates four factors:

$$RVI = F_s \cdot F_r \cdot F_m \cdot F_g \tag{6.20}$$

where F_s is a stress control factor, F_r is a rock petrophysical factor, F_m is a rockmass system stiffness factor, F_g is a geologic structure factor. This comprehensive index aims to account for stress conditions, rock properties, excavation geometry, and geological structures [47].

6. **Advanced Prediction Methods**

In recent years, researchers have developed more sophisticated methods for rockburst prediction, leveraging advancements in data analysis, numerical modeling, and artificial intelligence.

(1) Fuzzy Comprehensive Assessment

Fuzzy logic methods have been applied to rockburst prediction to handle the uncertainty and imprecision inherent in many input parameters. For example, an interval fuzzy comprehensive assessment method has been developed. This method incorporates multiple indicators and handles uncertainty through interval numbers and fuzzy set theory [48] (Fig. 6.24).

(2) Machine Learning Approaches

Various machine learning techniques have been applied to rockburst prediction, including: *ANN*, *SVM*, Random Forests, Gradient Boosting Machines. These methods can capture complex, non-linear relationships between input parameters and rockburst potential. They often show improved predictive accuracy compared to traditional criteria, especially when trained on large datasets [37].

(3) Hybrid Models

Researchers have developed hybrid models that combine multiple techniques to leverage the strengths of different approaches. For example, combining fuzzy logic with neural networks or integrating machine learning with numerical modeling [37].

(4) Real-Time Prediction Methods

Advancements in monitoring technologies and data analysis have enabled the development of real-time prediction methods. These approaches often integrate data from microseismic monitoring, stress measurements, and excavation parameters to provide continuous assessment of rockburst risk [49].

Advanced prediction methods offer significant capabilities, including the ability to capture complex non-linear relationships between parameters and rockburst potential, handle data uncertainty and imprecision, and, particularly in machine learning approaches, improve predictions through continuous data incorporation. However, these methods face notable limitations: they typically require extensive datasets for

6.2 Precursors and Prediction Methods of Rock Burst

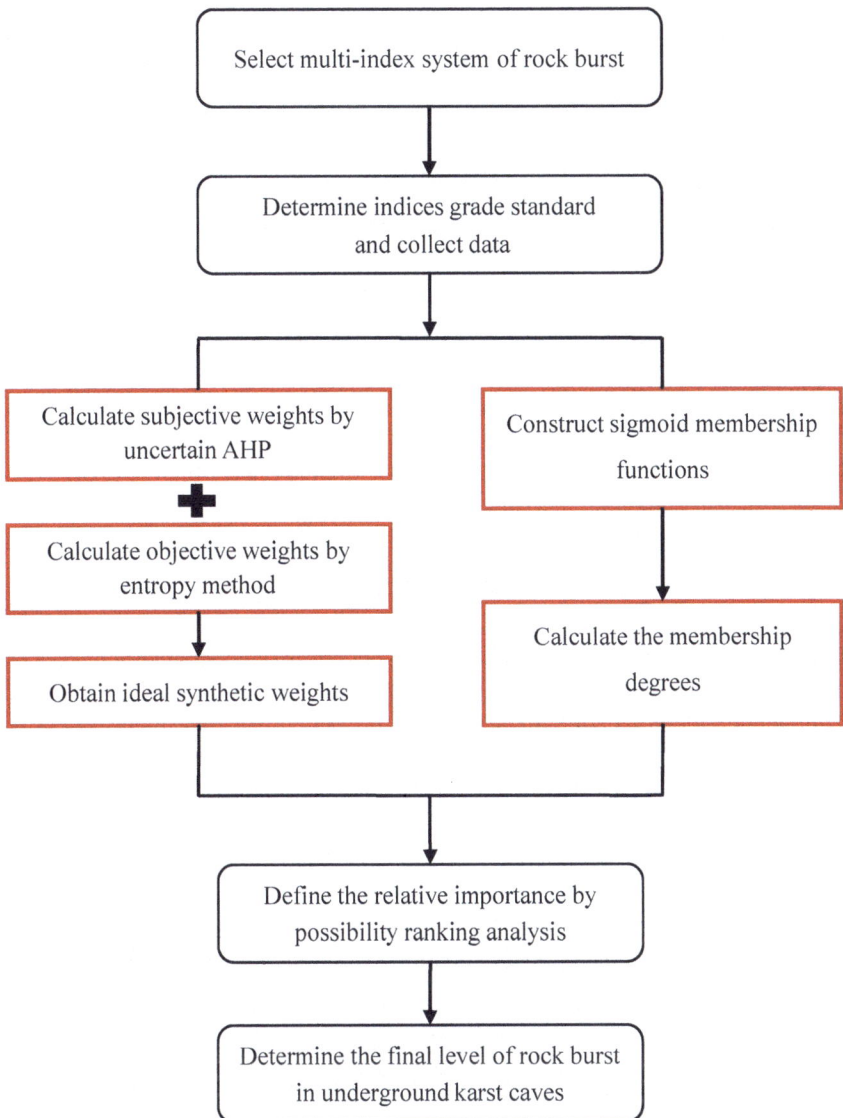

Fig. 6.24 The predicted process of rock burst hazard in underground caverns [48]

training and validation, may lack interpretability (especially in neural networks), and their performance is heavily dependent on the quality and representativeness of training data.

6.3 Rockburst Dynamic Mechanism

6.3.1 Rockburst Classification and Initiation Mechanisms

Rockbursts are one of the most severe and unpredictable hazards in deep underground excavations, posing significant risks to personnel safety and project stability. As mining and tunneling activities reach greater depths, understanding the classification and initiation mechanisms of rockbursts becomes increasingly crucial for effective risk assessment and mitigation. This section provides a comprehensive review of rockburst classification schemes and initiation mechanisms.

1. **Rockburst Classification**

Rockbursts can be classified based on various criteria, including their triggering mechanisms, damage characteristics, and timing relative to excavation. Several classification schemes have been proposed in the literature, each offering unique insights into the nature of these complex phenomena.

(1) Classification Based on Triggering Mechanism

One of the most fundamental ways to classify rockbursts is based on their primary triggering mechanism. This approach helps in understanding the underlying causes and prevention strategies. Two main categories are widely recognized [37, 50, 51]: (a) Strain bursts, also known as stress-induced rockbursts, occur due to the excessive concentration of stress in the rock mass surrounding an excavation. These rockbursts are typically localized and occur in close proximity to the excavation boundary. (b) Impact-induced bursts are triggered by external dynamic loads or disturbances that cause sudden stress changes in the rock mass.

(2) Classification Based on Damage Characteristics

Another approach to classifying rockbursts is based on the observed damage patterns and severity. This classification is particularly useful for post-event analysis and risk assessment. A commonly used classifcation includes the following categories [28, 37]: (a) None to minor: characterized by minor spalling or slabbing of rock surfaces; no significant impact on excavation stability or operations. (b) Moderate: localized ejection of rock fragments; visible damage to support systems; may require some additional support installation. (c) Strong: substantial rock ejection and support damage; potential for equipment damage and production delays; requires extensive rehabilitation of the affected area. (d) Severe: massive rock ejection and complete failure of support systems; significant damage to equipment and infrastructure; may result in partial or complete closure of the excavation; poses severe safety risks and requires major rehabilitation efforts.

(3) Classification Based on Timing Relative to Excavation

The timing of rockburst occurrence relative to the excavation process is another important aspect of classification. This temporal classification helps in understanding

6.3 Rockburst Dynamic Mechanism

the evolution of rockburst risk over time and informs monitoring and support strategies. Two main categories are recognized [52, 53]: (a) Immediate rockbursts: occur during excavation or within hours to 3 days after excavation; typically result from the immediate stress redistribution caused by excavation. (b) Time-delayed Rockbursts (TDRs): occur outside the immediate stress adjustment range of the tunnel face, typically 6–30 days after excavation and within 80 m behind the face; Result from the combined effects of long-term stress redistribution and external disturbances; often associated with complex geological conditions, such as areas rich in joints, fractures, and intercalations.

2. **Rockburst Initiation Mechanisms**

Understanding the fundamental mechanisms that initiate and drive rockburst events is crucial for developing effective prediction and prevention strategies. This section explores the key physical processes and conditions that lead to rockburst initiation.

(1) Energy Accumulation and Release

At its core, a rockburst represents a rapid release of stored strain energy in the rock mass. The energy accumulation process is primarily driven by three factors. (a) High in-situ stresses: particularly in deep underground environments, contribute to the storage of elastic strain energy in the rock mass. As excavation proceeds, this stored energy can be released violently if the rock's strength is exceeded. (b) Stress Redistribution: excavation causes a redistribution of stresses in the surrounding rock mass. This redistribution can lead to stress concentrations in certain areas, particularly near excavation boundaries, corners, and pillars. (c) Rock Properties: elastic modulus, which correlates positively with energy storage potential; brittleness, which determines the violence of failure; and strength, which dictates the maximum energy accumulation before failure occurs.

(2) Stress-Induced Failure Mechanisms

The specific failure mechanisms that initiate rockbursts can vary depending on the stress conditions and rock properties. Common mechanisms include: Tensile fracturing: In highly stressed rock near excavation boundaries, tensile stresses can develop perpendicular to the maximum compressive stress. This can lead to the formation and propagation of tensile cracks, which may coalesce and result in rock ejection. Shear fracturing: When shear stresses exceed the rock's shear strength, fractures can develop along planes of maximum shear stress. This mechanism is particularly important in fault slip-induced rockbursts. Mixed-mode fracturing: in many cases, rockbursts involve a combination of tensile and shear fracturing. The specific mode of failure depends on the local stress state and rock properties.

High stress concentrations play a central role in rockburst initiation. The redistribution of stresses around underground excavations can lead to areas of high stress concentration that exceed the rock's strength. This process can be described by the following stress-based criteria [54]:

Tangential stress criterion:

$$\frac{\sigma_\theta}{\sigma_c} > 0.4 \pm 0.1 \qquad (6.21)$$

where σ_θ is the maximum tangential stress around the opening, and σ_c is the uniaxial compressive strength of the rock.

Maximum principal stress criterion:

$$\frac{\sigma_1}{\sigma_c} > 0.15 \pm 0.05 \qquad (6.22)$$

where σ_1 is the maximum principal stress in the rock mass.

These criteria suggest that rockbursts are likely to occur when the stress levels exceed certain thresholds relative to the rock strength. However, it's important to note that these thresholds can vary depending on rock type and geological conditions.

(3) Role of Structural Features

Geological structures critically influence rockburst initiation and propagation through complex stress interactions and failure mechanisms. Pre-existing faults, shear zones, joints, and fractures serve as potential planes of weakness where stress concentrations can trigger sudden energy release. In sedimentary and metamorphic rocks, bedding planes and foliation create anisotropic strength properties that can facilitate rockburst initiation when unfavorably oriented relative to the stress field. Rock mass heterogeneity, characterized by variations in rock type, strength, and stiffness, further contributes to stress concentration at material interfaces. These structural features can induce diverse failure mechanisms, including shear failure along discontinuities, tensile fracturing in intact rock bridges, and block rotation and ejection. Notably, higher quality rock masses (RMR > 60 or Q < 10) demonstrate increased susceptibility to rockbursts due to their capacity to sustain high stresses before catastrophic failure [55, 56].

(4) Dynamic Triggering Mechanisms

Dynamic triggers play a crucial role in rockburst events, with both natural and anthropogenic sources initiating complex stress interactions. Seismic events, including earthquakes and mining-induced seismicity, generate stress waves that transiently increase rock mass stress states, apply dynamic loads to pre-existing fractures, and modify pore pressure conditions affecting rock strength. Blasting operations similarly trigger rockbursts by generating stress waves, creating new fractures, and dynamically loading discontinuities, particularly in multi-tunnel environments. In mining contexts, roof strata collapse contributes to rockburst potential through rapid load transfers, stress wave propagation, and comprehensive alterations of the local stress distribution, demonstrating the intricate mechanisms that can precipitate catastrophic rock mass failures.

6.3 Rockburst Dynamic Mechanism

(5) Influence of Excavation Method and Support Systems

Excavation methods and support systems play crucial roles in rockburst initiation mechanisms. While drill and blast techniques can introduce additional dynamic stresses and fractures, mechanical excavation methods typically provide more uniform stress redistribution, though both can trigger rockbursts under high-stress conditions. The selection of support systems significantly influences failure patterns: rigid supports may induce stress concentrations at rock-support interfaces, whereas yielding supports accommodate deformation, potentially mitigating sudden failures. Additionally, the temporal relationship between support installation and excavation critically affects stress redistribution within the rock mass.

(6) Thermomechanical Effects

Thermomechanical effects significantly influence rockburst initiation in deep underground environments through multiple mechanisms. Temperature gradients between excavations and surrounding rock masses generate thermal stresses that contribute to the overall stress state. Additionally, temperature fluctuations induce differential expansion or contraction of mineral grains, promoting microcrack initiation and propagation, thereby weakening the rock mass. In saturated conditions, thermal effects modify pore fluid pressures, altering the effective stress state. While these thermomechanical processes are generally secondary to mechanical factors, they warrant consideration in comprehensive rockburst mechanism assessments, particularly in deep excavations or environments with substantial temperature variations.

3. Quantitative Models for Rockburst Initiation

To better understand and predict rockburst initiation, researchers have developed various quantitative models. These models aim to capture the complex interplay of factors that lead to rockburst events. Some key approaches include:

(1) Energy-Based Models

Energy-based models focus on the balance between stored strain energy and the rock's capacity to dissipate energy. A general form of these models can be expressed as [57]:

$$\Delta E = E(\sigma_{1c}) - E(\sigma_c) > 0 \qquad (6.23)$$

where ΔE is the excess energy; $E(\sigma_{1c})$ is the energy at maximum principal stress; $E(\sigma_c)$ is the energy at uniaxial compressive strength. This criterion suggests that rockbursts occur when the energy stored in the rock exceeds its capacity to dissipate energy through stable deformation (Fig. 6.25).

(2) Stress-Based Criteria

Stress-based criteria employ various methodological approaches to predict rockburst potential by analyzing the relationship between induced stresses and rock strength.

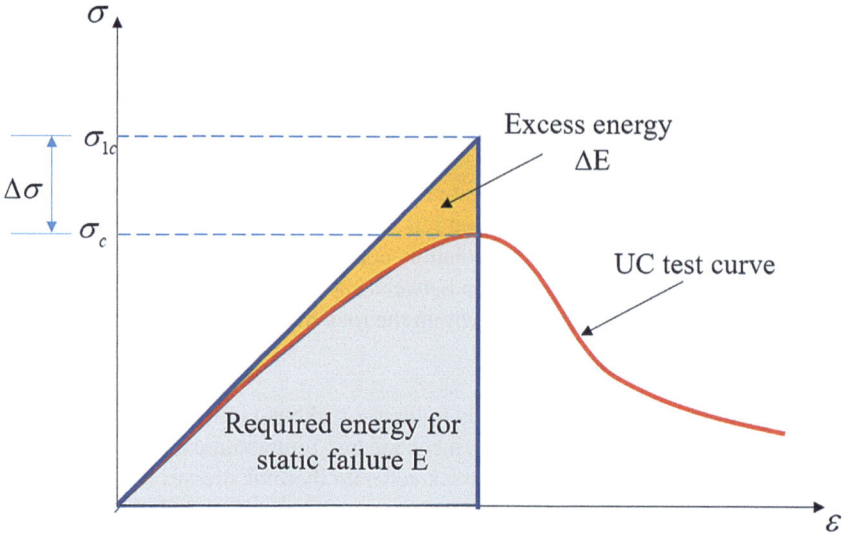

Fig. 6.25 Energy model of strainburst [4]

These approaches encompass three primary methods: stress ratio methods, which evaluate the relationship between induced and inherent rock stresses; the Hoek–Brown failure criterion, which considers the non-linear nature of rock mass strength; and fault slip criteria, which assess the potential for slip-induced rockbursts along existing discontinuities.

(3) Strain-Based Approaches

Strain-based approaches analyze rock mass deformation behavior by focusing on three fundamental deformation characteristics. These methods incorporate critical strain thresholds that define various failure modes, account for strain rate effects on rock strength and failure behavior, and monitor the accumulation of plastic strain as a precursor to failure. This integrated analysis of deformation parameters enables comprehensive assessment of rock mass stability.

(4) Probabilistic and Statistical Models

Probabilistic approaches to rockburst risk assessment employ three key analytical methods to address the inherent complexity and unpredictability of these events. These methods integrate statistical analysis of historical rockburst occurrences, utilize Bayesian approaches to dynamically update risk assessments as new data becomes available, and implement Monte Carlo simulations to account for uncertainties in input parameters. This comprehensive probabilistic framework enables more robust and adaptive risk assessment in complex geological environments.

(5) Machine Learning and Data-Driven Models

The emergence of enhanced computational capabilities and abundant monitoring data has facilitated the application of machine learning approaches to rockburst prediction

[37]. These advanced models demonstrate superior capability in integrating multiple parameters and identifying complex patterns that may elude traditional analytical methods. The primary machine learning techniques employed in rockburst prediction include Artificial Neural Networks (ANN), Support Vector Machines (SVM), Random Forests (RF), and Gradient Boosting Machines (GBM), each offering unique advantages in pattern recognition and prediction accuracy [37].

(6) Hybrid and Multi-scale Models

Hybrid modeling approaches in rockburst prediction integrate multiple techniques to provide a more comprehensive understanding across different scales. These models combine numerical-statistical analysis, multi-scale finite element modeling with microstructural detail incorporation, and integrated geomechanical-seismic modeling methodologies. This integrated approach enables a more thorough investigation of rockburst mechanisms, spanning from microscopic fracture processes to macroscopic rock mass behavior, thereby overcoming the limitations of individual prediction methods.

6.3.2 Rockburst Grading and Prediction

Rockbursts are complex and dangerous phenomena that pose significant challenges in deep underground excavations. Understanding the grading of rockbursts is crucial for effective risk assessment and mitigation strategies. This section provides a comprehensive review of current knowledge on rockburst classification systems.

1. **Rockburst Classification Systems**

Rockburst classification systems are essential tools for assessing the severity and potential impact of these events. Several classification schemes have been proposed over the years, each focusing on different aspects of rockburst phenomena. Here, we discuss some of the most widely used and recently developed classification systems.

(1) Severity-Based Classification

Rockbursts are commonly classified into four distinct levels based on their observable severity and potential consequences. The classification ranges from "No rockburst," characterized by the absence of visible damage or audible noise, to "Weak rockburst," which involves minor spalling and small rock fragment ejection with accompanying cracking sounds. "Moderate rockburst" represents a more significant event, featuring substantial rock fragment ejection, noticeable seismic activity, and potential minor equipment damage. The most critical category, "Strong or severe rockburst," involves violent rock mass ejection, intense seismic events, extensive excavation and equipment damage, and poses a serious risk of injuries or fatalities [36, 40, 55]. This classification system provides a systematic approach to understanding and characterizing the progressive levels of rock mass failure and associated hazards.

(2) Mechanism-Based Classification

Rockbursts are classified into four distinct types based on their underlying source mechanisms, each representing a unique mode of rock mass failure. Strain bursts involve the sudden release of stored elastic strain energy in rock masses surrounding excavations, characterized by high stress concentrations and brittle rock behavior. Fault-slip bursts emerge from abrupt movements along pre-existing geological discontinuities, typically triggered by stress redistribution during mining activities. Pillar bursts occur when load-bearing rock pillars exceed their structural capacity, resulting in sudden and violent failure, a phenomenon frequently observed in room-and-pillar mining operations. Buckling bursts represent a specialized mechanism involving the buckling failure of thin, plate-like rock layers under high compressive stresses, commonly encountered in layered sedimentary rock formations [37, 56]. This classification provides critical insights into the diverse mechanical processes underlying rock mass instability and failure.

(3) Composite Index-Based Classification

Recent research has focused on developing more comprehensive classification systems that incorporate multiple factors influencing rockburst potential. One such approach is the composite rockburst index (K_{rb}). This index combines three key parameters:

$$K_{rb} = K_u \times K_w \times K_\sigma \tag{6.24}$$

where K_u is the brittleness index, defined as the ratio of total deformation to permanent deformation before peak strength; K_w is the elastic strain energy index, representing the ratio of elastic strain energy to critical elastic energy for ejection; K_σ is the stress drop index, defined as the ratio of stress drop after peak strength to critical stress drop for ejection. The K_{rb} index provides a more nuanced assessment of rockburst potential, with the following classification criteria:

$K_{rb} < 3.5$: No rock burst potential;
$3.5 \leq K_{rb} < 15$: Weak rock burst potential;
$15 \leq K_{rb} < 50$: Moderate rock burst potential;
$K_{rb} \geq 50$: Strong rock burst potential.

This composite index approach offers a more comprehensive evaluation of rockburst risk by considering multiple factors that contribute to the phenomenon.

(4) Energy-Based Classification

An energy-based classification approach for rockbursts utilizes ejection velocity as a critical quantitative metric to categorize event severity. The classification ranges from "No ejection" (velocity < 0.1 m/s) to "Strong ejection" (velocity ≥ 5 m/s), with intermediate categories of "Weak ejection" (0.1–1 m/s) and "Moderate ejection" (1–5 m/s). This method provides a precise, measurable framework for characterizing the dynamic energy release during rock mass failure, offering researchers and practitioners an objective means of assessing the potential destructive potential of rockburst events. By quantifying the velocity of rock fragment ejection, this approach bridges the gap between observational and energy-based understanding of rock mass instability.

2. Integrated Approaches to Rockburst Prediction

Modern rockburst prediction strategies leverage advanced machine learning techniques like artificial neural networks and support vector machines to analyze complex relationships across diverse input parameters. These data-driven approaches integrate multiple factors including rock properties, in-situ stress conditions, excavation characteristics, and microseismic indicators, enabling more sophisticated predictive capabilities [58, 59].

Complementing machine learning methods, numerical modeling techniques such as finite element and discrete element methods provide critical insights into rock mass behaviors by simulating intricate stress distributions and failure mechanisms. These approaches generate key predictive indicators like energy release rate, excess shear stress, and failure approaching index, offering a comprehensive framework for evaluating rockburst potential that transcends traditional empirical and analytical approaches [56, 60].

6.4 Rockburst Engineering Protection Method

6.4.1 Emergency Plans for Rockburst Hazards

Effective emergency planning is crucial for mitigating the potentially catastrophic consequences of rockburst events in underground excavations. As rockbursts are characterized by their suddenness and destructiveness [58], having well-prepared emergency plans can significantly reduce the risk to personnel safety and minimize damage to equipment and infrastructure. This section discusses key components of emergency plans for rockburst hazards, including preparedness measures, response protocols, and post-event recovery strategies.

1. **Risk Assessment and Preparedness**

The foundation of any effective emergency plan lies in a thorough understanding of the rockburst risk. This involves comprehensive risk assessment techniques that consider various factors contributing to rockburst potential.

(1) Bow-tie Analysis

One valuable tool for rockburst risk assessment is the bow-tie analysis method [35]. This approach provides a visual representation of the potential causes and consequences of rockburst events, as well as the controls in place to prevent or mitigate them (Fig. 6.26).

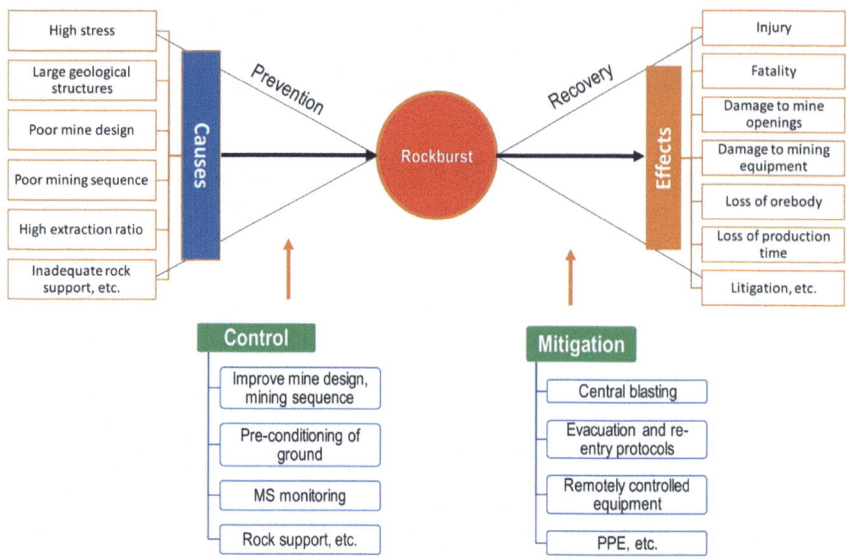

Fig. 6.26 Bow-tie analysis of rockburst risk [35]

(2) Microseismic Monitoring Systems

Implementing robust microseismic monitoring systems is a crucial aspect of rockburst preparedness [35]. Modern microseismic monitoring systems often integrate advanced data processing algorithms and machine learning techniques to improve their predictive capabilities [61]. When designing emergency plans, it is essential to establish clear protocols for interpreting and responding to microseismic data, including defined warning thresholds that trigger specific actions.

2. **Emergency Response Protocols**

When a rockburst event occurs or is imminent, having clear and well-rehearsed response protocols is essential for minimizing harm and facilitating an effective recovery. Key elements of these protocols include:

(1) Centralized Blasting and Exclusion Zones

To reduce the risk of personnel exposure to rockburst events, many deep mines implement centralized production blasting schedules [35]. This approach involves: coordinating all blasting activities to occur at predetermined times, establishing exclusion zones around active mining areas during and immediately after blasting, implementing strict re-entry protocols based on microseismic monitoring data and visual inspections. By concentrating potentially triggering activities (like blasting) into specific time windows and clearing personnel from high-risk areas, the potential for injuries due to rockbursts can be significantly reduced.

6.4 Rockburst Engineering Protection Method

Fig. 6.27 Rockburst risk exposure management [35]

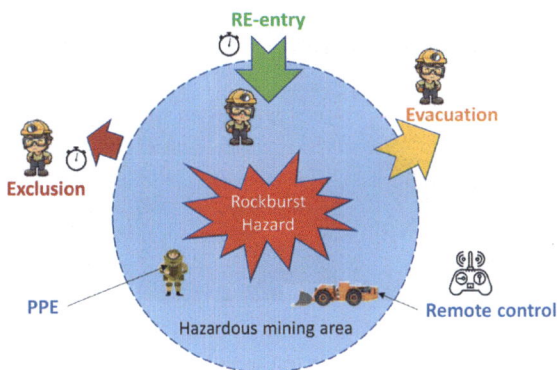

(2) Evacuation and Re-entry Procedures

Comprehensive safety protocols for rockburst events encompass two critical components: evacuation and re-entry procedures. The evacuation protocol requires established routes, safe assembly points, effective communication systems, and personnel accounting measures. Re-entry procedures involve a systematic stability assessment, including microseismic activity monitoring, professional geotechnical inspections, and thorough evaluation of ground support systems' integrity, ensuring safe resumption of operations only after necessary repairs are completed (Fig. 6.27).

(3) Emergency Communication Systems

Emergency response to rockbursts necessitates redundant communication systems to ensure uninterrupted information relay during critical situations. The communication infrastructure encompasses multiple complementary systems, including underground phones, wireless devices, leaky feeder radio systems, and visual-audible alarms. Regular testing and maintenance protocols are essential to guarantee system reliability during emergencies.

(4) Rescue and First Aid Provisions

Comprehensive emergency response plans for rockburst events must incorporate robust provisions for rescue operations and medical assistance, despite preventive measures. These provisions encompass strategically positioned first aid stations and emergency medical supplies, trained mine rescue teams with specialized equipment, and established protocols for expedited evacuation of injured personnel to medical facilities.

3. **Post-Event Recovery and Analysis**

After a rockburst event, a systematic approach to recovery and analysis is essential for both immediate safety concerns and long-term risk mitigation.

(1) Damage Assessment and Rehabilitation

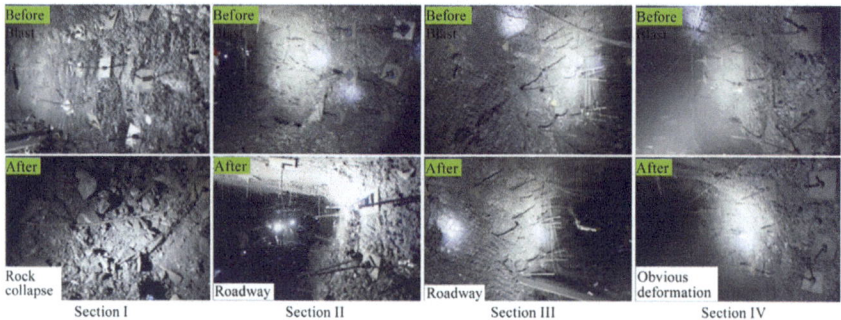

Fig. 6.28 Comparison of the test sections before and after blasting [51]

Following a rockburst, a thorough assessment of the affected area must be conducted to determine the extent of damage and necessary rehabilitation measures. This typically involves: geotechnical inspections to assess rock mass stability, evaluation of ground support systems and determination of required repairs or upgrades, assessment of any damage to infrastructure, equipment, or utilities.

Based on this assessment, a rehabilitation plan should be developed and implemented before normal operations can resume. This may involve installing additional ground support, scaling loose rock, or even modifying the mine design to reduce future rockburst risk [51] (Fig. 6.28).

(2) Event Analysis and Lessons Learned

Each rockburst event provides valuable data for improving future prediction and prevention efforts. A comprehensive post-event analysis should be conducted, including detailed examination of microseismic data leading up to and during the event, analysis of any precursor signals that may have been overlooked, review of the effectiveness of existing control measures and emergency response procedures, identification of any gaps in the emergency plan or areas for improvement. The insights gained from this analysis should be used to refine rockburst prediction models, update risk assessments, and improve emergency plans for future events.

(3) Updating Emergency Plans

Based on the lessons learned from each rockburst event and ongoing research in the field, emergency plans should be regularly reviewed and updated. This process should involve: incorporating new technologies and best practices in rockburst prediction and control; refining communication and evacuation procedures based on real-world experiences; updating training programs to address any identified gaps in knowledge or skills.

Effective emergency planning for rockburst hazards requires a multifaceted approach that combines thorough risk assessment, well-defined response protocols, and continuous improvement based on real-world experiences and emerging technologies. By implementing comprehensive emergency plans and staying abreast of advances in rockburst research and technology, mine operators can significantly reduce the risks associated with these unpredictable and potentially catastrophic events [35, 51, 58, 61].

6.4.2 Comprehensive Mitigation Measures for Rockburst Hazards

Rockbursts pose significant challenges to the safety and efficiency of deep underground excavations. As mining and tunneling projects venture into greater depths, the risk of rockburst events increases, necessitating the development and implementation of comprehensive mitigation strategies. This section examines the various approaches and techniques employed to prevent, control, and mitigate rockburst hazards in deep underground environments.

1. **Strategic Engineering Controls**

Strategic engineering controls form the foundation of a robust rockburst mitigation plan. These measures are implemented during the planning and design phases of a project to minimize the likelihood and potential impact of rockburst events.

(1) Mine Design Considerations

The strategic design of underground excavations is fundamental in rockburst risk management through three primary aspects. Mining method selection significantly impacts rock mass stress redistribution, with mechanized cut-and-fill and longwall mining typically preferred over room-and-pillar methods in high-stress conditions. Excavation layout optimization requires careful consideration of opening geometry and orientation, emphasizing alignment with maximum principal stress directions and elimination of geometric discontinuities. Additionally, effective pillar design incorporates critical parameters such as width-to-height ratio and pillar strength relative to in-situ stress, with yield pillar designs in deep hard rock mining facilitating controlled deformation and energy dissipation to mitigate violent failure risks [35, 58].

(2) Mining Sequence Optimization

The optimization of excavation sequencing is crucial for managing stress redistribution and rockburst potential in underground mining operations. Three primary techniques have been developed: the stress shadow technique, which creates protective zones through initial large excavations in low-stress areas, as demonstrated in South African gold mines [35]; retreat mining, which progresses from lower to higher stress zones to effectively manage stress concentrations, particularly in longwall coal mining operations [58]; and de-stress mining, which prioritizes ore extraction in less stressed areas before advancing to high-stress zones, proving especially effective in tabular orebodies [35].

(3) Rockburst Hazard Assessment

Effective rockburst risk management necessitates comprehensive hazard assessment methodologies encompassing three key components. The process begins with detailed geological and geotechnical characterization, including systematic mapping and analysis of rock mass properties, discontinuities, and in-situ stress conditions. This is complemented by advanced three-dimensional numerical simulations that predict stress distributions, potential failure modes, and energy release rates associated with planned excavations. Additionally, emerging machine learning and artificial intelligence technologies are being employed to analyze extensive datasets for identifying patterns and precursors associated with rockburst events [17, 35].

2. **Tactical Engineering Controls**

Tactical engineering controls are specific measures implemented during the excavation and support phases to directly address rockburst risks. These controls aim to modify the rock mass behavior, dissipate energy, and enhance the resilience of underground openings.

(1) Pre-conditioning of Ground

Pre-conditioning techniques, primarily including de-stress blasting and hydrofracturing, are implemented to alter the rock mass stress state and mechanical properties prior to excavation. De-stress blasting has demonstrated efficacy in deep South African gold mines and Canadian hard rock mines, with its success contingent upon rock properties, in-situ stress conditions, and blast design parameters [35, 58]. Hydrofracturing presents an advantageous alternative by generating more controlled and extensive fracture networks compared to blasting methods, proving effective in deep mines for rock mass pre-conditioning and rockburst risk mitigation [35] (Fig. 6.29).

(2) Backfill Systems

Implementing appropriate backfill systems can significantly contribute to rockburst mitigation by (a) Providing confinement to surrounding rock masses; (b) Reducing void spaces and limiting rock mass deformation; (c) Improving overall mine stability and stress distribution. The choice of backfill system depends on factors such as mining method, rock mass characteristics, and economic considerations. Proper backfill design and implementation can significantly reduce the risk of strain bursts and pillar bursts by providing confinement and limiting rock mass deformation [35].

(3) Dynamic Rock Support Systems

Developing and implementing effective dynamic rock support systems is crucial for mitigating the impacts of rockburst events. These systems are designed to absorb energy, accommodate large deformations, and maintain stability under dynamic

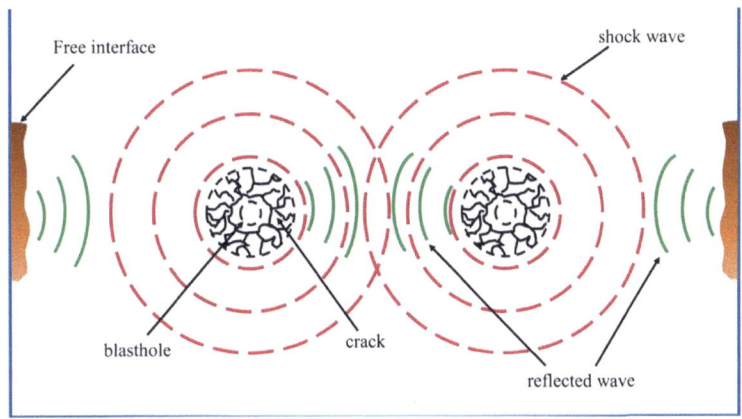

Fig. 6.29 Principle of de-stress blasting [62]

loading conditions. Dynamic support systems comprise several key elements: Energy-Absorbing Rockbolts, Meshes and Straps, Shotcrete, Yielding Support Systems, and Constant-Resistance Large-Deformation (CRLD) Support. The implementation of these systems necessitates thorough evaluation of loading conditions, rock mass characteristics, and project-specific requirements, often requiring the integration of multiple support elements to establish a comprehensive and effective support strategy [35, 63] (Fig. 6.30).

(4) Reduced Excavation Size and Advance Rate

Stress redistribution and energy release rates can be effectively managed through modified excavation practices, specifically through the implementation of Smaller Excavation Sizes and Controlled Advance Rates. However, these approaches require careful economic analysis to balance project productivity requirements against potential schedule extensions.

(5) Drilling Pressure Relief (DPR)

DPR has emerged as an effective technique for preventing rockbursts, particularly in deep tunneling projects. The method involves drilling boreholes around the excavation perimeter to create a stress relief zone (Fig. 6.31).

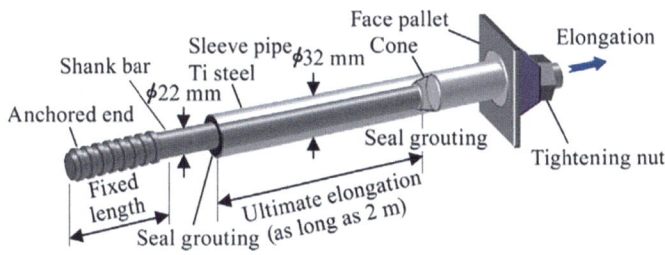

Fig. 6.30 Schematic diagram of the three-dimensional structure of the CRLD bolt [51]

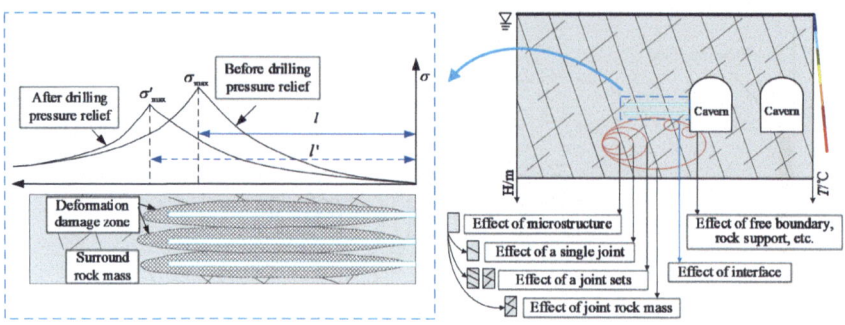

Fig. 6.31 Schematic of drilling pressure relief (DPR) [64]

3. **Administrative Controls**

While engineering measures form the core of rockburst mitigation strategies, administrative controls play a crucial role in managing risks and ensuring the effectiveness of technical measures. These controls focus on operational procedures, personnel management, and organizational aspects of rockburst mitigation.

(1) Use of Mechanized and Remote-Controlled Equipment

Remote-controlled and automated mining equipment, including LHD machines, tele-remote drilling systems, and automated rock bolting systems, along with advanced tunneling equipment such as remotely operated TBMs and robotic lining systems, can significantly reduce personnel exposure to rockburst hazards. The implementation of centralized control rooms and sophisticated sensor networks for equipment monitoring further enhances both safety and operational efficiency in challenging ground conditions [35].

(2) Personal Protective Equipment (PPE)

While engineering controls remain the primary rockburst mitigation strategy, PPE serves as a crucial final safeguard. The comprehensive PPE system encompasses high-energy impact-resistant helmets with full-face shields, impact-resistant body protection, reinforced footwear, integrated high-visibility elements, and communication devices. Regular training programs and strict compliance monitoring ensure proper PPE utilization, which, although not preventing rockbursts, can significantly reduce injury severity during rockburst events [35].

(3) Training and Awareness Programs

Comprehensive training and awareness programs are fundamental to implementing effective rockburst mitigation strategies. These programs encompass rockburst hazard recognition, emergency response protocols, specialized equipment operation, and dynamic support installation techniques. Continuous education through regular refresher courses and dissemination of new research findings, coupled with clear risk communication protocols and promotion of open reporting culture, helps

6.4 Rockburst Engineering Protection Method

develop a safety-conscious workforce capable of actively participating in rockburst risk management [35].

4. **Multiple Line Defense System**

A comprehensive rockburst mitigation strategy requires an integrated multiple line defense system incorporating strategic, tactical, and administrative controls. This system addresses three key aspects: hazard reduction through optimized mine design, pre-conditioning techniques, and real-time monitoring; vulnerability reduction via dynamic support systems, backfill strategies, and yielding support designs; and exposure reduction through automated equipment, exclusion zones, and enhanced PPE and training programs. This multi-faceted approach creates redundancy and enhances overall system resilience against rockburst hazards [35] (Fig. 6.32).

5. **Emerging Technologies**

Recent advancements in rockburst mitigation technologies have significantly enhanced the capability to predict, prevent, and manage rockburst events. Advanced monitoring systems, including Distributed Fiber Optic Sensing, LiDAR and Photogrammetry, and Micro-Electro-Mechanical Systems (MEMS), enable high-resolution data collection of rock mass behavior, facilitating early detection of rockburst precursors [17, 35]. This technological evolution is complemented by the development of intelligent support systems, encompassing Active Support Elements, Self-Diagnosing Support, and Adaptive Support Strategies, which provide dynamic response to changing ground conditions in dynamic underground environments [35].

The integration of Artificial Intelligence and machine learning techniques has revolutionized rockburst prediction and risk assessment capabilities. These advanced

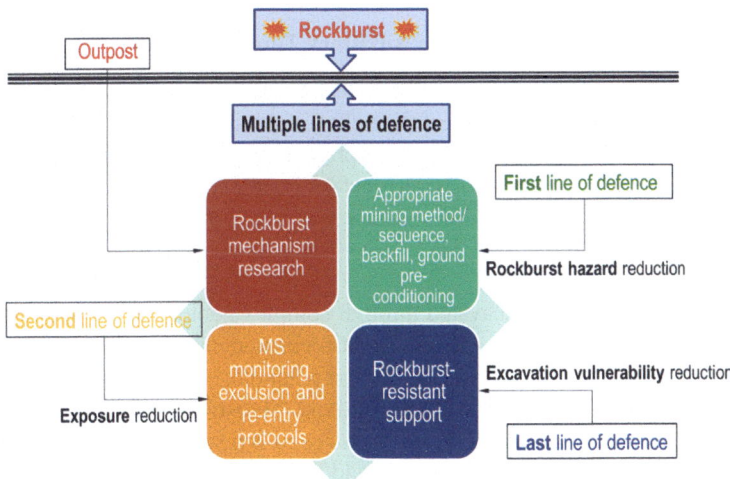

Fig. 6.32 Multiple lines of defense to deal with rockburst problems in deep mines [35]

computational methods, including Pattern Recognition and Real-Time Risk Assessment, demonstrate significant potential for enhancing decision-making processes in complex underground environments [17, 35]. Concurrently, novel research directions in rock mass property modification, such as Microwave Rock Breaking, Hydraulic Fracturing Enhancements, and Chemical Rock Softening, are being explored to provide more controlled and effective methods for altering rock mass behavior in burst-prone environments [35, 65].

The future of rockburst mitigation lies in the development of comprehensive modeling approaches that integrate multiple scales of analysis. These approaches, including coupled numerical models, data-driven modeling, and large-scale simulations, enable more accurate prediction of rockburst potential and optimization of mitigation strategies across various spatial and temporal scales [35, 65]. This multi-scale integration represents a significant advancement in understanding and managing rockburst phenomena in underground mining environments.

6.4.3 Rockburst Mitigation Engineering Case Studies

The study of rockburst mitigation through engineering measures has been a critical area of research in rock dynamics, particularly as mining and tunneling operations venture into deeper and more challenging environments.

This section presents a comprehensive review of several engineering case studies that demonstrate the practical application of various rockburst mitigation strategies. These case studies not only illustrate the effectiveness of different approaches but also highlight the complexities and challenges involved in managing rockburst risks in diverse geological settings.

1. **Jin-Ping II Hydropower Station Auxiliary Tunnels, China**

The Jin-ping II Hydropower Station project in China provides a compelling case study for rockburst mitigation in deep tunnel excavation. With depths ranging from 1500 to 2300 m, the auxiliary tunnels of this project experienced numerous rockburst events during construction, making it an ideal setting for studying and implementing rockburst prevention strategies [66].

(1) Geological and Stress Conditions

The Jin-ping tunnels were excavated through complex geological formations characterized by high in-situ stresses. The ratio of rock strength to maximum stress (R_b/σ_m) emerged as a critical indicator of rockburst potential, with values below 2 corresponding to zones of intensive rockburst activity. Statistical analysis revealed that rockbursts affected 18.7 and 16.9% of tunnels A and B respectively, with high-grade events predominantly occurring in deeper middle sections. Notably, most rockburst events were observed within 6–12 m from the excavation face and manifested 5–20 h

6.4 Rockburst Engineering Protection Method

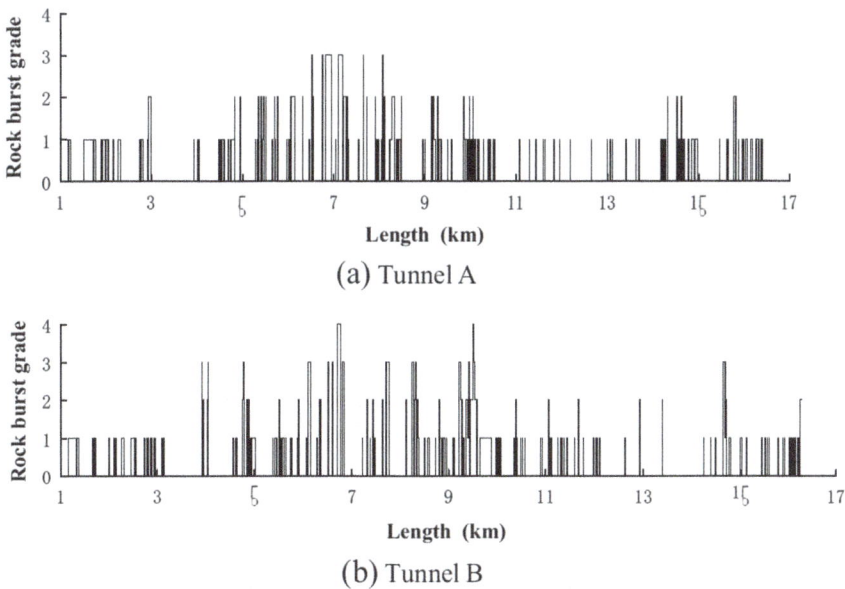

Fig. 6.33 Distribution of rock bursts along the axes of the auxiliary tunnels: **a** Tunnel A and **b** Tunnel B [66]

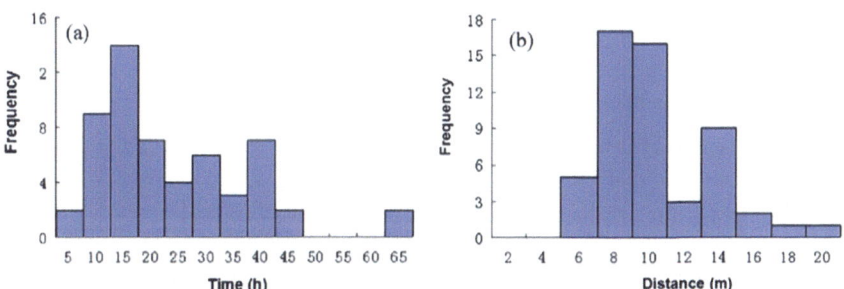

Fig. 6.34 The space–time distribution of rock bursts in Jin-ping auxiliary tunnels: **a** the rock bursts frequency changes with the time after excavation, and **b** the rock bursts changes [66]

post-excavation, highlighting the significant influence of both in-situ stress conditions and excavation-induced disturbances on rockburst occurrence [66] (Figs. 6.33 and 6.34).

(2) Stress Relief Blasting Technique

To address rockburst challenges, researchers developed and evaluated two stress relief blasting techniques: a 9-hole scheme utilizing perimeter blast holes, and an enhanced 14-hole scheme with increased explosive charges. Comparative analysis of their effectiveness, based on the estimated radius of cracked zones, demonstrated that the 9-hole scheme achieved a radius of 1.27–1.44 m per hole, while the 14-hole scheme produced larger zones of 1.27–1.92 m per hole. The 14-hole scheme proved superior in mitigating rockburst risk due to enhanced interconnection of cracked zones and more effective stress relief in the surrounding rock mass [66] (Fig. 6.35).

(3) Control of Blasting Excavation-Induced Disturbance

The research findings on excavation-induced disturbances revealed key strategies for rockburst risk reduction. Numerical simulations demonstrated the inverse relationship between excavation footage and Energy Release Rate (ERR), leading to the recommendation of limiting excavation advances to less than 2 m in high-risk areas. The implementation of controlled blasting techniques, including smooth blasting, pre-splitting methods, and step-by-step excavation approaches, combined with careful regulation of charge weight and blast hole parameters, was identified as essential for minimizing stress concentration and subsequent rockburst risk [66].

(4) Microseismic Monitoring for Rockburst Prediction

The Jin-ping project incorporated a microseismic monitoring system alongside stress relief blasting techniques, enabling real-time detection of precursory microcracking processes preceding catastrophic rockbursts [33]. This comprehensive case study of Jin-ping II Hydropower Station highlights the effectiveness of integrating multiple mitigation strategies in managing rockburst risks in deep tunneling projects, while emphasizing the necessity of site-specific engineering approaches and continuous refinement of prediction and mitigation techniques based on real-time monitoring data.

2. **Coeur D'Alene Mining District, Northern Idaho, USA**

The Coeur D'Alene mining district in northern Idaho, USA, provides a valuable long-term case study of rockburst management in deep metal mining. With over 60 years of experience dealing with rockbursts, this district offers important insights into the evolution of rockburst control strategies and their effectiveness over time [2].

(1) Geological and Mining Context

The Coeur D'Alene mining district exhibits distinctive geological characteristics that include deep mining operations exceeding 2000 m depth, predominantly hard and brittle rock formations consisting of quartzites and argillites, intricate geological structures with abundant faults and folds, and significant in-situ stress conditions particularly prevalent in quartzitic strata [67].

(2) Types of Rockbursts

The rockburst phenomena in the Coeur D'Alene district were categorized into three primary types: strain bursts, characterized by sudden failures of highly stressed rock near excavation surfaces; pillar bursts, manifesting as violent failures of mine pillars under excessive load; and slip bursts, involving sudden movements along pre-existing geological discontinuities [67].

6.4 Rockburst Engineering Protection Method

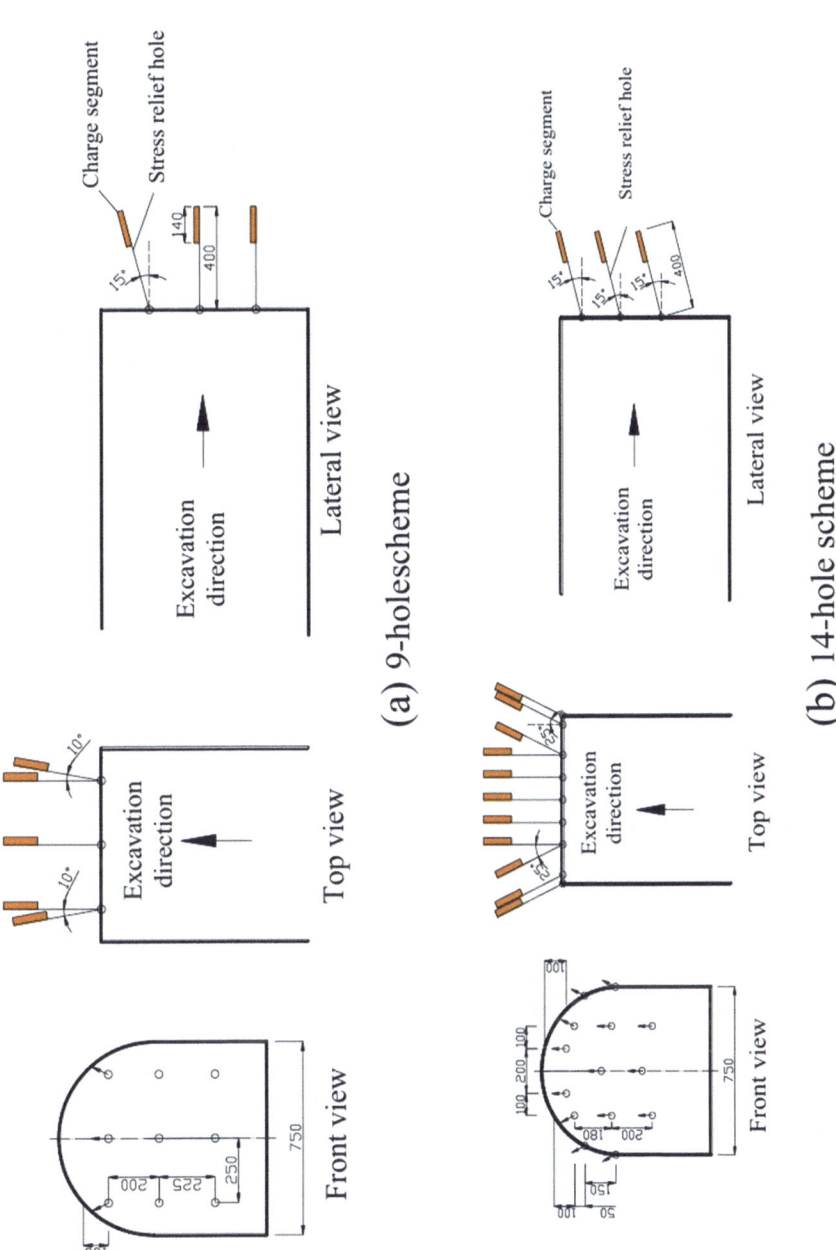

Fig. 6.35 Blast-hole distributions from the stress-relief blasting designs in the Jin-ping auxiliary tunnels: **a** 9-hole scheme and **b** 14-hole scheme (unit: mm) [66]

(3) Seismicity Characteristics

Seismic event analysis in the Coeur D'Alene district revealed that while mining-induced seismicity is predominantly benign, significant variations exist in relation to geological conditions, in-situ stress states, and depth. Notable seismic events recorded at the Lucky Friday Mine demonstrated magnitudes ranging from 2.5 to 4.2 ML (local magnitude) [67].

(4) Tactical Measures for Rockburst Control

Over the years, several tactical measures have been developed and implemented to mitigate rockburst hazards: Ground Support, Destress Blasting, and Mining Rate Control. The integrated application of these tactical measures in rockburst control can effectively manage and reduce the risk of rockbursts (Fig. 6.36).

(5) Strategic Methods for Rockburst Hazard Control

Rockburst hazard control in the mining environment involves dual strategic approaches focusing on mining method modifications and geological considerations. Mining adaptations include strategic backfilling of stopes for stress confinement, optimized stope sequencing for stress management, and implementation of under-hand cut-and-fill techniques for risk reduction. The geological component reveals that rockburst hazards predominantly occur in hard quartzitic strata under high in-situ stress conditions, while the orientation of excavations relative to discontinuities and faults requires careful consideration, with high-angle intersections being preferred when possible.

The Coeur D'Alene mining district case study provides valuable long-term insights into rockburst management in deep metal mines. It demonstrates the importance of combining tactical and strategic approaches, adapting to site-specific conditions, and continuously evolving mitigation strategies based on operational experience and scientific understanding.

3. **Dongguashan Copper Mine (DCM), China**

The DCM in China presents an important case study in the application of seismological methods for predicting and managing rockbursts in deep mining environments. As China's deepest metal mine, DCM offers unique insights into the challenges of rockburst mitigation at extreme depths [68].

(1) Geological and Mining Conditions

The DCM operation exhibits distinctive characteristics, including its location at approximately 1000 m depth, structural control by an anticline formation, and mineralization predominantly consisting of cupriferous skarn. The deposit features a marble roof and a floor composed of siltstone and quartz diorite, with mining operations conducted through a panel and stope system utilizing cemented tailings backfill [68].

6.4 Rockburst Engineering Protection Method

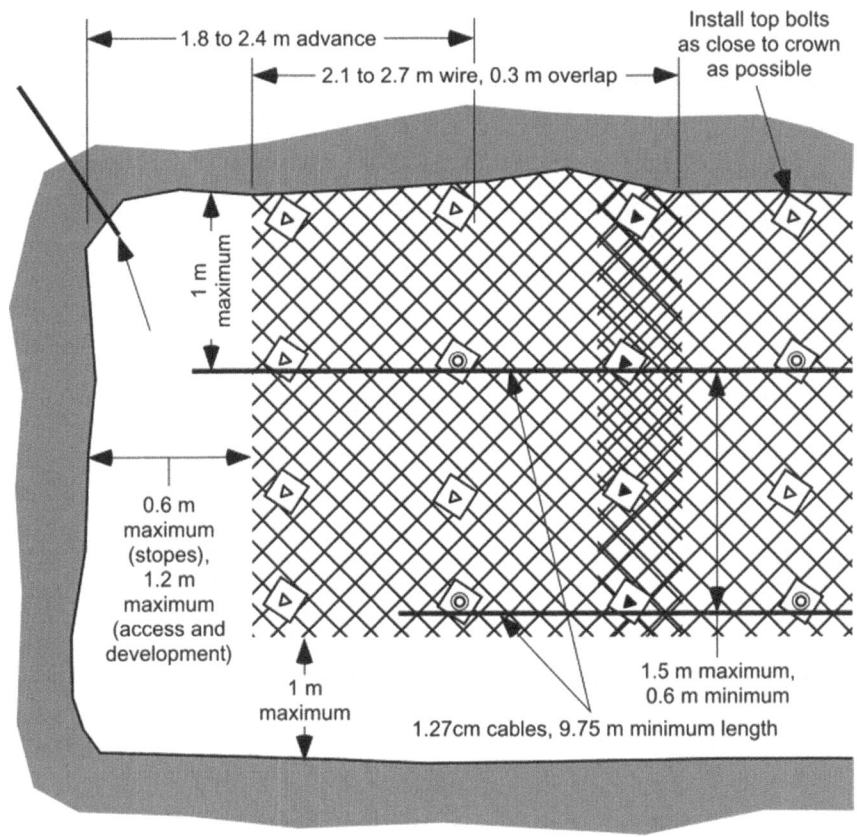

Fig. 6.36 Typical support scheme employed in rockbursting ground [67]

(2) Seismic Monitoring System

DCM established a comprehensive seismic monitoring system for rockburst prediction, featuring an ISS system equipped with 24 channels and 16 sensors, providing coverage across four panels in the initial mining area [68].

(3) Location of Hazardous Seismic Areas

The spatio-temporal analysis of seismic events at DCM revealed critical insights: panel barrier pillars and temporary pillars frequently developed into zones of persistent stress concentration, leading to the development of a conceptual model for hazardous seismic nucleation based on asperity theory [68] (Fig. 6.37).

(4) Mine Seismic Stiffness Method for Rockburst Prediction

A novel rockburst prediction methodology was developed based on unstable failure theory and mine stiffness theory, introducing key parameters d (nucleation area stiffness) and b (loading system stiffness). The approach establishes $S = d/b$ as a relative stiffness ratio and proposes dS/dt as a predictive index, where positive values indicate increasing rockburst possibility and negative values suggest decreasing risk [68].

(5) Case Study Results

The practical implementation of the seismic stiffness method at DCM between April 2006 and April 2007 demonstrated its effectiveness as a medium-term prediction tool, successfully identifying periods of varying rockburst probability approximately 60 days in advance through the application of the dS/dt index [68] (Figs. 6.38, 6.39 and 6.40).

The Dongguashan Copper Mine study showcases advanced seismological methods' effectiveness for rockburst prediction in deep mining, stressing multidisciplinary integration and site-specific approaches. It shows progress in medium-term predictions but also persistent challenges in reliable short-term forecasts in complex geology, highlighting the need for ongoing research and methodology refinement.

Fig. 6.37 Conceptual model of seismic nucleation: **a** Section perpendicular to strike; **b** Section along strike [68]

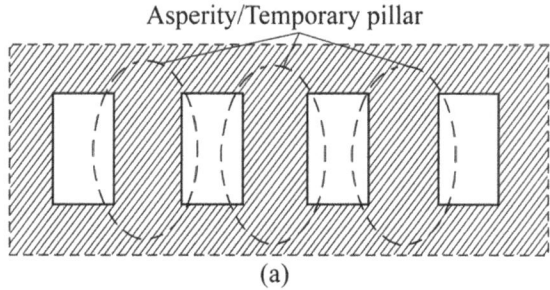

(a)

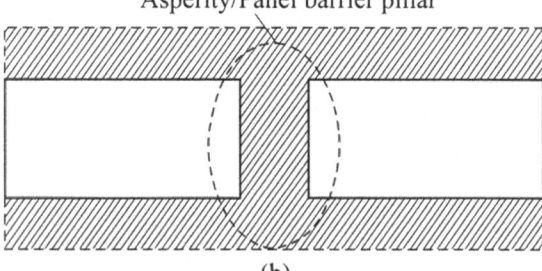

(b)

6.4 Rockburst Engineering Protection Method

Fig. 6.38 Distribution of seismicity in panel barrier pillar along 54# exploratory line: **a** Apparent stress contour map; **b** Displacement nephogram [68]

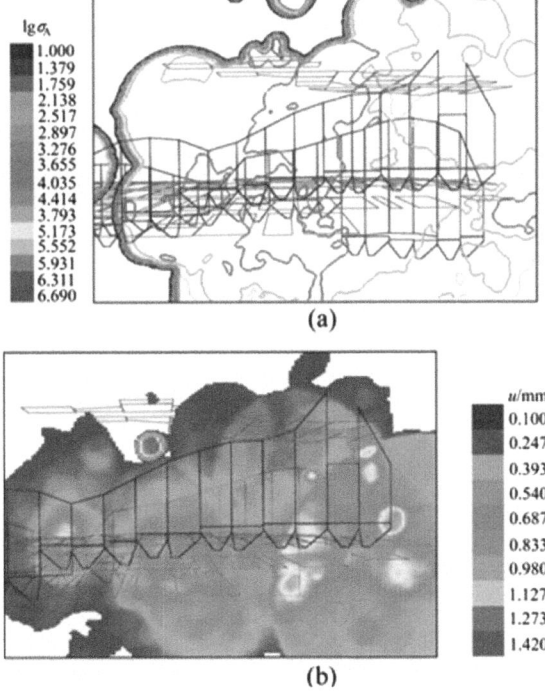

Fig. 6.39 B in loading system area and d in nucleation area [68]

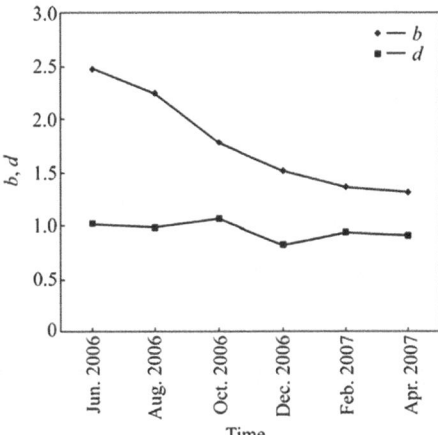

4. **Canadian Hard Rock Mines**

The experience of Canadian hard rock mines in managing rockburst risks provides valuable insights into practical strategies for rockburst mitigation in deep mining environments. This case study draws from the collective experience of multiple

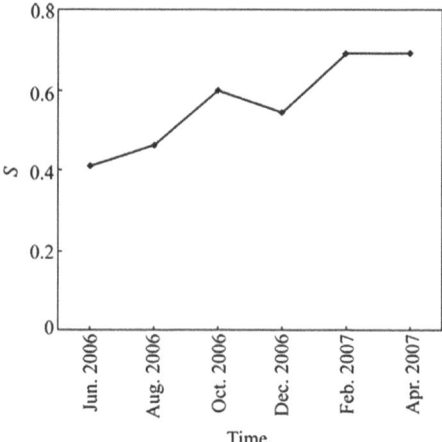

Fig. 6.40 Ratio of stiffness of load system area to that of nucleation area [68]

Canadian mines and highlights the multi-faceted approach developed over years of dealing with rockburst challenges [69].

(1) Geological and Mining Context

The Canadian Shield hosts numerous hard rock mines notable for their extreme operating conditions. These mines are characterized by exceptional depths, frequently surpassing 2000 m, and consist predominantly of hard, brittle rock formations. The geological environment is further distinguished by high in-situ stress conditions and intricate geological structures, including extensive fault systems and dyke intrusions [69].

(2) Rock Mechanics Tools for Rockburst Risk Mitigation

The Canadian approach to rockburst management implements a comprehensive strategy that integrates five essential components. This methodology encompasses strategic mine design to minimize stress concentrations, specialized burst-prone support systems featuring dynamic bolts and mesh-reinforced shotcrete, comprehensive seismic monitoring networks for real-time hazard assessment, systematic exclusion protocols to protect personnel, and advanced numerical modeling for predictive analysis. This multi-faceted approach enables mines to effectively manage rockburst risks through both preventive and reactive measures while maintaining operational efficiency.

(3) Field Observations and Lessons Learned

Studies of rockburst incidents in Canadian mines revealed critical insights into support system performance and geological influences [17]. Traditional support systems, comprising weld mesh and rebar, demonstrated limitations in containing ejected material, with particular vulnerability at mesh-plate interfaces. Dynamic support enhancements, including dome plates and mesh straps, exhibited superior

6.4 Rockburst Engineering Protection Method

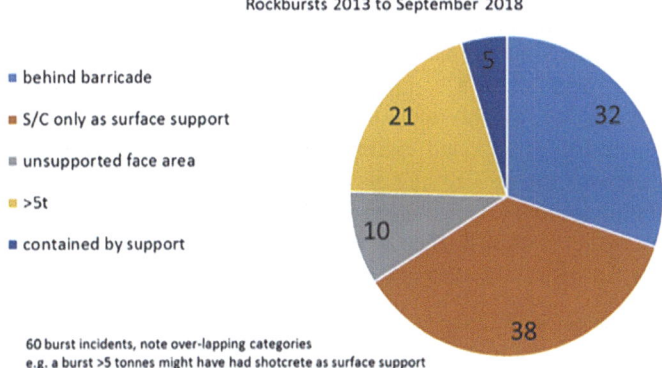

Fig. 6.41 Summary of 60 rockburst damage locations over a 69-month period from a relatively deep hard rock mine. Each category in the pie chart is expressed as a total number of the incidents. There is overlap between the categories so that the total of the pie is greater than 60 [69]

performance. Geologically, strong, fine-grained dyke material showed heightened susceptibility to strain energy storage and sudden release, resulting in localized rockbursts [69].

(4) Rockburst Database Analysis

Analysis of rockburst data spanning 69 months and encompassing 60 incidents at a Canadian mine yielded significant insights into ground support effectiveness and risk management. The transition from shotcrete to mesh outer support demonstrated notable success in reducing small-scale failures. The study revealed that development areas and primary stope access points were particularly susceptible to rockburst events. Implementation of comprehensive risk mitigation measures, including mesh installation, strategic barricading of high-risk zones, and strict ground control protocols, could have potentially reduced the incident count from 60 to 2 [69] (Fig. 6.41).

The Canadian hard rock mining case study demonstrates the effectiveness of a comprehensive, multi-faceted approach to rockburst management. It highlights the importance of integrating various strategies, from mine design to operational practices, and emphasizes the need for continuous improvement and adaptation as mining progresses to greater depths.

5. **Jinchuan Nickel Mine, China**

The Jinchuan Nickel Mine in China presents a unique case study focusing on the mechanism of large-scale roof caving events, which can be considered a special type of rockburst in mining environments. This case study provides insights into the complex interactions between geological structures, mining methods, and stress redistribution that can lead to catastrophic events in deep mines [70].

(1) Mine Background and Event Description

The Jinchuan Nickel Mine, China's largest metalliferous operation employing backfilling methods, operates in a complex geological environment characterized by multiple fault systems, notably including the major Fault F8. The mine utilizes a large-scale underhand horizontal cutting and filling method at depths exceeding 1000 m. A catastrophic roof caving event occurred on March 13, 2016, resulting in extensive damage that affected approximately 11,000 m^2 of underground workings and induced surface subsidence across 19,000 m^2. This event was accompanied by significant air blasts, acoustic emissions, and ground vibrations [70].

(2) Field Investigation Findings

Post-incident analysis of the 2016 Jinchuan Mine collapse revealed extensive structural damage concentrated at the 1650 and 1610 m level roadways. Surface manifestations included significant ground deformation characterized by crack formation and step-like features, with the most pronounced fracturing observed adjacent to Fault F8. Monitoring data indicated a dramatic acceleration in vertical displacement following the collapse event, with the subsidence center migrating towards the vicinity of Fault F8 [70] (Fig. 6.42).

(3) Numerical Modeling and Failure Mechanisms

A FLAC3D numerical simulation of the Jinchuan Mine collapse identified three critical failure mechanisms that culminated in the catastrophic event. Initially, the system behaved as a double-sided embedded rock beam, with support provided by intact rock mass on one side and Fault F8 on the other. The progressive advancement of mining operations toward Fault F8 induced stress redistribution, ultimately triggering fault activation when the supported pressure exceeded frictional resistance. Following fault activation, the structure transformed into a cantilever-articulated rock beam configuration, which proved insufficient to withstand the overburden pressure adjacent to Fault F8 [70] (Figs. 6.43 and 6.44).

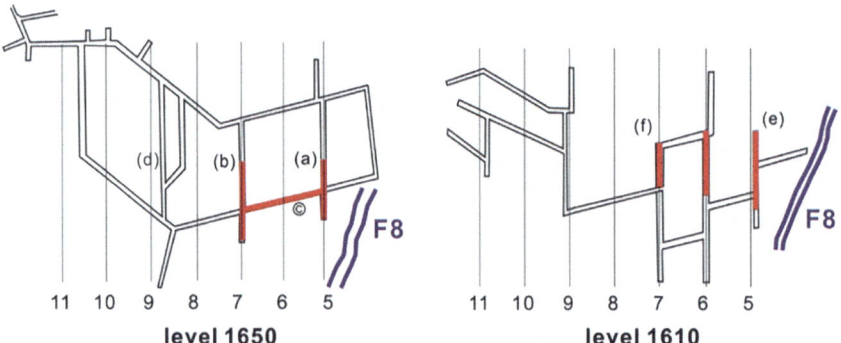

Fig. 6.42 Schematic diagram of the roadway failure at level 1650 and 1610 [70]

6.4 Rockburst Engineering Protection Method

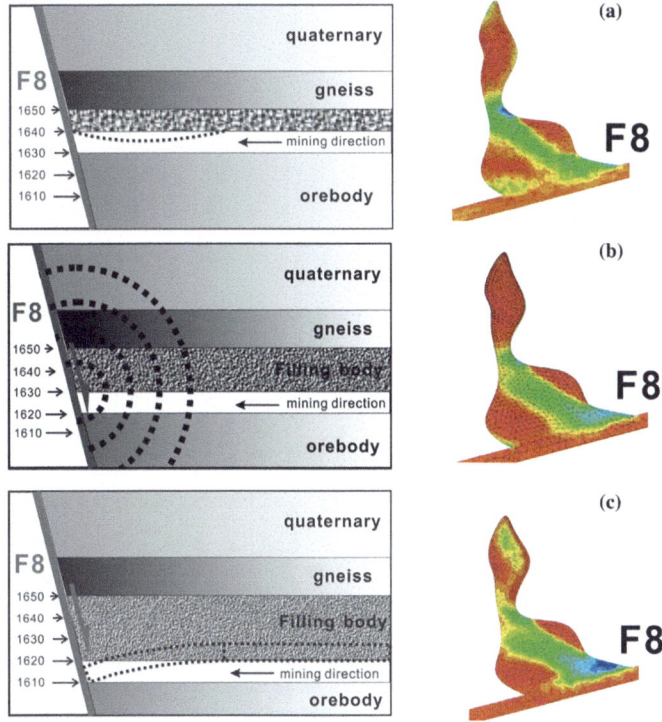

Fig. 6.43 Schematic diagram of the deformation and failure mechanisms at three mining stages. The right side is the transfer diagram of the vertical stress [70]

(4) Surface Deformation Characteristics

The observed surface deformation patterns exhibited a strong correlation with the three-stage failure mechanism identified in the numerical model. The sequence began with gradual deformation concentrated above the filling body, followed by accelerated deformation near Fault F8 during the fault activation phase. The final stage was marked by extensive subsidence following the cantilever beam failure [70].

(5) Influencing Factors

The catastrophic failure at Jinchuan Mine was primarily attributed to two critical factors. First, the stability analysis of Fault F8, which tracked normal and shear stress evolution, revealed a complex six-stage stress progression culminating in fault slip. Second, discrete element modeling demonstrated that hexagonal sections utilizing mosaic filling exhibited lower stability compared to rectangular sections when subjected to uniform loading conditions [70].

The Jinchuan Nickel Mine collapse illustrates how the interaction of geological structures, mining methods, and filling techniques can trigger catastrophic failures in deep mining operations. The case demonstrates the essential role of numerical

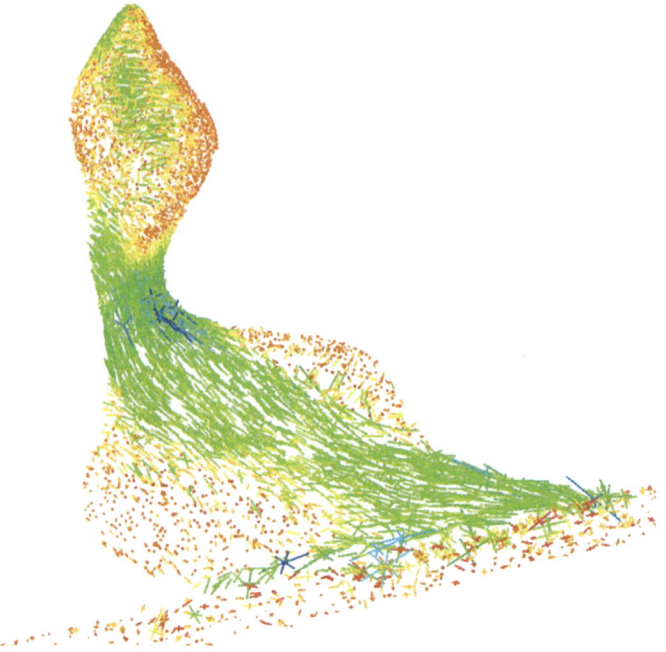

Fig. 6.44 Principal stress trace map [70]

modeling and monitoring systems in understanding complex failure mechanisms, providing valuable insights for improving stability in deep mines with challenging geological conditions.

6. **Synthesis of Case Studies and General Recommendations**

The analysis of global mining and tunneling case studies reveals critical strategies for rockburst management in deep underground excavations. Effective rockburst control requires an integrated approach encompassing detailed geological understanding, engineering design optimization, and comprehensive monitoring systems, particularly microseismic monitoring for risk assessment and early warning. This comprehensive strategy emphasizes the importance of both proactive measures, such as destress blasting, and reactive approaches, including dynamic support systems [66, 67, 69].

While rockburst management principles maintain universal applicability, their implementation demands site-specific adaptation based on local geological conditions, structural features, and stress regimes [66–68, 70]. The case studies demonstrate that excavation depth significantly influences management approaches due to varying stress conditions, while the selection and design of excavation methods directly impact rockburst potential [66, 67, 69, 70]. This adaptation is continually refined through the analysis of historical rockburst incident data from specific sites [67, 69].

Advanced monitoring and prediction techniques, coupled with innovative support systems, form the cornerstone of effective rockburst management [33, 69]. Comprehensive microseismic monitoring systems, exemplified by the seismic stiffness method at Dongguashan Copper Mine, provide essential insights into rock mass behavior and enhance medium-term prediction capabilities [68]. These systems are complemented by energy-absorbing support elements designed to withstand dynamic loading events, with their effectiveness continuously evaluated and refined based on performance during actual rockburst events [67, 69].

6.5 Conclusion

This chapter focuses on the phenomenon of rockbursts, providing a systematic explanation of rockburst dynamics and engineering protection measures. It offers comprehensive guidance for addressing rockburst issues in underground engineering.

Phenomenon and Causes of Rockbursts: Rockburst is a complex and dangerous phenomenon in underground excavations, characterized by the sudden violent failure of rock and ejection of rock fragments. Its occurrence is influenced by geological conditions, excavation methods, dynamic disturbances, and environmental factors. Hard, brittle rocks, high tectonic stresses, blasting excavation methods, and seismic activity all contribute to the increased risk of rockbursts. Understanding these factors is crucial for predicting and preventing rockbursts.

Rockburst Precursors and Prediction Methods: Monitoring rockburst precursors is key to prediction. Techniques such as microseismic monitoring, electromagnetic radiation, and stress–strain measurements can capture pre-rockburst changes. Prediction methods include theoretical guidelines, case-based intelligent methods, on-site monitoring, and comprehensive methods. Theoretical guidelines are simple but limited in scope, while intelligent methods can handle complex relationships but rely heavily on data. On-site monitoring provides real-time data, and comprehensive methods improve prediction accuracy by combining various approaches.

Rockburst Dynamics: Rockbursts can be categorized based on triggering mechanisms, damage characteristics, and timing. They are primarily caused by the accumulation and release of energy, with factors such as stress concentration, structural characteristics, dynamic triggering, excavation support, and thermal–mechanical interactions influencing their occurrence. Quantitative models involving energy, stress, and strain help in understanding and predicting rockbursts. The rockburst grading system assesses the severity of rockbursts from different perspectives, with integrated methods enhancing predictive capabilitie.

References

1. Wu M, Ye Y, Wang Q, et al. Development of rockburst research: a comprehensive review. Appl Sci. 2022;12(3):974.
2. Nussbaumer MM. A comprehensive review on rock burst. Massachusetts Institute of Technology;2000.
3. Cortés N, Hekmatnejad A, Pan P, et al. Empirical approaches for rock burst prediction: a comprehensive review and application to the new level of El Teniente Mine. Chile: Heliyon; 2024.
4. He M, Cheng T, Qiao Y, et al. A review of rockburst: experiments, theories, and simulations. J Rock Mech Geotech Eng. 2023;15(5):1312–53.
5. Rong H, Li N, Zhang H, et al. Insights into fundamental problems of rockburst under the modern structure stress field. Sci Rep. 2022;12(1):20299.
6. Kabwe E, Wang Y. Review on rockburst theory and types of rock support in rockburst prone mines. Open J Saf Sci Technol. 2015;5(04):104.
7. Gong FQ, Pan JF, Jiang Q. Analysis of the differences between rockburst and rockburst-induced ground pressure and key mechanisms of geological hazards in deep engineering. J Eng Geol. 2021;29(4):933–61 [in Chinese].
8. Zhao H, Liu C, Huang G, et al. Experimental investigation on rockburst process and failure characteristics in trapezoidal tunnel under different lateral stresses. Constr Build Mater. 2020;259: 119530.
9. Jiang J, Su G, Zhang X, et al. Effect of initial damage on remotely triggered rockburst in granite: an experimental study. Bull Eng Geol Env. 2020;79:3175–94.
10. Sun F, Guo J, Fan J, et al. Experimental study on rockburst fragment characteristic of granite under different loading rates in true triaxial condition. Front Earth Sci. 2022;10: 995143.
11. Su G, Yan X, Zheng Z, et al. Experimental study on the influence of a small-scale single structural plane on rockburst in deep tunnels. Rock Mech Rock Eng. 2023;56(1):669–701.
12. Qiu J, Xie H, Zhu J, et al. Dynamic response and rockburst characteristics of underground cavern with unexposed joint. Int J Rock Mech Min Sci. 2023;169: 105442.
13. Hu S, Su G, Qin Y, et al. Influence of the loading rate on the evolution characteristics of AE and MS signals during granite failure. Eng Fail Anal. 2023;152: 107428.
14. Liu DQ, Hu TX, Wang Y, et al. Experimental study on influence of dynamic load frequency on impact rockburst of sandstone. Chin J Rock Mech Eng. 2022;41(7):1310–24 [in Chinese].
15. Liu X, Liang Z, Zhang Y, et al. Experimental study on the monitoring of rockburst in tunnels under dry and saturated conditions using AE and infrared monitoring. Tunn Undergr Space Technol. 2018;82:517–28.
16. Su G, Jiang J, Zhai S, Zhang G. Influence of tunnel axis stress on strainburst: an experimental study. Rock Mech Rock Eng. 2017;50:1551–67.
17. Feng XT, Xiao YX, Feng GL, et al. Study on the development process of rockbursts. Chin J Rock Mech Eng. 2019;4:649–73 [in Chinese].
18. Wang Y, Liu DQ, Ren FQ, et al. Experimental study on influence of relationship between dynamic load and long–axis position on impact rockburst of surrounding rock of elliptical cavern. Rock Soil Mech. 2022;43(9):2347–59 [in Chinese].
19. Lu C-P, Liu G-J, Liu Y, et al. Microseismic multi-parameter characteristics of rockburst hazard induced by hard roof fall and high stress concentration. Int J Rock Mech Min Sci. 2015;76:18–32.
20. Li X, Wang E, Li Z, et al. Rock burst monitoring by integrated microseismic and electromagnetic radiation methods. Rock Mech Rock Eng. 2016;49:4393–406.
21. Xiao YX, Feng XT, Hudson JA, et al. ISRM suggested method for in situ microseismic monitoring of the fracturing process in rock masses. Rock Mech Rock Eng. 2016;49:343–69.
22. Ma C, Li T, Zhang H. Microseismic and precursor analysis of high-stress hazards in tunnels: a case comparison of rockburst and fall of ground. Eng Geol. 2020;265: 105435.
23. Konicek P, Waclawik P. Stress changes and seismicity monitoring of hard coal longwall mining in high rockburst risk areas. Tunn Undergr Space Technol. 2018;81:237–51.

24. Xu N, Li T, Dai F, et al. Microseismic monitoring and stability evaluation for the large scale underground caverns at the Houziyan hydropower station in Southwest China. Eng Geol. 2015;188:48–67.
25. Feng GL, Feng XT, Chen BR, et al. Effects of structural planes on the microseismicity associated with rockburst development processes in deep tunnels of the Jinping-II Hydropower Station, China. Tunnel Undergr Space Technol. 2019;84:273–80.
26. Ma T, Tang C, Tang L, et al. Rockburst characteristics and microseismic monitoring of deep-buried tunnels for Jinping II hydropower station. Tunn Undergr Space Technol. 2015;49:345–68.
27. Feng X-T, Liu J, Chen B, et al. Monitoring, warning, and control of rockburst in deep metal mines. Engineering. 2017;3(4):538–45.
28. Xu C, Liu X, Wang E, et al. Rockburst prediction and classification based on the ideal-point method of information theory. Tunn Undergr Space Technol. 2018;81:382–90.
29. Pu Y, Apel DB, Xu H. Rockburst prediction in kimberlite with unsupervised learning method and support vector classifier. Tunn Undergr Space Technol. 2019;90:12–8.
30. Liang W, Zhao G, Wu H, et al. Risk assessment of rockburst via an extended MABAC method under fuzzy environment. Tunn Undergr Space Technol. 2019;83:533–44.
31. Cai W, Dou L, Zhang M, et al. A fuzzy comprehensive evaluation methodology for rock burst forecasting using microseismic monitoring. Tunn Undergr Space Technol. 2018;80:232–45.
32. Lu CP, Liu Y, Zhang N, et al. In-situ and experimental investigations of rockburst precursor and prevention induced by fault slip. Int J Rock Mech Min Sci. 2018;108:86–95.
33. Wang J, Zhang J. Preliminary engineering application of microseismic monitoring technique to rockburst prediction in tunneling of Jinping II project. J Rock Mech Geotech Eng. 2010;2(3):193–208.
34. Ma TH, Tang CA, Tang SB, et al. Rockburst mechanism and prediction based on microseismic monitoring. Int J Rock Mech Min Sci. 2018;110:177–88.
35. Cai M. Rockburst risk control and mitigation in deep mining. Deep Resourc Eng 2024; 100019.
36. Yian T, Guangzhong S, Zhi G. A composite index Krb criterion for the ejection characteristics of the burst rock. Chinese J Geol. 1991;26(2):193–200.
37. Zhou J, Li X, Mitri HS. Evaluation method of rockburst: state-of-the-art literature review. Tunn Undergr Space Technol. 2018;81:632–59.
38. Hoek E, Brown ET. Underground Excavations in Rock. Institution of Mining and Metallurgy, London 1980; 527.
39. Wen J, Li H, Jiang F, et al. Rock burst risk evaluation based on equivalent surrounding rock strength. Int J Min Sci Technol. 2019;29(4):571–6.
40. Barton N, Lien R, Lunde J. Engineering classification of rock masses for the design of tunnel support. Rock Mech. 1974;6:189–236.
41. Zhao G, Wang D, Gao B, et al. Modifying rock burst criteria based on observations in a division tunnel. Eng Geol. 2017;216:153–60.
42. Kidybinski A. Bursting liability indices of coal. Int J Rock Mech Min Sci Geomech Abstr 1981;18(4):295–304.
43. Hucka V, Das B. Brittleness determination of rocks by different methods. Int J Rock Mech Min Sci Geomech Abstr. 1974;11(10):389–392.
44. Su G, Chen Y, Jiang Q, et al. Spalling failure of deep hard rock caverns. J Rock Mech Geotech Eng. 2023;15(8):2083–104.
45. Heal D, Potvin Y, Hudyma M. Evaluating rockburst damage potential in underground mining. In: Yale, D.P. et al. (Eds.), Proceedings of 41st U.S. Symposium on Rock Mechanics (USRMS). USA, Curran Associates, Colorado School of Mines, 2006; pp. 322–329.
46. Feng XT, Kong R, Yang C, Zhang X, Wang Z, Han Q, Wang G. A three-dimensional failure criterion for hard rocks under true triaxia 2020
47. Qiu SL, Feng XT, Zhang CT, et al. Development and validation of rockburst vulnerability index(RVI) in deep hard rock tunnels. J Rock Mech Geotech Eng. 2011;30(6):1126–41 [in Chinese].

48. Wang X, Li S, Xu Z, et al. An interval fuzzy comprehensive assessment method for rock burst in underground caverns and its engineering application. Bull Eng Geol Environ. 2019;78:5161–76.
49. Ma C, Chen W, Tan X, et al. Novel rockburst criterion based on the TBM tunnel construction of the Neelum-Jhelum (NJ) hydroelectric project in Pakistan. Tunn Undergr Space Technol. 2018;81:391–402.
50. Zhou J, Li X, Mitri HS. Classification of rockburst in underground projects: comparison of ten supervised learning methods. J Comput Civ Eng. 2016;30(5):04016003.
51. He M, Ren F, Liu D. Rockburst mechanism research and its control. Int J Min Sci Technol. 2018;28(5):829–37.
52. Feng XT, Chen BR, Ming HJ, et al. Evolution law and mechanism of rockbursts in deep tunnels:immediate rockburst. Chin J Rock Mech Eng. 2012;31(3):433–44.
53. Chen BR, Feng XT, Ming HJ, et al. The incubation law and mechanism of rockburst in deep-buried tunnels: time-lag rockburst. Chin J Rock Mech Eng. 2012;31(3):561–9.
54. Wang CH, Song CK, Liu LP. Study of stress characteristics of brittle failures of rock around underground openings. Rock Soil Mech 2012; 33.
55. Lee SM, Park BS, Lee SW. Analysis of rockbursts that have occurred in a waterway tunnel in Korea. Int J Rock Mech Min Sci. 2004;41:911–6.
56. Akram MS, Mirza K, Zeeshan M, et al. Geotechnical investigation and prediction of rock burst, squeezing with remediation design by numerical analyses along headrace tunnel in Swat Valley, Khyber Pakhtunkhwa, Pakistan. Open J Geol. 2018;8(10):965–86.
57. He M, Xia H, Jia X, et al. Studies on classification, criteria and control of rockbursts. J Rock Mech Geotech Eng. 2012;4(2):97–114.
58. Waqar MF, Guo S, Qi S. A comprehensive review of mechanisms, predictive techniques, and control strategies of rockburst. Appl Sci. 2023;13(6):3950.
59. Li CC, Mikula P, Simser B, et al. Discussions on rockburst and dynamic ground support in deep mines. J Rock Mech Geotech Eng. 2019;11(5):1110–8.
60. Zhang G, Chen J, Hu B. Prediction and control of rockburst during deep excavation of a gold mine in China. Chin J Rock Mech Eng. 2003;22(10):1607–12.
61. Zhou J, Zhang Y, Li C, et al. Rockburst prediction and prevention in underground space excavation. Undergr Space. 2024;14:70–98.
62. Zhang Q, Huo J, Yuan L, et al. A review of rockburst prevention and control methods in tunnels: graded and classified prevention and control. Bull Eng Geol Env. 2024;83(3):83.
63. Ghorbani M, Shahriar K, Sharifzadeh M, et al. A critical review on the developments of rock support systems in high stress ground conditions. Int J Min Sci Technol. 2020;30(5):555–72.
64. Gong F, He Z, Si X. Experimental study on revealing the mechanism of rockburst prevention by drilling pressure relief: status-of-the-art and prospects. Geomat Nat Haz Risk. 2022;13(1):2442–70.
65. Gong FQ, He ZC. Progress and prospect of experimental research on the mechanism of rockburst prevention and control by drilling pressure relief. Haz Control Tunnel Undergr Eng. 2023;5(2):1–23 [in Chinese].
66. Yan P, Zhao Z, Lu W, et al. Mitigation of rock burst events by blasting techniques during deep-tunnel excavation. Eng Geol. 2015;188:126–36.
67. Whyatt J, Blake W, Williams T, et al. 60 years of rockbursting in the Coeur d'Alene District of Northern Idaho, USA: lessons learned and remaining issues. 2002.
68. Tang LZ, Xia K. Seismological method for prediction of areal rockbursts in deep mine with seismic source mechanism and unstable failure theory. J Cent South Univ Technol. 2010;17(5):947–53.
69. Simser B. Rockburst management in Canadian hard rock mines. J Rock Mech Geotech Eng. 2019;11(5):1036–43.
70. Ding K, Ma F, Guo J, Zhao H, et al. Investigation of the mechanism of roof caving in the Jinchuan nickel mine, China. Rock Mech Rock Eng. 2018;51:1215–26.
71. He MC, Wu YY, Gao YB, et al. Progress in rock mechanics of deep mining. J China Coal Soc. 2024;49(1):75–99.

References

72. He S, Lai J, Zhong Y, et al. Damage behaviors, prediction methods and prevention methods of rockburst in 13 deep traffic tunnels in China. Eng Fail Anal. 2021;121: 105178.
73. Chen Y, Zhang J, Zhang J, et al. Rockburst precursors and the dynamic failure mechanism of the deep tunnel: a review. Energies. 2021;14(22):7548.
74. Gong FQ, Wang YL, Luo S. Rockburst proneness criteria for rock materials: review and new insights. J Central South Univ. 2020;27(10):2793–821.
75. Dou L, Chen T, Gong S, et al. Rockburst hazard determination by using computed tomography technology in deep workface. Saf Sci. 2012;50(4):736–40.
76. Sun XM, Ren C, Liu DQ, et al. Mechanism analysis and type determination of the rockburst of the gaoloushan tunnel based on a study of rockburst fragments. Chinese J Eng 2023; 45(3):337–48.
77. Qian QH. Definition, mechanism, classification and quantitative forecast model for rockburst and pressure bump. Rock Soil Mech. 2014;35(1):1–6.
78. Li TB, Xiao XP. Comprehensively integrated methods of rockburst prediction in underground engineering. Adv Earth Sci. 2008;23(5):533.
79. Zhang CQ, Yu J, Chen Q, et al. Evaluation method for potential rockburst in underground engineering. Rock Soil Mech 2016.
80. Gong F, He Z, Jiang Q. Internal mechanism of reducing rockburst proneness of rock under high stress by real-time drilling pressure relief. Rock Mech Rock Eng. 2022;55(8):5063–81.
81. Yan S, Liu R, Zhang Y, et al. Investigation and application of data balancing and combined discriminant model in rock burst severity prediction. Sci Rep. 2024;14(1):29657.
82. Gao M-t, Song Z-q, Duan H-q, et al. Mechanical properties and control rockburst mechanism of coal and rock mass with bursting liability in deep mining. Shock Vib. 2020;2020(1):8833863.
83. Feng XT, Xiao YX, Feng GL, et al. Study on the development process of rockbursts. J Rock Mech Geotech Eng 2019; 4:649–73
84. Lu Y, Lu WB, Jin XH, Chen M, Yan P. Mechanism study of slot prevention of delayed rockburst. Rock Soil Mech. 2011;32(10):3125–30 [in Chinese].
85. Zhang Y, Wang G, Tan L, et al. Research on occurrence law and the prevention of rockbursts in main roadways affected by mining activities: two case studies from Gaojiapu and Cuimu Coal Mines, Shaanxi, China. Appl Sci. 2024;14(22):10172.

Open Access This chapter is licensed under the terms of the Creative Commons Attribution-NonCommercial-NoDerivatives 4.0 International License (http://creativecommons.org/licenses/by-nc-nd/4.0/), which permits any noncommercial use, sharing, distribution and reproduction in any medium or format, as long as you give appropriate credit to the original author(s) and the source, provide a link to the Creative Commons license and indicate if you modified the licensed material. You do not have permission under this license to share adapted material derived from this chapter or parts of it.

The images or other third party material in this chapter are included in the chapter's Creative Commons license, unless indicated otherwise in a credit line to the material. If material is not included in the chapter's Creative Commons license and your intended use is not permitted by statutory regulation or exceeds the permitted use, you will need to obtain permission directly from the copyright holder.

Chapter 7
Key Techniques for Numerical Simulation of Rock Dynamics

7.1 High-Performance Computational Technique

7.1.1 Parallel Computing Methods

In rock dynamics, numerical simulations of complex models often demand extensive computational resources. Parallel computing methods have emerged as a critical solution for improving computational efficiency and enabling more sophisticated rock dynamic simulations. This section examines effective parallel computing implementations in rock dynamics, emphasizing their principles and performance characteristics. The approach distributes computational workload across multiple processors, with effectiveness dependent on problem characteristics, parallelization strategy, and hardware architecture.

1. Domain Decomposition Method (DDM)

A fundamental method for parallel computing in rock dynamics is the Domain Decomposition Method (DDM). The DDM partitions the problem domain into subdomains for distributed processing. This approach is particularly effective for finite element method (FEM) simulations in rock dynamics, including stress analysis, wave propagation, and fracture mechanics studies [1]. In electromagnetic applications like ground-penetrating radar, the domain Ω is divided into p sub-domains ($\Omega1$, $\Omega2$, ..., Ωp) with balanced node distribution [1]. The nodes in domain decomposition are categorized as: internal nodes within subdomains, border nodes at subdomain interfaces, and external nodes that connect with border nodes to maintain inter-subdomain coupling.

Domain decomposition in rock dynamics simulations follows a systematic workflow. The domain is first partitioned into discrete sections using graph partitioning or geometric methods to ensure balanced load distribution while minimizing inter-subdomain communication. Each processor then independently constructs the local stiffness matrix and load vector for its designated subdomain. Global matrix assembly follows, requiring controlled inter-processor communication for shared boundary nodes. The solution phase employs parallel linear algebra techniques, typically using the parallel conjugate gradient (PCG) algorithm for symmetric positive definite matrices common in rock mechanics.

The DDM for rock dynamics simulations consists of essential sequential stages [1]. The domain is first partitioned into sections using graph partitioning or geometric algorithms to optimize load distribution and minimize inter-subdomain communication. Each processor then assembles its local stiffness matrix and load vector independently. Global matrix assembly follows, requiring inter-processor communication for shared boundary nodes. The final stage implements parallel linear algebra techniques, typically using the parallel conjugate gradient (PCG) algorithm for symmetric positive definite matrices in rock mechanics applications.

The algorithm enables parallel processing of matrix–vector products step 2a and vector updates 2c, 2d, 2g, while dot products 2b, 2f require minimal global communication. Incomplete Cholesky factorization [1] serves as an effective preconditioner for PCG convergence in rock dynamics simulations, with parallelization effectiveness dependent on problem characteristics and decomposition strategy.

Parallel DDM implementations in rock dynamics simulations are evaluated through speedup metrics, defined as the ratio between single-processor and multi-processor execution times:

$$S(n) = \frac{T_1}{T_n} \tag{7.1}$$

where $S(n)$ is the speedup for n processors, T_1 is the execution time on a single processor, and T_n is the execution time on n processors.

Parallel efficiency is then calculated as:

$$E(n) = \frac{S(n)}{n} \tag{7.2}$$

Linear speedup ($S(n) = n$) and 100% efficiency ($E(n) = 1$) represent theoretical ideals, but communication overhead, load imbalance, and algorithmic limitations typically yield sublinear performance.

Experimental results [1] from comparable electromagnetic field simulations demonstrate parallel DDM effectiveness: a 1,043,000-node problem achieved 5.85 speedup and 97.5% efficiency with 6 processors, while a 4,155,000-node problem reached 5.89 speedup and 98.2% efficiency, indicating strong scalability at larger problem sizes.

2. Parallelization of Monte Carlo Methods

Monte Carlo methods in rock dynamics simulations provide valuable analysis of uncertainties in material properties and fracture networks. The method's independent sampling nature enables efficient parallelization. A parallel implementation example from polymer chain modeling [2], while not directly related to rock dynamics,

7.1 High-Performance Computational Technique

demonstrates adaptable parallelization principles for rock mechanics applications. The implementation uses Message Passing Interface (MPI) for sample distribution across processor cores. The process involves MPI environment initialization, equal distribution of Monte Carlo samples among cores, independent sample generation and simulation, and final result consolidation for statistical analysis (Fig. 7.1).

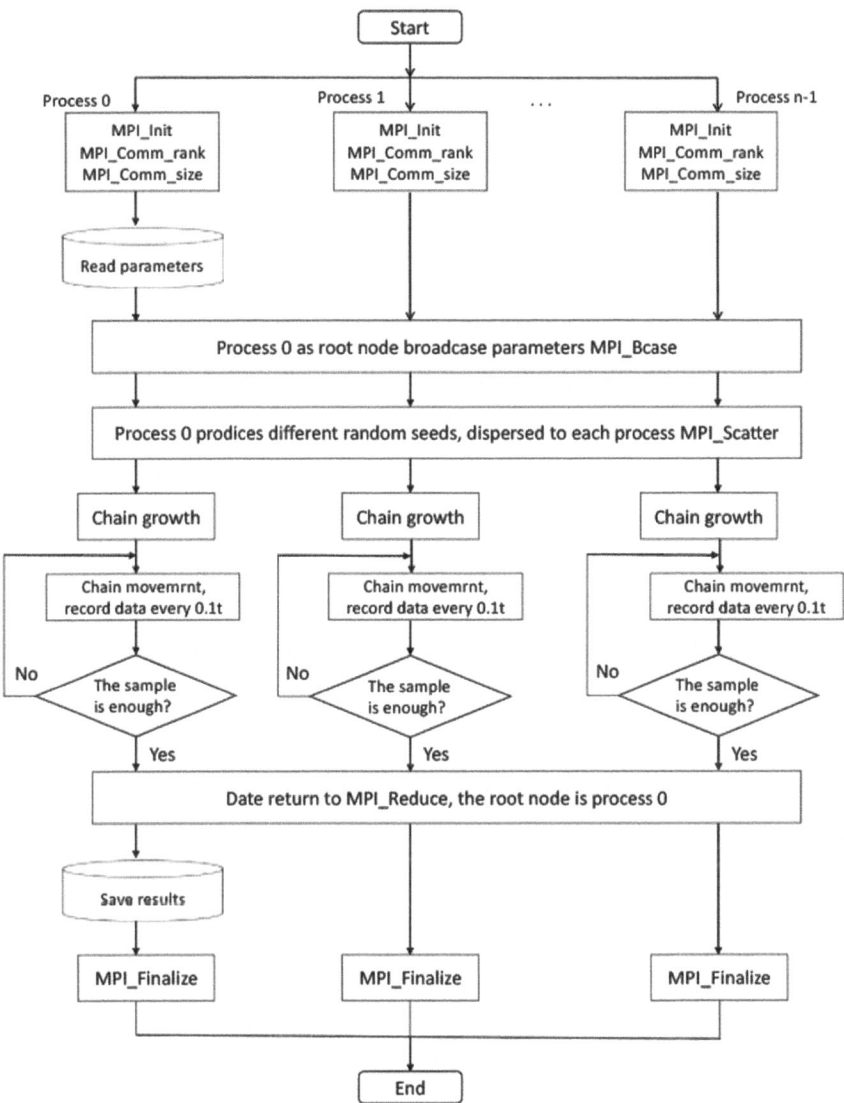

Fig. 7.1 The process of parallel program execution [2]

The parallel Monte Carlo approach demonstrates near-linear speedup, as shown in [2]. For a polymer chain (N = 100) with 480 samples and 10^6 Monte Carlo steps, linear speedup was achieved up to 480 cores, with 4×10^{-7} s average execution time per monomer move. At N = 400 with 960 samples, computation time reduced from 10.61 years (single core) to 4 days (960 cores), enabling analysis of larger systems. Similar performance gains apply to rock dynamics problems like fracture propagation in heterogeneous rock masses, due to minimal inter-processor communication requirements.

3. Numerical Manifold Method (NMM)

Beyond Monte Carlo methods, parallel computing extends to iterative solutions for large-scale linear systems in rock dynamics. Reference [3] presents a parallel Numerical Manifold Method (NMM) implementation using OpenMP for shared-memory systems, particularly effective for discontinuous media problems. The implementation parallelizes the Jacobi iterative method, which comprises 60–95% of total computing time, using OpenMP's "parallel for" directive for workload distribution. The pseudocode for the parallel Jacobi iteration is as follows:

```
#pragma omp parallel for private(i,j,sum) shared(x,b,a) reduction (+:err)
for(i=0; i<n; i++) {
  sum = 0.0;
  for(j=0; j<n; j++)
    if(j != i) sum += a[i][j] * x[j];
  x_new[i] = (b[i] - sum) / a[i][i];
  err += fabs(x_new[i] - x[i]);
}
```

The parallel NMM implementation achieves a 2.5 speedup on a quad-core Xeon 5420 platform for rock engineering problems, with improved performance at larger scales. The parallel Jacobi method shows enhanced convergence compared to sequential SOR: in one test case, parallel Jacobi reached 10^{-6} error in 400 iterations, while serial SOR required 600 iterations for 10^{-5} error with a 1.9 relaxation factor. These results demonstrate both reduced computation time and improved convergence, though optimal method selection depends on problem characteristics.

4. Parallelization of Collision Detection Algorithms

Parallel collision detection algorithms are essential for discontinuous rock mass simulations involving block systems or fractured rock. The self-avoiding walk (SAW) simulation method offers adaptable principles for parallel collision detection in rock dynamics, as its self-intersection prevention parallels block overlap constraints in discontinuous rock mass modeling.

7.1 High-Performance Computational Technique

In the context of rock dynamics, the implementation of a parallel collision detection algorithm follows a structured domain decomposition approach. The process begins with the division of the simulation domain into distinct spatial subdomains, followed by the strategic assignment of each subdomain to a dedicated processor or core. This spatial partitioning enables parallel execution of local collision detection processes within individual subdomains. To ensure comprehensive collision detection across the entire domain, the algorithm facilitates the exchange of boundary information between adjacent subdomains, thereby effectively handling collisions that occur at subdomain interfaces. The final phase involves the systematic resolution of all detected collisions, culminating in the appropriate adjustment of rock block positions based on the collision outcomes.

Parallel collision detection reduces computational complexity for large-scale rock block simulations, though efficient processor load balancing is crucial, particularly when blocks frequently cross subdomains.

5. Parallelization of Irregular Terrain Model

For electromagnetic phenomena in rock masses (ground-penetrating radar, fracture-associated emissions), reference [4] presents a parallel computing method using the Irregular Terrain Model (ITM). The method employs a 3D scalar field with static task allocation based on terrain reusability. The algorithm defines electromagnetic calculation boundaries, constructs the 3D scalar field, distributes workload across nodes based on terrain characteristics, executes parallel ITM power intensity calculations, and integrates results for comprehensive field analysis.

The parallel electromagnetic method achieves >70% efficiency across varying node counts, with efficiency improving at higher sampling densities. These principles are adaptable to electromagnetic phenomena in rock dynamics applications.

6. General High-Performance Computing Techniques

GPU acceleration offers significant performance benefits for rock dynamics simulations, particularly for operations with high data parallelism like explicit time integration and Monte Carlo methods. GPU implementation requires algorithm restructuring: identifying parallel components, optimizing data structures for GPU memory architecture, programming using CUDA or OpenCL frameworks, and managing CPU-GPU data transfer to minimize communication overhead.

GPU acceleration benefits vary across rock dynamics applications, with optimal performance for regular computations but potential limitations for algorithms requiring complex control flow, frequent synchronization, or irregular memory access patterns.

7. Distributed Computing Frameworks

Distributed computing frameworks (Apache Hadoop, Apache Spark) enhance rock dynamics simulations, particularly for large dataset analysis and multi-parameter studies. Hadoop can parallelize Monte Carlo simulations for rock slope stability analysis through map-reduce tasks, while Spark's in-memory computing accelerates iterative algorithms for nonlinear equations and model parameter optimization.

Parallel computing scalability in rock dynamics manifests as strong scaling (fixed problem size with varying processor count) and weak scaling (proportionally increasing both). Strong scaling ideally yields solution times inversely proportional to processor count, while weak scaling maintains constant solution times as both problem size and processors increase proportionally.

Effective scalability in rock dynamics simulations requires careful consideration of key constraints: load imbalance from uneven computational distribution, communication overhead between processors, sequential bottlenecks limiting parallelization per Amdahl's law, and memory bandwidth restrictions in memory-bound computations. These factors can significantly impact simulation performance as processor counts increase.

The optimization of scalability in parallel rock mechanics simulations can be achieved through the implementation of several sophisticated techniques. Parallel rock mechanics simulations achieve optimized scalability through several advanced techniques. Dynamic load balancing adaptively allocates tasks to processors, addressing unpredictable computational demands. Communication-avoiding algorithms enhance scalability by minimizing inter-processor communication through redundant computation and hierarchical methods. Hybrid parallelization combines MPI (distributed memory) with OpenMP (shared memory) across and within nodes. Parallel-optimized algorithms, such as multigrid methods, provide superior scalability compared to traditional iterative solvers for specific problems.

Platform-specific optimization strategies are essential for rock dynamics simulations across different architectures. Cluster computing requires minimized inter-node communication and optimized load balancing through domain decomposition. Shared-memory multicore CPU systems focus on thread-level parallelism via OpenMP or Intel TBB, emphasizing cache utilization and memory access patterns. Many-core architectures (Intel Xeon Phi) prioritize vectorization and SIMD instruction optimization.

GPU implementations demand optimized memory access, minimal CPU-GPU data transfer, and maximized GPU occupancy. Heterogeneous systems combining CPUs and GPUs require effective workload distribution and load balancing between processor types for optimal performance. Hardware characteristics significantly influence parallel algorithm selection, as demonstrated by the Jacobi iteration method [3]. While effective on multicore CPUs, its memory access pattern may be suboptimal for GPUs, where methods like colored Gauss–Seidel or block-iterative approaches offer more suitable memory access patterns.

Problem scale also determines optimal parallelization strategy: shared-memory parallelism (OpenMP) suffices for problems fitting single-node memory, while distributed-memory approaches (MPI) become necessary for larger problems. Hybrid implementations combining both paradigms—using MPI across nodes and OpenMP within nodes—can minimize communication overhead while maintaining scalability.

In conclusion, parallel computing has revolutionized rock dynamics, enabling large-scale complex simulations from domain decomposition in finite element analysis to Monte Carlo simulations for uncertainty quantification. Method selection depends on problem characteristics, computational resources, and accuracy-speed-scalability requirements. Hybrid approaches combining multiple parallelization strategies and heterogeneous computing resources are becoming increasingly prevalent.

7.1.2 Advanced Numerical Methods

Advanced numerical simulation techniques in rock dynamics have evolved to predict complex rock behavior under dynamic loading conditions, supporting mining, civil engineering, and geophysical applications. This section examines high-performance computational methods for rock dynamics simulation.

Finite Element Method (FEM) Enhancements

The Finite Element Method has long been a staple in rock mechanics simulations. However, recent advancements have significantly improved its capabilities in modeling dynamic rock behavior.

1. Extended Finite Element Method (X-FEM)

The Extended Finite Element Method (X-FEM) has emerged as a powerful tool for modeling crack propagation and fracture mechanics in rock dynamics. X-FEM extends the classical FEM by incorporating discontinuous basis functions to represent cracks and voids within elements, eliminating the need for remeshing as cracks propagate [5, 6].

In X-FEM, the displacement field $u(x)$ is approximated as:

$$u(x) = \sum_i N_i(x)u_i + \sum_j N_j(x)H(x)a_j + \sum_k N_k(x)\sum_\alpha F_\alpha(x)b_{k\alpha} \tag{7.3}$$

where $N_i(x)$ are the standard shape functions, u_i are the nodal displacements, $H(x)$ is the Heaviside function, a_j are the enriched nodal degrees of freedom for modeling discontinuities, $F_\alpha(x)$ are the crack-tip enrichment functions, and $b_{k\alpha}$ are the corresponding nodal enriched degrees of freedom.

The X-FEM enables modeling of arbitrary crack paths without domain remeshing, while providing accurate stress intensity factor calculations and efficient simulation of interacting cracks in rock dynamics.

The X-FEM has demonstrated effectiveness in rock dynamics applications, including dynamic crack propagation in impact-loaded brittle rocks [5], hydraulic fracturing simulation in unconventional reservoirs [6], and rock fragmentation analysis during blasting [7].

2. Isogeometric Analysis (IGA)

Isogeometric Analysis (IGA) bridges Computer-Aided Design (CAD) and Finite Element Analysis (FEA) by using Non-Uniform Rational B-Splines (NURBS) for both geometry description and solution approximation. In rock dynamics simulations, IGA provides exact geometry representation for complex geological structures and maintains higher-order continuity across element boundaries, offering improved accuracy per degree of freedom compared to traditional FEM.

The displacement field in IGA is approximated as:

$$u(\xi) = \sum_i R_{i,p}(\xi) P_i \qquad (7.4)$$

where $R_{i,p}(\xi)$ are the NURBS basis functions of degree p, and P_i are the control points.

Isogeometric Analysis (IGA) has demonstrated effectiveness in rock dynamics applications, including wave propagation modeling through heterogeneous rock masses, dynamic fracture simulation in layered rock structures [8], and seismic response analysis of underground caverns [9].

Meshless Methods

Meshless methods have gained popularity in rock dynamics simulations due to their ability to handle large deformations and discontinuities without the need for a predefined mesh.

1. Smoothed Particle Hydrodynamics (SPH)

Smoothed Particle Hydrodynamics is a Lagrangian particle method that approximates continuous fields using a set of discrete particles. In SPH, field variables and their derivatives are approximated using kernel functions [5, 8].

The approximation of a field variable $A(r)$ in SPH is given by:

$$A(r) \approx \sum_j \left(\frac{m_j}{\rho_j}\right) A_j W(r - r_j, h) \qquad (7.5)$$

where m_j and ρ_j are the mass and density of particle j, A_j is the value of A at particle j, W is the smoothing kernel function, and h is the smoothing length.

Advanced numerical methods in rock mechanics effectively handle large deformations and material separation, complex free-surface flows for fluid-rock interactions, and diverse constitutive models representing varied rock behavior patterns.

Advanced computational methods in dynamic rock mechanics simulate phenomena from high strain rate fragmentation [5] to large-scale debris flows and avalanches [3], while analyzing rock-structure interactions [7] for engineering applications.

7.1 High-Performance Computational Technique

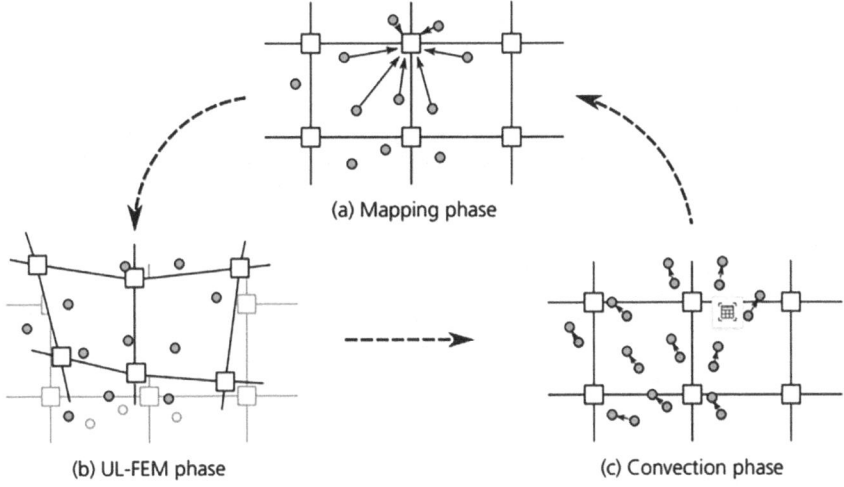

Fig. 7.2 Computational cycle of MPM

2. Material Point Method (MPM)

The Material Point Method is a hybrid Eulerian–Lagrangian approach that combines the strengths of particle methods and grid-based methods. In MPM, material points carry the physical properties and state variables, while a background grid is used for solving the governing equations [6, 9].

The MPM algorithm sequentially maps particle information to grid nodes, solves equations of motion, updates particle positions and velocities, and computes particle stresses and state variables for each time step [10] (Fig. 7.2).

The MPM effectively handles large deformations and material failure in rock dynamics, while avoiding mesh entanglement issues and efficiently managing contact and impact problems between multiple bodies.

The MPM has been applied to analyze dynamic rock slope failures and runout behavior [6], underground excavation dynamics with time-dependent deformation [9], and rock blasting processes including wave propagation and fragmentation [7].

Discrete Element Method (DEM) Advancements

The Discrete Element Method has been widely used in rock mechanics for modeling discontinuous media. Recent advancements have further improved its capabilities for dynamic simulations.

1. Bonded Particle Model (BPM)

The Bonded Particle Model is an extension of DEM that represents rock as an assembly of bonded circular or spherical particles. In BPM, inter-particle bonds can break under tensile or shear loading, allowing for the simulation of crack initiation and propagation [5].

The force–displacement law for bonded particles in BPM is typically given by:

$$F_n = k_n U_n \tag{7.6}$$

$$F_s = -k_s U_s \tag{7.7}$$

where F_n and F_s are the normal and shear forces, k_n and k_s are the normal and shear stiffnesses, and U_n and U_s are the normal and shear displacements.

Recent BPM advancements include rate-dependent bond strength models for time-dependent rock behavior, fluid–solid coupling mechanisms for hydraulic fracturing simulation, and particle clustering techniques for accurate rock fabric representation.

The BPM has been effectively applied to simulate dynamic rock failure under impact loading [5], seismic wave propagation through fractured rock masses [8], and rockburst mechanisms in deep underground excavations [7], demonstrating advantages over the Combined Finite-Discrete Element Method (FDEM).

2. Combined Finite-Discrete Element Method (FDEM)

The Combined Finite-Discrete Element Method (FDEM) integrates FEM's continuum representation with DEM's discontinuum capabilities, where finite elements can transform into discrete elements during fracture development [9, 11].

FDEM enables unified modeling of continuum deformation and discontinuous fracturing, handles complex contact interactions, and simulates progressive rock failure through continuum-discontinuum transition. The FDEM integrates equations of motion for discrete bodies with finite element formulations for internal deformation, incorporating contact detection and resolution algorithms for inter-body interactions.

The FDEM has been applied to simulate dynamic rock fracturing and fragmentation [11], earthquake-induced landslides under seismic loading [9], and rock support system performance under dynamic loading [7].

Multiscale and Multiphysics Approaches

Advanced numerical methods for rock dynamics increasingly incorporate multiscale and multiphysics approaches to capture the complex, coupled phenomena occurring across different spatial and temporal scales.

1. Hierarchical Multiscale Modeling

Hierarchical multiscale modeling couples different scale models to analyze phenomena from microscale fractures to macroscale rock mass behavior [2, 5], employing concurrent methods for simultaneous fine-coarse scale coupling, sequential methods for stepwise information transfer, and adaptive methods for dynamic resolution refinement based on local conditions.

Recent hierarchical multiscale modeling advances address microcrack-induced damage evolution under dynamic loading [5], seismic wave propagation in fractured porous media [6], and strain localization in heterogeneous rock masses [7], facilitating Coupled Thermo-Hydro-Mechanical-Chemical (THMC) analysis across multiple scales and physical processes.

7.1 High-Performance Computational Technique

2. Coupled Thermo-Hydro-Mechanical-Chemical (THMC) Modeling

Coupled THMC modeling aims to capture the complex interactions between thermal, hydraulic, mechanical, and chemical processes in rock dynamics simulations [11].

The general form of coupled THMC equations can be expressed as:

$$\begin{bmatrix} K_{uu} & K_{u\theta} & K_{up} & K_{uc} \end{bmatrix} \begin{bmatrix} u \\ \theta \\ p \\ c \end{bmatrix} = \begin{bmatrix} F_u \end{bmatrix} \tag{7.8}$$

$$\begin{bmatrix} K_{\theta u} & K_{\theta\theta} & K_{\theta p} & K_{\theta c} \end{bmatrix} \begin{bmatrix} \theta \end{bmatrix} = \begin{bmatrix} F_\theta \end{bmatrix} \tag{7.9}$$

$$\begin{bmatrix} K_{pu} & K_{p\theta} & K_{pp} & K_{pc} \end{bmatrix} \begin{bmatrix} p \end{bmatrix} = \begin{bmatrix} F_\theta \end{bmatrix} \tag{7.10}$$

$$\begin{bmatrix} K_{cu} & K_{c\theta} & K_{cp} & K_{cc} \end{bmatrix} \begin{bmatrix} c \end{bmatrix} = \begin{bmatrix} F_c \end{bmatrix} \tag{7.11}$$

where u, θ, p, and c represent displacement, temperature, pore pressure, and chemical concentration fields, respectively. K matrices represent coupling coefficients, and F vectors represent external forces or source terms.

Coupled THMC modeling requires accurate constitutive relationships for multi-physics interactions, efficient solutions for coupled equation systems, and numerical techniques for handling disparate time scales. Recent applications include thermally-induced fracturing in geothermal reservoirs [11], coupled seismic-electromagnetic wave propagation in saturated rocks [8], and chemically-induced stress corrosion cracking under dynamic loading [9].

Verification, Validation, and Uncertainty Quantification

As advanced numerical methods for rock dynamics become increasingly complex, rigorous verification, validation, and uncertainty quantification (VVUQ) procedures are essential for ensuring the reliability and applicability of simulation results.

1. Verification and Benchmarking

Rock dynamics numerical modeling verification includes analytical solution comparisons, code-to-code benchmarking, and convergence studies for error assessment. Standard test cases encompass dynamic fracture propagation [5], wave propagation in fractured media [6], and rate-dependent constitutive model comparisons [7], enabling systematic evaluation of computational approaches.

2. Validation Against Experimental Data

Validation compares simulation results with experimental observations to assess model accuracy. In rock dynamics, validation requires representative laboratory experiments to capture essential physical phenomena, while addressing scale relationships between laboratory and field behavior. Statistical analysis of model-data discrepancies quantifies prediction accuracy. Recent studies demonstrate this approach through dynamic rock fracturing validation using high-speed imaging, field-scale seismic response comparisons in underground excavations [8], and rockburst prediction verification using microseismic monitoring data [9].

3. Uncertainty Quantification

Uncertainty Quantification (UQ) assesses how various uncertainty sources impact simulation results in rock dynamics through sophisticated techniques. Monte Carlo methods propagate input uncertainties through numerical models, while sensitivity analysis identifies influential parameters for focused investigation. Bayesian inference enables parameter estimation and model calibration by combining prior knowledge with experimental data. Applications include probabilistic assessment of dynamic slope stability [5], seismic hazard analysis for underground structures [6], and optimization of rock support systems under dynamic loading [7].

7.1.3 Optimization Techniques

In rock dynamics numerical simulation, optimization techniques enhance computational efficiency and solution accuracy. These techniques address the computational demands of dynamic simulations in rock mechanics. Optimization encompasses two key areas: simulation process optimization (algorithm efficiency and parallel computing) and specific problem-solving techniques. Both are essential for high-performance rock dynamics computation.

Particle Swarm Optimization (PSO) and Hybrid PSO Techniques

Particle Swarm Optimization is a population-based stochastic optimization technique inspired by the social behavior of bird flocking or fish schooling. In rock dynamics simulations, PSO has found widespread application due to its simplicity, effectiveness, and ability to handle non-linear, non-convex optimization problems.

The basic PSO algorithm can be described mathematically as follows:

$$v_i^{k+1} = wv_i^k + c_1 r_1 (p_i - x_i^k) + c_2 r_2 (p_g - x_i^k) \tag{7.12}$$

$$x_i^{k+1} = x_i^k + v_i^{k+1} \tag{7.13}$$

where
 v_i^k is the velocity of particle i at iteration k
 x_i^k is the position of particle i at iteration k
 w is the inertia weight
 c1 and c2 are cognitive and social parameters
 r1 and r2 are random numbers between 0 and 1
 p_i is the best position found by particle i
 p_g is the global best position found by the swarm

7.1 High-Performance Computational Technique

PSO applications in rock dynamics include parameter identification, support system optimization, and mechanical property prediction. A PSO-ANN hybrid model [12] for TBM advance rate prediction in granitic rocks achieved superior performance (R^2: 0.958 training, 0.961 testing) compared to standalone ANNs. The model's success stems from PSO's optimization of ANN parameters (hidden neurons, learning rate, momentum coefficient).

To address premature convergence in complex rock dynamics optimization, researchers developed enhanced techniques like PSO-GP-FDM [13], combining PSO's global search capability with Gaussian Process learning and Finite Difference Method accuracy for rock mass parameter back-analysis in underground caverns (Fig. 7.3).

The PSO-GP-FDM algorithm can be summarized in the following steps: The optimization process begins with the initialization of the PSO population and its associated parameters. Subsequently, the fitness of each particle is evaluated through the implementation of the FDM model. Following this evaluation, the algorithm updates both the personal best (pbest) and global best (gbest) positions based on the obtained

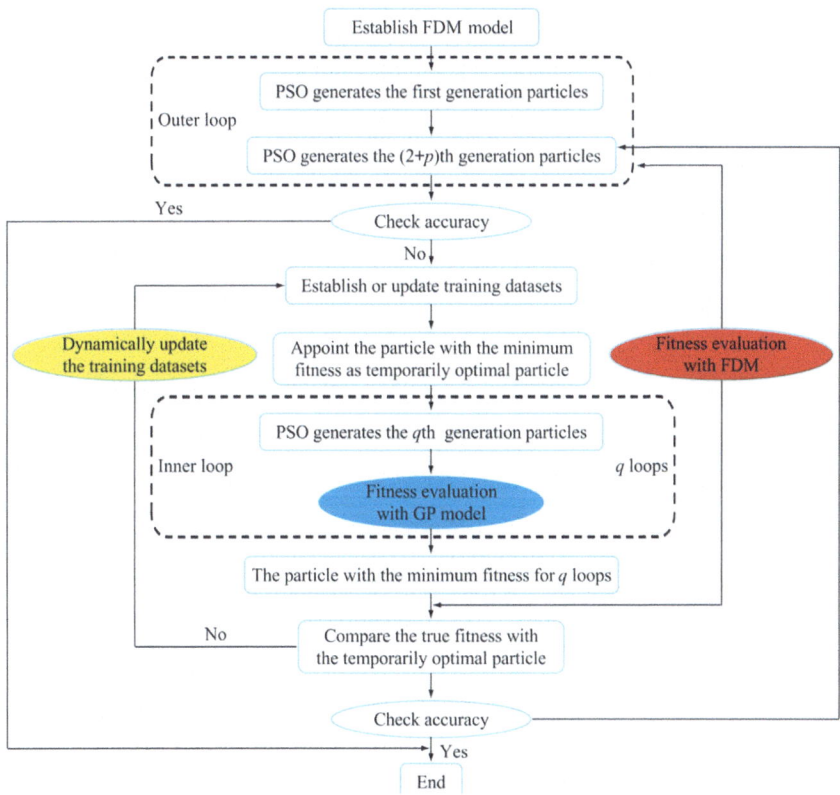

Fig. 7.3 Flow chart of the PS-GP-FDM [13]

fitness values. The process then advances to employ a Gaussian Process model, which predicts fitness values for newly generated positions. Using the standard PSO equations, the particle velocities and positions are systematically updated. These steps, from fitness evaluation through position updating, are iteratively executed until predefined convergence criteria are satisfied.

The PSO-GP-FDM hybrid method integrates Gaussian Process regression to minimize computational FDM evaluations while preserving accuracy. Applied to the Tai'an Pumped Storage Power Station [13], this approach successfully back-analyzed seven geomechanical parameters (Young's moduli, cohesion values, internal friction angle) using third-stage excavation displacement data from seven monitoring points. The optimized parameters matched in-situ displacements with 5.28% mean relative error.

Genetic Algorithms (GA) and Evolutionary Strategies

Genetic Algorithms (GAs), derived from natural selection principles, serve as effective optimization tools for rock dynamics simulations, particularly in solving non-linear problems with extensive search spaces. The GA process initializes a population of solution-representing chromosomes, evaluates their fitness, selects parent chromosomes for reproduction, and applies crossover and mutation operators to generate subsequent generations. This iteration continues until meeting termination criteria.

In rock dynamics, GAs have been applied to various optimization problems, including parameter identification, design optimization, and prediction of rock properties. For instance, [14] employed GA to develop predictive models for estimating rock brittleness. The GA-quadratic model developed in this study outperformed other models, including PSO-based models, in terms of accuracy and reliability.

The GA-quadratic model for predicting normalized rock brittleness (Y_n) *as a function of normalized rock density* (X_{1n}), uniaxial compressive strength (X_{2n}), *and Brazilian tensile strength* (X_{3n}) is given by:

$$Y_n = -0.0361 X_{1n} + 2.1728 X_{2n} - 1.1340 X_{3n} + 0.2748 X_{1n}^2 + 0.6479 X_{2n}^2 \\ + 0.5439 X_{3n}^2 - 0.7859 X_{1n} X_{2n} + 0.5775 X_{1n} X_{3n} - 1.4771 X_{2n} X_{3n} + 0.2479 \tag{7.14}$$

This model achieved impressive performance metrics, including: The model's performance was evaluated using multiple statistical metrics, yielding robust validation results. The coefficient of determination (R^2) achieved a value of 0.932, indicating strong correlation between predicted and observed values. This high predictive accuracy is further supported by a Root Mean Square Error (RMSE) of 2.64, demonstrating relatively low prediction deviation. Additionally, the Variance Account For (VAF) reached 93.06%, confirming the model's excellent capability in explaining data variability. The overall effectiveness of the model is further quantified by a Performance Index (PI) of 1.708, providing a comprehensive measure of model efficiency. The GA-quadratic model's success in predicting rock brittleness showcases genetic algorithms' effectiveness in capturing complex, non-linear relationships between rock properties, particularly valuable for rock dynamics simulations where material behavior depends on multiple interacting factors.

Genetic algorithms excel in rock dynamics optimization by naturally handling discrete variables and constraints, making them ideal for support system selection and excavation sequence optimization in rock engineering. However, like PSO, they can face premature convergence issues and high computational demands in large-scale applications.

To address these limitations, researchers have developed hybrid approaches, notably the GA-ANN (Genetic Algorithm–Artificial Neural Network) method for parameter identification in rock dynamics models. This hybrid system optimizes ANN architecture and initial weights through GA while modeling rock behavior parameters through the neural network.

The GA-ANN hybridization process involves initializing ANN populations with random architectures and weights, evaluating fitness through training data performance, selecting top performers for reproduction, and applying genetic operators to generate and train new networks until convergence or generation limit is reached.

This hybrid approach combines the global search capability of GA with the powerful function approximation abilities of ANNs, potentially leading to more accurate and robust models for rock dynamics simulations.

Simulated Annealing (SA)

Simulated Annealing (SA), based on metallurgical annealing principles, excels at finding global optima in complex rock dynamics problems with multiple local optima.

The SA algorithm operates by initializing a random solution at temperature T and iteratively generating neighboring solutions. The change in objective function (ΔE) determines solution acceptance: solutions with $\Delta E < 0$ are accepted automatically, while those with $\Delta E > 0$ are accepted with probability $\exp(-\Delta E/T)$. The temperature T decreases according to a cooling schedule until stopping criteria are met.

The acceptance probability for a worse solution is given by:

$$P(\text{accept}) = \exp\left(-\frac{\Delta E}{T}\right) \tag{7.15}$$

where ΔE is the change in objective function value, and T is the current temperature.

In rock dynamics simulations, SA optimizes support systems, excavation sequences, and constitutive model parameters, with its local optima escape capability particularly valuable for irregular objective function landscapes. As exemplified in tunnel support design, SA optimizes rock bolt layouts by balancing factors including system cost, induced rock mass stress, and safety factors. A key challenge in SA application is selecting appropriate cooling schedules, which govern the temperature parameter's decrease rate and solution space exploration effectiveness.

Common cooling schedules include:

Linear cooling:

$$T_k = T_0 - \beta k \tag{7.16}$$

Geometric cooling:

$$T_k = \alpha T_{k-1} \tag{7.17}$$

Logarithmic cooling:

$$T_k = \frac{T_0}{\log(k+1)} \tag{7.18}$$

where T_k is the temperature at iteration k, T_0 is the initial temperature, and α, β are cooling parameters.

The choice of cooling schedule and its parameters can significantly impact the performance of the SA algorithm in rock dynamics optimization problems. Researchers often need to experiment with different schedules and parameters to find the best configuration for their specific problem.

Artificial Neural Networks (ANN) in Optimization

Artificial Neural Networks (ANNs) in rock dynamics simulations serve dual optimization roles: as surrogate models reducing computational costs and as components in hybrid optimization algorithms.

The basic structure of a feedforward ANN can be described mathematically as:

$$y = f\left(\sum_{i=1}^{n} w_i x_i + b\right) \tag{7.19}$$

where
  y is the output
  x_i are the inputs
  w_i are the weights
  b is the bias
  f is the activation function

In rock dynamics optimization, ANNs create surrogate models for rapid simulation approximation within optimization algorithms. A notable application in [15] used a PSO-optimized ANN to predict Tunnel Boring Machine (TBM) advance rates, achieving R^2 values of 0.958 and 0.961 for training and testing datasets. The three-layer ANN architecture featured eight input neurons processing rock mechanics parameters (UCS, BTS, RQD), twelve hidden neurons, and one output neuron for TBM advance rate prediction.

7.1 High-Performance Computational Technique

This $8 \times 12 \times 1$ architecture was determined to be optimal through the PSO optimization process, which searched for the best combination of network parameters.

ANNs can also be used in conjunction with other optimization techniques to solve complex rock dynamics problems. For instance, a hybrid approach combining ANNs with genetic algorithms (GA-ANN) could be used for parameter identification in rock mechanics models. In this approach, the GA would optimize the architecture and initial weights of the ANN, while the ANN itself would model the relationship between input parameters and rock behavior.

The training process of an ANN for rock dynamics applications typically involves minimizing an error function, such as the mean squared error (MSE):

$$MSE = \frac{1}{n} \sum_{i=1}^{n} (y_i - \hat{y}_i)^2 \qquad (7.20)$$

where y_i are the actual values and
$\hat{y}_i$ are the predicted values.

Various optimization algorithms, including gradient descent, conjugate gradient, and Levenberg–Marquardt, can be used to minimize this error function and train the ANN. The choice of optimization algorithm can significantly impact the training speed and final performance of the ANN.

One of the challenges in using ANNs for rock dynamics optimization is the risk of overfitting, especially when dealing with limited datasets. Techniques such as regularization, early stopping, and cross-validation are often employed to improve the generalization ability of ANN models in rock mechanics applications.

Multi-Objective Optimization Techniques

Rock dynamics problems frequently involve multiple conflicting objectives, such as balancing safety, cost, and constructability in rock support system design. The Non-dominated Sorting Genetic Algorithm II (NSGA-II) addresses these challenges through Pareto dominance and crowding distance concepts.

The NSGA-II process initializes a parent population P (size N), generates offspring population Q, and combines them into population R (size 2N). Non-dominated sorting identifies Pareto fronts (F1, F2, ...), constructing the next generation P′ sequentially. When |P′|<N, additional solutions are selected from the last front using crowding distance calculations. The process iterates with new offspring Q′ generation until termination criteria are met.

In rock dynamics simulations, NSGA-II and other multi-objective optimization techniques have been applied to various problems, including the optimization of tunnel support systems, rock slope stability analysis, and blast design optimization.

The tunnel support design optimization can be formulated as a multi-objective problem with three primary objectives to be minimized simultaneously. These objectives comprise F1(x), representing the total cost of the support system; F2(x), denoting the maximum induced stress in the rock mass; and F3(x), quantifying the tunnel convergence. The optimization is bounded by three fundamental constraints:

g1(x) ≤ 0, which ensures an adequate factor of safety against rock mass failure; g2(x) ≤ 0, which addresses constructability requirements; and g3(x) ≤ 0, which accounts for. Where x is a vector of decision variables representing the support system design parameters (e.g., bolt length, spacing, shotcrete thickness).

The result of such a multi-objective optimization is not a single optimal solution, but a set of non-dominated solutions known as the Pareto front. Each solution on the Pareto front represents a different trade-off between the objectives, allowing engineers to make informed decisions based on their specific project requirements and constraints.

Hybrid and Ensemble Optimization Techniques

Hybrid and ensemble optimization techniques address complex rock dynamics problems by combining multiple algorithms' strengths.

This hybrid approach leverages the global search capability of PSO, the efficient approximation of the GP model, and the accuracy of FDM simulations. By using the GP model as a surrogate for expensive FDM evaluations, the method significantly reduces computational cost while maintaining accuracy.

Another example of a hybrid approach is the combination of Genetic Algorithms (GA) and Artificial Neural Networks (ANN) for rock property prediction, as demonstrated in [14]. In this study, both GA and PSO were used to optimize ANN models for predicting rock brittleness. The GA-quadratic model outperformed other models, including:

GA-linear:

$$Y_n = 0.2322X_{1n} + 1.8351X_{2n} - 0.9156X_{3n} + 0.1839 \tag{7.21}$$

PSO-linear:

$$Y_n = 0.2992X_{1n} + 1.0000X_{2n} - 0.2140X_{3n} + 0.1455 \tag{7.22}$$

PSO-quadratic:

$$Y_n = 0.4510X_{1n} + 1.0000X_{2n} - 0.3811X_{3n} - 0.0148X_{1n}^2 - 0.2608X_{2n}^2 + 0.7477X_{3n}^2 \\ + 0.6439X_{1n}X_{2n} - 0.2678X_{1n}X_{3n} - 0.8701X_{2n}X_{3n} + 0.1233 \tag{7.23}$$

The GA-quadratic model achieved superior performance ($R^2 = 0.932$, RMSE = 2.64) in capturing complex rock mechanics relationships.

Ensemble optimization techniques further enhance problem-solving through parallel implementation of multiple algorithms (PSO, GA, SA) with interactive information sharing. The process executes algorithms simultaneously, exchanges best-found solutions periodically, and updates search strategies based on shared information until convergence, determining final solutions through performance-based selection or collective voting.

7.1 High-Performance Computational Technique

This ensemble approach can help overcome the limitations of individual algorithms and increase the robustness of the optimization process, particularly for complex rock dynamics problems with irregular objective function landscapes.

Machine Learning-Enhanced Optimization

The integration of machine learning techniques with traditional optimization algorithms has opened new avenues for solving complex rock dynamics problems. Machine learning models can be used to create surrogate models, guide the search process, or even learn optimal optimization strategies.

One promising approach is the use of Gaussian Process Regression (GPR) in optimization, as demonstrated in the PSO-GP-FDM method [13]. GPR can create efficient surrogate models that capture the uncertainty in predictions, allowing for more informed decision-making in the optimization process.

The Gaussian Process model can be described by its mean function $m(x)$ and covariance function $k(x, x')$:

$$f(x) \sim \mathcal{GP}(m(x), k(x, x')) \tag{7.24}$$

where

$$m(x) = \mathbb{E}[f(x)] \tag{7.25}$$

$$k(x, x') = \mathbb{E}[(f(x) - m(x))(f(x') - m(x'))] \tag{7.26}$$

In rock dynamics optimization, Gaussian Process Regression (GPR) models design parameter-objective function relationships, enabling efficient design space exploration without extensive numerical simulations.

Reinforcement Learning (RL) offers an alternative approach, learning optimal solution space exploration strategies through five components: state space S representing problem conditions, action space A containing possible solution modifications, reward function $R(s,a)$ quantifying immediate benefits, policy π(a|s) determining action probabilities, and value function $V(s)$ calculating expected cumulative rewards.

The goal of RL is to learn an optimal policy π* that maximizes the expected cumulative reward:

$$\pi^* = \arg\max_{\pi} \mathbb{E}_{\pi}\left[\sum_{t=0}^{\infty} \gamma^t R(s_t, a_t)\right] \tag{7.27}$$

where γ is a discount factor for future rewards.

In rock dynamics optimization, RL could be used to learn optimal strategies for adjusting design parameters or selecting promising regions of the solution space to explore. This approach could be particularly valuable for problems with large, complex solution spaces where traditional optimization methods struggle to find good solutions efficiently.

7.2 Multi-Physics Coupling Simulation

7.2.1 Thermo-Hydro-Mechanical (THM) Coupling in Rock Dynamics

Thermo-Hydro-Mechanical (THM) coupling is a critical aspect of rock dynamics that plays a fundamental role in various geotechnical and geoenvironmental applications. This complex phenomenon involves the intricate interplay between thermal, hydraulic, and mechanical processes occurring within rock masses. The understanding and accurate modeling of THM coupling are essential for addressing challenges in fields such as geothermal energy extraction, nuclear waste disposal, carbon dioxide sequestration, and underground construction. In this section, we will explore the key concepts, governing equations, numerical modeling approaches, and applications of THM coupling in rock dynamics.

Fundamentals of THM Coupling

THM coupling refers to the simultaneous occurrence and interaction of thermal (T), hydraulic (H), and mechanical (M) processes within a porous medium, such as rock. These processes are inherently linked and can significantly influence each other, leading to complex behavior that cannot be adequately described by considering each process in isolation.

1. Thermal Processes

Thermal processes involve heat transfer within the rock mass and between the rock and fluids present in its pores. The primary mechanisms of heat transfer in rocks are conduction, convection, and radiation. In most cases, conduction and convection dominate, while radiation plays a minor role except at very high temperatures.

Heat conduction is described by Fourier's law:

$$q = -k\nabla T \tag{7.28}$$

where q is the heat flux vector, k is the thermal conductivity tensor, and ∇T is the temperature gradient.

Heat convection occurs due to fluid movement and is governed by the energy balance equation:

$$\rho c_p v \cdot \nabla T = \nabla \cdot (\nabla T) + Q \tag{7.29}$$

7.2 Multi-Physics Coupling Simulation

where ρ is the fluid density, c_p is the specific heat capacity, v is the fluid velocity vector, and Q is a heat source or sink term.

2. Hydraulic Processes

Hydraulic processes involve fluid flow through the interconnected pore spaces and fractures within the rock mass. The flow of fluids in porous media is typically described by Darcy's law:

$$v = -K\nabla h \tag{7.30}$$

where v is the Darcy velocity vector, K is the hydraulic conductivity tensor, and ∇h is the hydraulic gradient.

3. Mechanical Processes

Mechanical processes involve the deformation and stress distribution within the rock mass in response to applied loads and changes in pore pressure. The mechanical behavior of rocks is typically described using the principles of continuum mechanics and elastoplasticity theory.

The stress–strain relationship for a linear elastic material is given by Hooke's law:

$$\sigma = D : \varepsilon \tag{7.31}$$

where σ is the stress tensor, D is the elasticity tensor, and ε is the strain tensor.

For more complex rock behavior, elastoplastic models or damage mechanics approaches may be employed to capture nonlinear and irreversible deformations.

4. Coupling Mechanisms

Thermal, hydraulic, and mechanical processes in rock systems interact through multiple coupling mechanisms. Thermal–hydraulic (TH) coupling manifests through temperature-induced changes in fluid density and viscosity affecting flow patterns, convective heat transfer, and temperature-dependent permeability. Thermal–mechanical (TM) coupling occurs via thermal expansion/contraction of rock matrix and fluids, temperature-dependent elastic moduli, and thermally induced stresses and deformations. In hydro-mechanical (HM) coupling, pore pressure modifies effective stress conditions, while stress-dependent permeability and fluid-induced deformations affect rock mass behavior. The fully coupled THM system combines these mechanisms, producing additional phenomena such as thermally induced pore pressure changes (Fig. 7.4).

Governing Equations for THM Coupling

The mathematical description of THM coupling requires a set of coupled partial differential equations that represent the conservation of mass, momentum, and

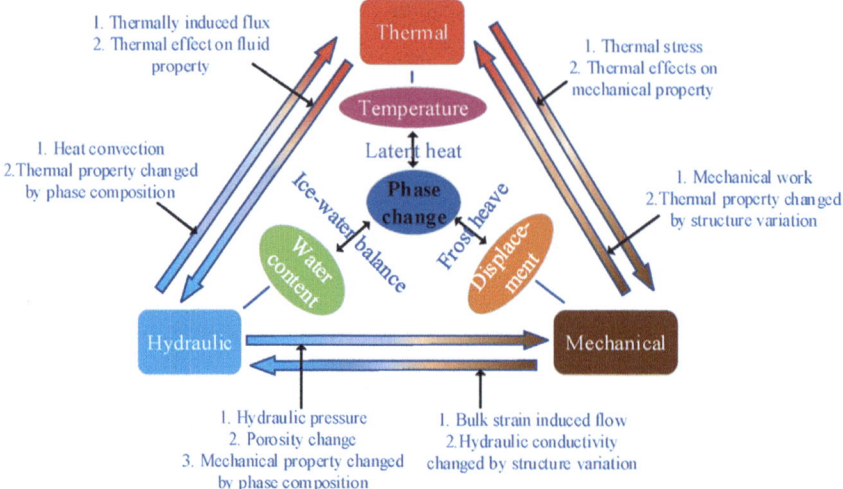

Fig. 7.4 Schematic representation of cryogenic THM coupling [16]

energy, along with appropriate constitutive relations. The following equations form the basis for most THM coupling models in rock dynamics:

1. Mass Conservation Equation

Mass conservation equations for water and gas phases [17]:

Methane (gas phase):

$$\frac{\partial(\phi \rho_g S_g)}{\partial t} + \nabla \cdot (\rho_g v_g) + (\phi \rho_g S_g)\frac{\partial \varepsilon v}{\partial t} = m_g \tag{7.32}$$

Water (liquid phase):

$$\frac{\partial(\phi \rho_w S_w)}{\partial t} + \nabla \cdot (\rho_w v_w) + (\phi \rho_w S_w)\frac{\partial \varepsilon v}{\partial t} = m_w + q_w \tag{7.33}$$

where ϕ is porosity, ρ_w and ρ_g are densities of water and gas, S_w and S_g are saturations of water and gas, v_w and v_g are Darcy velocities of water and gas, and Q_w and Q_g are source/sink terms for water and gas. q_w is source-sink term of water and m_w is water production rate.

The momentum conservation for the solid phase is represented by the quasi-static equilibrium equation:

$$\nabla \cdot \sigma + \rho g = 0 \tag{7.34}$$

7.2 Multi-Physics Coupling Simulation

where σ is the total stress tensor, ρ is the bulk density of the rock-fluid system, and g is the gravitational acceleration vector.

2. Energy Conservation Equation

The energy conservation equation, accounting for heat conduction, convection, and source terms, can be expressed as [18]:

$$(\rho c_p)_{\text{eff}} \frac{\partial T}{\partial t} + \nabla \cdot \left(\rho_w c_{pw} v_w T + \rho_g c_{pg} v_g T \right) = \nabla \cdot (k_{\text{eff}} \nabla T) + Q \quad (7.35)$$

where $(\rho c_p)_{\text{eff}}$ is the effective heat capacity of the rock-fluid system, c_{pw} and c_{pg} are specific heat capacities of water and gas, k_{eff} is the effective thermal conductivity tensor, Q is a heat source/sink term and T is temperature.

3. Constitutive Relations

The effective stress principle governs rock-fluid system behavior:

$$\sigma' = \sigma - \alpha P I \quad (7.36)$$

where σ' is the effective stress tensor, α is the Biot coefficient, P is the pore pressure, and I is the identity tensor.

Thermoelastic behavior follows [19]:

$$\sigma = D : (\varepsilon - \varepsilon T - \varepsilon M) - \alpha P I \quad (7.37)$$

where D is the elasticity tensor, ε is the total strain tensor, εT is the thermal strain tensor, and εM is the moisture-induced strain tensor.

Fluid flow follows generalized Darcy's law:

$$\begin{aligned} v_g &= -\frac{K_{rg} K}{\mu_g} \nabla P_g \\ v_w &= -\frac{K_{rw} K}{\mu_w} \nabla P_w \end{aligned} \quad (7.38)$$

The modified van Genuchten (VG) model describes saturation-suction relationship [3]:

$$S_{ew} = \left\{ \frac{1}{1 + \left[(\alpha_0 + \lambda_\alpha T - \lambda_\alpha T_0) e^b H_c \right]^{n_0 + \lambda_m (T - T_0)}} \right\} \quad (7.39)$$

where S_{ew} is the effective saturation, α_0, n_0, and m_0 are fitting parameters at the reference temperature T_0, λ_α, λ_n, and λ_m are temperature-dependent fitting parameters, e is the void ratio, b is a new proposed fitting parameter, H_c is the matrix suction head, and T is temperature.

Numerical Modeling Approaches for THM Coupling

The complexity of THM coupling processes in rock dynamics necessitates the use of advanced numerical modeling techniques. Several approaches have been developed to simulate these coupled phenomena, with the most common being:

1. Fully Coupled Approach

In the fully coupled approach, all governing equations for thermal, hydraulic, and mechanical processes are solved simultaneously as a single system of equations. This method provides the most accurate representation of the coupled processes but can be computationally intensive, especially for large-scale problems.

2. Sequential Coupling Approach

The sequential coupling approach provides a systematic method for solving complex rock mechanics problems by addressing governing equations separately through iterative information exchange. This approach encompasses two distinct implementation strategies: explicit and implicit coupling. In explicit implementation, the solution proceeds with a single calculation per time step, where results are passed sequentially between different physical processes. The implicit implementation, conversely, employs an iterative solution methodology within each time step, continuing calculations until convergence is achieved. This iterative process ensures higher accuracy in capturing the complex interactions between different physical phenomena, albeit at a higher.

3. One-Way Coupling Approach

One-way Coupling Approach Applied when one process dominantly influences others, particularly in thermal–hydraulic (TH) coupling where mechanical effects are minimal. Thermal and hydraulic equations are solved first, then used to update mechanical properties.

4. Numerical Methods

The implementation of these coupling approaches relies on various numerical methods for discretizing and solving the coupled THM equations. The FEM is widely adopted due to its versatility in handling complex geometries and material heterogeneities. The FDM offers computational efficiency and simplicity, making it particularly valuable in reservoir simulation applications. The FVM proves especially suitable for fluid flow problems owing to its conservative nature. For problems involving infinite or semi-infinite domains, the BEM provides an effective solution strategy. Additionally, the DEM serves as a specialized tool for simulating discontinuous rock masses and fracture propagation [16–19].

7.2.2 Numerical Methods for Coupled Multi-Physics Field Simulations

The study of rock dynamics often involves complex interactions between multiple physical phenomena, such as mechanical deformation, fluid flow, heat transfer, and chemical reactions. To accurately simulate these coupled processes, advanced numerical methods have been developed to solve the governing equations simultaneously or in a tightly coupled manner. This section provides a comprehensive overview of the state-of-the-art numerical methods for coupled multi-physics field simulations in rock dynamics.

FEM for Coupled Problems

The FEM has emerged as one of the most powerful and versatile numerical techniques for simulating coupled multi-physics problems in rock dynamics [11]. Its ability to handle complex geometries and material heterogeneities makes it particularly suitable for modeling realistic geological formations and engineering structures.

1. Formulation of Coupled FEM Equations

In a coupled multi-physics problem, the governing equations for different physical fields are typically expressed as a system of partial differential equations (PDEs). For example, in a THM coupled problem, the governing equations may include:

(1) Linear momentum balance (mechanical field)

$$\nabla \cdot \sigma + \rho b = 0 \tag{7.40}$$

where σ is the total Cauchy's stress, ρ denotes the body force vector and b is weighted average density of the mixture.

(2) Fluid flow continuity (hydraulic field)

$$\alpha \nabla \cdot \dot{\mathbf{u}} + \nabla \cdot \left(\frac{k_f}{\mu_f}(-\nabla p + \rho_f \mathbf{b}) \right) + \frac{1}{Q_t}\dot{p} - \beta_t \dot{T} = 0 \tag{7.41}$$

where α is a medium—property coefficient, ∇u is the divergence of velocity u (volumetric strain rate), k is permeability, μ_f is fluid dynamic viscosity, p is pressure, ρ_f is fluid mass density, b is body—force vector, Q_t is a time—related parameter, $\dot{p}$ is pressure's time derivative, B_T is thermal expansion coefficient, $\dot{T}$ is temperature's time derivative.

(3) Heat transfer (thermal field)

$$(\rho C)_{\text{eff}} \dot{T} + \rho_f C_f \left(\frac{k_f}{\mu_f}(-\nabla p + \rho_f \mathbf{b}) \right) \cdot \nabla T - \nabla \cdot (\lambda_{\text{eff}} \nabla T) = 0 \tag{7.42}$$

where $(\rho C)_{\text{eff}}$ is effective heat capacity, ρ_f is fluid mass density, C_f is fluid specific heat, k is permeability, μ_f is fluid dynamic viscosity, p is pressure, $\rho_f b$

is fluid body—force term, V_T is temperature gradient, λ_{eff} is effective thermal conductivity.

To solve these coupled equations using FEM, the domain is discretized into finite elements, and the field variables (u, p, T) are approximated using shape functions. The weak form of the governing equations is then derived using the principle of virtual work or the Galerkin method. This results in a system of algebraic equations that can be written in matrix form as:

$$\begin{bmatrix} K_{uu} & K_{up} & K_{uT} \\ K_{pu} & K_{pp} & K_{pT} \\ K_{Tu} & K_{Tp} & K_{TT} \end{bmatrix} \begin{bmatrix} \Delta u \\ \Delta p \\ \Delta T \end{bmatrix} = \begin{bmatrix} F_u \\ F_p \\ F_T \end{bmatrix} \qquad (7.43)$$

2. Solution Strategies for Coupled FEM Equations

Several solution strategies have been developed to address the computational challenges of solving coupled FEM equations efficiently. The monolithic approach represents the most direct method, wherein all field variables are solved simultaneously within a single system of equations. While this approach ensures strong coupling between fields, it often proves computationally expensive when applied to large-scale problems. In contrast, the staggered approach offers improved computational efficiency by solving each field separately in a sequential manner, facilitating information exchange between fields at each time step or iteration. However, this method may require smaller time steps to maintain stability and accuracy. Additionally, operator splitting methods have emerged as an effective alternative, decomposing the coupled problem into simpler subproblems that can be solved sequentially. These techniques, particularly the Isothermal Split and the Drained Split, have demonstrated particular utility in addressing THM problems [20, 21].

3. Advanced FEM Techniques for Coupled Problems

Several advanced FEM techniques have been developed to address specific challenges in coupled multi-physics simulations. Among these approaches, Mixed Finite Element Methods employ different interpolation functions for different field variables, thereby improving stability and accuracy, particularly when dealing with incompressible materials or fluid flow in porous media. Furthermore, Stabilized Finite Element Methods, such as the Streamline Upwind Petrov–Galerkin (SUPG) method and the Galerkin Least Squares (GLS) method, have been implemented to stabilize numerical solutions in cases involving advection-dominated problems or sharp gradients. Additionally, Adaptive Mesh Refinement has emerged as a dynamic technique that enhances accuracy while maintaining computational efficiency by refining the mesh in regions characterized by high solution gradients or error [20, 22].

XFEM for Discontinuous Problems

The XFEM is an extension of the classical FEM that allows for the modeling of discontinuities, such as cracks or material interfaces, without the need for mesh

7.2 Multi-Physics Coupling Simulation

conformity [20]. This makes XFEM particularly useful for simulating fracture propagation and fluid flow through fractured rock masses.

1. XFEM Formulation for Coupled Problems

In XFEM, the standard finite element approximation is enriched with discontinuous functions to represent the jump in field variables across discontinuities. For a coupled THM problem, the XFEM approximation can be written as:

$$\boldsymbol{u}(\boldsymbol{x}) = \boldsymbol{u}^c(\boldsymbol{x}) + H_{\Gamma_d}(\varphi(\boldsymbol{x}))\boldsymbol{u}^d(\boldsymbol{x}) \tag{7.44}$$

$$p(\boldsymbol{x}) = p^c(\boldsymbol{x}) + H_{\Gamma_d}(\varphi(\boldsymbol{x}))p^d(\boldsymbol{x}) \tag{7.45}$$

$$T(\boldsymbol{x}) = T^c(\boldsymbol{x}) + H_{\Gamma_d}(\varphi(\boldsymbol{x}))T^d(\boldsymbol{x}) \tag{7.46}$$

where H_{Γ_d} is the Heaviside enrichment function, $\varphi(\boldsymbol{x})$ is the level set function describing the discontinuity geometry, and the superscripts c and d denote continuous and discontinuous parts, and u^c, p^c, T^c and u^d, p^d, T^d are associated with the standard (continuous) and enriched (discontinuous) approximations of the displacement, pressure and temperature fields, respectively [20].

2. Implementation of XFEM for Coupled Problems

The implementation of XFEM for coupled multi-physics problems encompasses several interconnected technical components that must be carefully considered. At the foundation of this implementation lies the enrichment strategy, which requires the careful selection of appropriate enrichment functions to effectively represent discontinuities across various physical fields. Following this crucial first step, the development of specialized integration schemes becomes necessary, particularly for elements intersected by discontinuities. These schemes typically involve either element partitioning or the application of high-order quadrature methods. Subsequently, the coupling terms within the governing equations must be modified to properly account for the discontinuous nature of the enriched approximation. Finally, the implementation process culminates in the development of robust algorithms for tracking the evolution of discontinuities, such as crack propagation, primarily through the application of level set methods.

3. Applications of XFEM in Rock Dynamics

The XFEM has demonstrated significant versatility and effectiveness in analyzing various rock dynamics challenges. In the context of hydraulic fracturing, XFEM enables the simulation of fluid-driven fracture propagation in reservoir rocks, effectively accounting for the complex coupled interactions between fluid pressure, rock deformation, and fracture mechanics. Furthermore, XFEM has proven valuable in modeling thermally induced fracturing, particularly in the analysis of crack initiation and propagation resulting from thermal stresses in geothermal reservoirs and nuclear waste repositories. Additionally, the method has been successfully applied to fault reactivation studies, where it facilitates the analysis of potential fault slip and induced seismicity in geological CO_2 storage and enhanced geothermal systems [20].

FVM for Coupled Problems

The FVM is particularly well-suited for problems involving fluid flow and heat transfer in porous media due to its inherent conservation properties [5]. In the context of rock dynamics, FVM is often used in combination with FEM to simulate coupled fluid–solid interactions.

1. FVM Formulation for Coupled Problems

In FVM, the computational domain is divided into control volumes, and the governing equations are integrated over each control volume. For a coupled fluid–solid problem, the discretized equations may take the form:

(1) Fluid mass conservation

$$\frac{\partial}{\partial t} \int_V \phi \rho_f dV + \int_S \rho_f \mathbf{v} \cdot \mathbf{n} dS = \int_V q_m dV \qquad (7.47)$$

q_m is the mass source/sink term per unit volume, V represents a control volume, and S is the surface of V.

(2) Fluid momentum conservation

$$\frac{\partial}{\partial t} \int_V \phi \rho_f \mathbf{v} dV + \int_S \rho_f \mathbf{v}\mathbf{v} \cdot \mathbf{n} dS = -\int_S p\mathbf{n} dS + \int_V \phi \rho_f \mathbf{g} dV + \int_V \mathbf{F}_d dV \qquad (7.48)$$

where ϕ is porosity, ρ_f is fluid density, v is fluid velocity, p is pressure, g is gravity, and F_d represents drag forces [21].

2. Coupling FVM with FEM

In the simulation of coupled fluid–solid interactions, the FVM for fluid flow can be effectively integrated with the FEM for solid mechanics through several coupling approaches. The implementation of this coupling can be achieved through three distinct methodologies. The first approach, sequential coupling, involves solving the fluid and solid problems separately while exchanging information at each time step. Alternatively, iterative coupling provides a more rigorous solution by resolving the fluid and solid problems repeatedly within each time step until convergence is achieved. The third methodology, monolithic coupling, takes a comprehensive approach by simultaneously solving the combined fluid–solid system as a unified problem.

3. Applications of FVM in Rock Dynamics

The FVM based approaches have demonstrated significant versatility in addressing various challenges in rock dynamics. In the context of reservoir simulation, these methods effectively model multiphase fluid flow in porous and fractured reservoirs, while accounting for complex well configurations and heterogeneous rock

7.2 Multi-Physics Coupling Simulation

properties. Furthermore, FVM-based approaches have proven valuable in analyzing coupled hydro-mechanical problems, particularly in simulating the intricate interaction between fluid flow and rock deformation during processes such as reservoir compaction and hydraulic fracturing. Additionally, these methods have been successfully applied to model heat transfer phenomena in geothermal systems, specifically in analyzing the transport of heat through porous rock masses and fracture networks within geothermal reservoirs [11].

DEM for Coupled Problems

The DEM is particularly useful for simulating the behavior of discontinuous rock masses, where the material is represented as an assembly of discrete particles or blocks [23]. DEM can be coupled with continuum-based methods to simulate multi-physics problems involving fluid flow, heat transfer, and chemical reactions.

1. DEM Formulation for Coupled Problems

In DEM, the equations of motion for each particle or block are solved explicitly:

$$m_i \frac{d^2 \mathbf{x}_i}{dt^2} = \sum_j \mathbf{F}_{ij} + \mathbf{F}_i^{ext} \tag{7.49}$$

$$I_i \frac{d\omega_i}{dt} = \sum_j \mathbf{M}_{ij} \tag{7.50}$$

where m_i and I_i are the mass and moment of inertia of particle i, x_i and ω_i are its position and angular velocity, F_{ij} and M_{ij} are the contact forces and moments between particles i and j, and F_i^{ext} represents external forces [23].

2. Coupling DEM with Continuum Methods

To simulate coupled multi-physics problems, DEM can be combined with continuum-based methods in several ways, each addressing specific aspects of rock mechanics phenomena. The first significant coupling approach involves combining DEM with CFD, which enables the simulation of fluid-particle interactions, particularly in applications such as hydraulic fracturing or proppant transport. Furthermore, the integration of DEM with FEM provides a robust framework for modeling the interaction between discontinuous and continuous domains, specifically at interfaces between intact rock and fractured zones. Additionally, thermo-mechanical coupling expands DEM's capabilities by incorporating heat transfer models, thereby facilitating the simulation of thermally induced fracturing and thermal expansion of granular materials [23].

3. Applications of Coupled DEM in Rock Dynamics

Coupled DEM approaches have demonstrated significant utility across multiple critical applications in rock dynamics. In the context of hydraulic fracturing, these methods enable the simulation of complex interactions between fluid pressure, rock deformation, and fracture propagation within naturally fractured reservoirs. Furthermore, DEM approaches have proven valuable in modeling blast-induced fracturing, where they effectively capture the coupled effects of stress waves and gas pressurization during rock mass fragmentation. Additionally, these numerical methods have been successfully applied to analyze thermally induced spalling phenomena, particularly in high-temperature environments such as deep geothermal reservoirs and underground nuclear waste repositories [23].

Meshless Methods for Coupled Problems

Meshless methods, such as the Element-Free Galerkin (EFG) method and Smoothed Particle Hydrodynamics (SPH), offer advantages in handling large deformations, moving boundaries, and discontinuities without the need for mesh generation or remeshing [5, 24]. These methods have gained popularity in simulating coupled multi-physics problems in rock dynamics.

1. EFG Method

The EFG method uses a set of nodes distributed over the problem domain without the need for element connectivity. The approximation of field variables is based on moving least squares (MLS) interpolants:

$$u^h(\mathbf{x}) = \sum_{I=1}^{n} \phi_I(\mathbf{x}) u_I \tag{7.51}$$

where $\phi_I(x)$ are the MLS shape functions and u_I are the nodal parameters [24].

For coupled problems, the EFG method can be formulated similarly to FEM, with the main difference being the use of MLS shape functions instead of polynomial shape functions. The resulting system of equations can be solved using similar techniques as those used in FEM.

2. SPH

SPH is a Lagrangian particle method where the continuum is represented by a set of particles carrying physical properties. The approximation of field variables and their derivatives is based on kernel functions:

$$f(\mathbf{x}) \approx \sum_j \frac{m_j}{\rho_j} f_j W(\mathbf{x} - \mathbf{x}_j, h) \tag{7.52}$$

where $f(x)$ represents the field variable approximated at a position defined by the vector x; m_j and ρ_j are the mass and density of particle j, f_j is the value of the function at particle j, W is the kernel function, and h is the smoothing length [21].

For coupled problems, SPH can be used to simulate both fluid and solid domains, with coupling achieved through interface conditions or unified formulations.

3. Applications of Meshless Methods in Rock Dynamics

Meshless methods have emerged as a powerful computational approach in rock dynamics, demonstrating particular effectiveness across several challenging application domains. In the realm of large deformation problems, these methods excel at simulating complex phenomena such as rock mass collapse, landslides, and debris flows—scenarios where traditional mesh-based methods often encounter significant limitations. Furthermore, meshless methods have proven valuable in fracture propagation analysis, enabling the modeling of crack initiation and propagation in heterogeneous rock masses without requiring remeshing or special enrichment functions. Additionally, these methods have been successfully applied to fluid–structure interaction problems, facilitating detailed analysis of complex scenarios such as dam-break floods and underwater explosions, where the interplay between rock masses and fluids requires sophisticated computational approaches [21, 24].

7.2.3 Multi-scale Methods for Coupled Problems

Many problems in rock dynamics involve processes occurring at different spatial and temporal scales. Multi-scale methods have been developed to efficiently capture these scale-dependent phenomena while maintaining computational feasibility [5, 11].

1. Hierarchical Multi-Scale Methods

Multiscale modeling approaches in rock mechanics employ hierarchical solution strategies that address phenomena at different scales sequentially. These methods facilitate the systematic transfer of information between scales, with two primary techniques emerging as particularly significant. The first approach utilizes homogenization techniques, which leverage detailed microstructural information to derive effective properties that can be subsequently implemented in macroscale simulations [21, 22]. Complementing this methodology, submodeling enables the application of detailed, high-resolution models within specific regions of interest that are situated within a larger, coarser model framework [21, 22]. These complementary approaches allow for efficient computational solutions while maintaining accuracy at critical scales of interest.

2. Concurrent Multi-Scale Methods

Modern computational approaches in rock mechanics employ various multi-scale methodologies to address complex mechanical problems. These methods solve the problem at different scales simultaneously through two primary approaches. The first approach, Domain Decomposition, involves dividing the problem domain into subdomains with different resolutions or physics [21, 22]. In parallel, the Heterogeneous Multi-Scale Method provides an alternative strategy by coupling macro-scale and micro-scale models, wherein micro-scale problems are solved on-the-fly to inform macro-scale simulations [21, 22]. Together, these complementary methods enable comprehensive analysis of rock mechanical behavior across multiple spatial scales.

3. Applications of Multi-Scale Methods in Rock Dynamics

Multi-scale methods have emerged as powerful tools for analyzing complex rock dynamics phenomena across various geological applications. In fractured reservoirs, these methods enable sophisticated modeling of fluid flow and transport mechanisms by simultaneously accounting for both matrix and fracture contributions. Building upon this foundation, multi-scale approaches have also proven invaluable in earthquake mechanics, where they facilitate the simulation of fault rupture dynamics while effectively capturing the influence of small-scale heterogeneities and friction laws. Furthermore, these methods have demonstrated particular utility in analyzing coupled THM processes, especially in contexts such as geothermal systems and nuclear waste repositories, where understanding the intricate interplay between thermal, hydraulic, and mechanical processes across different scales is crucial [21, 22].

7.2.4 Multi-Physics Coupling Simulation Softwares and Engineering Applications

The simulation of rock mass behavior under dynamic loading conditions requires advanced computational tools capable of handling coupled multi-physics processes. Modern software packages integrate thermal, hydraulic, mechanical, and chemical interactions within a unified simulation framework [25], enabling accurate prediction of rock behavior under complex environmental conditions. This integration is particularly crucial for analyzing dynamic loading scenarios and extreme conditions.

1. COMSOL Multiphysics

COMSOL Multiphysics stands as a leading multi-physics simulation platform for rock dynamics, employing FEM to solve coupled physical processes [26]. The software integrates structural mechanics, heat transfer, and fluid flow interfaces for comprehensive rock behavior modeling under dynamic loading. Its adaptive meshing algorithms optimize domain discretization, while advanced solvers handle nonlinear and time-dependent problems characteristic of rock dynamics. The platform enables modeling across scales, from microstructure to full-scale systems, with capabilities for complex geometries and custom material models.

The general form of the coupled equations solved by COMSOL can be expressed as:

$$\frac{\partial(\rho u)}{\partial t} + \nabla \cdot (-c\nabla u - \alpha u + \gamma) + \beta \cdot \nabla u + au = f \qquad (7.53)$$

where ρ is the density, u is the dependent variable (e.g., displacement, temperature, pressure), t is time, c is the diffusion coefficient, α is the conservative flux convection coefficient, γ is the conservative flux source term, β is the convection coefficient, a is the absorption coefficient, and f is the source term.

This general form can be adapted to represent various physical processes relevant to rock dynamics, such as stress wave propagation, heat conduction, and fluid flow through porous media.

7.2 Multi-Physics Coupling Simulation

2. ABAQUS

ABAQUS serves as a powerful multi-physics simulation platform for rock dynamics, excelling in nonlinear structural analysis with comprehensive element types and material models [27, 28]. Its modular architecture, separating pre-processing, analysis, and post-processing stages, enables efficient management of large-scale simulations. The software's extensive material library includes crucial rock behavior models, notably the Drucker-Prager Cap model for pressure-dependent yield behavior. The yield surface is described by:

$$F = t - p\tan\beta - d = 0 \quad (7.54)$$

where t is the deviatoric stress measure, p is the equivalent pressure stress, β is the friction angle in the p–t plane, and d is the cohesion of the material.

ABAQUS enables custom material model implementation through user subroutines, essential for modeling nonlinear rock behavior under dynamic loading. The software incorporates specialized tools for soil-structure interaction analysis, notably the Domain Reduction Method (DRM) and Perfectly-Matched-Layers (PMLs) [28]. DRM operates through prescribed effective nodal forces, while PMLs are implemented via user-defined element (UEL) subroutines. This coupled DRM-PML system efficiently simulates wave propagation through rock masses and their interaction with structures, particularly for applications like tunnel seismic response and blast-induced vibrations.

3. TOUGH-FLAC

TOUGH-FLAC is a prominent coupled simulator integrating TOUGH2's multiphase fluid flow and heat transfer capabilities with FLAC3D's geomechanical analysis functionality [25]. The software employs sequential coupling, exchanging information between the codes at each time step to model THM processes in fractured rock masses. This approach enables simulation of stress-dependent porosity and permeability changes, thermal expansion, and fluid pressure effects on mechanical deformation. The coupled system is governed by the following equations:

Mass balance:

$$\frac{\partial(\phi\rho)}{\partial t} + \nabla \cdot (\rho v) = Q \quad (7.55)$$

Energy balance:

$$\frac{\partial(\phi\rho u)}{\partial t} + \nabla \cdot (\rho u v - \lambda \nabla T) = Q_h \quad (7.56)$$

Momentum balance:

$$\nabla \cdot \sigma + \rho g = 0 \quad (7.57)$$

where ϕ is porosity, ρ is fluid density, v is Darcy velocity, Q is mass source/sink term, u is specific internal energy, λ is thermal conductivity, T is temperature, Q_h is heat source/sink term, σ is stress tensor, and g is gravitational acceleration.

An extension of TOUGH-FLAC, known as TOUGHREACT-FLAC3D, incorporates geochemical reactions into the THM framework, enabling the simulation of fully coupled THMC processes in rock masses. This advanced capability is particularly relevant for studying long-term behavior of rock repositories for nuclear waste disposal or predicting the evolution of geothermal reservoirs.

In addition to these commercial software packages, several specialized codes have been developed specifically for rock mechanics applications. The RFPA (Rock Failure Process Analysis) software series, developed in China, employs a statistical damage model and FEM to simulate damage and fracture processes in heterogeneous materials under multi-field coupling conditions [25]. RFPA's distinctive capability lies in its representation of material heterogeneity through statistical distribution of element properties, enabling realistic simulation of progressive damage accumulation, strain localization, and fracture coalescence. The damage evolution law in RFPA is expressed as:

$$D = 1 - \exp\left[-\int \left(\frac{\varepsilon}{\varepsilon_0}\right)^m dt\right] \tag{7.58}$$

where D is the damage variable, ε is the strain, ε_0 is a reference strain, and m is a material parameter.

This damage model is coupled with the mechanical, thermal, and hydraulic field equations to provide a comprehensive simulation of rock behavior under complex loading conditions.

To illustrate the application of multi-physics coupling simulation software in rock dynamics, consider the following example of a coupled thermo-hydro-mechanical analysis of a rock mass subjected to thermal loading, such as might occur in a geothermal reservoir or nuclear waste repository:

The problem involves a fractured rock mass with a heat source (e.g., a waste canister or geothermal injection well) placed within it. The simulation aims to predict the evolution of temperature, pore pressure, stress state, and potential fracture propagation over time. This problem can be approached using a software package like TOUGH-FLAC or COMSOL Multiphysics.

The key governing equations for this coupled THM problem include:

(1) Heat conduction equation

$$\rho c_p \frac{\partial T}{\partial t} + \nabla \cdot (-k \nabla T) = Q_T \tag{7.59}$$

where ρ is rock density, c_p is specific heat capacity, T is temperature, k is thermal conductivity and Q_T is heat source term.

(2) Fluid flow equation (based on Darcy's law)

7.2 Multi-Physics Coupling Simulation

$$\phi\frac{\partial(\rho_f S)}{\partial t} + \nabla \cdot \left(-\frac{\rho_f k}{\mu}\nabla p\right) = Q_f \tag{7.60}$$

where φ is porosity, ρf is fluid density, S is saturation, k is permeability tensor, μ is fluid viscosity, p is pore pressure and Q_f is fluid source/sink term.

(3) Mechanical equilibrium equation

$$\nabla \cdot \sigma + \rho g = 0 \tag{7.61}$$

where σ is the stress tensor, ρ is bulk density and g is gravitational acceleration.

THM numerical simulation implementation follows a systematic workflow encompassing model construction, parameter definition, and computation. The process begins with establishing the rock mass geometry and mesh generation, incorporating pre-existing fractures and essential material properties, particularly temperature-dependent and stress-dependent parameters. Initial and boundary conditions are then specified, including ambient temperature distribution, pore pressure state, in-situ stress conditions, and far-field stress states. The model configuration concludes with heat source characterization and coupled physics interface setup. The computational phase involves selecting appropriate numerical solvers and time-stepping schemes for solution stability. Post-processing analysis examines temperature distributions, pore pressure evolution, stress field modifications, and potential fracture formation zones.

This example demonstrates the complexity of multi-physics coupling in rock dynamics problems and the need for advanced simulation software capable of handling such coupled phenomena.

In conclusion, multi-physics field coupling simulation software plays a crucial role in advancing our understanding of rock dynamics. These sophisticated computational tools enable researchers and engineers to model the complex interplay between thermal, hydraulic, mechanical, and chemical processes that govern rock behavior under dynamic loading conditions. Software packages such as COMSOL Multiphysics, ABAQUS, TOUGH-FLAC, and specialized codes like RFPA offer a range of capabilities for tackling diverse rock dynamics problems.

Applications of THM Coupling in Rock Dynamics

THM coupling plays a crucial role in various geotechnical and geoenvironmental applications. Some key areas where THM coupling is particularly important include:

1. Geothermal Energy Systems

In geothermal energy systems, THM coupling within enhanced geothermal systems (EGS) manifests through three key permeability evolution stages in hot dry rocks (HDR). The initial slow increase stage occurs between 25 and 400 °C, where microscopic parameters demonstrate a steady rise while being constrained by triaxial stress. This is followed by a fluctuation stage (400–550 °C) characterized by parameter variations driven by thermo-elastic–plastic fluidity and dynamic interactions between pores and fissures. The final rapid increase stage (550–650 °C) involves significant rock mass deterioration, leading to dramatic permeability growth. Understanding these stages is essential for optimizing reservoir performance while maintaining stability.

2. Nuclear Waste Disposal

Nuclear waste disposal applications rely heavily on THM coupling for ensuring repository safety. The process involves three primary mechanisms: thermal effects from radioactive decay heat causing expansion and property changes; hydraulic effects where heat-induced pore pressure variations influence fluid flow and radionuclide transport; and mechanical effects where thermal stresses lead to fracturing and permeability changes.

3. Carbon Dioxide Sequestration

In carbon dioxide sequestration, THM processes are critical for successful implementation. These processes affect formation injectivity through mechanical-chemical interactions, determine storage capacity through pore space modifications, and influence caprock integrity through thermal–mechanical stress effects on sealing capacity.

4. Underground Construction

Underground construction projects demonstrate THM coupling effects across multiple applications. In tunneling, heat generation and groundwater patterns significantly impact structural stability. Storage facilities require careful consideration of long-term cavern stability for various substances, while deep mining operations must account for rock mass stability, ventilation requirements, and groundwater management.

This comprehensive understanding of THM coupling across different applications enables engineers to develop more effective and safer geotechnical solutions.

Case Studies and Numerical Examples

To illustrate the application of THM coupling in rock dynamics, we present several case studies and numerical examples from recent research:

1. Cryogenic THM Model for Frozen Media

A study by [16] developed an integrated cryogenic THM model implemented in FDEM code to simulate freezing processes in porous media. The model's framework encompasses energy conservation for heat conduction, advection, and latent heat; mass conservation for variably saturated water flow; and mechanical equilibrium with stress–strain constitutive relationships. Critical coupling parameters include unfrozen water content, permeability, and thermal conductivity.

Model validation against experimental data demonstrated accurate predictions of temperature evolution and frost heave strain across multiple scenarios, including sandstone freezing, water redistribution, and frost heave near chilled pipelines. The results confirmed the model's ability to reproduce the three-zone distribution (frozen,

7.2 Multi-Physics Coupling Simulation

freezing, unfrozen) while capturing cryogenic suction-driven water migration and its interaction with insulation effects on frost processes.

The validated model provides a foundation for simulating cryogenic processes in cold regions, with applications in permafrost engineering and artificial ground freezing.

2. THM Model for Hydrate-Bearing Sediments

Research by [17] developed a coupled THM model to investigate hydrate phase transition effects on reservoir damage during thermal fluid injection. The mathematical framework comprises fundamental governing equations: hydrate decomposition kinetics and phase equilibrium equations describing thermodynamic behavior, mass conservation equations for methane, water, and hydrate components, and the temperature field equation capturing thermal dynamics. The model incorporates dynamic equations for key reservoir properties:

Permeability change:

$$K = K_0 \exp\left[-\left(d \cdot S_h + e \cdot \bar{\sigma}' \cdot 10^{-6}\right)\right] \quad (7.62)$$

where K is the dynamic permeability of hydrate reservoir, K_0 is the permeability of hydrate reservoir without hydrate, $\bar{\sigma}'$ is average effective stress, S_h is hydrate saturation, d = 6.8364, e = 0.106.

Cohesion change:

$$C = C_0 \times \left[1 - 1.2 \cdot \left(\phi \times \left(S_{h_i} - S_h\right)\right)\right] \quad (7.63)$$

where C is dynamic cohesion of hydrate reservoir, C_0 is initial cohesion of hydrate reservoir, φ is porosity of hydrate reservoir, S_{hi} is initial hydrate saturation, Sh is hydrate saturation.

Elastic modulus change:

$$E = (n_s E_s + n_h E_h) \cdot \left[a \cdot \left(10^{-6}\sigma\prime\right) + b \cdot \left(10^{-6}\sigma\prime\right) + c\right] \quad (7.64)$$

where n_s is volume fraction of reservoir rock, n_h is volume fraction of hydrate, E_s and E_h are the elastic modulus of reservoir rock and hydrate, respectively.

Model simulations revealed significant coupled THM processes during thermal fluid injection. Hydrate decomposition in the near-wellbore region reduces hydrate saturation, generating increased pore pressure as solid hydrate dissociates into methane and water. This pressure elevation modifies the formation's effective stress state and permeability characteristics. The decomposition process weakens particle cementation, decreasing cohesion and elastic modulus, while pressure increases can form hydrate secondary formation zones away from the wellbore.

Experimental results demonstrated temperature-dependent effects on reservoir damage mechanisms, with complete hydrate dissociation observed at 303.15K (30 °C above initial reservoir temperature). Injection rate analysis revealed that increasing rates from 0.05×10^{-5} m/s to 7.5×10^{-5} m/s resulted in: 28.79% permeability decrease, 11.25% elastic modulus increase, and 2.58% cohesion increase. Higher injection rates elevated pore pressure, inhibiting hydrate decomposition processes.

3. Geothermal Reservoir Modeling

Recent advances [18] examined advanced numerical tools (TOUGH2, FEHM, OpenGeoSys, FLAC3D) that integrate coupled THMC processes in geothermal reservoirs. These platforms demonstrate diverse capabilities in multi-phase flow simulations, reactive transport modeling, and stress-deformation coupling, incorporating fundamental equations for mass-momentum conservation, energy balance, force equilibrium, and reactive transport phenomena.

The investigation identified distinct temporal hierarchies in coupling effects: poro-elastic effects manifest over hours to days, thermo-elastic processes operate within days to months, and chemical effects evolve over months to years. Despite these varying timescales, THMC processes interact simultaneously, with reservoir heterogeneity, fracture characteristics, and mineral composition significantly influencing heat extraction efficiency. Chemical reactions modify porosity and permeability, while thermal stresses may induce fracturing and seismic risks.

Based on these findings, the review identified several critical areas requiring further research attention. Critical research priorities include advancing experimental and theoretical understanding of complex reservoir dynamics, developing numerical codes for field-scale THMC simulations, and investigating induced seismicity risks and transmissivity changes. Additional focus areas encompass complex fracture networks, rough fracture surfaces, and multi-mineral systems.

4. THM Model for Unsaturated Soil

A significant advancement in unsaturated soil modeling has been achieved through an enhanced THM coupling model [20]. This model extends the classical VG approach by incorporating void ratio and temperature effects on the soil–water characteristic curve (SWCC). Implemented via COMSOL Multiphysics, the model employs coupled differential equations to describe moisture migration, heat transfer, and mechanical deformation while accounting for both water and gas phases.

Validation against experimental data demonstrated superior performance compared to the classical VG approach, particularly in predicting volumetric water content increment during heating and cooling cycles. The model revealed that higher heat source temperatures correlate with greater saturation at specific measurement points under thermal stability. Volumetric strain exhibited a complex interplay between wet and thermal expansion effects, with wet expansion dominating near the heat source. Temperature gradient analysis showed progressive changes over time, while total displacement patterns were primarily governed by z-direction displacement.

This sophisticated THM model serves as a valuable tool for analyzing and predicting unsaturated soil behavior under various thermal and mechanical loading conditions, offering crucial insights into the complex interactions between thermal, hydraulic, and mechanical processes.

7.3 Data-Driven Modelling

7.3.1 Machine Learning for Rock Dynamics

Machine learning (ML) techniques have emerged as powerful tools for addressing complex problems in rock dynamics. These data-driven approaches offer new capabilities for analyzing, modeling, and predicting rock behavior under dynamic loading conditions. This section provides a comprehensive overview of recent advances in applying machine learning to various aspects of rock dynamics research and engineering.

Prediction of Rate-Dependent Rock Strength

One of the key challenges in rock dynamics is understanding and predicting how rock strength varies with loading rate. Traditional empirical or analytical models often struggle to capture the complex, nonlinear relationships between rock properties, loading conditions, and dynamic strength.

Zhang et al. developed a suite of machine learning models to predict the rate-dependent compressive strength of rocks based on various rock properties and loading conditions [29]. They compared the performance of Support Vector Machine (SVM), Back-Propagation Neural Network (BPNN), and Random Forest (RF) algorithms. The study utilized a comprehensive dataset comprising 11 types of hard rocks, with strain rates ranging from $5 \times 10^{-6}/s$ to 223/s. Input parameters included static compressive strength, P-wave velocity, specimen dimensions (length and diameter), grain size, bulk density, and strain rate.

The sensitivity analysis identified strain rate as the dominant factor affecting prediction accuracy, with excluding it causing the most significant decline in model performance. The study also revealed key behavioral patterns in rock strength dynamics. A critical threshold of 20/s was identified, below which compressive strength remained constant, and above which it increased sharply. An inverse relationship between specimen diameter and strength was observed, with smaller specimens showing higher strength values. Dynamic Increase Factor curves displayed distinctive behavior above 50/s, especially for larger diameter specimens, which exhibited more pronounced increases.

These findings advance our understanding of the relationship between strain rates and rock strength, particularly in transitional loading zones. The study demonstrates how machine learning can capture complex, nonlinear relationships and process multiple variables simultaneously, complementing traditional experimental and numerical methods in rock mechanics. It establishes a solid foundation for future dynamic rock strength prediction work and underscores the potential of machine learning in geotechnical engineering.

Fluid Flow in Rough Rock Fractures

Understanding fluid flow behavior in fractured rock masses is critical for many engineering applications, including oil and gas production, geothermal energy, and underground water control. The roughness and tortuosity of natural rock fractures make fluid transport much more complex than in idealized smooth parallel-plate models.

Wu et al. developed an innovative approach combining an electrical circuit (EC) model with machine learning to efficiently simulate fluid flow in rough rock fractures [30]. Their method introduced three key advances: they created an "aperture reduction ratio" concept showing that fracture void spaces contribute differently to fluid transport, developed a neural network to predict this ratio from fracture characteristics, and created an algorithm linking EC model resistances to rough fracture aperture distributions.

The EC model represents each rough fracture as a network of distributed electrical resistances, with voltage and current analogous to pressure and flow rate. This approach exploits the physical similarity between electrical current (I) and fluid flow rate (Q):

$$\frac{V_{ec}}{I_{ec}} = R \tag{7.65}$$

$$\frac{\Delta P}{Q} = C \frac{1}{h_x^3} \tag{7.66}$$

Where V_{ec} is electrical potential, I_{ec} is electrical current, R is electrical resistance, ΔP is pressure difference, C is a constant depending on fluid and fracture properties, and h_x is the local fracture aperture.

The electrical resistance is related to the local fracture aperture as:

$$R = C h_x^{-3} \tag{7.67}$$

Accounting for aperture reduction:

$$h_r = (1 + r_c) h_m \tag{7.68}$$

where h_r is the reduced aperture used in the EC model, r_c is the aperture reduction ratio, and h_m is the mean mechanical aperture.

The machine learning model (ANN) was trained to predict r_c based on seven input parameters characterizing the fracture and flow conditions. The ANN showed good predictive performance, with $R^2 = 0.8787$ between ANN predictions and CFD simulation results for test cases.

7.3 Data-Driven Modelling

The study achieved remarkable accuracy and efficiency. The electrical circuit (EC) model closely matched computational fluid dynamics (CFD) simulations, with an R^2 of 0.9995 at its optimal configuration. It maintained high accuracy in most scenarios, with around 90% of predictions having relative errors under 10%. When the aperture reduction component was excluded, errors increased significantly, ranging from 48 to 85%, highlighting the importance of the machine learning-derived aperture reduction ratio.

Beyond flow predictions, the EC model also accurately replicated pressure distributions along fracture cross-sections when compared to CFD results. Notably, the EC model provided these results in seconds, offering a significant improvement in computational efficiency over traditional CFD methods.

This study's integrated EC-ML approach, combining electrical circuit modeling with machine learning, brings significant benefits to the field of fracture fluid flow modeling. The method stands out primarily for its computational speed, completing calculations in seconds compared to the minutes required for traditional CFD simulations. It also shows remarkable versatility in handling challenging geometries, including fractures with minimal apertures and extensive contact areas. Looking forward, the framework shows promise for scaling up to three-dimensional fracture networks [30].

Integration of Physics-Driven Dynamics Simulation and Machine Learning

While machine learning offers powerful predictive capabilities, integrating ML with physics-based simulations can provide even deeper insights into complex rock dynamics problems. This approach combines the flexibility and pattern recognition abilities of ML with the physical constraints and process understanding of traditional numerical models [31].

Liu et al. demonstrated this integrated approach for mineral deposit prediction in the Tongguanshan (TGS) skarn ore field in China [31].

The methodology combined advanced techniques, beginning with 3D geological models based on exploration data. Physics-driven simulations of magmatic intrusion cooling and rock deformation were then used. Machine learning models, specifically Random Forest algorithms, were developed to predict mineralization locations, incorporating simulation results as input features for a hybrid approach that blended physical insights and data-driven predictions.

The physics-driven dynamics simulation modeled parameters such as temperature, fluid pressure, differential stress, shear strain, and volume strain (dilatant deformation). These simulated parameters were then used as input features for the Random Forest models, along with geological and geophysical data.

This study highlighted the effectiveness of machine learning for predicting mineral deposits. Researchers developed an optimal Random Forest model that integrated eight key features: four geological characteristics, three dynamic measurements, and one geophysical measurement (resistivity). The combination of geological and dynamic features enhanced the model's predictive power.

SHAP analysis revealed the relative importance of various factors in ore formation. Lithology (rock characteristics and layering) was the most significant, followed by volume strain (rock deformation under pressure). Other key factors included spatial measurements, such as distance to the contact zone and the Devonian-Carboniferous interface, as well as characteristics of the contact zone itself (curvature, direction, slope), pressure, and temperature. Notably, traditional geophysical measurements like resistivity were less influential than dynamic features like volume strain.

Most importantly, when the model was tested using only dynamic features, it identified mineralization patterns that differed from the training data, suggesting that dynamics-based modeling could uncover new mineral deposits often overlooked by conventional methods. This analysis underscores the potential of combining geological data and machine learning to revolutionize mineral exploration.

This integrated approach offers several advantages. It improves predictive performance by combining diverse data sources and modeling techniques. Additionally, it provides insights into the relative importance of different physical processes and geological factors. Most importantly, it holds the potential to identify new exploration targets or mineralization styles not captured in the training data.

Elastic Rock Properties Estimation and Fracturability Evaluation

Accurate estimation of elastic rock properties and evaluation of rock fracturability are crucial for many rock engineering applications, particularly in the oil and gas industry for hydraulic fracturing design.

Kadeethum et al. developed a machine learning approach for estimating Young's modulus and evaluating rock fracturability using a combination of well log data, mineral composition, microstructural information from rock images, and geomechanical properties [32].

Their methodology involved several key steps. Data fusion was performed by integrating multiple sources, including well log data (caliper, resistivity, porosity, gamma ray, density, compressional and shear velocity logs), mineral composition from XRD-calibrated mineralogy logging, microstructural data from thin section rock images, and geomechanical properties from laboratory measurements. Feature selection was carried out using regularization techniques such as LASSO, Ridge, and Elastic Net.

Young's modulus was predicted using both linear and non-linear regression models. Linear models included LASSO, Ridge Regression, and Elastic Net, while non-linear models featured ANN and Multivariate Adaptive Regression Splines (MARS). Fracturability analysis involved clustering techniques, including K-Means and Hierarchical Clustering with Support Vector Machine, alongside fracture propagation modeling using GOHFER software.

Key findings from the study include that non-linear regression methods, specifically ANN and MARS, outperformed linear regularized methods in predicting Young's modulus. The most important features for prediction were the ratio of compressional to shear velocity (V_p/V_s), bulk density, effective porosity, clay content, pyrite content, and the gamma ray log.

Clustering analysis revealed three distinct geomechanical facies, each with different fracturability characteristics. Facies 1 exhibited an intermediate Young's modulus, the highest Poisson's ratio, and the lowest minimum horizontal stress.

7.3 Data-Driven Modelling

Facies 2 had the highest Young's modulus, an intermediate Poisson's ratio, and the lowest elastic anisotropy. Facies 3 had the lowest Young's modulus, the lowest Poisson's ratio, and the highest elastic anisotropy.

Fracture propagation modeling showed varying behaviors for each facies. Facies 1 exhibited short but smooth fracture geometries, while Facies 2 demonstrated promising fracture height and length when continuous. Facies 3, however, showed restricted fracture height growth.

Finally, microstructural analysis from thin-section images strongly correlated with the classified geomechanical and fracturability facies, providing physical validation for the machine learning-derived classifications.

Machine Learning for Rock Mechanics Problems: An Overview

Machine learning techniques have found applications across a wide range of rock mechanics problems, from property estimation to complex simulations. Liang et al. provided a comprehensive overview of these applications, highlighting the diverse ways in which ML is transforming rock mechanics research and practice [33]. Key areas of application include:

1. Rock Mass Properties

ML methods have been applied to classify rock masses, estimate parameters like UCS, and predict shear behavior. Various algorithms like ANNs, SVM, and decision trees have been used. Input parameters often include physical and mechanical properties of rocks. Comparison studies show ML models often outperform traditional regression methods in predicting rock mass properties.

2. Back-Analysis

ML is used for inverse analysis to determine rock mass parameters and in-situ stresses from measured displacements. Genetic algorithms and neural networks have been applied to improve the efficiency and reliability of back-analysis. This approach helps overcome challenges in obtaining accurate rock mass properties in situ.

3. Determining Constitutive Behaviors

ML, especially ANNs, is used to model complex constitutive relationships of rocks and soils. These models can capture nonlinear elastic and elasto-plastic behaviors, often using stress–strain data, loading conditions, and material properties as inputs. ML-based constitutive models offer advantages in handling complex, nonlinear behaviors that are difficult to capture with traditional analytical approaches.

4. Rock Mass Stability Research

ML is applied to predict displacements of tunnel surrounding rock and assess slope stability. SVMs and ANNs have shown good performance in displacement prediction. For slope stability analysis, input parameters typically include geometry, soil properties, and hydrological factors. These ML approaches can integrate multiple factors and capture complex relationships that influence stability.

5. Fracture Mechanics

ML is used to predict fracture toughness, crack propagation, and crack identification. Various algorithms like decision trees, ANNs, and SVMs have been applied. These methods can handle high-dimensional data and complex relationships involved in fracture processes, offering new insights into rock fracture behavior.

6. Pore-Scale Analysis

ML is applied to digital rock modeling for fluid flow simulations and mechanical property predictions. Convolutional neural networks (CNNs) are used for automated image segmentation and property prediction. These techniques can overcome limitations of traditional numerical simulations in terms of computational cost and sample size, enabling more detailed and efficient analysis of pore-scale processes.

7. Physics-Informed Neural Networks (PINNs)

PINNs combine data-driven machine learning with physical models, offering a promising approach for integrating ML with domain knowledge in rock mechanics. They have been applied to solve complex problems in porous media flow and deformation. PINNs show particular promise in handling inverse problems and incorporating physical constraints into ML models.

This overview demonstrates the broad impact of machine learning across rock mechanics, from basic property estimation to complex multi-physics simulations.

Machine learning techniques offer several advantages in rock mechanics research. They can handle complex, nonlinear relationships and integrate diverse data sources. Additionally, they have the potential for real-time analysis and prediction, while also improving computational efficiency compared to traditional numerical methods.

However, challenges remain, including the need for large, high-quality datasets and the interpretability of complex ML models. Future research directions may focus on developing more widely applicable methods, improving model interpretability, and further integrating ML with physical understanding of rock mechanics processes [33].

Prediction of Shear Strength Parameters

Accurate prediction of shear strength parameters (cohesion c and angle of internal friction φ) is crucial for many rock engineering applications. Machine learning approaches offer new capabilities for estimating these parameters based on more easily measurable rock properties. Qi et al. conducted a comprehensive study comparing various ML algorithms for this task [34].

The study developed and compared four artificial intelligence (AI) models:

- Group Method of Data Handling (GMDH)
- Gene Expression Programming (GEP)
- Bidirectional Long Short-Term Memory (BILSTM)
- Random Forest (RF)

7.3 Data-Driven Modelling

These models were trained and tested on a database of 199 sets of experimental data from 4 rock types (limestone, slate, quartzite, quartz-mica schist). Input variables included P-wave velocity, density, UCS, and tensile strength (TS). The dataset was split with 80% (159 samples) used for training and 20% (40 samples) for testing.

Key findings from the study reveal that all four AI models demonstrated strong performance, with R^2 values greater than 0.9 for predicting both cohesion (c) and the angle of internal friction (φ). Among these models, the Random Forest (RF) algorithm outperformed the others.

To further enhance the RF model, six meta-heuristic optimization algorithms were applied to optimize its hyperparameters. These included Particle Swarm Optimization (PSO), Bald Eagle Search (BES), Marine Predators Algorithm (MPA), Northern Goshawk Optimization (NGO), Golden Jackal Optimization (GJO), and Dung Beetle Optimizer (DBO). The DBO-RF hybrid model delivered the best overall performance.

Finally, sensitivity analysis of the RF model indicated that the input parameters influencing cohesion and friction angle in order of importance were density, P-wave velocity, UCS, and tensile strength.

This study highlights several key insights into the application of machine learning (ML) to rock mechanics problems. First, the choice of ML algorithm plays a critical role in prediction accuracy, with ensemble methods such as Random Forest often yielding strong results for rock property prediction. Second, hyperparameter optimization through meta-heuristic algorithms can further enhance model performance, though this comes with increased complexity.

The study also demonstrates that ML models can uncover unexpected relationships between input and output parameters. For example, density had a significant influence on shear strength parameters, despite showing a weak correlation in mutual information tests. Lastly, ML approaches can achieve high accuracy in predicting complex rock properties, such as shear strength parameters, based on more easily measurable properties.

The hybrid DBO-RF model developed in this study offers a powerful tool for estimating rock shear strength parameters. However, it's important to note that the model's applicability may be limited to rock types similar to those in the training dataset. Further work to expand the database and validate the model across diverse geological settings would enhance its generalizability [34].

Real-Time Rock Characterization During Drilling

Real-time characterization of rock properties during drilling operations is crucial for optimizing drilling performance and ensuring safety. Machine learning techniques offer new possibilities for rapid rock characterization based on measurable drilling parameters. Cao et al. developed an innovative approach using ML models trained on drill-bit acceleration data to characterize downhole rock formations during vibro-impact drilling (VID) [35].

The study methodology involved simulating bit-rock impacts using an impact oscillator model and extracting features from acceleration signals. These features included average impact duration, impact duration statistics, and raw data statistics.

Regression models were then developed, including Multilayer Perceptron (MLP), Support Vector Regression (SVR), and Gaussian Process Regression (GPR).

Key findings from the study indicate that all three machine learning models performed well in predicting rock strength from acceleration signals, with R^2 values greater than 0.9. The MLP networks generally delivered the best performance in terms of accuracy, error distribution, training time, and consistency. Using raw data statistics as features led to exceptional results, particularly with MLP and GPR models.

The machine learning approach also demonstrated several advantages over existing methods. It allows for quantitative rather than qualitative rock characterization, is not limited to specific impact categories, and operates autonomously with minimal human interaction. Additionally, it uses readily measurable drill-bit vibrations and is less computationally intensive than some alternative methods.

This ML-based approach to real-time rock characterization during drilling offers several potential benefits for rock dynamics applications. It enables the real-time tuning of VID operating parameters for optimal performance, potentially improving drilling efficiency and reducing costs. Additionally, it provides continuous rock property data that could enhance geomechanical models and deepen the understanding of dynamic rock behavior during drilling.

7.3.2 Big Data Analysis in Rock Testing

The field of rock mechanics and engineering has experienced a significant transformation in recent years with the advent of big data analytics and artificial intelligence (AI) technologies. These advancements have revolutionized the way we approach rock testing, allowing for more comprehensive, accurate, and efficient analysis of complex geological systems. In this section, we will explore the application of big data analysis in rock testing, focusing on its implications for numerical simulation of rock dynamics and data-driven modeling.

Big data analysis in rock testing refers to the collection, processing, and interpretation of large volumes of data generated during various rock testing procedures. This approach allows researchers and engineers to extract meaningful insights from vast datasets, uncovering patterns and relationships that may not be apparent through traditional analysis methods. The integration of big data analytics with rock testing has opened up new possibilities for understanding rock behavior, predicting geological conditions, and optimizing engineering designs.

One of the primary advantages of big data analysis in rock testing is its ability to handle the inherent complexity and heterogeneity of rock masses. Rocks are natural materials with intricate structures, varying compositions, and complex mechanical behaviors that can be challenging to characterize using conventional methods. Big data approaches allow for the simultaneous consideration of multiple parameters and their interactions, providing a more holistic understanding of rock properties and behaviors.

7.3 Data-Driven Modelling

The application of big data analysis in rock testing spans various aspects of rock mechanics and engineering, including:

- Mechanical Property Characterization
- Geological Condition Prediction
- Rock Mass Classification
- Failure Mechanism Analysis

Let us delve into each of these aspects in detail, exploring the methodologies, challenges, and recent advancements in big data analysis for rock testing.

Mechanical Property Characterization

The characterization of mechanical properties is a fundamental aspect of rock testing, providing essential information for engineering design and analysis. Traditional approaches often rely on a limited number of tests, which may not fully capture the variability and scale-dependent nature of rock properties. Big data analysis has enabled a more comprehensive and statistically robust characterization of rock mechanical properties.

One notable application of big data analysis in mechanical property characterization is the use of nanoindentation techniques combined with advanced statistical methods. Luo et al. [36] conducted a study using big data nanoindentation to investigate the mechanical properties of shale and sandstone rocks. Their approach involved performing approximately 500 indentations per sample to depths of 6–8 μm, generating a large dataset of depth-dependent hardness and Young's modulus values.

The key innovation in their methodology was the development of an improved data analytics approach using Gaussian mixture modeling (GMM) to analyze the two-dimensional datasets of hardness and Young's modulus. This technique allowed for the transformation of scattered property-depth curves into distinct lines corresponding to different mineral phases within the rocks.

The study revealed several key findings. First, the big data analysis identified 5–7 mechanically distinct phases in shales and 5 phases in sandstone, offering a more detailed understanding of rock composition and its impact on mechanical properties. Second, it showed a multi-scale characterization, where properties progressively merged and homogenized from individual phases at the nanoscale to bulk rock at the macroscale. This insight helps address long-standing challenges in rock mechanics research.

Additionally, the method allowed for the quantification of phase properties, including characteristic sizes, volume fractions, and mechanical properties, which is critical for developing accurate constitutive models and predicting bulk rock behavior. Finally, the data enabled upscaling from nano to macro scales, providing an empirical basis for predicting bulk rock properties without relying on theoretical models.

The mathematical framework for the Gaussian mixture modeling approach used in this study can be expressed as follows:

For a given dataset $X = \{x_1, x_2, ..., x_n\}$, where each x_i represents a vector of mechanical properties (e.g., hardness and Young's modulus), the probability density function for a Gaussian mixture model with K components is given by:

$$p(x) = \sum_{j=1}^{k} \pi_j \mathcal{N}(x|\mu_j, \Sigma_j) \tag{7.69}$$

where

π_j represents the mixture weight for the jth component. $\mathcal{N}(x|\mu_j, \Sigma_j)$ is the probability density of a multivariate Gaussian distribution with mean μ_j and covariance matrix Σ_j.

The parameters of the GMM (π_j, μ_j, Σ_j) are typically estimated using the Expectation–Maximization (EM) algorithm, which iteratively refines the parameter estimates to maximize the likelihood of the observed data.

This approach to mechanical property characterization using big data nanoindentation offers several advantages over traditional methods. It provides statistical robustness by performing hundreds of indentations, ensuring a statistically significant dataset that captures the inherent variability in rock properties. The ability to distinguish between different mineral phases within the rock allows for a more detailed understanding of how individual components influence overall rock behavior.

Additionally, the continuous depth-dependent data bridges scales, enabling the study of how properties evolve from the nanoscale to the macroscale, addressing a critical gap in rock mechanics. The empirical upscaling data also serves to validate and refine existing theoretical models, improving their accuracy and applicability.

However, it is important to note that this approach also has limitations. The nanoindentation technique is primarily applicable to surface measurements and may not fully capture the three-dimensional variability of rock properties. Additionally, the method requires specialized equipment and expertise, which may limit its widespread adoption in routine rock testing practices.

Another important aspect of mechanical property characterization using big data analysis is the integration of multiple testing methods to provide a more comprehensive understanding of rock behavior. For example, combining nanoindentation data with results from uniaxial compression tests, triaxial tests, and acoustic emission measurements can provide a multi-scale, multi-physics characterization of rock properties.

This integrated approach can be particularly valuable for understanding the relationship between microstructural features and macroscopic rock behavior. By correlating nanoindentation results with bulk mechanical properties and failure mechanisms observed in larger-scale tests, researchers can develop more accurate predictive models for rock behavior under various loading conditions.

Furthermore, the application of machine learning techniques to these integrated datasets can reveal complex relationships that may not be apparent through traditional analysis methods. For instance, neural networks or support vector machines could be trained on the combined dataset to predict bulk rock properties based on nanoindentation measurements and other easily obtainable parameters (e.g., mineralogy, porosity).

7.3 Data-Driven Modelling

The mathematical framework for a machine learning approach in rock property prediction involves several steps. First, feature selection is performed to identify relevant features from various testing methods, such as nanoindentation hardness, Young's modulus, mineral composition percentages, and porosity. Next, data preprocessing is carried out to normalize and scale the input features, ensuring they are on comparable scales. Finally, a machine learning model, such as a neural network, is trained to predict bulk rock properties based on the selected features.

This integrated, machine learning-based approach to mechanical property characterization offers several key advantages. It enables comprehensive characterization by combining multiple testing methods, providing a more complete understanding of rock behavior across various scales and loading conditions. The trained model also has strong predictive power, allowing for the estimation of bulk rock properties based on easily obtainable parameters, which reduces the need for extensive and costly large-scale testing.

In addition, many machine learning techniques offer uncertainty quantification, which is vital for risk assessment in engineering applications. The model benefits from continuous improvement, with predictions refined as more data is incorporated over time.

However, it is important to note that the success of this approach relies heavily on the quality and representativeness of the training data. Ensuring that the dataset covers a wide range of rock types and conditions is crucial for developing a robust and generalizable model.

In conclusion, the application of big data analysis to mechanical property characterization in rock testing has opened up new possibilities for understanding and predicting rock behavior. By leveraging advanced statistical techniques, machine learning algorithms, and integrated multi-scale testing approaches, researchers and engineers can develop more accurate and comprehensive models of rock mechanical properties. These advancements have significant implications for various fields, including geotechnical engineering, mining, and petroleum engineering, potentially leading to more efficient and safer design practices in rock engineering applications.

Geological Condition Prediction

The prediction of geological conditions during tunneling and excavation projects is a critical aspect of rock engineering that has been significantly enhanced by the application of big data analysis. Traditional methods of geological prediction often rely on limited borehole data and expert interpretation, which can lead to uncertainties and potential safety risks. The integration of big data analytics with real-time monitoring systems has revolutionized the way we approach geological condition prediction, particularly in tunnel boring machine (TBM) operations.

A notable example of this approach is presented in the study by Chenet al. [37], which focused on predicting potential tunnel collapses during TBM construction by analyzing TBM performance data using deep learning techniques. The research was conducted on the Yinsong Water Diversion Project in northeast China, involving a 20 km tunnel section drilled by TBM.

The methodology employed in this study demonstrates key aspects of big data analysis in geological condition prediction. Real-time monitoring of TBM performance generated continuous data at 1-s intervals, capturing key parameters such as penetration rate, cutterhead rotation speed, torque, and thrust force. This high-frequency data collection creates a rich dataset, capturing subtle variations in machine performance that may indicate changes in geological conditions.

To simplify the analysis, the authors introduced a parameter called the drilling efficiency index (TPI), defined as:

$$TPI = \frac{T \cdot \text{RPM}}{N \cdot P_r} \tag{7.70}$$

where

T is the Torque per disc cutter
RPM is the Rotation speed
N is the Number of disc cutters
P_r is the Penetration rate

This data simplification step is crucial in big data analysis, as it reduces the dimensionality of the dataset while retaining essential information about the TBM's performance in relation to the geological conditions.

The results of this study demonstrated the effectiveness of big data analysis in geological condition prediction. The model achieved high accuracy, correctly identifying 14 out of 16 documented collapse sections as "unhealthy suspected" based on the proposed criteria. It also enables real-time prediction, continuously updating the model as new data becomes available, which allows for ongoing assessment of geological conditions during TBM operation. Additionally, the model's ability to detect deviations between predicted and observed TPI values provides an early warning mechanism for potential tunnel collapses.

However, this method, like many data-driven approaches, has certain limitations. Its performance may vary between different tunneling projects due to differences in geological conditions and TBM specifications. The accuracy of predictions is also highly dependent on the quality and consistency of the input data. The proposed criteria for identifying unstable sections require further validation across various geological settings and TBM types to ensure their robustness.

Building on this approach, future research could focus on integrating additional data sources to enhance prediction accuracy. For example, incorporating geological survey data, geophysical measurements, and historical information about the local geology could provide a more comprehensive basis for prediction.

Another promising direction for big data analysis in geological condition prediction is the use of multi-modal data fusion techniques. This approach involves combining data from various sources, such as TBM operational parameters, seismic measurements, and ground penetrating radar, to create a more comprehensive picture of the subsurface conditions.

A mathematical framework for multi-modal data fusion could be represented as follows:

7.3 Data-Driven Modelling

Let $X = \{X_1, X_2, \ldots, X_n\}$ be a set of n data sources, where each X_i represents a different modality (e.g., TBM data, seismic data, etc.). The goal is to find a joint representation Z that captures the complementary information from all sources:

$$Z = f(X_1, X_2, \ldots, X_n; \theta) \tag{7.71}$$

where f is a fusion function (e.g., a neural network) and θ represents the parameters of the fusion model.

The fusion model can be trained to optimize a task-specific objective, such as minimizing the prediction error for geological conditions:

$$\min L(y, g(Z)) \tag{7.72}$$

where y represents the true geological conditions, g is a prediction function, and L is a loss function (e.g., mean squared error).

This multi-modal fusion approach offers several advantages. It enables a comprehensive analysis by integrating data from multiple sources, providing a more complete understanding of subsurface conditions. The approach is also robust, as the use of diverse data streams helps mitigate the impact of noise or errors in any single source. Furthermore, the framework is flexible, allowing for easy adaptation to incorporate new data sources as they become available.

However, implementing such a system also presents challenges, including data synchronization, handling missing data, and dealing with varying scales and units across different data modalities.

Another important aspect of big data analysis in geological condition prediction is the incorporation of uncertainty quantification. Given the inherent variability and uncertainty in geological systems, it's crucial to provide not just point predictions but also estimates of prediction uncertainty.

Bayesian methods offer a natural framework for uncertainty quantification in geological prediction models. For example, a Bayesian neural network (BNN) could be used to predict geological conditions while also providing uncertainty estimates:

$$p(y|x, D) = \int p(y|x, w) p(w|D) \, dw \tag{7.73}$$

where
  y is the predicted geological condition

x is the input data

D is the training dataset

w represents the model parameters

$p(w|D)$ is the posterior distribution of the model parameters given the data

$p(y|x, w)$ is the likelihood of the prediction given the input and model parameters

- The integral is typically approximated using techniques like Markov Chain Monte Carlo (MCMC) or variational inference.

This Bayesian approach offers several benefits. It provides uncertainty estimates by generating a distribution of possible predictions, which supports more informed decision-making. The model also allows for the incorporation of prior knowledge, enabling the inclusion of geological expertise through informative priors on model parameters. Additionally, the Bayesian framework facilitates continuous learning, allowing predictions to be updated as new data becomes available.

In conclusion, the application of big data analysis to geological condition prediction represents a significant advancement in rock engineering. By leveraging large datasets, advanced machine learning techniques, and multi-modal data fusion, these approaches offer the potential for more accurate, real-time prediction of subsurface conditions. This can lead to improved safety, efficiency, and cost-effectiveness in tunneling and other underground construction projects. However, it's important to note that these data-driven methods should complement, rather than replace, traditional geological expertise. The integration of big data analysis with domain knowledge and experience will likely yield the most robust and reliable prediction systems for geological conditions.

Rock Mass Classification

Rock mass classification is a fundamental aspect of rock engineering that has traditionally relied on expert judgment and simplified rating systems. However, the advent of big data analysis has opened up new possibilities for more comprehensive, objective, and data-driven classification approaches. These advanced methods can capture the complex interrelationships between various rock mass parameters and provide more accurate and nuanced classifications.

One of the key advantages of applying big data analysis to rock mass classification is its capacity to handle large volumes of diverse data. Traditional classification systems are typically limited to a small number of parameters, which may fail to capture the full complexity of rock mass behavior. Big data approaches, on the other hand, can integrate a wide array of data sources, including geotechnical parameters like RQD, joint spacing, and joint condition; geophysical measurements such as seismic velocity and electrical resistivity; laboratory test results like uniaxial compressive strength and point load index; in-situ stress measurements; geological mapping data; borehole imaging logs; and excavation performance data, including TBM penetration rates and support requirements. This comprehensive approach allows for a more holistic understanding of rock mass characteristics, leading to more accurate and reliable classification models.

By incorporating this diverse range of data, big data analysis can provide a more holistic and accurate representation of rock mass conditions.

A key example of the application of big data analysis in rock mass classification is presented in the study by Zhang et al. [38], which focuses on the use of machine learning techniques for rock mass classification. While the paper doesn't provide

7.3 Data-Driven Modelling

specific details on rock mass classification, the principles of machine learning and big data analysis discussed can be applied to this domain.

A rock mass classification approach using big data analysis typically follows a systematic process. First, data is collected from various sources, such as geotechnical, geophysical, laboratory tests, and field measurements. Missing data is handled through imputation techniques, and the features are normalized and scaled to ensure comparability across the dataset.

Next, dimensionality reduction techniques like Principal Component Analysis (PCA) or t-distributed Stochastic Neighbor Embedding (t-SNE) are used to reduce the complexity of the dataset while retaining important information. Feature selection algorithms are also applied to identify the most relevant parameters for classification.

Once the data is processed, unsupervised learning methods are employed to identify patterns. Clustering algorithms such as K-means or DBSCAN help detect natural groupings in the data, which may correspond to different rock mass classes. Additionally, Self-Organizing Maps (SOM) can be used to visualize high-dimensional data and uncover key patterns.

For the classification itself, supervised machine learning models like Random Forests, Support Vector Machines, or Neural Networks are trained using the patterns identified in the data and expert-labeled classifications. Cross-validation is conducted to evaluate the model's performance and ability to generalize to unseen data.

Finally, the model's predictions are interpreted using SHAP (SHapley Additive exPlanations) values, which help to understand the contribution of individual features to the classification outcomes. The predictions are validated by comparing them with expert classifications and field observations to ensure their reliability and accuracy.

This approach integrates unsupervised and supervised learning techniques, leveraging the full potential of big data to develop a more robust and data-driven rock mass classification model.

The mathematical framework for some of these techniques can be described as follows:

- Principal Component Analysis (PCA): PCA transforms the original features into a new set of uncorrelated variables (principal components) that capture the maximum variance in the data. The transformation can be represented as:

$$Z = XW \tag{7.74}$$

where X is the original data matrix, W is the matrix of eigenvectors of the covariance matrix of X and Z is the transformed data in the new principal component space.

- K-means Clustering: K-means aims to partition n observations into k clusters, minimizing the within-cluster sum of squares:

$$\mathrm{argmin} \sum_{j=1}^{k} \sum_{i \in S_j} ||x_i - \mu_j||^2 \tag{7.75}$$

where S_j is the set of points in the j-th cluster and μ_j is the mean of points in S_j.

- Random Forest Classification: Random Forests construct multiple decision trees and merge them to get a more accurate and stable prediction. The probability of a rock mass belonging to class k can be estimated as:

$$P(y = k|x) = \frac{1}{B} \sum_{b=1}^{B} P^b(y = k|x) \qquad (7.76)$$

where B is the number of trees in the forest; $P^b(y = k \mid x)$ is the prediction of the b-th tree.

Machine learning approaches offer several advantages over traditional rock mass classification systems. One of the key benefits is objectivity. These methods rely on data-driven algorithms rather than subjective expert judgment, leading to more consistent and reproducible classifications. Another significant advantage is their ability to handle complexity. Machine learning models can capture intricate, non-linear relationships between rock mass parameters, which are often difficult to represent using simpler rating systems.

Moreover, these models are highly adaptable. As new data becomes available, they can be retrained and updated, ensuring that the models remain accurate and relevant in varying geological conditions. Lastly, uncertainty quantification is an important benefit of machine learning techniques. Many models provide probabilistic outputs, allowing for the estimation and communication of uncertainties inherent in rock mass classifications, which is essential for making informed engineering decisions.

While data-driven approaches offer numerous advantages, they also come with some limitations. One key challenge is data requirements. Big data analysis typically necessitates large datasets to train reliable models, and such datasets may not always be available, especially for new projects or in unique geological settings. Another limitation is interpretability. Some advanced machine learning models, such as deep neural networks, can be difficult to interpret, making it challenging to understand the rationale behind specific classifications or predictions.

Additionally, extrapolation is a concern when applying models trained on data from one geological setting to a different one. The relationships between rock mass parameters may vary significantly between settings, and models that work well in one context may not necessarily produce accurate results when applied to another. These limitations highlight the need for careful consideration and expert validation when using machine learning in rock mass classification and other geotechnical applications.

To address these limitations, hybrid approaches that combine data-driven methods with expert knowledge and traditional classification systems are often most effective. For example, a machine learning model could be used to provide an initial classification, which is then reviewed and adjusted by expert geologists or engineers based on their domain knowledge and site-specific considerations.

One promising direction for future research in this area is the development of transfer learning techniques for rock mass classification. Transfer learning allows

7.3 Data-Driven Modelling

models trained on data from one domain or project to be adapted and applied to a new, related domain with limited data. This could be particularly valuable in rock engineering, where data from previous projects could be leveraged to improve classifications in new geological settings.

A mathematical framework for transfer learning in rock mass classification might involve the following steps:

- Pre-train a deep neural network on a large dataset from multiple rock engineering projects

$$L_{\text{source}}(\theta_s) = \frac{1}{N_s} \sum_{i=1}^{N_s} l(f_s(x_i; \theta_s), y_i) \tag{7.77}$$

where

L_{source} is the loss function for the source domain,
θ_s is the model parameters, f_s is the model function,
(x_i, y_i) is the input–output pairs from the source domain,
Ns is the number of samples in the source domain.

- Fine-tune the pre-trained model on a smaller dataset from the target project

$$L_{\text{target}}(\theta_t) = \frac{1}{N_t} \sum_{i=1}^{N_t} l(f_t(x_i; \theta_t), y_i) + \lambda R(\theta_t, \theta_s) \tag{7.78}$$

where

L_{target} is the loss function for the target domain,
θ_t is the model parameters for the target domain,
f_t is the model function for the target domain,
N_t is the number of samples in the target domain,
$R(\theta_t, \theta_s)$ is a regularization term that encourages the target model to stay close to the source mode,
λ is a hyper-parameter controlling the strength of the regularization.

This approach allows the model to leverage knowledge gained from a large dataset of diverse rock masses while adapting to the specific characteristics of the target geological setting.

Another important aspect of big data analysis in rock mass classification is the integration of spatial and temporal information. Traditional classification systems often treat each measurement point independently, ignoring the spatial relationships between different locations in a rock mass. By incorporating spatial statistics and geostatistical methods, big data approaches can provide a more cohesive and realistic representation of rock mass variability.

For example, kriging techniques could be used to interpolate rock mass properties between measurement points, providing a continuous spatial model of rock mass quality. The kriging estimator can be expressed as:

$$Z^*(x_0) = \sum_{i=1}^{N} \lambda_i Z(x_i) \tag{7.79}$$

where

$Z^*(x_0)$ is the estimated value at location x_0

$Z(x_i)$ are the known values at locations x_i

λ_i are the kriging weights, determined by solving a system of equations that minimize the estimation variance

By combining these geostatistical techniques with machine learning approaches, it's possible to create sophisticated 3D models of rock mass classification that account for spatial continuity and anisotropy.

In conclusion, the application of big data analysis to rock mass classification represents a significant advancement in rock engineering. These methods offer the potential for more accurate, objective, and comprehensive classifications that can capture the complex nature of rock masses. By leveraging large datasets, advanced machine learning techniques, and spatial statistics, these approaches can provide valuable insights for engineering design, risk assessment, and decision-making in rock engineering projects. However, it's crucial to remember that these data-driven methods should complement, rather than replace, geological expertise and site-specific knowledge. The most effective rock mass classification systems will likely be those that successfully integrate big data analysis with traditional engineering judgment and domain expertise.

Failure Mechanism Analysis

The analysis of failure mechanisms in rock masses is a critical aspect of rock engineering that has been significantly enhanced by the application of big data analysis techniques. Traditional approaches to failure mechanism analysis often rely on simplified models and limited datasets, which may not fully capture the complex, multiscale nature of rock failure processes. Big data analysis offers the potential to integrate large volumes of diverse data, enabling a more comprehensive understanding of the factors contributing to rock mass failure.

One of the key advantages of applying big data analysis to failure mechanism analysis is the ability to identify subtle precursors and complex interactions that may not be apparent through conventional methods. By analyzing large datasets of monitoring data, laboratory test results, and field observations, it's possible to detect patterns and correlations that can provide early warning of impending failures and insights into the underlying mechanisms.

A notable example of the application of big data analysis in failure mechanism analysis is presented in the study by Wang et al. [39], which focused on predicting tunnel collapses during TBM construction. While this study primarily addressed prediction rather than mechanism analysis, the principles and techniques used can be extended to understand failure mechanisms more broadly.

7.3 Data-Driven Modelling

The approach used in this study can be adapted for failure mechanism analysis by integrating data from multiple sources, applying advanced machine learning techniques, and combining empirical data with physical models. The process begins with data collection and integration, gathering information from various sources such as in-situ monitoring (e.g., displacement measurements, microseismic activity), laboratory test results (e.g., strength parameters, failure modes), geophysical surveys (e.g., seismic tomography, electrical resistivity), construction records (e.g., excavation sequence, support installation), and environmental data (e.g., temperature, precipitation).

Next, feature engineering involves developing relevant features that capture key aspects of rock mass behavior and potential failure mechanisms. These features could include stress concentration factors, damage indices based on microseismic activity, time-dependent deformation rates, and pore pressure evolution.

Time series analysis is used to identify trends, patterns, and anomalies in monitoring data that may indicate the development of failure mechanisms. Techniques like ARIMA (AutoRegressive Integrated Moving Average) models or deep learning approaches, such as Long Short-Term Memory (LSTM) networks, are employed for this purpose.

In the next step, machine learning for pattern recognition is applied. Unsupervised learning techniques, such as clustering or dimensionality reduction, are used to identify natural groupings in the data that may correspond to different failure mechanisms or stages of the failure process. Supervised learning algorithms are then applied to classify observed behaviors into known failure mechanism categories or predict the likelihood of different failure modes.

To improve prediction accuracy, physical model integration combines data-driven insights with physics-based models. This hybrid approach leverages both empirical data and theoretical understanding of rock mechanics. Techniques like physics-informed neural networks (PINNs) ensure that machine learning models adhere to fundamental physical laws and constraints.

Finally, causal inference is applied to move beyond correlation and identify causal relationships between factors and failure mechanisms. Methods such as causal forests or structural equation modeling can be employed for this purpose, providing a deeper understanding of the interactions driving failure mechanisms.

This integrated approach allows for a more comprehensive analysis of failure mechanisms and can significantly improve predictive accuracy in rock mass stability assessments. The mathematical framework for some of these techniques can be described as follows:

- ARIMA Model

$$\left(1 - \sum_{i=1}^{p} \phi_i L^i\right)(1-L)^d y_t = \left(1 + \sum_{j=1}^{q} \theta_j L^j\right) \varepsilon_t \qquad (7.80)$$

where

L is the lag operator
φ_i are the parameters of the autoregressive part
θ_j are the parameters of the moving average part
d is the degree of differencing
ε_t is white noise

- LSTM Network

$$f_t = \sigma\left(W_f \cdot [h_{t-1}, x_t] + b_f\right) \tag{7.81}$$

$$i_t = \sigma\left(W_i \cdot [h_{t-1}, x_t] + b_i\right) \tag{7.82}$$

$$\tilde{C}_t = \tanh\left(W_c \cdot [h_{t-1}, x_t] + b_c\right) \tag{7.83}$$

$$C_t = f_t * C_{t-1} + i_t * \tilde{C}_t \tag{7.84}$$

$$o_t = \sigma\left(W_o \cdot [h_{t-1}, x_t] + b_o\right) \tag{7.85}$$

$$h_t = o_t * \tanh(C_t) \tag{7.86}$$

where
f_t, i_t, o_t are the forget, input, and output gates respectively,
C_t is the cell state,
h_t is the hidden state,
W and b are weight matrices and bias vectors,
σ is the sigmoid function.

- Physics-Informed Neural Networks (PINNs): PINNs incorporate physical laws into the loss function of neural networks. For example, in the context of rock mechanics, we might include a term in the loss function that enforces conservation of momentum

$$L = \text{MSE}\left(y_{\text{pred}}, y_{\text{true}}\right) + \lambda \cdot \text{MSE}(\nabla \cdot \sigma + \rho g, 0) \tag{7.87}$$

where
MSE is the mean squared error,
y_{pred} and y_{true} are the predicted and true values,
σ is the stress tensor,
ρ is density,
g is the gravity vector,

These advanced analytical techniques offer several advantages for failure mechanism analysis. One key benefit is multi-scale integration, where big data analysis can merge information from different scales, ranging from microscopic crack propagation to macroscopic deformation. This enables a more comprehensive understanding of failure processes that traditional methods may overlook.

Another advantage is the ability to identify precursors to failure. By analyzing large volumes of monitoring data, subtle indicators of impending failure can be detected that might not be visible through conventional analysis. This early detection can be crucial for improving safety and making timely interventions.

Lastly, complex interaction modeling is a significant benefit. Machine learning techniques excel at capturing complex, non-linear relationships between various factors contributing to rock mass failure. These models can provide insights into how different variables interact, helping to predict failure more accurately and understand underlying processes.

7.3.3 Predictive Models for Rock Failure

In recent years, the field of rock dynamics has seen significant advancements in the development and application of predictive models for rock failure. These models, particularly those leveraging data-driven approaches and machine learning techniques, have revolutionized our ability to understand, predict, and mitigate rock failure in various geotechnical and mining engineering contexts. This section provides a comprehensive review of the predictive models for rock failure, focusing on their theoretical foundations, methodologies, applications, and limitations.

Rock failure prediction is a critical aspect of numerous engineering disciplines, including civil engineering, mining, petroleum engineering, and geological disaster prevention. Accurate prediction of rock failure is essential for ensuring the safety and stability of underground excavations, slopes, tunnels, and other rock engineering structures. Traditional approaches to rock failure prediction often relied on simplified analytical models or empirical correlations, which may not fully capture the complex, nonlinear behavior of rock masses under various loading conditions. The advent of data-driven techniques and machine learning algorithms has opened up new possibilities for more accurate and robust predictive models.

This section will explore various aspects of predictive models for rock failure, including:

- Fundamental concepts and challenges in rock failure prediction
- Data-driven approaches to rock failure modeling
- Machine learning techniques for rock property estimation
- Predictive models for specific failure modes
- Integration of multi-scale and multi-physics approaches
- Uncertainty quantification and model validation
- Applications in rock engineering and geohazard assessment
- Future trends and research directions

Fundamental Concepts and Challenges in Rock Failure Prediction

Rock failure prediction stands as one of the most complex challenges in geomechanical engineering, involving the interplay of multiple physical, geological, and environmental factors. The fundamental complexity stems from several interconnected challenges that researchers and engineers must address.

The inherent heterogeneity and anisotropy of rock masses present a primary obstacle. Rock materials, shaped by their geological history and internal structure, exhibit varying properties in different directions and locations within the same mass. This natural variability makes it exceptionally difficult to develop generalized predictive models that can accurately represent behavior across different conditions and locations [40].

Scale effects introduce another layer of complexity. The behavior of rock samples studied in laboratory conditions often differs significantly from the behavior of large rock masses in the field. This scale-dependent variation in properties and behavior creates a substantial challenge in translating laboratory findings to practical field applications [40].

Time-dependent phenomena add further complexity to prediction models. Many rocks exhibit creep behavior and fatigue effects that evolve over time, making it necessary to account for these temporal changes in any comprehensive prediction model. These time-dependent characteristics can significantly affect long-term stability and safety assessments [41].

The presence of coupled processes in rock failure mechanisms presents additional challenges. Hydro-mechanical and thermo-mechanical interactions often occur simultaneously, creating complex feedback loops that affect rock behavior. These coupled processes require sophisticated modeling approaches that can account for multiple interacting physical phenomena [42].

Data limitations pose a practical challenge to model development and validation. Obtaining comprehensive, high-quality data about rock properties and failure conditions, particularly in-situ measurements, requires significant investment in time and resources. This constraint often leads to incomplete datasets that may not fully capture the range of conditions affecting rock behavior [43].

The nonlinear nature of rock failure presents a fundamental modeling challenge. Traditional linear models often fail to capture the complex stress–strain relationships and failure mechanisms observed in real rock masses. This nonlinearity necessitates more sophisticated mathematical approaches and computational methods [44].

To address these challenges, researchers increasingly employ advanced numerical techniques and data-driven approaches. These modern methods aim to better capture the complex behavior of rock masses and provide more reliable predictions under various conditions. The integration of machine learning techniques, coupled with traditional mechanical models, shows particular promise in addressing these fundamental challenges.

7.3 Data-Driven Modelling

Data-Driven Approaches to Rock Failure Modeling

Data-driven approaches have gained significant traction in rock mechanics and rock engineering due to their ability to capture complex, nonlinear relationships without relying on simplified assumptions. These methods leverage large datasets of rock properties, stress conditions, and failure observations to develop predictive models.

Data-driven approaches offer several important advantages in modern analysis and decision-making. When researchers apply these methods, they can capture complex nonlinear relationships that might be missed by simpler models. These approaches reduce the need for oversimplified assumptions about how variables interact, leading to more nuanced and accurate results. The models can continuously improve as new data becomes available, adapting to evolving conditions and patterns. However, these benefits come with a significant challenge: the need for high-quality, comprehensive datasets. This has led researchers to develop sophisticated strategies for collecting, cleaning, and enriching their data. They employ various preprocessing techniques to handle missing values and outliers, while data augmentation methods help expand limited datasets in meaningful ways. The detailed data-driven approach are summarized as follow steps:

1. Data Collection and Preprocessing

Effective data-driven modeling in rock mechanics relies on diverse, representative datasets capturing rock behavior and properties. Key sources include laboratory tests (e.g., uniaxial/triaxial compression, direct shear tests), field measurements (stress state and geophysical surveys), continuous monitoring (time-series data on deformation and pressure), geological/geotechnical logs, and numerical simulations. Preprocessing these datasets is crucial for robust predictive models. Key steps include normalization and standardization to align scales, handling missing/erroneous data through imputation or removal, and feature engineering to select relevant parameters. Data augmentation techniques can also expand limited datasets while preserving rock mechanics constraints. For example, in the study by Lei et al. [40], the authors used a database of 168 test results from direct shear tests on discontinuous joint specimens with different wall strengths. The data was preprocessed using minimum–maximum normalization, which was found to perform best among the tested normalization methods.

2. Feature Selection and Importance Analysis

Linear methods help identify relationships between rock properties and failure indicators. For example, UCS may strongly correlate with factors like density, porosity, and mineral composition, though non-linear relationships are often overlooked.

Principal Component Analysis (PCA) is valuable for analyzing interconnected rock properties. By combining related measurements (e.g., grain size, mineral content, texture) into principal components, PCA reduces data dimensionality while retaining key variance.

Recursive Feature Elimination is effective for rock mechanics data, iteratively removing less important features. It helps identify the most influential parameters (e.g., stress conditions, rock composition, environmental factors) for predicting failure, particularly with complex datasets from various testing methods.

Mutual Information Analysis captures non-linear relationships, such as how discontinuities, water content, and stress history interact to influence failure probability—insights linear methods may miss.

Combining methods often enhances rock failure prediction. For example, using mutual information analysis to capture non-linear relationships, followed by recursive feature elimination to optimize the feature set, ensures both physical relevance and statistical significance.

Lei et al. [40] conducted sensitivity analyses on their hybrid data-driven model for predicting the shear strength of discontinuous rock materials. They found that the compressive strength ratio of joint walls ($\sigma ch/\sigma cs$) and the normal stress (σn) were the most influential factors in their model.

In another study on predicting rock slope failure [43], multi-factor sensitivity analysis using the mutual information (MI) technique revealed that friction angle (φ) and slope height (H) had the most significant impact on the factor of safety (FOS). This type of analysis can guide the selection of input parameters for predictive models and provide insights into the underlying physical mechanisms of rock failure.

3. Hybrid Modeling Approaches

Hybrid modeling approaches combine data-driven techniques with physics-based models or expert knowledge to leverage the strengths of both paradigms. These approaches can improve model accuracy, interpretability, and generalization capabilities. Some examples of hybrid modeling in rock failure prediction include:

Physics-informed Neural Networks (PINNs): PINNs integrate physical laws into the loss function or network structure, ensuring predictions align with rock mechanics principles. For example, they enforce stress–strain relationships or failure criteria during training, maintaining physically meaningful predictions even with limited data and improving extrapolation beyond the training set [45].

Expert-guided Feature Engineering: Domain expertise shapes the model's input space by identifying critical parameter combinations (e.g., stress ratios, strength indices), guiding the creation of composite features that capture failure mechanisms, and ensuring features align with the physical understanding of rock behavior. Experts also define appropriate value ranges and constraints for features [44].

Multi-fidelity Modeling: This approach combines data from various sources: high-fidelity data from precise tests, lower-fidelity data from empirical correlations, and historical data with varying accuracy. The goal is to establish relationships between different data fidelities to leverage all available information effectively [42].

Ensemble Methods: These combine different model types to improve prediction robustness: machine learning models for data patterns, physics-based models for fundamental mechanics, and empirical models based on engineering relationships. The ensemble can weight models according to their reliability in specific

7.3 Data-Driven Modelling

scenarios. Hybrid approaches balance theoretical understanding with data-driven insights, resulting in models that are both accurate and physically meaningful [43].

For instance, the hybrid data-driven model proposed by Lei et al. [40] for predicting the shear strength of discontinuous rock materials combines Extreme Gradient Boosting (XGBoost) with the Levy-Sparrow Search Algorithm (LSSA) for hyperparameter optimization. This hybrid approach outperformed both the original XGBoost and SSA-XGBoost models in terms of accuracy and computational efficiency.

Machine Learning Techniques for Rock Property Estimation

Machine learning techniques have proven particularly effective in estimating rock properties and predicting failure behavior. These methods can capture complex, nonlinear relationships between input variables and output parameters, often outperforming traditional empirical correlations. Some of the most commonly used machine learning techniques in rock mechanics include:

1. Artificial Neural Networks (ANNs)

ANNs are versatile machine learning models inspired by biological neural networks. They consist of interconnected nodes (neurons) organized in layers, capable of learning complex patterns from data.

Estimation of Rock Strength: ANNs are highly effective in predicting UCS and tensile strength by learning complex relationships between rock properties. Models can forecast UCS using inputs like porosity, density, P-wave velocity, and mineral composition, reducing the need for costly and time-consuming direct testing.

Prediction of Elastic Properties: ANNs model the nonlinear relationships between rock characteristics and elastic behavior. With sufficient data, ANNs can predict Young's modulus (stiffness) and Poisson's ratio (lateral-to-axial strain ratio) based on parameters like mineral content, texture, and stress conditions, helping engineers assess rock deformation without extensive lab testing.

Forecasting Rock Mass Behavior: ANNs predict failure mechanisms by analyzing historical data on variables such as joint spacing, weathering, groundwater conditions, and stress states. This is particularly valuable in mining and tunneling, where understanding rock mass behavior is essential for safety and design optimization.

For example, in the study on predicting rock slope failure [43], a Backpropagation Neural Network (BPNN) model was developed to predict the factor of safety (FOS) for rock slopes. The BPNN model achieved high accuracy, with an R^2 of 0.86 and a Mean Absolute Percentage Error (MAPE) of 0.005.

2. Support Vector Machines (SVMs)

SVMs are powerful machine learning algorithms that can be used for both classification and regression tasks. In rock mechanics, SVMs have been applied to:

Rock Type Classification: SVMs are highly effective for classifying rock types by handling high-dimensional data and finding optimal boundaries between different classes. By mapping input features (e.g., mineral composition, texture, grain size) into a higher-dimensional space, SVMs make it easier to separate rock types, especially in complex geological formations where traditional methods may struggle.

Prediction of Rock Mass Rating (RMR) and Geological Strength Index (GSI): SVMs excel in predicting rock mass classification parameters. For RMR, they process inputs like rock strength, RQD, joint spacing, joint condition, and groundwater conditions. For GSI, they analyze rock structure, surface conditions, and discontinuity characteristics. SVMs' ability to handle nonlinear relationships makes them ideal for these complex predictions.

Estimation of Rock Strength and Deformability: SVMs are effective in predicting rock strength parameters, such as UCS, tensile strength, Young's modulus, and Poisson's ratio. Their ability to manage noisy data and avoid overfitting is crucial when dealing with the variability of geological materials. SVMs capture complex relationships between input parameters (e.g., density, porosity, mineral content) and strength or deformability values.

The study on predicting rock slope failure [43] also implemented a Least-Square Support Vector Machine (Ls-SVM) model, which showed good performance in predicting the factor of safety for rock slopes.

3. Random Forests and Decision Trees

Random Forests are ensemble learning methods that construct multiple decision trees and combine their predictions. These methods have gained popularity in rock mechanics due to their ability to handle high-dimensional data and provide feature importance rankings. Applications include:

Rock Mass Classification: Random Forests are effective for rock mass classification by processing multiple parameters simultaneously and accounting for their relative importance. They handle both numerical and categorical variables commonly used in rock mass systems. Their ensemble approach reduces classification errors, manages noisy or missing data, and ranks the most influential parameters.

Prediction of Rock Strength and Deformability: Random Forests predict rock mechanical properties by learning from various input parameters. They capture complex nonlinear relationships, handle interactions between parameters, and provide uncertainty estimates. Their ability to maintain accuracy with limited data makes them especially useful in this context.

Estimation of Rock Mass Behavior in Tunneling and Mining: Random Forests predict rock mass behavior in tunneling and mining by analyzing multiple factors, including ground conditions, support requirements, and stability issues. Their ability to rank feature importance helps engineers understand key parameters affecting excavation performance, supporting informed decision-making in project planning and execution.

In the study on predicting rock slope failure [43], a Random Forest model was developed and showed excellent performance, achieving an R^2 of 0.86 in predicting the factor of safety for rock slopes.

7.3 Data-Driven Modelling 399

4. Extreme Gradient Boosting (XGBoost)

XGBoost is an advanced gradient boosting algorithm known for its high performance and efficiency. In rock mechanics, it has been successfully applied to several prediction tasks, including the estimation of rock strength parameters, prediction of rock mass behavior, and forecasting of geohazards.

Lei et al. [40] used XGBoost as the base model in their hybrid approach for predicting the shear strength of discontinuous rock materials. The XGBoost model showed superior performance compared to other machine learning algorithms, achieving high accuracy and computational efficiency.

5. Genetic Programming and Evolutionary Algorithms

Genetic programming and other evolutionary algorithms have been applied in rock mechanics to develop predictive models and optimize design parameters. These techniques are particularly effective for deriving empirical equations for rock property estimation, optimizing rock support systems, and creating hybrid models that combine machine learning with physical insights.

For instance, in the study on predicting rock slope failure [43], a Gene Expression Programming (GEP) model was developed and showed excellent performance in predicting the factor of safety for rock slopes, achieving an R^2 of 0.86 and a Mean Absolute Percentage Error (MAPE) of 0.004.

6. Deep Learning Approaches

Deep learning techniques, such as deep neural networks and convolutional neural networks (CNNs), have shown significant promise in rock mechanics, especially for image analysis and pattern recognition tasks. Key applications include automated rock mass classification from borehole images or photogrammetry, prediction of rock fracture patterns and failure modes, and real-time monitoring and analysis of rock mass behavior.

While deep learning approaches offer powerful capabilities, they often require large datasets and significant computational resources. As more data becomes available and computing power increases, these methods are likely to play an increasingly important role in rock failure prediction.

Predictive Models for Specific Failure Modes

Different types of rock failure require specialized predictive models that account for the unique characteristics and mechanisms involved. This section discusses predictive models for several important failure modes in rock mechanics.

1. Shear Failure of Rock Joints

Shear failure along rock joints is a key factor in rock mass stability. Predictive models for shear failure estimate the shear strength and failure behavior of rock joints under various loading conditions. Lei et al. [40] developed a hybrid data-driven model to predict the shear strength of discontinuous rock materials with varying joint wall strengths. Their approach combined Extreme Gradient Boosting (XGBoost) with the Levy-Sparrow Search Algorithm (LSSA) for hyperparameter optimization.

The model used input variables such as the compressive strength ratio of joint walls (σ_{ch}/σ_{cs}), shear strength of discontinuities with identical joint wall strength (τ_s), joint roughness coefficient (JRC), and normal stress (σ_n). The target variable was the shear strength of discontinuous joint specimens (τ_i). The hybrid LSSA-XGBoost model achieved a high coefficient of determination (R^2) of 0.972 on the test set.

The key findings from Lei et al. [46] include: (1) The hybrid model outperformed conventional empirical correlations and other machine learning approaches; (2) The compressive strength ratio of joint walls and normal stress were found to be the most influential factors in predicting shear strength; (3) The model demonstrated improved computational efficiency, reducing runtime by 54.7% compared to the SSA-XGBoost model.

This approach provides a more accurate and efficient method for predicting the shear strength of rock joints, which is crucial for assessing the stability of rock slopes, tunnels, and other geotechnical structures.

2. Brittle Rock Failure

Brittle rock failure is a complex phenomenon that can result in sudden and catastrophic collapse in underground excavations. Predicting its severity and extent is crucial for ensuring the safety and stability of these structures. A study by [42] introduced a new approach for classifying and predicting brittle rock failure, focusing on failure behaviors and damage depth. The authors classified brittle rock failure into four types based on the mechanical properties of the rupture surface: Type I, stress-induced spalling (depth < 30 cm); Type II, stress-induced buckling (depth < 60 cm); Type III, stress-induced shear (depth > 60 cm); and Type IV, stress-structure coordinated failure, with depth depending on the coupling between stress and structure. And the severity of brittle rock failure was classified into four grades based on the ratio of failure depth to tunnel radius: Minor (<0.1); Moderate (0.1–0.2); Major (0.2–0.3); Severe (>0.3).

The method is built around two key rock mass characterization parameters—K_v and GSI—and three critical stress thresholds. The Modified Rock Mass Integrity Coefficient (K_v), expressed as the squared ratio of P-wave velocity between the rock mass (v_{pm}) and intact rock (v_{pr}), quantifies rock mass degradation. The Geological Strength Index (GSI), derived from visual geological observations, assesses overall rock mass quality by considering rock structure and surface conditions. The three critical stress thresholds include the Maximum Tangential Stress (σ_{max}), which occurs near the tunnel surface and signals potential failure; the Crack Initiation Stress (σ_{ci}), where microcracks begin to form, marking the start of failure progression; and the Crack Damage Stress (σ_{cd}), the point at which unstable crack growth begins, signaling severe damage. Together, these parameters form a comprehensive framework that integrates rock mass quality (K_v and GSI) with induced stress conditions (σ_{max}, σ_{ci}, and σ_{cd}).

The method showed good agreement with actual brittle rock failure observations in case studies at the Shuangjiangkou Hydropower Station in China. This

7.3 Data-Driven Modelling

approach provides a more comprehensive prediction process for brittle rock failures in geological engineering, considering both stress conditions and rock mass characteristics.

3. Rock Slope Failure

Predicting rock slope failure is essential for infrastructure safety and preventing geohazards. A study by [43] explored the use of multiple machine learning algorithms to predict the factor of safety (FOS) in rock slope stability analysis, using a dataset of 344 rock slope cases from Iran, analyzed with PLAXIS software. The input parameters included cohesion, slope angle, unit weight, slope height, friction angle, and pore pressure ratio.

Six machine learning models were developed and compared: Multi-variable Regression (MVR), Backpropagation Neural Network (BPNN), Feed-Forward Neural Network (FFNN), Takagi–Sugeno Fuzzy System (TSF), Gene Expression Programming (GEP), and Least-Square Support Vector Machine (Ls-SVM). The results showed that all models produced satisfactory FOS predictions, with BPNN and GEP achieving the highest precision, both with an R^2 of 0.86. Feature importance analysis revealed that the friction angle was the most influential parameter, while slope height had the least impact. Additionally, a new empirical formula for calculating FOS was developed based on the machine learning results.

The study also introduced a user-friendly Graphical User Interface (GUI) that simplifies the implementation of these ML models, enabling quick, cost-effective, and accurate slope FOS estimations.

This research demonstrates the potential of machine learning techniques to enhance rock slope stability analysis, providing more accurate and efficient tools for risk management and hazard prevention in geotechnical engineering projects.

4. Cyclic Loading and Fatigue Failure

Predicting rock failure under cyclic loading is crucial for understanding the long-term stability of geotechnical structures exposed to repeated loads, such as those in earthquake zones or near heavy machinery. A study by [41] developed a 1D bounding surface model integrated into a 4D lattice spring model (4D-LSM) to predict rock failure under cyclic loading.

The model incorporates key components such as a simplified 1D bounding surface model, a microcrack model to represent nonlinear rock behavior, a macro strength criterion and cohesive zone model for failure processes, and GPU-based parallel computing for acceleration. The model successfully simulates the deformation and failure process of rock under various cyclic loading paths, hysteresis loops, irreversible deformation accumulation, higher tangent modulus during unloading, and fatigue life under different stress levels and confining pressures.

Key features of the approach include a combination of the strengths of both continuum-based and discrete models, the ability to model both intact rock and rock with preexisting macro joints, and the reproduction of the three phases of rock deformation under cyclic loading: initial, stable, and accelerated.

The model was validated against experimental data for granite and sandstone samples, showing good agreement in reproducing stress–strain behavior under various cyclic loading paths and predicting fatigue life trends.

This modeling approach provides valuable insights into rock deformation and failure mechanisms under cyclic loading, with applications in predicting long-term stability of geotechnical structures and modeling progressive failure and fatigue life of rock materials.

5. Progressive Failure of Rock-Like Materials

Understanding the progressive failure of rock-like materials is essential for predicting the behavior of rock masses in engineering applications. A study by [40] introduced a new microplane model to characterize progressive failure based on the traction vector.

The model's key features include the use of the traction vector to intuitively capture directional failure modes under high stress, accounting for the distinct effects of tensile and compressive traction on mechanical failure. It also incorporates an efficient implicit return mapping algorithm for numerical implementation and derives constitutive equations at both the micro-plane and macroscopic scales.

The model was validated by simulating stress–strain curves and anisotropic damage evolutions for rock under tensile, compressive, and shear stress paths. It showed good agreement with experimental results for sandstone under uniaxial tensile, triaxial compressive, and direct shear stress conditions.

Key findings and advantages of this approach include the accurate capture of multi-scale, anisotropic, and progressive failure behavior of rock, the prediction of nonlinear deformation, strength, and post-peak softening under different stress paths, and the identification of anisotropic, multi-scale failure mechanisms under tensile, compressive, and shear stresses. The model provides a clear physical interpretation of the traction vector in describing rock defect failures, offering a unified framework for both tensile and compressive-shear stress states, with few model parameters and an intuitive understanding of directional failure.

This traction-based microplane model provides an effective approach for describing and predicting rock's complex progressive failure characteristics across scales, showing promise for further development to simulate more complex stress paths and multi-field coupled conditions in rock mechanics applications.

Integration of Multi-Scale and Multi-Physics Approaches

Predicting rock failure often requires consideration of phenomena occurring at different scales and involving multiple physical processes. Integrating multi-scale and multi-physics approaches into predictive models can provide a more comprehensive understanding of rock failure mechanisms and improve prediction accuracy.

1. Multi-Scale Modeling

Rock masses exhibit varying behaviors across different scales, from microscopic defects to macroscopic discontinuities. Multi-scale modeling approaches aim to integrate these scales, offering a comprehensive view of rock failure processes. At the

7.3 Data-Driven Modelling

microscale, models consider the effects of grain boundaries, microcracks, and mineral composition on rock strength and deformation. Mesoscale modeling focuses on the influence of small-scale discontinuities, like joints and bedding planes, on rock mass behavior. At the macroscale, models address large geological structures and their impact on stability. Scale-bridging techniques, such as homogenization and hierarchical modeling, enable the transfer of information between these scales, enhancing the accuracy of rock failure predictions.

The traction-based microplane model presented by [43] is an example of a multi-scale approach that captures both microscopic and macroscopic aspects of rock failure. By using the traction vector to characterize directional failure modes, the model can represent anisotropic multi-scale failure mechanisms under various stress conditions.

2. Multi-Physics Coupling

Rock failure often involves the interaction of multiple physical processes, including mechanical deformation, fluid flow, heat transfer, and chemical reactions. Integrating these coupled processes into predictive models enhances their accuracy and relevance for complex real-world scenarios. Key multi-physics couplings in rock failure prediction include hydro-mechanical coupling, which considers the interaction between pore fluid pressure and mechanical deformation, crucial for phenomena like hydraulic fracturing and fluid-induced seismicity. Thermo-mechanical coupling accounts for the effects of temperature changes on rock strength and deformation, important in geothermal energy extraction and nuclear waste disposal. Chemo-mechanical coupling incorporates the influence of chemical reactions on rock strength and permeability, particularly significant in CO_2 sequestration and acid mine drainage. Electro-magnetic coupling, which involves the effects of electromagnetic fields on rock behavior, is relevant in geophysical exploration and electromagnetic stimulation of reservoirs.

Developing predictive models that incorporate these multi-physics couplings remains a challenging task, often requiring advanced numerical techniques and high-performance computing resources. However, such models can provide valuable insights into complex rock failure mechanisms and improve the accuracy of predictions in challenging engineering scenarios.

3. Data Assimilation and Model Updating

Integrating real-time monitoring data with predictive models enhances their accuracy and reliability by enabling continuous updates through data assimilation techniques. These methods allow model parameters and predictions to be refined as new observations are collected. Key aspects of data assimilation in rock failure prediction include the use of sensor networks, which deploy arrays of sensors to monitor rock mass behavior, such as deformation, stress changes, and microseismic activity. Inverse modeling uses observed data to infer rock mass properties and update model parameters. Machine learning algorithms can efficiently process large volumes of real-time monitoring data to update predictive models. Additionally, uncertainty quantification is crucial for assessing and communicating the uncertainties in model predictions, considering both aleatory (inherent randomness) and epistemic (knowledge-based) uncertainties.

Implementing effective data assimilation and model updating strategies can enhance the predictive capabilities of rock failure models and improve their applicability to real-world engineering problems.

Uncertainty Quantification and Model Validation

Uncertainty quantification and model validation are crucial aspects of developing reliable predictive models for rock failure. These processes help assess the accuracy and reliability of model predictions and provide guidance for model improvement and application.

1. Sources of Uncertainty

In rock failure prediction, uncertainties can arise from multiple sources. Data uncertainty refers to errors and variability in input parameters, such as rock properties and in-situ stress measurements. Model uncertainty stems from simplifications and assumptions in the mathematical formulation of the predictive model. Parameter uncertainty involves uncertainties in model parameters, which are often estimated or calibrated from limited data. Aleatory uncertainty reflects the inherent randomness in rock mass behavior and failure processes, while epistemic uncertainty arises from a lack of knowledge or understanding of certain aspects of rock failure mechanisms. Identifying and characterizing these sources of uncertainty is essential for developing robust predictive models and interpreting their results.

2. Uncertainty Quantification Techniques

Engineers and geologists use several advanced methods to quantify uncertainties in rock failure predictions. Sensitivity analysis helps identify key factors by systematically varying input parameters, such as determining that joint orientation influences slope stability more than rock mass strength in certain formations.

Monte Carlo simulation provides a probabilistic approach by running thousands of calculations with randomly sampled inputs, producing a distribution of potential outcomes. This allows engineers to understand the range of possible failure scenarios, like the probability of deformation values when assessing tunnel stability.

Bayesian inference refines predictions by updating initial estimates with new monitoring data. Instruments such as extensometers or inclinometers provide ongoing insights, which enhance prediction accuracy over time, particularly for long-term projects.

Fuzzy set theory addresses uncertainties that are difficult to quantify, like vague or qualitative descriptions of rock properties. For example, terms like "highly weathered" or "moderately jointed" are represented mathematically using fuzzy logic, enabling more precise analysis.

Interval analysis helps define the bounds of possible outcomes by propagating input value ranges through the model. This method helps engineers assess both worst-case and best-case scenarios, ensuring safety factors account for uncertainties in rock mass properties.

7.3 Data-Driven Modelling

3. Model Validation Approaches

Model validation in rock mechanics employs several complementary approaches to ensure prediction accuracy and reliability.

Cross-validation is a key method, where portions of the data are held back to test model performance. This helps confirm that the model isn't just memorizing patterns but is effectively learning underlying relationships in rock behavior. Advanced cross-validation schemes, such as multiple folds or stratified sampling, further account for the heterogeneous nature of rock formations.

Field validation offers the most convincing evidence, as it tests predictions against actual rock mass behavior in real-world conditions. This can include monitoring slope movements, tunnel convergence, or pillar deformation in mines. Field validation not only confirms model predictions but also identifies unexpected behaviors, which can lead to model refinement.

Blind predictions are rigorous tests where predictions are made without prior knowledge of outcomes, such as forecasting displacement patterns around a planned tunnel before construction. The comparison between predicted and measured results offers unbiased insights into model performance.

Multi-model comparison provides a deeper understanding of different modeling approaches. By comparing empirical, numerical, and hybrid models on the same rock failure problem, it becomes clear which methods excel under specific conditions. Some models may capture progressive failure better, while others handle brittle behavior more effectively.

Benchmarking against known solutions provides a solid foundation for validation, testing models against cases with analytical solutions to ensure consistency and reliability.

4. Performance Metrics

Various performance metrics are used to assess the accuracy and reliability of predictive models for rock failure.

The Coefficient of Determination (R^2) measures how much of the variance in the dependent variable is explained by the independent variables. For example, rock slope failure models developed by [43] achieved R^2 values of 0.86 for their best-performing models.

Root Mean Square Error (RMSE) quantifies the average deviation between predicted and observed values. For instance, Lei et al.'s hybrid model for predicting shear strength in discontinuous rock materials achieved an RMSE of 31.081 MPa on the test set.

Mean Absolute Percentage Error (MAPE) offers a percentage-based measure of prediction accuracy. The Gene Expression Programming model for rock slope failure prediction [43] achieved a MAPE of 0.004.

The Area Under the Receiver Operating Characteristic Curve (AUC-ROC) is particularly useful for classification models, especially for binary outcomes like failure/no-failure predictions.

A confusion matrix gives a detailed breakdown of correct and incorrect predictions for classification problems, while the F1 score combines precision and recall, making it valuable for models with imbalanced datasets.

Selecting appropriate performance metrics depends on the specific prediction task and the nature of the available data. Using multiple complementary metrics can provide a more comprehensive assessment of model performance.

Applications in Rock Engineering and Geohazard Assessment

Predictive models for rock failure have numerous applications in rock engineering and geohazard assessment. These models play a crucial role in ensuring the safety and efficiency of various engineering projects and in mitigating natural hazards associated with rock masses. Some key applications include:

1. Tunnel and Underground Excavation Design

Predicting brittle rock failure is crucial in underground engineering, as stress concentrations around new openings can lead to spalling, slabbing, or more severe failure mechanisms. These predictions influence key aspects of the project, including initial support design and long-term monitoring strategies. For example, in deep mining, understanding the depth and extent of the damaged zone helps determine the necessary shotcrete thickness and the length of rock bolts required for stable anchoring.

Accurate predictive models are essential for support system design, as they account for both immediate and long-term loading conditions. Engineers use these models to optimize bolt spacing, shotcrete thickness, and the timing of support installation. The models must also consider dynamic conditions, such as seismic events, and how different support systems interact. For instance, understanding the synergy between bolting and steel sets is critical for creating a robust support system.

Predictive modeling also plays a vital role in optimizing excavation sequences, particularly in complex projects with multiple openings. By analyzing how the stress field redistributes during tunneling, engineers can better plan face advances and support installation, ensuring stability throughout the excavation process.

Time-dependent behaviors, such as creep, swelling, or deterioration, add further complexity. In conditions like squeezing ground or areas with significant water interaction, rock masses may undergo long-term deformation. Predictive models help engineers anticipate these behaviors, enabling the design of yielding support systems or the planning of periodic reinforcement.

Together, these predictive elements provide a comprehensive understanding of underground behavior. By integrating time-dependent deformation with excavation sequence planning, engineers can ensure both short-term construction safety and long-term operational stability of underground structures.

2. Rock Slope Stability Analysis

Stability assessment of natural and engineered slopes requires advanced predictive modeling to account for complex geological factors. Engineers evaluate discontinuity patterns, groundwater conditions, and rock mass strength to identify potential failure

7.3 Data-Driven Modelling

mechanisms and calculate factors of safety under various scenarios. For example, in highway cut slopes, predictive models can uncover hidden risks, such as wedge failures or toppling, that may not be apparent from visual inspection alone.

Predictive models also enhance the design of reinforcement systems. By evaluating different support measures like rock bolts, anchors, and drainage systems, engineers can determine optimal configurations, such as anchor lengths, bolt patterns, and capacities. These models can, for instance, show how a combination of horizontal drains and tensioned anchors might improve slope stability more effectively than either method alone.

With climate change impacting rainfall patterns and development expanding into more challenging terrain, analyzing external factors has become increasingly critical. Predictive models help engineers assess how intense rainfall or seismic shaking might influence slope stability. These models simulate water pressure buildup in discontinuities during storms and how dynamic loads from earthquakes might affect different parts of a slope.

Risk assessment for nearby infrastructure also relies on predictive modeling. Engineers use these models to estimate failure volumes, runout distances, and impact energies, which are crucial for designing protective measures and establishing safety zones. For example, when evaluating the risk to a railway line beneath a rock slope, models can predict not only the likelihood of failure but also the potential debris reach, guiding decisions on monitoring systems and emergency response plans.

3. Mining Engineering

In mining engineering, predictive models for rock failure are crucial for various applications. They are used to design stable underground mine openings and pillars, ensuring safety while minimizing material costs. These models also assess the risk of rockbursts and other dynamic failure events, helping engineers anticipate and mitigate sudden failures. Additionally, predictive models optimize open-pit slope angles, balancing safety concerns with economic considerations. Finally, they evaluate the stability of tailings dams and waste rock piles, ensuring long-term structural integrity and preventing environmental hazards.

4. Petroleum and Gas Engineering

In petroleum and gas engineering, predictive models are essential for several key applications. They assess wellbore stability during drilling operations to prevent collapses and optimize drilling efficiency. These models also predict sand production in weakly consolidated formations, helping to mitigate potential reservoir damage. Additionally, they evaluate caprock integrity for underground gas storage and CO_2 sequestration, ensuring long-term containment. Predictive models further optimize hydraulic fracturing operations by forecasting fracture propagation and its interaction with natural fractures, improving extraction efficiency and minimizing environmental risks.

5. Geohazard Assessment and Mitigation

Predictive models are essential in assessing and mitigating geohazards. They help identify potential landslide and rockfall failure mechanisms, evaluating risks to infrastructure and communities. In earthquake-prone regions, these models predict the response of rock masses to seismic loading, including liquefaction potential. For subsidence, they estimate ground surface deformation resulting from underground excavations or fluid extraction. Additionally, predictive models assess the stability of volcanic edifices and forecast potential collapse scenarios, providing crucial information for disaster preparedness and mitigation.

6. Nuclear Waste Disposal

In nuclear waste disposal, predictive models for rock failure are critical for ensuring the long-term safety and effectiveness of underground repositories. These models help assess the stability of repository sites over extended periods, evaluate the potential migration of radionuclides through fractured rock masses, and design effective engineered barriers and sealing systems. Additionally, they predict how heat generation from radioactive decay might influence rock mass behavior, ensuring that repository conditions remain stable and secure over time.

7. Geothermal Energy Development

In geothermal energy development, predictive models for rock failure are essential for ensuring the stability and efficiency of operations. These models help assess the stability of geothermal wells and reservoirs, predict how fracture networks will evolve during reservoir stimulation, and evaluate the risk of induced seismicity associated with geothermal activities. They also assist in optimizing well placement and reservoir management strategies to enhance resource extraction while minimizing environmental impacts.

8. Hydroelectric Projects

In hydroelectric project design and operation, predictive models are crucial for ensuring the stability and safety of infrastructure. These models are used to assess the stability of dam foundations and abutments, evaluate the potential for reservoir-induced seismicity, and design underground powerhouses and water conveyance tunnels. Additionally, they predict the long-term behavior of rock masses subjected to cyclic loading, such as the effects of fluctuating reservoir levels, which is essential for maintaining the integrity of structures over time.

By providing more accurate and reliable predictions of rock failure behavior, advanced predictive models enable engineers and geoscientists to make better-informed decisions, optimize designs, and improve the safety and efficiency of various rock engineering projects. As these models continue to evolve and incorporate more sophisticated data-driven and multi-physics approaches, their applicability and accuracy in addressing complex real-world problems are expected to further improve.

7.4 Big Data Visualization and Mining

7.4.1 Visualization Techniques for Stress Distribution

In recent years, the field of rock dynamics has seen significant advancements in numerical simulation techniques, particularly in the area of big data visualization and mining. One crucial aspect of this progress is the development and refinement of visualization techniques for stress distribution. These techniques play a vital role in understanding the complex behavior of rock materials under various loading conditions and are essential for both research and practical applications in rock engineering. This section provides a comprehensive review of the state-of-the-art visualization techniques for stress distribution in rock dynamics, focusing on their principles, applications, and recent developments.

Mechanoluminescence Imaging for Stress Visualization

One of the most innovative and promising techniques for visualizing stress distribution in materials is mechanoluminescence (ML) imaging. This method leverages the phenomenon of light emission from certain materials when subjected to mechanical stress, allowing for real-time, full-field visualization of stress patterns [47].

1. Principle and Material Development

ML imaging effectiveness relies on advanced ML materials, with recent research focusing on intensity and sensitivity enhancement. A key advancement has been the development of Ho co-doped $SrAl_2O_4$:Eu (SAO) [47], which achieved an order of magnitude increase in ML intensity compared to conventional SAO materials, significantly improving the technique's sensitivity and resolution.

ML sensing film preparation involves two key stages: ML powder synthesis and film application. The synthesis combines $SrCO_3$, Al_2O_3, $Eu(NO_3)_3 \cdot 2H_2O$, and Ho_2O_3, calcined at 1300 °C under reducing conditions. The film is then prepared using paste and screen-printing methods to create an adherent coating on metal substrates.

2. Experimental Setup and Methodology

The ML imaging setup consists of three key components: a material testing machine (MTS 810) for precise load application, a high-speed camera for ML image capture, and a computer system for data acquisition and processing (Fig. 7.5).

The experimental protocol comprises three sequential phases: specimen preparation (coating metal plates with ML sensing film), controlled mechanical loading

Fig. 7.5 ML image system [47]

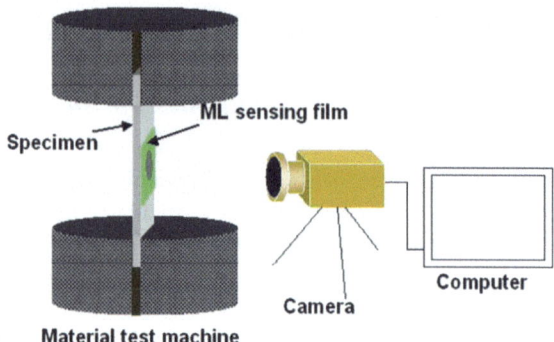

(tensile, compressive, or cyclic), and data analysis of captured ML images to determine stress distribution.

3. Calibration and Quantitative Analysis

To obtain quantitative stress information from ML images, a calibration process is necessary. This typically involves establishing a relationship between ML intensity and applied stress. Research has shown that for many ML materials, there is a linear relationship between ML intensity (I_{ML}) and stress (σ) within a certain range [47]:

$$I_{ML} = k * \sigma + c \tag{7.88}$$

where k is the slope and c is the intercept of the calibration curve.

This linear relationship allows for the conversion of ML intensity values to stress values, enabling quantitative analysis of stress distribution.

4. Applications and Advantages

ML imaging enables real-time, full-field stress visualization with significant advantages over conventional methods. Its non-contact nature prevents measurement interference, while high material sensitivity allows detection of microscopic stress variations.

The technique excels in visualizing stress concentrations around geometric discontinuities, as demonstrated in studies of circular holes in metal plates under tensile loading [47]. The stress concentration factor (K_t) quantifies this effect:

$$K_t = \frac{\sigma_{max}}{\sigma_{nom}} \tag{7.89}$$

where σ_{max} is the maximum stress at the hole edge and σ_{nom} is the nominal stress away from the hole.

5. Limitations and Future Directions

7.4 Big Data Visualization and Mining

ML imaging faces three key limitations: temperature sensitivity affecting measurement accuracy, material specificity (primarily limited to metal substrates), and restriction to surface stress visualization.

Current research addresses these challenges through: development of temperature-stable ML materials with broader substrate compatibility, integration with digital image correlation for comprehensive analysis, and creation of ML materials for extreme conditions (high-temperature and high-pressure) in rock engineering applications.

Organic Mechanoresponsive Luminogens for Stress/Strain Visualization

Another promising approach for stress visualization in materials is the use of organic mechanoresponsive luminogens with aggregation-induced emission (AIE) characteristics. This technique offers unique advantages in terms of sensitivity and versatility, particularly for visualizing stress/strain distribution and fatigue crack propagation in metal specimens [48].

1. Principle and Material Development

A key advancement in stress visualization is the development of 1,1,2,2-tetrakis(4-nitrophenyl)ethane (TPE-4N), an AIE luminogen [48] combining ultra-sensitive mechanochromic response with superior film-forming properties. TPE-4N exhibits facile crystallization upon heating and demonstrates a 130-fold increase in photoluminescence (PL) intensity in its aggregated versus isolated state (Fig. 7.6).

The stress visualization operates through state transformation: mechanical stress converts the non-fluorescent crystalline state to a fluorescent amorphous state, enabling quantifiable "turn-on" fluorescence response.

2. Experimental Setup and Methodology

The stress visualization setup using organic luminogens comprises three core components: specimen preparation (dip-coating with TPE-4N), imaging system (UV light source, CCD camera, computer), and loading apparatus (in situ fatigue testing machine).

The protocol follows three phases: application of uniform crystalline luminogen film, controlled mechanical loading (tensile, compressive, or cyclic), and UV-illuminated fluorescence imaging for stress–strain distribution analysis.

3. Quantitative Analysis

Stress–strain distribution analysis using organic luminogens relies on two key metrics. The Grayscale Enhancement Ratio is defined as (Fig. 7.7).

where G is the current grayscale value and G_0 is the initial value. The Fluorescence Intensity Ratio is expressed as:

$$I_a mor / I_c ryst$$

where $I_a mor$ and $I_c ryst$ represent photoluminescence intensities in amorphous and crystalline states, respectively. These metrics quantify stress/strain-induced changes and correlate fluorescence patterns with mechanical behavior.

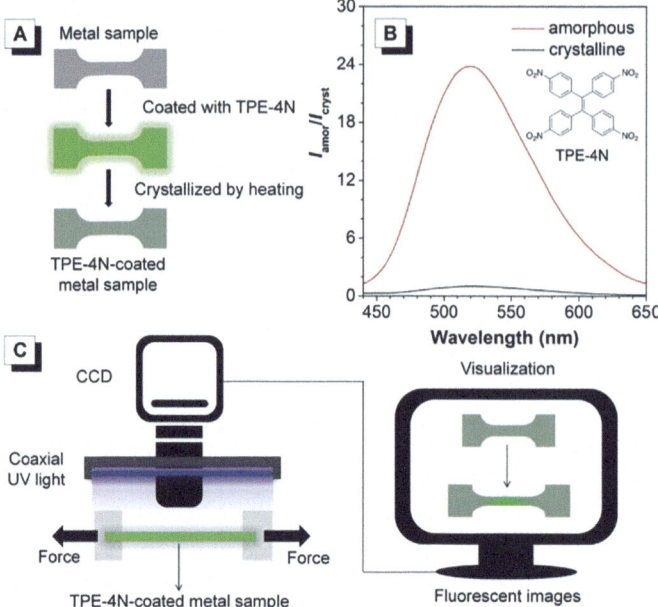

Fig. 7.6 a Illustration of sample preparation of TPE-4N-coated metal specimen. **b** PL spectra of TPE-4N film on SUS316L austenitic stainless steel in amorphous and crystalline states. Iamor and Icryst are the PL intensity of TPE-4N in amorphous and crystalline states, respectively. Inset: molecular structure of TPE-4N. **c** Illustration of the experiment setup [48]

4. Applications and Advantages

Organic mechanoresponsive luminogens provide advanced stress visualization in materials testing through their ultra-sensitive mechanical response and real-time, full-field visualization capabilities. The technique offers simplified setup and operation, with fluorescence patterns often visible to the naked eye. Its proven effectiveness across steel and aluminum alloy substrates demonstrates broad applicability.

In specialized applications, particularly crack monitoring, the technology enables predictive analysis of crack propagation in cyclically loaded metal specimens through accumulated fluorescence patterns, revealing potential failure paths before visible damage occurs [48].

5. Limitations and Future Directions

Current limitations of organic mechanoresponsive luminogens include material specificity (primarily developed for metals rather than rocks), challenges in quantitative stress measurement, and environmental sensitivity to temperature and humidity. Research priorities focus on developing luminogens with enhanced sensitivity, stability, and broader material compatibility, while integrating complementary stress analysis methods and exploring applications in complex loading scenarios relevant to rock engineering.

7.4 Big Data Visualization and Mining

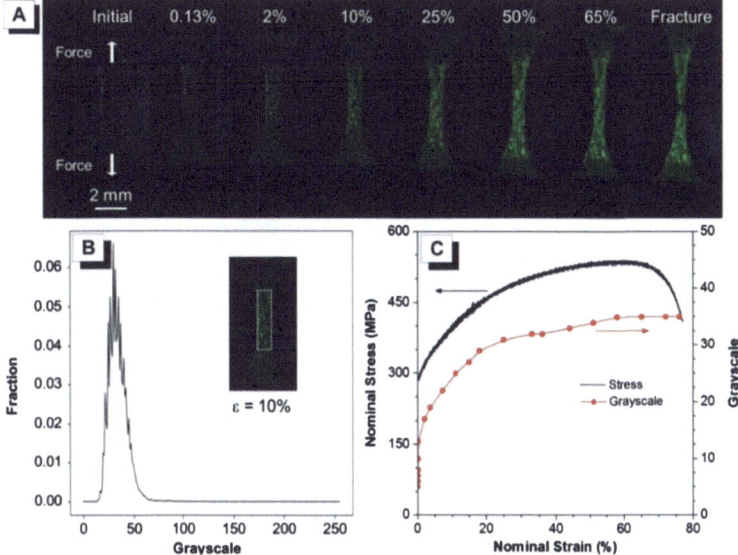

Fig. 7.7 **a** Fluorescence images of TPE-4N-coated steel tensile specimen at different strains (ε, %). Direction of stretching force: vertical. **b** Grayscale distribution of the selected area at ε = 10%. Inset: fluorescence image of TPE-4N-coated steel tensile specimen at ε = 10% and the selected area for grayscale analysis. **c** Plots of strain against stress and grayscale of the TPE-4N-coated steel tensile specimen [48]

X-ray CT Imaging Combined with 3D Printing and Photoelasticity

A novel and powerful approach to visualizing stress distribution in heterogeneous materials, such as rocks, combines X-ray computed tomography (CT) imaging, 3D printing, and photoelastic testing techniques [49, 50]. This integrated method allows for unprecedented visualization and analysis of internal stress distributions in complex geomaterials, validating and complementing numerical modeling approaches.

1. Principle and Methodology

The internal stress visualization process in rock samples integrates CT imaging, 3D printing, and experimental mechanics. High-resolution micro-focus X-ray CT scanning generates a digital 3D model of the rock's mesostructure, mapping particles, matrix, and void distributions. The model is then 3D printed using transparent photosensitive polymer (VeroClear) for matrix and opaque material (RGD525) for particles, maintaining void spaces. Photoelastic testing under uniaxial compression reveals stress distribution patterns, supplemented by FEM stress field simulations and DEM fracture analysis.

2. Quantitative Analysis

While primarily qualitative, the technique enables quantitative analysis through key metrics. The Stress Concentration Factor (SCF) is one of these factors.

The photoelastic fringe order relationship:

$$N = (\sigma_1 - \sigma_2) * \frac{t}{f_\sigma} \tag{7.90}$$

where N is fringe order, σ_1 and σ_2 are principal stresses, t is model thickness, and $f\sigma$ is the stress-optic coefficient, can be rearranged to determine stress differences:

$$\sigma_1 - \sigma_2 = N * \frac{f_\sigma}{t} \tag{7.91}$$

This relationship allows for quantitative estimation of stress differences from observed fringe patterns.

3. Applications and Advantages

The integrated CT-3D printing-photoelastic methodology enables comprehensive stress visualization in heterogeneous materials through physical representation of internal structures and full-field photoelastic analysis. The technique reveals stress concentrations around structural features and provides insights into deformation processes and failure mechanisms, while validating numerical simulations.

Studies [49, 50] have demonstrated that stress concentrations develop predominantly around voids and irregular particles, with higher gradients near void spaces. Particle distribution patterns significantly influence stress field configuration and failure patterns. Three-dimensional models show reduced stress concentrations compared to two-dimensional versions due to lateral inertial confinement effects.

4. Limitations and Future Directions

The integrated CT-3D printing-photoelastic approach faces three primary challenges: material property replication between printed models and real rocks, quantitative analysis of three-dimensional stress fields, and scale limitations of current 3D printing technologies affecting feature resolution.

Research priorities focus on developing advanced 3D printable materials to better replicate rock properties, enhancing quantitative analysis techniques for 3D photoelastic stress fields, and integrating complementary imaging and analysis methods. Efforts also aim to expand applications across diverse geomaterials and loading conditions.

Color-Based Visualization Techniques

Color-based visualization techniques play a crucial role in representing stress distributions in rock dynamics simulations. These methods leverage the human visual system's ability to quickly perceive and interpret color variations, making them particularly effective for conveying complex stress patterns. Several color-based approaches have been developed and refined for stress visualization in rock dynamics (Fig. 7.8).

7.4 Big Data Visualization and Mining

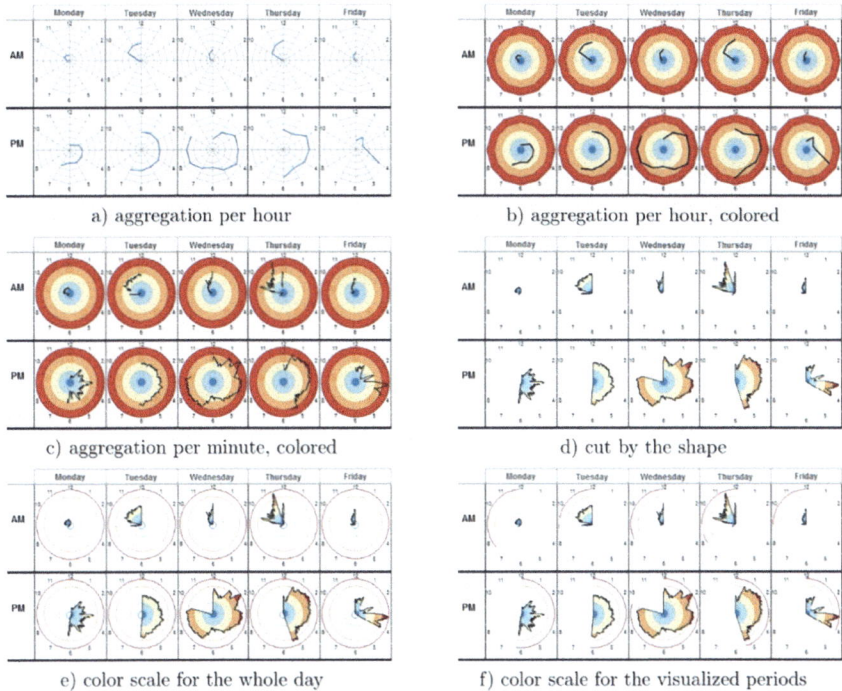

Fig. 7.8 Star visualization with the stress value aggregated per minute [49]

1. Principles of Color-Based Stress Visualization

Color-based stress visualization maps continuous stress fields to color scales through systematic relationships. Key design elements include color space selection and scale design, which determine the effectiveness of stress data translation into visual information.

Color space options include RGB (aligned with cone cell response), HSV (matching human perception), and perceptually uniform spaces (providing linear relationships between values and perceived differences). Color scales serve distinct purposes: sequential scales for ordered progression, diverging scales for emphasizing critical midpoints, and qualitative scales for categorical stress states.

Color transfer functions mathematically link stress values to colors through linear, logarithmic, or custom mappings, enabling emphasis on specific stress ranges.

2. Techniques and Applications

Color-based stress visualization in rock dynamics employs several specialized techniques. Color bar visualization with diverging schemes (red-yellow-blue) effectively represents temporal stress changes [51], while 3D surface color mapping enables stress visualization on complex geological structures [52].

Volume rendering with color transfer functions allows analysis of three-dimensional stress fields and selective highlighting within rock masses. Hybrid contour plots with color gradients provide clear visualization of stress boundaries and gradients in both 2D and 3D applications [52].

3. Quantitative Analysis

Color-based visualization techniques integrate quantitative elements through numerical color scale labels and mathematical analysis methods. The stress gradient calculation based on color variations is expressed as:

$$\nabla \sigma \approx \frac{\Delta C}{\Delta x} \tag{7.92}$$

where $\nabla \sigma$ is the stress gradient, ΔC is the change in color value, and Δx is the spatial distance.

Histogram equalization optimizes stress variation visualization through:

$$h_{eq}(v) = floor\left(\frac{(cdf(v) - cdf_{min})}{(N - cdf_{min}) * (L - 1)}\right) \tag{7.93}$$

where cdf is the cumulative distribution function, N is pixel count, and L is possible intensity levels.

4. Advantages and Considerations

Color-based stress visualization provides intuitive interpretation of complex patterns with high information density, flexible emphasis capabilities, and natural compatibility with other visualization methods. However, effectiveness varies with individual color perception, cultural interpretations, and color vision deficiencies.

Best practices include using color scales robust against vision deficiencies, incorporating redundant visual elements (patterns/textures), and offering customizable schemes to accommodate individual needs.

5. Future Directions

Current research in color-based stress visualization focuses on developing perceptually optimized color scales for rock stress patterns, implementing machine learning for adaptive color mapping, and exploring novel color spaces. Research extends to virtual/augmented reality applications while integrating complementary visualization methods like glyphs and streamlines.

Vector and Tensor Field Visualization Techniques

In rock dynamics, stress states are often characterized by complex vector and tensor fields. Visualizing these multidimensional data presents unique challenges and has led to the development of specialized techniques. These methods aim to effectively represent both the magnitude and directional aspects of stress fields in rocks.

1. Vector Field Visualization

7.4 Big Data Visualization and Mining

Vector field visualization in rock dynamics employs three main approaches for representing displacement, velocity, and principal stress directions:

Direct methods use arrow plots (vector magnitude/direction with color coding) and hedgehog plots (line segments for dense 3D fields).

Indirect methods employ streamlines (tangent paths showing flow-like behavior), pathlines (particle tracking in dynamic processes), and streaklines (point loci visualization in time-varying fields).

Feature-based methods identify critical points (sources, sinks, saddles) and adapt vortex core extraction for rotational stress components.

2. Tensor Field Visualization

Stress tensor visualization in rock dynamics employs specialized methods to address higher dimensionality challenges. Glyph-based approaches use ellipsoid and superquadric shapes aligned with eigenvectors and scaled by eigenvalues to represent stress anisotropy. Hyperstreamlines visualize continuous tensor variations by following one eigenvector while representing two others in cross-section.

Tensor field topology analysis identifies degenerate points and separatrices, while color-coding methods include RGB mapping of eigenvalues and barycentric schemes for tensor shape visualization. These techniques extend to fabric tensor analysis for visualizing crack and grain boundary orientations.

3. Quantitative Analysis

Vector and tensor fields in rock mechanics enable quantitative analysis through specific measures. For vector fields, key quantities include divergence:

$$\nabla v = \frac{\partial vx}{\partial x} + \frac{\partial vy}{\partial y} + \frac{\partial v_z}{\partial z} \tag{7.94}$$

And Curl:

$$\nabla \times v = \left(\frac{\partial vz}{\partial y} - \frac{\partial vy}{\partial z}, \frac{\partial vx}{\partial z} - \frac{\partial vz}{\partial x}, \frac{\partial vy}{\partial x} - \frac{\partial vx}{\partial y} \right) \tag{7.95}$$

Tensor analysis employs three invariants: first invariant (trace):

$$J_1 = \sigma_{11} + \sigma_{22} + \sigma_{33} \tag{7.96}$$

Second invariant:

$$J_2 = \sigma_{11}\sigma_{22} + \sigma_{22}\sigma_{33} + \sigma_{33}\sigma_{11} - \sigma_{12}^2 - \sigma_{23}^2 - \sigma_{31}^2 \tag{7.97}$$

And third invariant (determinant):

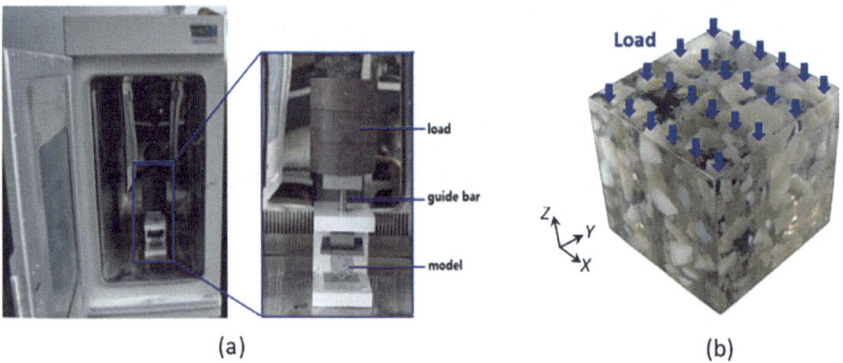

Fig. 7.9 Device and loading scheme for the frozen stress test: **a** equipment and loading device, **b** loading and boundary conditions [50]

$$J_3 = det(\sigma ij) \tag{7.98}$$

These invariants provide scalar measures that are independent of coordinate system orientation.

The Lode angle characterizes stress tensor shape

$$\theta = \left(\frac{1}{3}\right) arcsin\left(\frac{\left(3\sqrt{3}J_3\right)}{\left(2J_2^{\frac{3}{2}}\right)}\right) \tag{7.99}$$

where J_2 and J_3 are the second and third invariants of the deviatoric stress tensor.

This measure can be used to quantify the shape of the stress tensor (e.g., axisymmetric compression, pure shear, axisymmetric extension) (Figs. 7.9 and 7.10).

4. Advantages and Considerations

Vector and tensor field visualization techniques effectively represent stress distributions in rock dynamics through their ability to capture directional stress fields and provide multi-scale representation via streamlines, enabling simultaneous visualization of local and global characteristics. However, these methods face challenges including visual complexity from dense fields, difficulties in representing 3D tensor fields on 2D displays, and high computational resource requirements.

Best practices to address these limitations include implementing interactive visualization systems for dynamic data exploration, integrating multiple visualization techniques, and applying feature extraction methods to highlight key stress field aspects while reducing complexity.

5. Future Directions

Recent advances in vector and tensor field visualization for rock dynamics focus on developing intuitive tensor glyphs, implementing machine learning for automated

7.4 Big Data Visualization and Mining

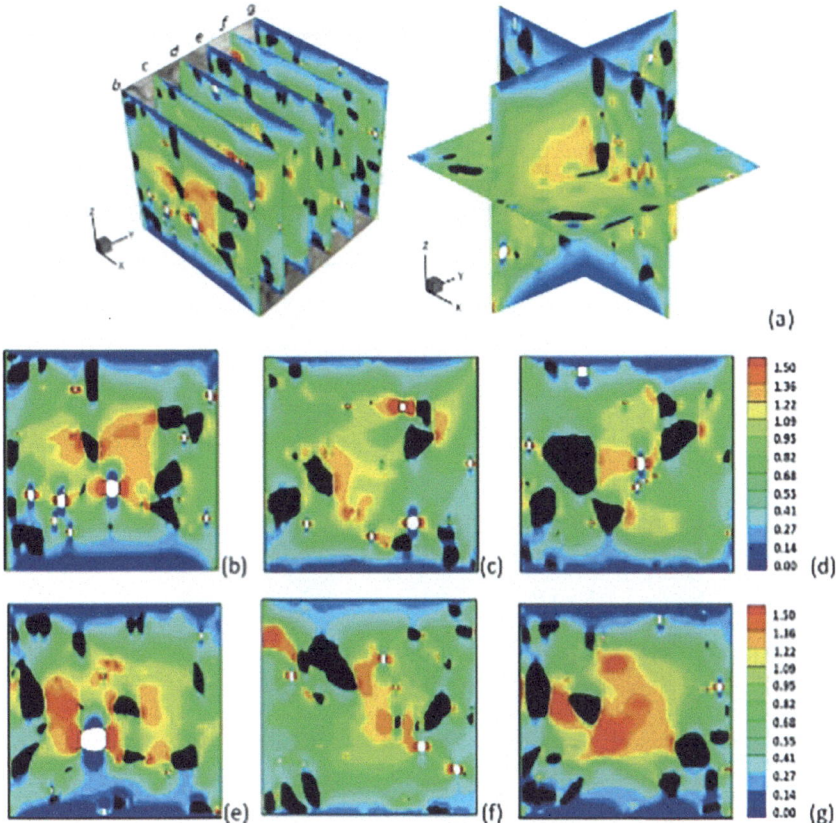

Fig. 7.10 Numerical results of the principal stress difference (r1 r2) distribution (unit MPa) in slices taken from different positions in printed concrete models under uniaxial compression using the frozen stress technique and 3D photoelastic method. **a** Spatial location of photoelastic slices taken at half-slice coordinates of: **b** y = 5 mm, **c** y = 11 mm, **d** y = 17 mm, **e** y = 23 mm, **f** y = 29 mm, **g** y = 35 mm [50]

feature detection, and utilizing VR/AR for immersive 3D field visualization. Progress includes multi-field visualization techniques integrating stress, strain, and other rock properties, alongside specialized interactive tensor field exploration tools for rock mechanics.

Integration of Visualization Techniques with Big Data and Machine Learning

As the field of rock dynamics continues to advance, the integration of visualization techniques with big data analytics and machine learning is becoming increasingly important. This integration allows for more comprehensive analysis of complex stress distributions and enables new insights into rock behavior under dynamic loading conditions (Figs. 7.11 and 7.12).

Fig. 7.11
Three-dimensional visualization of the same problem

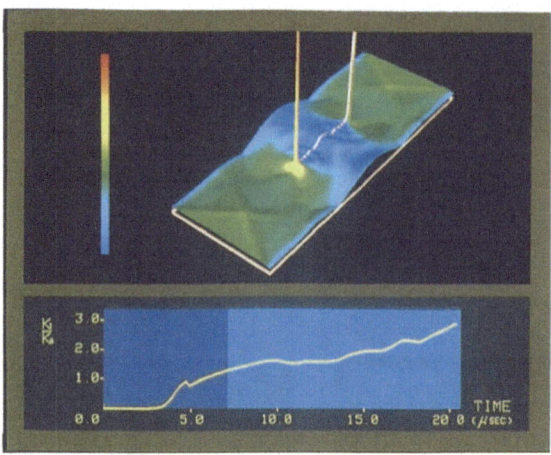

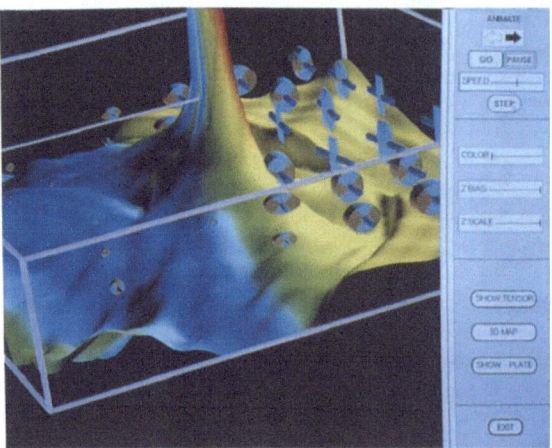

Fig. 7.12 Interactive visualization with tensor glyphs (1989). This system uses parallel processing to compute the visualization mappings in real time during the animation. The user can interactively adjust the color transfer function applied to the strain energy density field and two parameters ("Z-BIAS" and "Z-SCALE") that determine the nonlinear mapping of the kinetic energy density to the height field. Tensor glyphs (a = 2.0) indicate the three-dimensional state of stress at selected locations [52]

Here, the square root of the kinetic energy density defines the height field and color indicates the log of the strain energy density. This idiom clearly reveals the elastic wave patterns. In this image, the longitudinal waves from the step function loading applied on the long edges of the structure have reached the opposite sides of the plate. The diagonal wave patterns result from the interaction of these waves with the free

7.4 Big Data Visualization and Mining

edge conditions on the short sides of the plate. The circular waves surrounding each crack tip are due to the initial scattering of the longitudinal waves and to the motion of the crack tips. The lower graph shows the history of the normalized dynamic stress intensity factor [50].

1. Big Data Approaches in Stress Visualization

Big data visualization challenges in rock dynamics span three domains: data management, real-time streaming visualization, and multi-scale representation. Efficient data management employs optimized storage strategies, including raw data archival in files, processed data in databases for rapid access, and user/device-based organization for scalability.

Real-time visualization of in-situ monitoring data requires adaptive sampling and progressive rendering techniques, while multi-scale visualization enables stress field analysis from microscopic rock fabric to macroscopic mass behavior through hierarchical data structures and level-of-detail rendering.

2. Machine Learning in Stress Visualization

Machine learning enhances stress field visualization in rock dynamics through automated feature detection using CNNs for identifying stress concentrations and fracture initiation points. Dimensionality reduction via PCA and t-SNE improves complex stress tensor visualization, while predictive models analyze patterns to visualize future stress states for real-time monitoring systems.

Adaptive rendering algorithms optimize visualization parameters based on stress field characteristics and user interactions. Unsupervised learning techniques, including autoencoders and isolation forests, enable efficient anomaly detection in large-scale rock dynamics simulations.

3. Integrated Visualization Frameworks

Integrated visualization frameworks for rock dynamics merge big data and machine learning through three advances: interactive visual analytics systems combining data management with ML algorithms, knowledge-based approaches incorporating domain expertise with adaptive rules, and collaborative platforms enabling multi-user stress field analysis with ML-supported insight sharing.

4. Quantitative Analysis in Integrated Approaches

Quantitative stress field analysis integrates big data and ML through complementary methods. Statistical analysis employs mean stress

$$\mu = \left(\frac{1}{N}\right) \sum \sigma_i \tag{7.100}$$

and variance

$$\sigma^2 = \left(\frac{1}{N}\right) \sum (\sigma_i - \mu)^2 \tag{7.101}$$

across large datasets to characterize global field properties. Correlation analysis uses Pearson coefficient

$$r = \frac{cov(X, Y)}{(\sigma X * \sigma Y)} \qquad (7.102)$$

(where X and Y are stress and another rock property).

to relate stress patterns to rock properties, while k-means clustering (min $\sum\sum \|x_i - \mu_j\|^2$) identifies pattern groups. Predictive modeling follows:

$$y = f(X) + \varepsilon \qquad (7.103)$$

to forecast stress field evolution

5. Advantages and Considerations

The integration of big data and ML with visualization techniques advances rock mechanics analysis by enabling massive dataset visualization, complex pattern identification, and stress field evolution prediction with adaptive, context aware visualization capabilities.

Implementation challenges include significant computational demands, data quality dependencies, model interpretability (especially for deep neural networks), and data security concerns. Best practices address these through robust data validation pipelines, interpretable ML models for rock mechanics, secure data management protocols, and clear documentation of ML-based visualizations.

7.4.2 Data Mining in Rock Dynamics

Data mining has revolutionized rock dynamics research by enabling advanced analysis of complex geological processes and improving engineering practices. The technique extracts meaningful patterns from large datasets generated through monitoring systems, laboratory experiments, and field observations to predict behaviors and optimize rock engineering decisions. The proliferation of big data in geotechnical engineering has made data mining essential.

Rock burst prediction represents a crucial application of data mining. Traditional single-index evaluations of these sudden, violent rock mass failures proved insufficient for underground mining and tunneling safety. Modern data mining approaches integrate multiple parameters for improved prediction accuracy.

Zhu et al. [53] demonstrated this advancement through their Intelligent Early Warning System for Rock Bursts (IEWSRB), which combines real-time monitoring with intelligent analysis. The system's core clustering algorithm, based on improved K-means methodology, classifies monitoring data into three categories: no rock burst, impending rock burst, and rock burst occurred (Fig. 7.13).

The IEWSRB employs the following key formulas for rock burst risk assessment:

7.4 Big Data Visualization and Mining

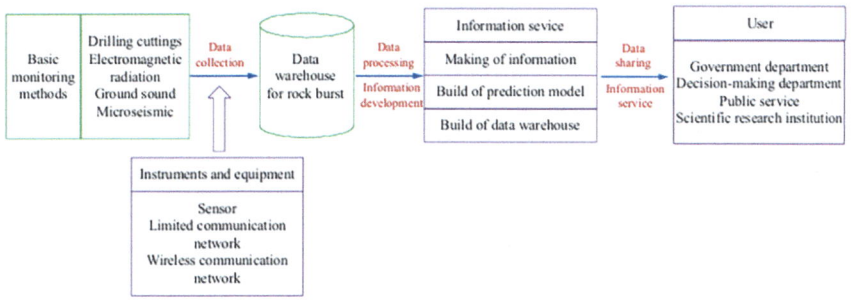

Fig. 7.13 Overview structure of the intelligent early warning system for rock bursts (IEWSRB) [53]

Critical energy density calculation:

$$E_{p\min} = \frac{1}{2}\rho V^2 \qquad (7.104)$$

$$E_{f\min} = \frac{\sigma_c^2}{2E} \qquad (7.105)$$

$$E_{\min} = E_{p\min} + E_{f\min} \qquad (7.106)$$

where

$E_{p\min}$ = Minimum kinetic energy
$E_{f\min}$ = Minimum energy to destroy coal-rock mass
$E_{\min}$ = Critical energy value for rock burst
ρ = Rock density
V = P-wave velocity
σ_c = Uniaxial compressive strength
E = Young's modulus

These formulas provide a quantitative basis for assessing the energy state of the rock mass and its potential for violent failure [53].

Data mining also enhances coal and gas outburst prediction in mining safety. The multivariate information coupling approach combines statistical process control (SPC) with logistic regression to improve prediction accuracy beyond single-index methods. The outburst risk probability model is:

$$p = \frac{1}{1 + \exp\left[-\left(\sum b_i x_i\right)\right]} = \frac{1}{1 + \exp(-z)} \qquad (7.107)$$

where p represents outburst probability, b_i are regression coefficients, and x_i denote input variables including drilling cuttings quantity, initial gas emission velocity, and dynamic phenomena.

The performance of this multivariate approach is evaluated using three key metrics:

- Prediction outburst rate

$$\eta_1 = \frac{100 n_t}{N} \qquad (7.108)$$

- Prediction outburst accuracy rate

$$\eta_2 = \frac{100 n_A}{n_t} \qquad (7.109)$$

- Prediction non-outburst accuracy rate

$$\eta_3 = \frac{100 n_B}{n_C} \qquad (7.110)$$

where n_t is the number of predicted outburst hazards, N is the total number of predictions, n_A is the number of actual outbursts among predicted outbursts, n_B is the number of predicted non-outbursts, and n_C is the number of actual non-outbursts among predicted non-outbursts.

Field studies have validated these data mining techniques, demonstrating significant improvements over traditional single-index methods. The multivariate approach achieved 95% accuracy in outburst prediction and 99% for non-outburst events [55]. Additionally, Artificial Neural Networks (ANNs) have proven effective in predicting rock strength properties from dynamic parameters obtained through in-situ tests, enabling non-destructive property estimation [56].

The ANN models developed for various rock strength parameters showed high correlations and low root mean square (RMS) errors.

These results demonstrate the potential of ANNs to accurately predict rock strength properties from easily obtainable dynamic parameters.

Data mining applications in rock dynamics encompass classification and characterization of rock masses, with decision trees enabling landslide hazard zoning and Support Vector Machines (SVMs) improving slope displacement prediction. These methods excel at capturing non-linear relationships in geological data.

However, data mining in rock dynamics faces notable challenges. Many techniques require extensive datasets for training and validation, which are often limited in rock engineering applications. Furthermore, geological system heterogeneity complicates model generalization across different sites.

Data mining results interpretation requires careful integration of statistical analysis with geological expertise. Effective implementation demands collaboration between data scientists and rock mechanics specialists to ensure both statistical validity and geological relevance.

7.4 Big Data Visualization and Mining

The field's future development focuses on integrating multiple data sources, combining GIS and remote sensing for comprehensive geological analysis, and developing real-time monitoring capabilities. Advanced artificial intelligence techniques, particularly deep learning, show promise in analyzing complex, high-dimensional datasets to reveal subtle patterns beyond traditional methods.

Data mining has transformed rock dynamics by enabling complex geological analysis, hazard prediction, and engineering optimization. Despite existing challenges, continued technological advancement expands its capabilities. Future progress depends on integrating data mining with emerging technologies and domain expertise to enhance rock engineering safety and efficiency [53–59].

7.4.3 Real-Time Simulation and Visualization

Real-time simulation and visualization techniques have become increasingly important in the field of rock dynamics, enabling researchers and engineers to gain deeper insights into complex geological processes and improve decision-making in various applications. This section explores the latest advancements in real-time simulation and visualization methods for rock dynamics, focusing on their implementation, challenges, and potential applications in geosciences and engineering.

Real-Time Visualization of 3D Terrains and Subsurface Geological Structures

One of the primary challenges in rock dynamics is the efficient representation and visualization of large-scale 3D terrains and subsurface geological structures. Recent research has made significant progress in addressing this issue through the development of novel data structures and rendering techniques.

1. Stack-Based Representation (SBRT)

One of the primary challenges in rock dynamics is the efficient representation and visualization of large-scale 3D terrains and subsurface geological structures. Recent research has made significant progress in addressing this issue through the development of novel data structures and rendering techniques [60]. The Stack-Based Representation (SBRT) advances geological surface–subsurface structure modeling [60] by extending Digital Elevation Models (DEMs) through vertical interval sequences with material attributes per cell, superseding single elevation values. SBRT demonstrates superior memory efficiency, reducing storage requirements by up to 95% compared to voxel representations and 74–90% versus octrees [60], enabling real-time processing of large-scale geological datasets. The approach accurately represents complex geological features (faults, folds, intrusions) while facilitating regional-scale modeling through enhanced dataset visualization capabilities.

2. GPU-Based Raycasting Algorithm

A GPU-based raycasting algorithm enables real-time visualization of SBRT data [60]. The rendering pipeline consists of four stages: geometry setup for defining viewing frustum and ray directions, optimized GPU memory access of SBRT data, surface normal vector calculation using a novel hybrid object-image space algorithm, and compositing of geological structures with lighting effects. This approach achieves interactive frame rates for large-scale geological models, enabling real-time exploration of complex subsurface structures.

3. GPU Memory Management Strategies

GPU memory management is essential for real-time visualization of large geological datasets. Five strategies for storing SBRT data in GPU memory have been investigated: separate 2D textures for indices and intervals, R-Super texels (single 2D texture with super texels), L-Super texels (2D texture with linear super texels), 3D texture approach, and Shader Storage Buffer Objects (SSBO). The SSBO approach demonstrates optimal rendering speed and memory efficiency [60] through its flexible data access capabilities.

4. Visual Operations for Geoscientific Applications

Real-time visualization systems for rock dynamics require four key visual operations: strata visualization with layer attenuation for exploring geological layers through transparency adjustment, cross-section visualization for real-time display of subsurface structures, borehole log visualization integrated with 3D geological models for subsurface condition analysis, and texture application to terrain surfaces for enhanced surface feature interpretation. These operations enable comprehensive geoscientific analysis and accurate structural interpretation.

5. Illumination Model

To enhance the visual quality and realism of rendered geological structures, real-time visualization systems often employ illumination models. A commonly used approach is the Lambertian diffuse model, which can be expressed as:

$$I_p = i_a + k_d\left(\vec{L} \cdot \vec{N}\right)i_d \tag{7.111}$$

where
  I_p is the final pixel color
  i_a and i_d are ambient and diffuse lighting intensities
  k_d is the diffuse Lambertian reflectance
  $\vec{L}$ is the light direction vector
  $\vec{N}$ is the surface normal vector

This model provides a good balance between visual quality and computational efficiency, making it suitable for real-time applications in rock dynamics visualization.

7.4 Big Data Visualization and Mining

Real-Time Visualization of Large-Scale Geological Models with Nonlinear Feature-Preserving Levels of Detail

As the complexity and scale of geological models continue to increase, maintaining real-time performance while preserving important features becomes increasingly challenging. Recent research has addressed this issue by developing advanced LOD techniques specifically tailored for geological models [61].

1. Decoupled Approach for LOD Computation

A framework for interactive visualization of large-scale geological models separates level-of-detail (LOD) computation from data compression [61]. This approach enables nonlinear filtering techniques to preserve geometric and attribute features across detail levels. The decoupled design maintains geological features at coarse resolutions, allows independent optimization of LOD generation and compression stages, and accommodates various geological data types and feature preservation requirements.

2. Nonlinear Feature-Preserving Filtering

The LOD generation process employs nonlinear feature-preserving filters through joint bilateral filters computed on high-dimensional permutohedral grids [61]. This technique preserves discontinuities in geometry and attributes, maintaining geological features like faults and stratigraphic boundaries. The method incorporates multiple attributes to identify features not apparent in geometry alone, while permutohedral grids enable efficient high-dimensional filter computation for real-time applications.

3. Wavelet-Based Encoding

The framework employs wavelet-based encoding for efficient representation and transmission of multi-resolution geological models [61]. The methodology uses wavelet transforms to encode detail levels separately, enabling resolution adaptation based on data smoothness. The system implements PROGRES coding for compact wavelet coefficient representation, while providing adaptive resolution that allocates higher detail to complex geological features and lower resolution to smoother regions.

4. GPU Data Structure for Direct Visualization

The framework introduces an innovative GPU data structure enabling real-time visualization without prior decoding [61]. The structure consists of three components: a bytestream buffer storing compressed wavelet coefficients, an offset buffer for random data access, and an adaptive resolution parameter encoding η for render-time resolution control. This GPU-optimized design reduces memory usage while enabling interactive exploration of large geological models.

5. Adaptive Level-Of-Detail Rendering

The framework features an adaptive LOD rendering system that adjusts model resolution based on viewing parameters and data characteristics [61]. The system utilizes task and mesh shaders for efficient on-the-fly geometry generation, while performing per-tile visibility and LOD calculations. Feature-preserving mechanisms maintain crucial geological details at coarser resolutions to ensure interpretability during exploration.

The mesh workgroup count (m) is determined by

$$m = \max\left(1, \frac{t_i}{m_i \cdot 2^\gamma} \times \frac{t_j}{m_j \cdot 2^\gamma}\right) \tag{7.112}$$

where
  m is the number of mesh workgroups.
  t_i and t_j are tile dimensions.
  m_i and m_j are mesh workgroup dimensions.
  γ is the LOD level.
This enables real-time visualization while preserving geological feature fidelity.

Real-Time AUV Simulation in Underwater Environments

Real-time simulation of Autonomous Underwater Vehicles (AUVs) in complex underwater environments presents unique challenges in the context of rock dynamics and subsea exploration. Recent advancements in integrating robotic simulation frameworks with physics engines have led to significant improvements in this area [62].

1. Rock-Gazebo Integration Architecture

The Rock framework integrates with the Gazebo simulator through a three-component system plugin architecture [62]. The RockBridge Class, inheriting from the Gazebo system plugin, enables Rock-Gazebo communication. The World Task exports Gazebo world resources to Rock, while the Model Task handles AUV model resources. This architecture allows flexible customization for underwater rock dynamics simulations.

2. Robot Description and Visualization

The Rock framework integrates with Gazebo simulator via a three-component system plugin architecture [62]. The RockBridge Class inherits from Gazebo system plugin to enable Rock-Gazebo communication. The World Task exports Gazebo world resources to Rock, while the Model Task handles AUV model resources. This modular architecture supports customization for underwater rock dynamics simulations.

7.4 Big Data Visualization and Mining 429

The conversion of link positions from model-relative to parent-relative coordinates is achieved using the following matrix transformation:

$$C' = P^{-1} \cdot C \tag{7.113}$$

where
  C' is the new link pose relative to the parent
  P is the parent link pose matrix relative to the model
  C is the child link pose matrix relative to the model
  Similarly, joint positions are converted from child-relative to parent-relative using:

$$J' = C' \cdot J \tag{7.114}$$

where
  J' is the joint pose relative to the parent link
  C' is the child link pose relative to the parent (from the previous equation)
  J is the joint pose relative to the child link
  These transformations ensure accurate representation of the AUV's structure and kinematics within the simulated underwater environment.

3. Underwater Environment Simulation

The Rock-Gazebo integration implements key physics components for underwater AUV simulation [62]. The system incorporates Archimedes' principle-based buoyancy calculation with a fixed center of buoyancy, computes viscous drag forces based on AUV-fluid relative velocity, and accounts for ocean current effects. A unidirectional force model for AUV thrusters, controlled via Rock component (Thruster_Task) and Gazebo topic, enables propulsion simulation.

4. Underwater Visualization

The Rock-Gazebo integration employs osgOcean library through the Vizkit3D Ocean Plugin (gui-vizkit3d_ocean) for underwater visualization [62]. The system simulates key underwater phenomena: sunlight penetration and attenuation, water turbidity, refraction and caustics, and underwater fog and color absorption.

Visual Simulation of Granular Rock Crushing

Understanding the behavior of granular rocks during crushing processes is crucial for various applications in rock dynamics, including mining operations and material processing. Recent advancements in visual simulation techniques have enabled more accurate and insightful representations of these complex phenomena.

1. SRCVS Architecture for Rock Crushing Visual Simulation

The Solution for Rock Crushing Visual Simulation (SRCVS) architecture [63] integrates three components: a VR System for process visualization and interaction, a

Modeling Subsystem using fractal geometry for rock model generation, and Physical Control Models that calculate crushing parameters. These components ensure accurate representation of rock crushing processes through mapping functions.

The SRCVS architecture separates phenomenon generation from physical computation, enabling independent optimization. Specialized mapping functions allow flexible subsystem coupling for various crushing scenarios, while integrated VR technology provides advanced visualization capabilities.

2. Fractal-Based Rock Geometry Generation

A key innovation in the SRCVS approach is the use of fractal geometry for generating realistic rock models. The iterative algorithm for rock geometry generation can be expressed as:

$$\{P_i^j\}_{i=1,\ldots,Nvj} \Rightarrow \{P_i^{j+1}\}_{i=1,\ldots,Nvj1} \tag{7.115}$$

where

P_i^j represents the vertex position at iteration j

N_{vj} is the number of vertices at iteration j

For existing vertices, the position update is calculated as:

$$P_G P_i^{j+1} = P_G P_i^j \cdot (1 + r_\varepsilon) + \frac{\sum_{kag=1}^{Kag} P_G \vec{P}_{kaj}^j \cdot r_\alpha}{Kaj} \tag{7.116}$$

For new midpoint vertices:

$$\begin{cases} P_G \vec{P}_{M,1} = \frac{1}{2} \cdot \left(P_G \vec{P}_1 + P_G \vec{P}_2\right) \cdot (1 + r_\varepsilon) + \frac{\left(P_G \vec{P}_3 + P_G \vec{P}_3'\right) \cdot r_q}{2} \\ P_G \vec{P}_{M,2} = \frac{1}{2} \cdot \left(P_G \vec{P}_2 + P_G \vec{P}_3\right) \cdot (1 + r_\varepsilon) + \frac{\left(P_G \vec{P}_1 + P_G \vec{P}_1'\right) \cdot r_q}{2} \\ P_G \vec{P}_{M,3} = \frac{1}{2} \cdot \left(P_G \vec{P}_3 + P_G \vec{P}_1\right) \cdot (1 + r_\varepsilon) + \frac{\left(P_G \vec{P}_2 + P_G \vec{P}_2'\right) \cdot r_q}{2} \end{cases} \tag{7.117}$$

where

$r_\varepsilon, r_\alpha, r_q$ are Gaussian random variables

P_G is the geometric center of the polyhedron

The fractal dimension D and parameters r_E and k control the rock shape characteristics, allowing for the generation of a wide variety of realistic rock geometries.

3. Real-Time Rendering of Large-Scale Rock Objects

The SRCVS implements a modified sort-first parallel rendering solution [63] based on frustum division into N partitions (matching rendering node count) for parallel processing. Geometric models are generated simultaneously across nodes, with rendered images assembled on output nodes.

7.4 Big Data Visualization and Mining

4. Rapid Texture Creation for Rock Models

The SRCVS implements an approximate texture mapping method [63] through sequential steps: calculating rock polyhedron bounding boxes, mapping textures to box faces, and assigning interpolated texture colors to vertices based on nearest box faces. This $O(n)$ complexity method (n = vertex count) enables efficient real-time texturing.

5. Mapping Between Physical Computation and Visual Phenomena

SRCVS maintains physical-visual consistency [63] through mapping functions that connect physical parameters (productivity, crushing levels, size distribution, shape coefficients) with modeling parameters (initial elements, iterations, random factors, rock quantities). These functions translate physical simulation changes into visual representation.

CT-Based Visualization of Rock Damage Evolution

Understanding the internal damage evolution of rocks under various conditions is crucial for predicting and mitigating potential hazards in rock engineering. Recent advancements in CT scanning and image analysis techniques have enabled detailed visualization and quantification of rock damage processes [64].

1. Experimental Setup and Methods

The sandstone damage evolution study under freeze–thaw cycles and uniaxial compression [64] utilized in-situ CT scanning for microstructure observation. Interactive threshold segmentation isolated pore and crack structures, while the maximum sphere method reconstructed 3D pore networks. Lattice Boltzmann simulations analyzed permeability and fluid transport changes.

2. Quantitative Analysis of Pore Structure

Sandstone pore structures were classified [64] into four categories by radius: small (<100 μm), medium (100–500 μm), large (500–1000 μm), and extra-large (>1000 μm).

Analysis of pore size evolution [64] revealed distinct damage patterns: low freeze–thaw cycles primarily affect small pores, while high cycle conditions generate interconnected crack networks and reduce isolated pore proportions.

3. Visualization of Damage Evolution Process

CT visualization [64] revealed progressive damage evolution in freeze–thaw sandstone under compression, with critical points marking sudden changes in damage progression. Pore volume exhibited three phases: initial decrease, gradual increase, and rapid expansion near peak strength.

4. Effects of Freeze–Thaw Cycles

Analysis of samples under 0–30 freeze–thaw cycles [64] showed failure mode transition from single shear plane to mixed tensile-shear failure with multiple surfaces. Freeze–thaw cycling modified pore characteristics: enhanced connectivity, increased maximum pore radius, higher pore quantity, and elevated porosity.

5. Pore Connectivity Analysis

Lattice Boltzmann simulations [64] showed pore connectivity develops through both freeze–thaw cycles and compressive loading, with pore channel quantity being more influential than dimensions. A critical threshold for structural changes occurred within the first 10 freeze–thaw cycles.

Virtual Reality-Based Rock Mechanics Simulation and Visualization

The integration of virtual reality (VR) technology with numerical simulation of rock mechanics offers new possibilities for analyzing and visualizing complex geological processes. This approach combines the immersive visualization capabilities of VR with the quantitative analysis power of numerical methods [65].

1. Advantages of VR-Numerical Simulation Integration

The VR-numerical simulation integration creates immersive geological environments with real-time visualization of stress fields, displacement fields, and safety parameters. This system enables dynamic monitoring of mining and excavation processes through interactive geological model exploration.

2. Key Components of the Integrated System

The integration of VR with numerical simulation provides essential capabilities for rock mechanics analysis. VR technology creates immersive digital environments replicating geological conditions, while numerical simulation generates quantitative data for stress fields and displacement distributions. Through intuitive VR interfaces, users can interact with geological models and observe real-time updates of simulation results during mining and excavation processes, enabling continuous monitoring.

3. Case Study: Xincheng Gold Mine

The VR-based simulation system was implemented at the Xincheng Gold Mine, featuring 3D models of surface topography, ore bodies, geological faults, and tunnel networks. The system integrated rock joint visualization through scanning and 3D reconstruction, while incorporating mining sequence simulations for operational planning. Stress field analysis utilized FLAC software to generate stress nephograms for tunnel sections, complemented by ShapeMetriX3D software for key block identification. Real-time deformation monitoring provided continuous updates on rock mass displacement.

4. Benefits and Applications

7.4 Big Data Visualization and Mining

The integration of VR and numerical simulation in rock mechanics provides critical operational and safety benefits [65]. The system's visualization capabilities enable improved interpretation of geological data and mining operations, enhancing hazard identification through combined visual and quantitative stress-deformation analysis. The VR environment serves dual functions: as a risk-free training platform for personnel and as a decision support tool for mining operations. This technological integration advances intelligent digital mining practices.

Visualized Numerical Simulation of Rock Breaking Processes

Accurate simulation and visualization of rock breaking processes are crucial for various applications in rock dynamics, including mining, tunneling, and excavation. Recent advancements in numerical methods, particularly the coupled finite-discrete element method, have enabled more realistic and detailed simulations of these complex phenomena.

1. Coupled Finite-Discrete Element Method

The coupled finite-discrete element method combines the advantages of both finite element and discrete element approaches, making it particularly effective for simulating rock breaking processes [66]. Key aspects of this method include:

- Dynamic Equilibrium Equation: For deformable bodies (finite element discretization)

$$\mathbf{M}\ddot{\mathbf{u}} + \mathbf{C}\dot{\mathbf{u}} = \mathbf{f}^{ext} - \mathbf{f}^{int}(\mathbf{u}) \tag{7.120}$$

where

$\mathbf{M}$ and $\mathbf{C}$ are mass and damping matrices.
$\mathbf{f}^{ext}$ is the external force vector including contact forces.
$\mathbf{f}^{int}(\mathbf{u})$ is the internal force vector due to deformation.
$\mathbf{u}, \dot{\mathbf{u}}, \ddot{\mathbf{u}}$ are displacement, velocity, and acceleration vectors.

- Equations of Motion: For rigid bodies (discrete element approach)

$$m_i a_i + c_i v_i = F_i - G_i \tag{7.121}$$

$$\mathrm{d}\frac{L_i}{\mathrm{d}t} = M_i \tag{7.122}$$

where

m_i, a_i, and v_i are the mass, acceleration, and velocity of rigid body i.
c_i is the damping coefficient.
F_i is the resultant force excluding gravity.
G_i is the gravity force.
L_i is the angular momentum.
M_i is the resultant moment.

2. Fracture Model

The coupled method introduces a hybrid fracture model combining virtual and single crack approaches [66]. The model comprises three key components: stress-based crack initiation, propagation governed by stress intensity factors and energy release rates, and multiple crack handling through virtual crack implementation. This approach enables efficient simulation of multiple crack phenomena.

3. Brazilian Disc Test Simulation

A Brazilian disc test validated the coupled finite-discrete element method [66]. The test used a 150 mm diameter disc specimen, with loading plates incorporating elastic cushion layers ($E = 60$ GPa, $\nu = 0.2$). The rock sample properties were: $E = 30$ GPa, $\nu = 0.2$, tensile strength $= 3.2$ MPa, density $= 2700$ kg/m3, and fracture energy $= 100$ N/m. The simulation used a 0.003 m/s loading rate and discretization of 1114 triangular elements with 3342 nodes under plane stress conditions.

4. Visualization and Analysis of Simulation Results

The simulation revealed detailed rock failure mechanics [66]. The specimen's center exhibited characteristic horizontal stress evolution: linear response until peak load, followed by rapid load capacity decline. The failure sequence progressed from central crack initiation at 38 ms to complete vertical separation at 55 ms. Stress distribution and crack pattern visualizations documented the progressive failure process.

5. Implications and Applications

The coupled finite-discrete element simulation method offers key advantages for rock dynamics applications. The method enables detailed analysis of rock failure progression, with demonstrated accuracy in replicating experimental failure patterns. Parameter sensitivity studies through variable material properties and loading conditions facilitate rock breaking optimization. Stress and crack propagation visualization enhances safety assessment and excavation planning, while simulation results improve rock breaking equipment design and efficiency.

Advanced numerical methods integrated with visualization techniques enhance rock breaking process analysis and prediction. These tools advance applications across rock dynamics, including mining, tunneling, hazard assessment, and equipment design. Continued computational advances will further bridge the gap between numerical simulation and physical rock mechanics behavior.

7.5 Conclusions

The detailed examination of advanced computational methods in rock dynamics, focusing on three main areas: parallel computing methods, advanced numerical methods, and optimization techniques are outlined.

The multi-physics coupling simulation in rock dynamics has evolved from the fundamental THM (Thermal-Hydro-Mechanical) coupling framework, through diverse numerical methods (including FEM, XFEM, FVM, and DEM enhanced by data-driven techniques), to the practical implementation in advanced software packages (such as COMSOL Multiphysics, ABAQUS, and TOUGH-FLAC), enabling effective solutions for critical engineering applications in geothermal energy, nuclear waste disposal, carbon sequestration, and underground construction.

Machine learning approaches in rock dynamics has been corroborated to capture complex, nonlinear relationships between rock properties and dynamic behaviors, and integrate diverse data sources to improve predictions. They also enable real-time analysis and decision-making, enhancing dynamic rock engineering applications, and offer improved computational efficiency compared to traditional numerical methods. Moreover, ML models have the capacity to reveal unexpected relationships and generate new insights into rock dynamic processes. Additionally, future research directions in ML for rock dynamics may include the development of hybrid models that combine physics-based simulations with data-driven ML techniques. In particular, integrating real-time ML-based characterization and prediction tools into dynamic rock engineering practices can significantly advance the field.

The integration of advanced numerical methods with sophisticated visualization techniques has greatly enhanced our ability to understand and predict rock dynamic processes. In addition, integration of data mining with other advanced technologies and domain-specific knowledge will be crucial in addressing the complex challenges of rock dynamics and improving the safety and efficiency of rock engineering projects. These tools provide valuable insights for a wide range of applications in rock dynamics, from mining and tunneling to geological hazard assessment and equipment design.

References

1. Li T, Wu T. Parallel computing of electromagnetic field based on domain decomposition method. In: 2015 5th International conference on electric utility deregulation and restructuring and power technologies (DRPT). IEEE; 2015. p. 1583–6.
2. Gao HP, Li H, Gong B. Parallel computing based on numerical simulation of self-avoiding walk. Appl Res Comp. 2014;31(4):1039–42 [in Chinese].

3. Miao Q, Huang M, Wei Q. Parallel computing of numerical manifold method with OpenMP. In: 2009 IEEE youth conference on information, computing and telecommunication. IEEE; 2009. p. 174–7.
4. Yu RH, Wu LD, Deng BS, et al. Parallel computing research of complex electromagnetic environment based on ITM. Syst Eng Electron 2012; 34(7):1339–43 [in Chinese].
5. Zhang YH, Wong LNY. A review of numerical techniques approaching microstructures of crystalline rocks. Comput Geosci. 2018;115:167–87.
6. Wang J, Apel DB, Pu Y, et al. Numerical modeling for rockbursts: a state-of-the-art review. J Rock Mech Geotech Eng. 2021;13(2):457–78.
7. Wang ZY, Cheng ZQ. Numerical simulation of damage of rock materials based on peridynamics. Chinese Quart Mech. 2019;40(1):22–31 [in Chinese].
8. Nikolić M, Roje-Bonacci T, Ibrahimbegović A. Overview of the numerical methods for the modelling of rock mechanics problems. Tehnički vjesnik. 2016;23(2):627–37.
9. Aydan O. Rock dynamics. CRC Press; 2017.
10. Wang B, Hicks MA, Vardon PJ. Slope failure analysis using the random material point method. Géotech Lett. 2016;6(2):113–8.
11. Jing L, Hudson JA. Numerical methods in rock mechanics. Int J Rock Mech Min Sci. 2002;39(4):409–27.
12. Armaghani DJ, Koopialipoor M, Marto A, et al. Application of several optimization techniques for estimating TBM advance rate in granitic rocks. J Rock Mech Geotech Eng. 2019;11(4):779–89.
13. Zhang Y, Su G, Li Y, et al. Displacement back-analysis of rock mass parameters for underground caverns using a novel intelligent optimization method. Int J Geomech. 2020;20(5):04020035.
14. Yagiz S, Ghasemi E, Adoko AC. Prediction of rock brittleness using genetic algorithm and particle swarm optimization techniques. Geotech Geol Eng. 2018;36:3767–77.
15. Chen CW, Huang KL, Lyuu YD. Accelerating the least-square Monte Carlo method with parallel computing. J Supercomput. 2015;71:3593–608.
16. Sun L, Tang XH, Aboayanah KR, Zhao Q, Liu QS, Grasselli G. A coupled cryogenic thermo-hydro-mechanical model for frozen medium: theory and implementation in FDEM. J Rock Mech Geotech Eng 2024; 16(11):4335–53.
17. Liu XQ, Qu ZQ, Guo TK, et al. A coupled thermo-hydrologic-mechanical (THM) model to study the impact of hydrate phase transition on reservoir damage. Energy 2021; 216.
18. Pandey SN, Vishal V, Chaudhuri A, et al. Geothermal reservoir modeling in a coupled thermo-hydro-mechanical-chemical approach: a review. Earth-Sci Rev 2018; 185:1157–69.
19. Chen PP, Zhang HW, Yan GC, et al. Thermo-hydro-mechanical coupling model of unsaturated soil based on modified VG model and numerical analysis. Front Earth Sci 2022; 10.
20. Jafari A, Vahab M, Broumand P, et al. An eXtended finite element method implementation in COMSOL multiphysics: thermos–hydro–mechanical modeling of fluid flow in discontinuous porous media. Comput Geotech. 2023;159: 105458.
21. Bég OA. Numerical methods for multi-physical magnetohydrodynamics. J Magnetohydrodyn Plasma Res 2013; 18(2/3):93–203.
22. Barbero EJ. Finite element analysis of composite materials using abaqus™ (1st ed.). CRC Press;2013.
23. Potyondy DO, Cundall PA. A bonded-particle model for rock. Int J Rock Mech Min Sci. 2004;41(8):1329–64.
24. Areias P. Finite element technology, damage modeling. Contact Constraints and Fracture Analysis: University of Porto; 2003.

25. Yan BQ, Ren FH, Cai MF, et al. Research review of rock mechanics experiment and numerical simulation under THMC multi–field coupling. Chinese J Eng. 2021;43(1):47–57.
26. Ghosh A. Modeling and simulation. In: Dynamic systems for everyone. Springer International Publishing;2015.
27. Hibbitt HD. ABAQUS/EPGEN–a general purpose finite element code with emphasis on nonlinear applications. Nucl Eng Des. 1984;77(3):271–97.
28. Zhang W, Seylabi EE, Taciroglu E. An ABAQUS toolbox for soil–structure interaction analysis. Comput Geotech. 2019;114: 103143.
29. Wei M, Meng W, Dai F, et al. Application of machine learning in predicting the rate-dependent compressive strength of rocks. J Rock Mech Geotech Eng. 2022;14:1356–65.
30. Xiao F, Shang J, Wanniarachchi A, et al. Assessing fluid flow in rough rock fractures based on machine learning and electrical circuit model. J Petrol Sci Eng. 2021;206: 109126.
31. Liu L, Zhou F, Cao W. Integrate physics-driven dynamics simulation with data-driven machine learning to predict potential targets in maturely explored orefields: a case study in Tongguangshan orefield, Tongling, China. J Geochem Explor. 2024;262: 107478.
32. Gong Y, El-Monier I, Mehana M. Machine learning and data fusion approach for elastic rock properties estimation and fracturability evaluation. Energy and AI. 2024;16: 100335.
33. Yu H, Taleghani A, Al Balushi F, et al. Machine learning for rock mechanics problems; an insight. Front Mech Eng 2022; 8.
34. Han D, Xue X. Machine learning-based prediction of shear strength parameters of rock materials. Rock Mech Rock Eng. 2024;57:8795–819.
35. Afebu KO, Liu Y, et a. Machine learning-based rock characterisation models for rotary-percussive drilling. Nonlinear Dyn 2022; 109:2525–45.
36. Luo S, Kim D, Wu Y, et al. Big data nanoindentation and analytics reveal the multi-staged, progressively-homogenized, depth-dependent upscaling of rocks' properties. Rock Mech Rock Eng. 2021;54(3):1–32.
37. Chen ZY, Zhang YP, Li JB, et al. Diagnosing tunnel collapse sections based on TBM tunneling big data and deep learning: a case study on the Yinsong Project, Chin. Tunnel Underground Space Technol. 2021;108: 103700.
38. Zhang Q, Liu Z, Tan J. Prediction of geological conditions for a tunnel boring machine using big operational data. Autom Constr. 2019;100:73–83.
39. Wang SF, Cai X, Zhou J, Song ZY, Li XF. Analytical, numerical and big-data-based methods in deep rock mechanics. Mathematics. 2020;10(18):3403–3.
40. Lei D, Zhang Y, Lu Z, et al. Hybrid data-driven model for predicting the shear strength of discontinuous rock materials. Mater Today Commun. 2024;41: 110327.
41. Qi S, Guo S, Waqar FM, et al. Prediction of brittle rock failure severity: an approach based on rock mass failure progress. J Rock Mech Geotech Eng. 2024;16(12):4852–65.
42. Miah IM. Predictive models and feature ranking in reservoir geomechanics: a critical review and research guidelines. J Nat Gas Sci Eng. 2020;82: 103493.
43. Kong L, Xie H, Li C. Traction-based microplane model for charactering the progressive failure of rock-like material. J Mech Phys Solids. 2025;194: 105910.
44. Hadi AF. Polyaxial rock failure criteria: insights from explainable and interpretable data-driven models. Rock Mech Rock Eng. 2022;55(4):1–19.
45. Liu BW, Wang ZW, Muhodir SH, et al. Prediction of rock slope failure using multiple ML algorithms. Geomech Eng. 2024;36(5):489–509.
46. Zeng G, Liu W. An iso-time scaling method for big data tasks executing on parallel computing systems. J Supercomput. 2017;73:4493–516.
47. Li C, Xu CN, Zhang L, et al. Dynamic visualization of stress distribution on metal by mechanoluminescence images. J Visual. 2008;11(4):329–35.
48. Qiu ZJ, Zhao WJ, Cao MK, et al. Dynamic visualization of stress/strain distribution and fatigue crack propagation by an organic mechanoresponsive AIE luminogen. Adv Mater. 2018;30(44):1803924.

49. Ju Y, Wang L, Xie HP, et al. Visualization and transparentization of the structure and stress field of aggregated geomaterials through 3D printing and photoelastic techniques. Rock Mech Rock Eng. 2017;50:1383–407.
50. Ju Y, Wang L, Xie HP, et al. Visualization of the three-dimensional structure and stress field of aggregated concrete materials through 3D printing and frozen-stress techniques. Constr Build Mater. 2017;143:121–37.
51. Semikina S. Stress data visualization. Degree Thesis of Master, Eindhoven University of Technology; 2014.
52. Haber RB. Visualization techniques for engineering mechanics. Comp Syst Eng 1990; 1(1):37–50.
53. Zhu XJ, Jin XN, Jia DD, et al. Application of data mining in an intelligent early warning system for rock bursts. Processes. 2019;7(2):55.
54. Hossein Alavi A, Hossein GA. A robust data mining approach for formulation of geotechnical engineering systems. Eng Comput. 2011;28(3):242–74.
55. Li XZ, Hao SG, Wu T, et al. Data mining technology and its applications in coal and gas outburst prediction. Sustainability. 2023;15(15):11523.
56. Terra GS, Costa MCA, Ebecken NFF. Prediction of rock strength properties by data mining approach. WIT Trans Inform Commun Technol. 2002;28:489–97.
57. Chen JH, Tang F. Visualized analysis of microscale rock mechanism research: a bibliometric data mining approach. Heliyon. 2024;10(20): e39160.
58. Zhang DC. On data mining technique and its research progress in rock projects. Shanxi Arch. 2010;36(29):91–2 [in Chinese].
59. Li ZH, Liu YF, Zhang LQ, et al. Application of data mining method in lithology identification using well log. Fault Block Oil Gas Field. 2019;26(6):713–8 [in Chinese].
60. Graciano A, Rueda AJ, Feito FR. Real-time visualization of 3D terrains and subsurface geological structures. Adv Eng Softw. 2018;115:314–26.
61. Sicat R, Ibrahim M, Ageeli A, et al. Real-time visualization of large-scale geological models with nonlinear feature-preserving levels of detail. IEEE Trans Visual Comput Graph. 2021;29(2):1491–505.
62. Watanabe T, Neves G, Cerqueira R, et al. The rock-gazebo integration and a real-time auv simulation. In: 2015 12th Latin American robotics symposium and 2015 3rd brazilian symposium on robotics (LARS-SBR). IEEE; 2015 Oct 29. p. 132–8.
63. Wu DL, Hu Y, Fan XM. Visual simulation for granular rocks crush in virtual environment based on fractal geometry. Simul Model Pract Theory. 2009;17(7):1254–66.
64. Liu H, Yang GS, Sheng YJ, et al. CT visual quantitative characterization of meso-damage evolution of sandstone under freeze-thaw-loading synergistic effect. Chin J Rock Mech Eng. 2023;42(5):1136–49 [in Chinese].
65. Liu XG, Yang TH, Zhang PH, et al. Numerical simulation of rock mechanics and 3D visualization based on virtual reality technology. Metal Mine. 2015;8:20–3 [in Chinese].
66. Qiu LC, Lu H, He SJ. Visualized numerical simulation of the process of rock breaking. Min Res Dev. 2010;5:62–3 [in Chinese].

References

Open Access This chapter is licensed under the terms of the Creative Commons Attribution-NonCommercial-NoDerivatives 4.0 International License (http://creativecommons.org/licenses/by-nc-nd/4.0/), which permits any noncommercial use, sharing, distribution and reproduction in any medium or format, as long as you give appropriate credit to the original author(s) and the source, provide a link to the Creative Commons license and indicate if you modified the licensed material. You do not have permission under this license to share adapted material derived from this chapter or parts of it.

The images or other third party material in this chapter are included in the chapter's Creative Commons license, unless indicated otherwise in a credit line to the material. If material is not included in the chapter's Creative Commons license and your intended use is not permitted by statutory regulation or exceeds the permitted use, you will need to obtain permission directly from the copyright holder.

Chapter 8
Engineering Applications in Rock Dynamics

8.1 Engineering Applications in China

China, with its vast and diverse geological landscape, has been at the forefront of rock dynamics research and application for decades. The country's rapid economic growth and infrastructure development have necessitated extensive underground construction and resource extraction, often in challenging geological conditions. This section examines the various engineering applications of rock dynamics in China, focusing on key areas such as deep mining, hydropower projects, and underground construction.

8.1.1 Deep Mining and Rock Burst Control

As China's mineral resources near the surface become depleted, mining operations are increasingly moving to greater depths, often exceeding 1000 m. This shift has led to significant challenges in rock mechanics, particularly in managing rock bursts and other dynamic rock failures. The mining industry in China has developed innovative approaches to understand, predict, and control these hazards.

One of the primary concerns in deep mining is the occurrence of rock bursts. These sudden, violent failures of the rock mass can pose severe risks to workers and equipment. Chinese researchers have made significant strides in understanding the mechanisms behind rock bursts and developing methods to mitigate their effects. A comprehensive study by He et al. [1] provides valuable insights into the classification, mechanisms, and control methods for various types of rock mass dynamic disasters in deep mining.

Rock mass dynamic disasters are complex phenomena occurring in high-stress underground environments, classified into four main categories based on their mechanisms and the materials involved. The first is the rock burst, a nonlinear dynamic phenomenon where elastic energy stored within rock masses is suddenly and violently released along free excavation surfaces, such as mine walls or tunnel faces. This often results in intense rock ejection and deformation, posing significant risks to infrastructure and safety. Rock bursts are commonly observed in deep mining operations and are influenced by in-situ stress levels, excavation geometry, and geological conditions.

Closely related is the coal burst, which shares many similarities with rock bursts but occurs specifically in coal mining environments. In this case, the accumulated energy is released rapidly from brittle coal seams along free surfaces. Coal bursts are distinct in their behavior due to the unique properties of coal, such as its lower strength and higher susceptibility to deformation under stress, making these events particularly hazardous in tectonically stressed or structurally compromised mining areas.

Another category is the mine pressure bump, which refers to a nearby dynamic failure caused by the sudden release of elastic energy stored in both coal and the surrounding rock masses. Unlike rock or coal bursts, pressure bumps are often localized and triggered by mining activities, such as blasting or rapid excavation, resulting in localized fracturing and movement that can lead to significant damage within a confined area.

Lastly, the mine earthquake is characterized by the release of elastic deformation energy in the form of seismic waves, typically without causing coal or rock bursts. These events occur due to large-scale structural failures, such as the breaking or sliding of critical rock formations, including fault planes or major geological discontinuities. While mine earthquakes involve broader stress redistribution and are less localized than other dynamic disasters, they can result in substantial ground motion and widespread structural damage, especially in seismically active mining regions.

This classification provides a foundation for understanding the mechanisms driving dynamic disasters, enabling better risk assessment, prediction, and mitigation strategies in high-stress mining environments. These distinctions are essential for the development of safer and more resilient practices in the field of underground engineering and resource extraction (Fig. 8.1).

Chinese researchers have developed several control theories to address rock mass stability challenges. The excavation compensation theory aims to mitigate stress loss from excavation using high pre-tensioned support systems, stabilizing the surrounding rock and reducing dynamic failure risks. The mining compensation theory relies on broken rock dilation to fill goaf areas and restore equilibrium stress. Key technologies implementing these theories include Negative Poisson's Ratio (NPR) support materials, which exhibit enhanced mechanical properties compared to traditional materials. NPR materials absorb more energy and provide superior confinement, reducing rock burst risk. Directional roof-cutting pressure relief uses bidirectional cavity blasting to create controlled cracks in the roof, managing stress release and minimizing violent failure. Additionally, broken rock dilation filling leverages rock dilation to fill goaf areas, reducing voids and mitigating stress concentrations that could lead to rock burst. These technologies have been applied in deep

8.1 Engineering Applications in China 443

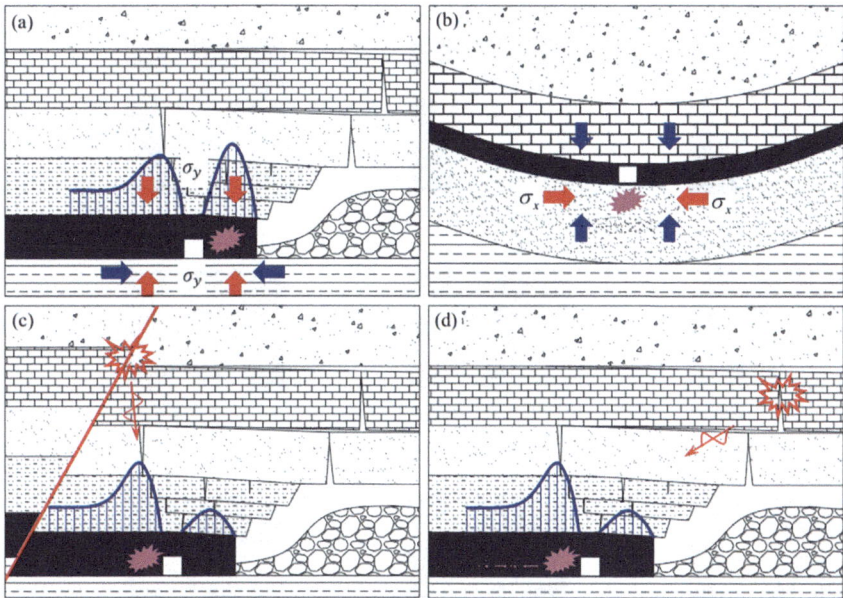

Fig. 8.1 Four types of rock bursts **a** pillar-induced rock burst; **b** fold structure-induced rock burst; **c** fault induced rock burst; **d** hard roof induced rock burst [2]

mines in China, reaching depths of over 1000 m, with promising results. Microseismic event energy was reduced by up to 92.5%, and support pressures decreased by 24.4% following compensation control [1].

The success of these applications demonstrates the effectiveness of the compensation control method in reducing the risk of rock mass dynamic disasters in deep mines. By integrating high pre-stress compensation, roof-cutting pressure relief, and broken rock dilation filling technologies, Chinese engineers have developed a comprehensive approach to managing rock dynamics in challenging deep mining environments.

Another critical aspect of rock dynamics in deep mining is the influence of geological structures on rock burst occurrence. A study investigated the characteristics of rock bursts in syncline regions and the use of micro seismic precursors based on energy–density clouds. This research provides valuable insights into the relationship between geological structures and rock burst hazards, as well as methods for predicting and mitigating these events.

The study found that the axis of synclines presents a higher rock burst hazard compared to the limbs and anticlines. Many coal mines experiencing rock bursts are located in tectonic stress zones with syncline structures. In-situ stress measurements showed high horizontal stresses (1.2–3.88 times the vertical stress) in these regions, indicating that tectonic stress plays a dominant role in rock burst occurrence [2].

To better understand and predict rock bursts, the researchers developed an innovative approach using energy–density clouds. This method provides a more effective way to visualize the evolution of micro seismic tremors preceding rock bursts. The

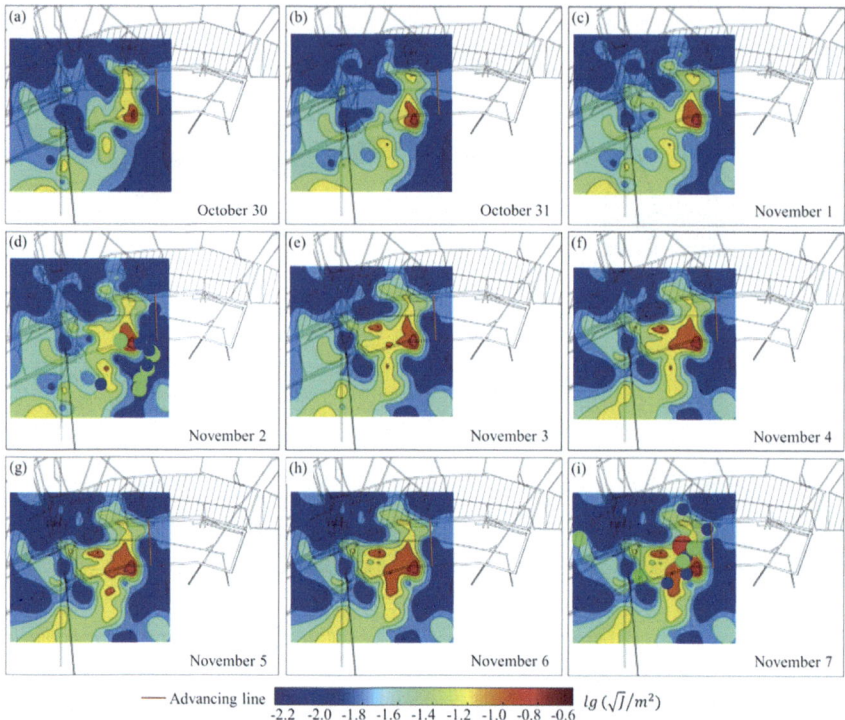

Fig. 8.2 Evolution of energy density clouds before the strong tremor [2]

analysis revealed that a nucleation area appears first, followed by a gradual extension of the cloud with increasing intensity. Strong tremors and rock bursts tend to occur at the edges of the energy–density cloud (Fig. 8.2).

Energy density cloud analysis, aligned with laboratory acoustic emission tests, reveals high-energy fracturing at the edges of microcrack zones, suggesting localized deformation and stress transfer before major ruptures. This provides insight into rock burst initiation and propagation. Seismic velocity tomography was used to identify high-stress regions in syncline structures, with high-velocity areas indicating stress concentration and guiding targeted stress relief, such as large-diameter hole drilling and deep-hole blasting. A case study demonstrated the effectiveness of these measures: after implementation, a strong tremor (1.49×10^5 J) did not trigger a rock burst, unlike a similar event before the intervention, where a rock burst occurred. Lower energy density before the tremor confirmed the success of the stress relief measures [2].

This research provides valuable tools for assessing and managing risks associated with rock bursts in complex geological settings. The energy density cloud analysis offers a promising new approach for rock burst prediction, while the identification of high-stress regions through seismic velocity tomography allows for more targeted and effective prevention measures.

8.1 Engineering Applications in China   445

Chinese researchers have explored hydraulic fracturing to initiate vertical fractures in hanging roofs for stress relief. Liu et al. [3] proposed a method that utilizes horizontal tensile stress from roof bending to enhance vertical crack propagation. The fracturing point is positioned at the location of maximum additional tensile stress, identified as the optimal position. A mechanical model was developed, treating the hanging roof as an elastic thin plate on an elastic foundation, subjected to overburden pressure, to calculate this optimal fracturing position.

The study found that the reasonable fracturing position is influenced by several factors, in order of importance: hanging roof length, hanging roof elastic modulus, hanging roof thickness, elastic cushion coefficient, and horizontal stress. A shorter hanging roof, thicker hanging roof, larger elastic modulus, and smaller elastic cushion coefficient all shift the reasonable fracturing position deeper into the coal wall [3].

The method was applied in roadway 5103 of Majiliang Coal Mine, resulting in vertical hydraulic cracks at the point of maximum horizontal tensile stress. This led to a 44% reduction in roof-to-floor convergence and a 47% reduction in side convergence. The implementation involved drilling holes at a 30° angle, 2.5 m above the floor, slot drilling the borehole, injecting high-pressure water, and monitoring water leakage through fractures and neighboring boreholes. A borehole camera was used to observe fracture formation [3]. This novel hydraulic fracturing method offers a promising solution for controlling stress and deformation in underground coal mines with hard-hanging roofs. The mechanical model developed provides a reliable guide for determining the optimal fracturing position, potentially improving safety and efficiency in coal mining operations.

8.1.2 Hydropower Projects and Rock Mass Stability

China's rapid economic growth has led to an increasing demand for energy, with hydropower playing a crucial role in meeting this demand. The construction of large-scale hydropower projects often involves extensive underground excavation in complex geological environments, presenting significant challenges in rock dynamics and stability control.

One of the most notable examples of such projects is the Jinping I hydropower station in Southwest China. Analyzed the surrounding rock mass stability in the underground caverns of this project. The underground powerhouse complex at Jinping I faces challenging geological conditions, including high in-situ stresses (up to 35.7 MPa), complex deformation and failure mechanisms, and large unloading depths during excavation.

Jinping I's underground structures include a main powerhouse (204.5 m × 25.6 m × 68.8 m), a transformer chamber, and two tailrace surge chambers, located in an area with three major faults (f13, f14, f18) intersecting the caverns, four groups of critically oriented joints, and a high in-situ stress field with maximum principal stress between 20 and 35.7 MPa [4]. The study combined field monitoring, borehole imaging, numerical simulations, and EDM. A key innovation was the EDM, which analyzes rock mass stability by incorporating total resistance energy, dissipated energy during softening/hardening, secant Young's modulus, unloading modulus, and an energy dissipation ratio criterion [4]. Field monitoring revealed significant deformation, with a maximum sidewall displacement of 246 mm in the main powerhouse, most of which occurred within two years after excavation. The downstream sidewall showed larger deformation than the upstream, with relaxation depths of 6.5 m upstream and 8 m downstream. Failure mechanisms included spalling and cracking of shotcrete, stress concentration and cracking around bolt holes, and large deformations during the 5th and 6th bench excavations. Energy dissipation analysis identified three zones based on the energy dissipation ratio (Rd): Failure zone (Rd > 0.90–0.95), Strong relaxation zone (0 < Rd < 0.90–0.95), and Weak relaxation zone (Rd = 0) [4] (Fig. 8.3).

The study combined energy dissipation theory with field measurements, borehole imaging, numerical simulations, and acoustic wave velocity analysis to assess rock mass stability in underground excavations. Using the EDM, the researchers found good agreement between numerical simulations and field data, particularly in displacement contours, stress distributions, and energy dissipation zones, which aligned with relaxation depths and extensometer data. Based on these findings, a

Fig. 8.3 Failure and deformation characteristics of the underground caverns. **a** expansion break of rock mass in the exploration adit. **b** rock core discing phenomenon in the exploration adit. **c** spalling phenomenon after concrete spraying in the underground powerhouse. **d** the outlook of the underground powerhouse after the second-floor excavation [4]

8.1 Engineering Applications in China

comprehensive criterion was proposed to evaluate surrounding rock mass stability, incorporating energy dissipation ratios, acoustic wave velocity, and elastic modulus variations. This approach provides a more quantitative method for determining support requirements compared to traditional elastic–plastic analysis, offering valuable insights for designing and analyzing large caverns in high-stress environments [4].

Another significant hydropower project that has contributed to the understanding of rock dynamics in China is the Dagangshan hydropower station. A study by Meng et al. [5] focused on the micro-seismic monitoring and stability analysis of the right bank slope at this station after the initial impoundment. This research is particularly important as it addresses the impact of reservoir impoundment on slope stability, a critical issue in many large-scale hydropower projects.

The Dagangshan hydropower station features a 210 m high arch dam, located on a steep right bank with a 600 m height difference. The main rock type is medium-grain granite, interspersed with diabase dikes and faults. To strengthen weak geological structures, anti-shear galleries were built. The study employed a comprehensive monitoring strategy, including micro-seismic monitoring with a 19-sensor system (recording 113 events from 2014 to 2016), piezometer data, and numerical modeling. Key findings showed that seepage pathways formed along diabase dikes and faults, with 2D modeling revealing damage progression along these structures. The Load/Unload Response Ratio (LURR) indicated slope instability when values exceeded 1, and 3D numerical modeling calculated a safety factor of 1.645, suggesting overall slope stability post-impoundment, with minimal impact from local porous rock masses [6] (Fig. 8.4).

The researchers developed several key equations to analyze the slope stability:

Energy Index:

$$EI = \frac{E_r}{\overline{E}(M_0)} \tag{8.1}$$

where E_r is the radiated seismic energy of an event, $\overline{E}(M_0)$ is the average radiated seismic energy for a given seismic moment M_0, and E_I is the energy index.

LURR:

$$Y = \frac{X_+}{X_-} = \frac{\frac{V_{A+}}{\overline{EI}_+}}{\frac{V_{A-}}{\overline{EI}_-}} \tag{8.2}$$

where X_+ and X_- are the loading response rate and unloading response rate, V_{A+} represents the increase of the apparent volume corresponding to $\overline{EI}_+$, V_{A-} represents the increase of the apparent volume corresponding to $\overline{EI}_-$, $\overline{EI}_+$ and $\overline{EI}_-$ represents the average EI_+ and EI_-.

Mohr-Coulomb Criterion with Effective Stress:

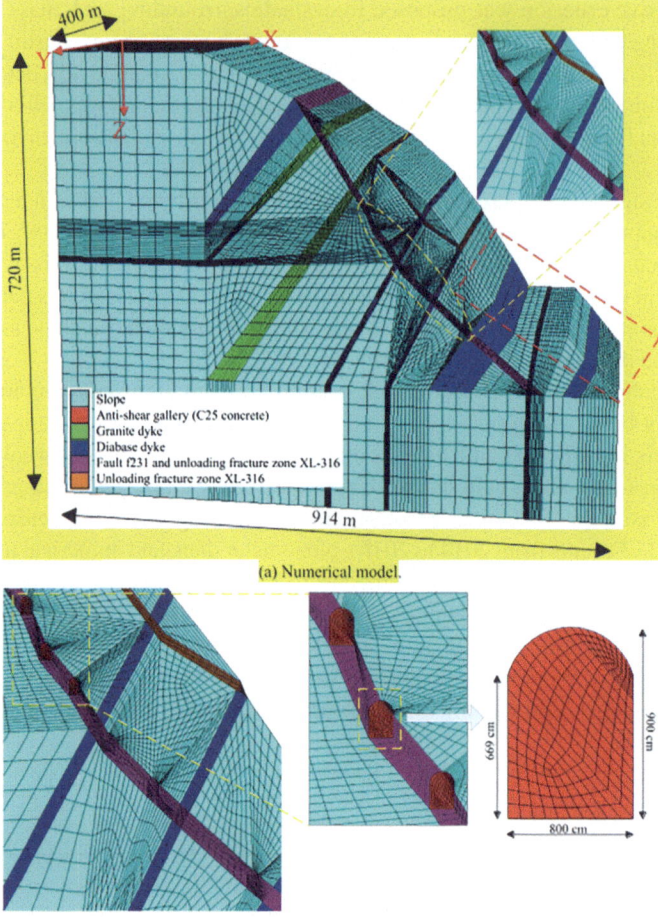

Fig. 8.4 3D numerical model based on finite element method (ANSYS) [6]

$$\tau = (\sigma - \lambda_p)\tan\varphi + c \quad (8.3)$$

where τ and σ are the shear stress and normal stress, φ is the friction angle, and c is the cohesion. λ_p is the effective stress coefficient.

Safety Factor Calculation:

$$c_F = \frac{c}{F_s} \quad (8.4)$$

$$\tan(\varphi_F) = \frac{\tan(\varphi)}{F_s} \quad (8.5)$$

8.1 Engineering Applications in China

Where F_s is the safety factor, c_F and φ_F are the cohesion and friction angles, when the slope is in a limit state condition [6].

This comprehensive study demonstrates the value of integrating multiple monitoring and modeling techniques to assess slope stability in complex hydropower projects. The approach provides insights into damage mechanisms and overall stability that can inform risk assessment and management strategies for large-scale hydropower projects in China and beyond.

8.1.3 Underground Construction and Tunnel Stability

China's rapid urbanization and infrastructure development have led to an increasing number of underground construction projects, including subway systems, road tunnels, and underground storage facilities. These projects often encounter challenging geological conditions, requiring innovative approaches to ensure stability and safety.

Tu et al. [7] studied the excavation and kinematic analysis of a shallow, large-span tunnel at Xing gong jie Station on Line 1 of the Dalian Metro, facing challenges from its large dimensions (21.5 m width, 18.11 m height, 344 m^2 excavation area) and shallow depth (7.18–11.75 m). They developed a kinematic mechanism model based on the nonlinear Hoek–Brown failure criterion and upper bound theorem, incorporating rock strength, density, subsidence, and tunnel geometry to calculate slip surface index and maximum subsidence. Three excavation sequences were compared using 3D finite element simulations (Fig. 8.5).

The simulation results for the Xing gong jie Station tunnel excavation showed that Case 1 resulted in the smallest surrounding rock displacement (5.21 cm crown settlement, 3.93 cm surface subsidence) and adhered to the 50 mm safety limit. Case 2 exhibited the largest deformations (8.75 cm crown settlement, 6.56 cm subsidence), while Case 3 showed intermediate results. The Case 1 method was implemented in construction, with field monitoring confirming its success: maximum pressure on surrounding rock was 125 kPa, with over 80% of the load supported by the initial lining. The maximum surface settlement was 70.13 mm. The theoretical model's

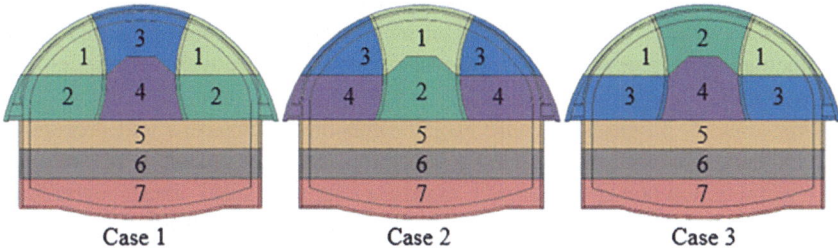

Fig. 8.5 Excavation sequences used in the numerical simulation (finite element method) [7]

predictions showed a small discrepancy with numerical results (4.72% for slip surface index, 4.84% for surface subsidence) [7].

This research provides valuable guidance for the design and construction of large-span subway tunnels in similar up-soft/low-hard geological conditions. The kinematic model and optimized excavation sequence offer an effective approach to predict and control deformations in challenging shallow tunnel projects.

A study by Lu et al. [8] focused on stress evolution and precursors related to hard roof fracturing and rock bursts in underground coal mines, particularly at the Junde coal mine (JCM) in China. The research employed seismic tomography, micro seismic (MS) monitoring, electromagnetic emission (EME), and acoustic emission (AE) measurements to analyze the stress dynamics. The study identified the triggering mechanism of rock bursts as a combination of high intrinsic stress concentration ahead of the working face and external dynamic stress transfer from roof fracturing. Pre-fracturing, the source region showed high P-wave velocity, velocity gradient, and stress concentration, while post-fracturing, the source region exhibited lower velocity, and surrounding areas showed increased velocity and stress due to dynamic stress transfer.

The researchers developed a P-wave velocity-stress relationship:

$$V_p = a \times \sigma^\lambda \tag{8.6}$$

where V_p is the P-wave velocity, σ is the imposed stress, and a and λ are optional parameters fitted from laboratory tests.

They also calculated the stress concentration coefficient using the formula:

$$\varphi = \frac{(V_p/\varphi)^{\frac{1}{\psi}}}{\sigma_p^a} \tag{8.7}$$

where φ is the stress concentration coefficient, V_p is the P-wave velocity obtained by inversion, φ and ψ are parameters from the experimental relational model, and σ_p^a is the estimated value based on mean V_p^a from the experimental model [8].

The study identified several precursors for rock bursts, including MS changes, EME drops, and AE characteristics. MS precursors showed a shift from hybrid to low-frequency events, indicating roof micro-fissure development and coalescence. EME precursors involved a significant amplitude drop below 30 mV, signaling rock burst danger, while AE data revealed an increase in event count post-rock burst and fluctuating AE energy patterns, suggesting ongoing micro-fissure formation and propagation in the coal-rock mass [8].

This research provides valuable insights into the mechanisms and precursors of rock bursts induced by hard roof fracturing, offering potential methods for early warning and prevention of such events in coal mines and other underground constructions.

The application of rock dynamics in Chinese engineering extends to the study of mining-induced fractures and their impact on surrounding strata. A study by Wang

8.1 Engineering Applications in China

et al. [9] proposed novel methods to quantitatively characterize the separation and fracturing of overlying strata disturbed by longwall mining in mineral deposit seams.

The researchers developed theoretical models to calculate the three-dimensional distribution of the void ratio of fractures (VRF) in mining-disturbed overburden. Two types of VRFs were modeled:

Transverse VRF: Quantifies bed separation between adjacent key strata.
Longitudinal VRF: Quantifies fracturing within individual strata.
The VRF models are based on analytical expressions for key strata subsidence:

$$S_i(x, y, z_i) = \frac{\left[M - (K_{pi} - 1)z_i\right]\left\{1 - \left[1 + \exp\left(\frac{2l_x - |2l_x - 4x|}{l_i} - 2\right)\right]^{-1}\right\}\left\{1 - \left[1 + \exp\left(\frac{2l_y - 4|y|}{l_i} - 2\right)\right]^{-1}\right\}}{1 - \left[1 + \exp\left(\frac{2l_y}{l_i} - 2\right)\right]^{-1}} \quad (8.8)$$

where $S_i(x, y, z_i)$ is the subsidence of key stratum, M is the thickness of deposit seam, l_x is the strike length of the mined-out area l_y is the dip width of the mined-out area, l_i is the length of rock block key stratum, z_i is the vertical height of key stratum, K_{pi} is the bulking factor.

The transverse VRF is calculated as:

$$TF_i = \frac{S_i - S_i + 1}{z_{i+1} - z_i + S_i - S_{i+1}} \quad (8.9)$$

where TF_i is the transverse VRF caused by bed separation between the ith and $(i+1)^{th}$ key strata.

The longitudinal VRF is calculated as:

$$LF_i = \sqrt{1 - \frac{1}{\sqrt{1 + (\frac{\partial S_i}{\partial x})^2 + (\frac{\partial S_i}{\partial y})^2}}} \quad (8.10)$$

where LF_i is the longitudinal VRF caused by the fracture of the ith key stratum.

The study used UDEC (Universal Distinct Element Code) for numerical simulations of strata movement and fracturing, applying novel image processing techniques to extract fracture networks and calculate VRF distributions and fractal dimensions. Results showed that VRFs peak at the edges of the mining-disturbed zone and decrease from deep to shallow strata. For a 200 m wide longwall panel, transverse VRF peaks at the edges (0–0.25), longitudinal VRF peaks at both the edges and center (0–0.15), and total VRF peaks at the edges (0–0.3). The fractal dimension of fractures decreased from 1.314 to 1.099 as mining progressed, indicating zone compaction. These findings can improve mining safety by enabling accurate assessments of fracture intensity, porosity, permeability, and gas/water flow, aiding

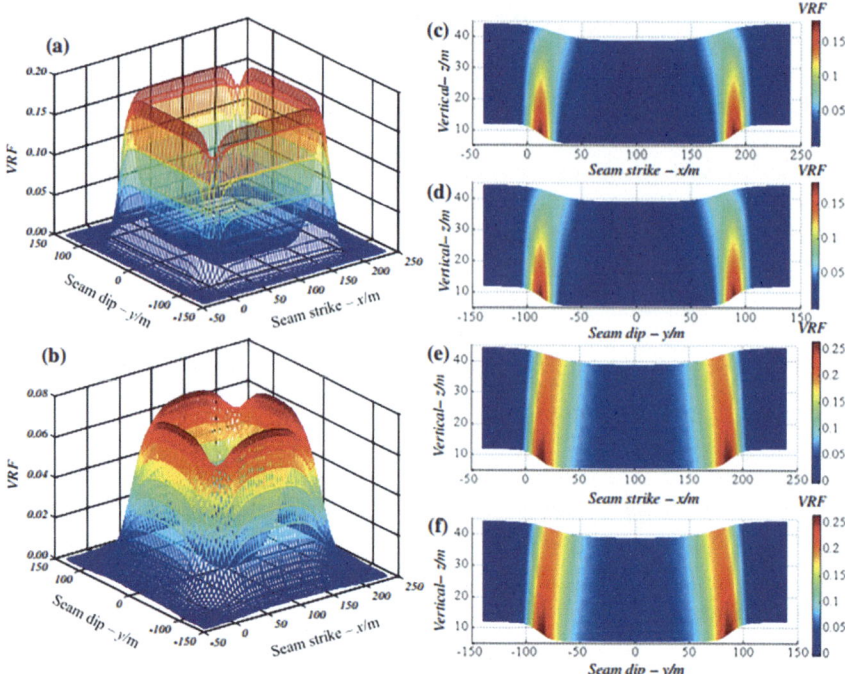

Fig. 8.6 Theoretical VRF distribution due to strata fracturing over 200 m. Includes: **a** VRF mesh in the first key stratum; **b** VRF mesh in the second; **c** longitudinal VRF contour at y = 0; **d** at x = 100 m; **e** total VRF contour at y = 0; **f** at x = 100 m [9]

in gas drainage and risk management for water inflow and gas outbursts. The developed image processing methods offer a powerful tool for analyzing mining-induced fracture networks [9] (Fig. 8.6).

The application of rock dynamics in Chinese engineering also extends to the study of deformation, failure, and permeability changes in coal-bearing strata during longwall mining. A comprehensive study by Meng et al. [5] investigated these phenomena at the daliuta coal mine in the Shen Dong mining area of China.

The researchers conducted laboratory experiments to obtain complete stress–strain-permeability curves for different rock types. Key findings include that the permeability coefficient reaches its maximum value after the stress reaches peak strength, typically during the strain-softening stage. Permeability decreases during initial loading and elastic deformation, then increases sharply after peak stress due to crack development. The magnitude of permeability changes varies significantly with rock lithology. Based on these observations, the researchers developed a six-stage conceptual model to describe the stress–strain-permeability relationship: (1) Closure of primary microcracks, (2) Elastic deformation, (3) Crack occurrence and expansion, (4) Rapid development of cracks to failure, (5) Strain softening, and (6) Residual strength (Fig. 8.7).

8.1 Engineering Applications in China

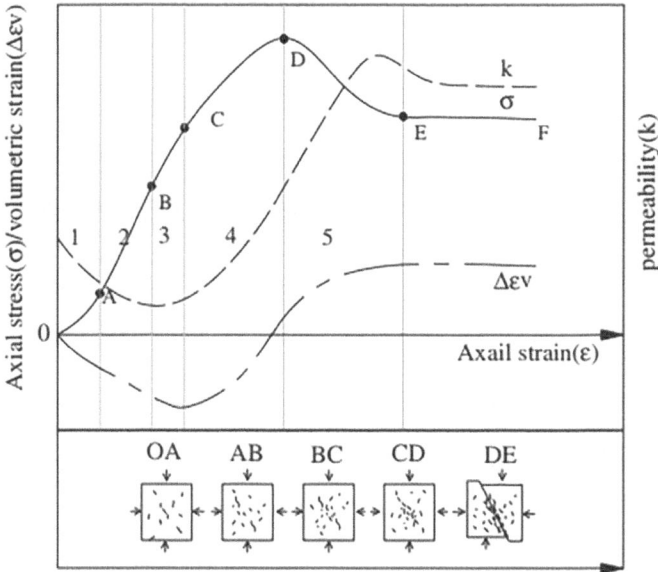

Fig. 8.7 A conceptual model between stress, permeability and axial strain [5]

A three-dimensional coupling model between stress and permeability was developed:

$$K_k = \frac{K_{k0}}{2}\left\{\left[1 - \left(\frac{1}{K_{ni}b_i} + \frac{1}{K_{ni}s_i} + \frac{1}{E_r}\right)(\Delta\sigma_i - \nu(\Delta\sigma_j + \Delta\sigma_k))\right]^3 \right.$$
$$\left. + \left[1 - \left(\frac{1}{K_{nj}b_j} + \frac{1}{K_{nj}s_j} + \frac{1}{E_r}\right)(\Delta\sigma_j - \nu(\Delta\sigma_i + \Delta\sigma_k))\right]^3\right\} \quad (8.11)$$

where K_k and K_{k0} are the permeability in the k-direction after and before stress variation, s is the original fracture spacing, b is the original fracture aperture, K_n is the fracture normal stiffness, E_r is Young's modulus of the rock matrix, $\Delta\sigma$ represents stress changes (Fig. 8.8).

Numerical simulations revealed important characteristics of the stress field, strain and failure, and permeability distribution in the mining-affected area. In the stress field, the advancing direction shows sections of initial stress, gentle stress concentration, sharp stress concentration, and stress relaxation, with a peak stress concentration coefficient of about 2.7 occurring approximately 10 m ahead of the working face. Three stress regimes are identified in the overlying strata: bidirectional tensile, tensile-compressive, and bidirectional compressive. For strain and failure, there is pronounced tensile deformation in the caving zone behind the working face, shear

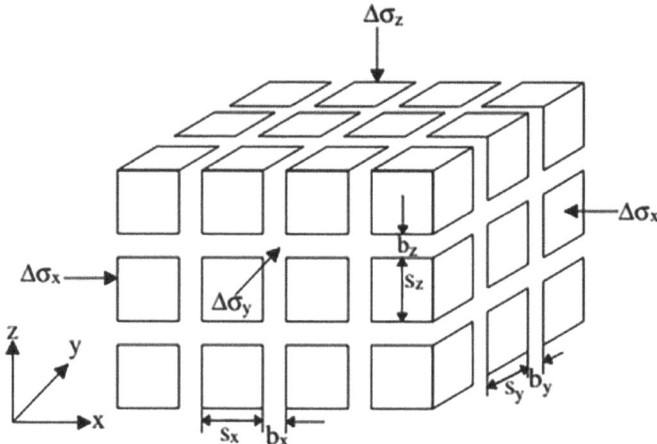

Fig. 8.8 A conceptual model for three sets of orthogonal fractured rock mass in a three-dimensional stress state [5]

deformation above pillars, and compressive deformation in the abutment stress zone. Three failure types are observed: tension, tensile-shear, and shear failure, with the failure area being saddle-shaped horizontally and higher on the sides than in the middle of the goaf. Regarding permeability distribution, permeability significantly increases around the goaf, diminishing with vertical distance from the coal seam, and is larger on the periphery of the goaf compared to the central area. Permeability decreases near coal pillar areas, with horizontally oval-shaped and vertically saddle-shaped permeability-increased zones. The increase in horizontal permeability is larger compared to vertical, but over a smaller range.

8.2 Engineering Applications in the US

The United States has been at the forefront of research and applications in rock dynamics, with numerous studies and practical implementations spanning various fields such as mining engineering, civil engineering, and earthquake engineering. This section delves into the diverse engineering applications of rock dynamics principles in the US, highlighting key research findings and practical case studies.

8.2.1 Mining Engineering Applications

Mining engineering in the US has significantly benefited from advancements in rock dynamics research, particularly in areas such as longwall mining, blast-induced fracturing, and mine stability analysis.

Longwall Mining Subsidence Prediction

One of the critical applications of rock dynamics in US mining engineering is the prediction and management of ground subsidence caused by longwall mining operations. A recent study applied the Knothe influence function method to predict both final and dynamic ground movements in an eastern US longwall coal mine [10]. This research expanded on previous work to develop equations for dynamic slope, horizontal displacement, and horizontal strain, providing a more comprehensive understanding of mining-induced ground deformations.

The study utilized a two-step calibration process, first for final subsidence values and then for dynamic subsidence values. The key equations developed for final subsidence (S^f) and dynamic subsidence (S^d) are as follows:

$$S^f(x_t, x_0, y_1, y_2, z) = \frac{S_{max}}{r_z^2} \int_{x_0}^{x_t} \int_{y_1}^{y_2} \exp\left(-\frac{\pi(x^2 + y^2)}{r_z^2}\right) dx dy \quad (8.12)$$

$$S^d(x_t, x_0, y_1, y_2, z, \Delta t) = S^f(x_t, x_0, y_1, y_2, z) - \exp\left(\frac{u_z^2}{4\pi}\right) \exp\left(\frac{u_z x_t}{r_z}\right)$$
$$S^f\left(x_t + \frac{r_z u_z}{2\pi}, x_0 + \frac{r_z u_z}{2\pi}, y_1, y_2, z\right) + \Delta S^f(x_t, x_0, y_1, y_2, z)[1 - \exp(-c_z \Delta t)]$$
$$(8.13)$$

where S_{max} is the maximum subsidence, r_z is the radius of influence, x_t is the current face position, y_1 and y_2 are the lateral extents of the panel, z is the seam depth, u_z is a dimensionless time factor, and c_z is a time constant (Fig. 8.9).

The study demonstrated that the influence function method effectively modeled vertical subsidence with low root mean square error (RMSE), though site-specific calibration was crucial for accurate predictions, particularly for parameters like the tangent of the influence angle and edge effect offset. A key finding was that the dynamic subsidence time factor was one order of magnitude larger than previously recommended for eastern US coalfields, suggesting slower temporal evolution of mining-induced deformations, which affects risk assessments. While vertical subsidence predictions were accurate, horizontal displacement predictions showed high relative root mean square error (RRMSE) due to factors like surface topography and secondary movements not included in the model. However, the model still accurately predicted transitions in horizontal displacement curves and specific portions of the displacement profile. The research underscores the need for further model refinement, including surface topography effects and secondary movements, to improve

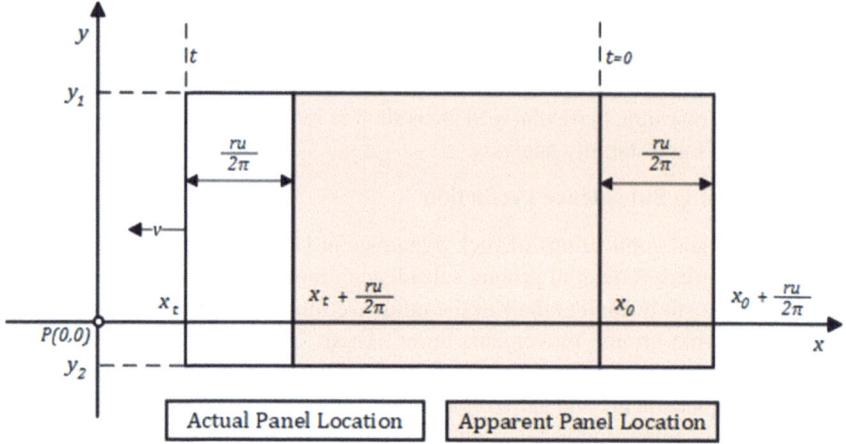

Fig. 8.9 Dynamic edge effect set shifting the extraction panel effect regarding point P (0, 0) [10]

horizontal displacement predictions in US mining applications and enhance safety and environmental management in mining operations [10].

The dynamic horizontal displacement (U_x^d) was calculated using the following equation:

$$U_x^d(x_t, x_0, y_1, y_2, z) = -B_s \cdot r_z \cdot T_x^d(x_t, x_0, y_1, y_2, z) \qquad (8.14)$$

where B_s is a dimensionless factor relating horizontal displacement to tilt, and T_x^d is the dynamic tilt.

Blast-Induced Fracturing and Rock Fragmentation

Another important application of rock dynamics in US mining engineering is the optimization of blast-induced fracturing and rock fragmentation. Understanding the dynamic behavior of rocks under high-strain-rate loading conditions is crucial for improving blasting efficiency, reducing energy consumption, and enhancing ore recovery.

Recent research by Braunagel and Griffith [11] investigated the microstructural controls on mixed-mode dynamic fracture propagation in crystalline and porous granular rocks, which has direct implications for blast design in US mining operations. The study compared the behavior of Westerly Granite (crystalline) and Berea Sandstone (porous granular) under dynamic loading conditions using notched semi-circular bend (NSCB) specimens in a split Hopkinson pressure bar (SHPB) apparatus. Key findings from this research include (Fig. 8.10).

The fracture initiation toughness of both rock types increased linearly with the loading rate under mode I (opening) conditions. For Berea Sandstone, the toughness values ranged from 0.83 to 2.64 MPa·m$^{1/2}$ across the tested loading rates. Similarly, for Westerly Granite, the range was between 3.29 and 8.80 MPa·m$^{1/2}$ [11]. This

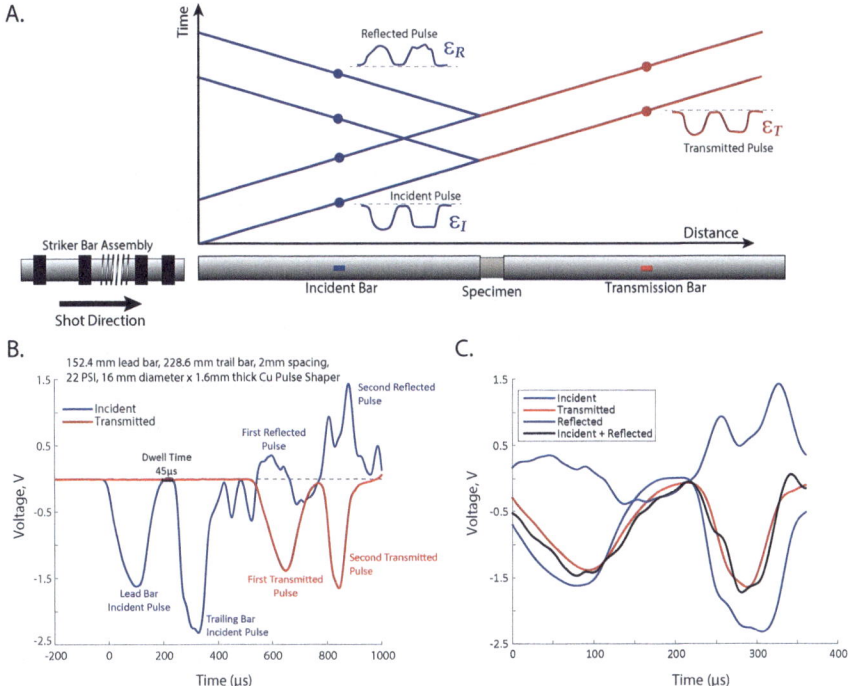

Fig. 8.10 a Conceptual illustration of SHPB apparatus with double striker bar and position-time diagram of wave propagation. **b** voltage changes recorded during the experiment. **c** dynamic force balance at specimen-bar interfaces [11]

trend provides critical insights for predicting rock breakage behavior under various blasting conditions in US mines.

$$K_{IC}^d = K_{IC}^{ss}\left(1 + \frac{\dot{K}_I}{K_{IC}^{ss}} \cdot 2 \cdot 10^{-5}\right) \quad (8.15)$$

where K_{IC}^d is the dynamic fracture toughness, K_{IC}^{ss} is the quasi-static fracture toughness, and $\dot{K}_I$ is the loading rate.

This study investigated fracture initiation and propagation dynamics in two rock types—Westerly Granite and Berea Sandstone—emphasizing their distinct behaviors in blasting operations. In Westerly Granite, the energy release rate increased with mode II (shear) contribution, from 247 J/m2 at a 0° notch angle to 424 J/m2 at a 30° angle. In contrast, Berea Sandstone exhibited a decrease in energy release with increasing shear, dropping from 800 J/m2 to 147 J/m2. Crack propagation velocities ranged from 180 to 760 m/s, with average velocities of 347 m/s (0.21cR) in Berea Sandstone and 383 m/s (0.12cR) in Westerly Granite. Fracture propagation toughness increased linearly with crack length in both rocks, with higher velocities in Westerly Granite correlating to increased toughness, while Berea Sandstone remained stable. Microstructural analysis revealed different fracture behaviors: in granite, fractures propagated through grains and along boundaries, while in sandstone, fractures were confined to cement bridges. The study also found that continuum-based models underestimate dynamic fracture toughness at lower velocities, suggesting that current blast design methodologies in US mining should be refined for more controlled blasting. These findings underscore the need for customized blast designs based on the specific rock types to optimize fragmentation and minimize damage to surrounding rocks [11].

The application of these advanced rock dynamics principles to blast design in US mining operations has the potential to improve ore recovery rates, reduce energy consumption, and minimize environmental impacts associated with blasting activities. Future research in this area should focus on developing more sophisticated numerical models that incorporate the observed microstructural effects and validating these models through large-scale blasting trials in diverse geological settings across the US.

Mine Stability Analysis

Rock dynamics principles play a crucial role in assessing and ensuring the stability of underground mine openings in the US. Recent research has provided new insights into the behavior of fractured rock masses under dynamic loading conditions, which has important implications for mine design and support systems. A study by Wood et al. [12] investigated the relationship between hydro-mechanical and east-dynamic properties of dynamically stressed fractured rock, using tightly controlled laboratory experiments on tensile-fractured Westerly granite samples. The experiments were conducted under conditions representative of varying depths and stresses in the Earth's crust, making the findings particularly relevant to US mining operations (Fig. 8.11).

The study found that relative permeability change ($\Delta k/k_0$) increases with increasing pore pressure oscillation amplitude but generally decreases with increasing normal stress on the fracture. This relationship is described by:

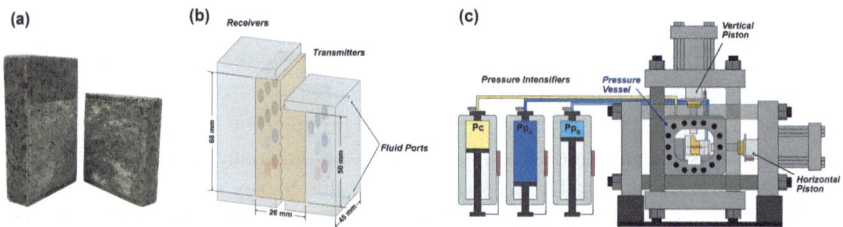

Fig. 8.11 a Pre-fractured L-shaped Westerly granite sample. b transmitter receiver pairs used in active source ultrasonic monitoring embedded inside loading block. c biaxial loading apparatus [12]

8.2 Engineering Applications in the US

$$\frac{\Delta k}{k_0} = f(P_{osc}, \sigma_n) \tag{8.16}$$

where P_{osc} is the pore pressure oscillation amplitude and σ_n is the normal stress.

This finding has implications for assessing the stability of mine openings in water-bearing rock masses, as changes in fluid pressure due to mining activities or natural seismic events could alter the permeability and strength characteristics of the rock mass [12].

The research also revealed that relative change in wave velocity (R_0) and wave velocity amplitude modulation (R_1) increase with oscillation amplitude but generally decrease with increasing effective stress after an initial increase. This relationship can be expressed as:

$$R_0, R_1 = f\left(P_{osc}, \sigma'_{eff}\right) \tag{8.17}$$

where σ'_{eff} is the effective stress.

These changes in elastic wave velocities could potentially be used as indicators of stress changes and damage accumulation in rock masses surrounding mine openings [12].

The study found that the real contact area of fractures increases with applied stress in a Hertzian-contact-like manner, following a cubic relationship:

$$S = k \cdot \sigma_n^{3/2} \tag{8.18}$$

where S is the real contact area and k is a constant (Fig. 8.12).

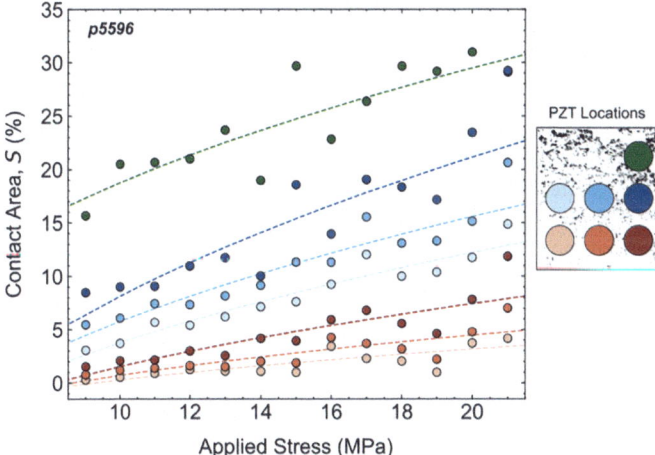

Fig. 8.12 Percent area of fracture in contact within each piezoelectric transducer "footprint area" as a function of applied stress for one experiment [12]

The research highlighted the importance of understanding the mechanical behavior of fractured rock masses, particularly in relation to shear strength and stability in mine environments. It found that permeability and wave velocity recovery after oscillations follows a logarithmic time dependence, which is crucial for assessing the long-term stability of mine openings after dynamic events like blasting or seismic activity. Additionally, hysteresis in permeability and nonlinear elastic parameters between loading and unloading phases must be considered when evaluating mine opening stability under cyclic loading conditions such as repeated blasting [12].

The study's findings are crucial for mine stability analysis and design, particularly in US mining operations. The relationship between permeability changes and stress state highlights the need for designers to account for stress redistribution, which can impact both excavation stability and dewatering system efficiency. Additionally, insights into the dynamic properties of fractured rock under dynamic loading provide a foundation for more precise rock mass characterization, guiding the development of improved support systems and excavation sequences. Time-dependent behavior, including recovery patterns and hysteresis, is important for long-term stability assessments, especially in abandoned or closing mines. These findings also suggest opportunities for advancing mine monitoring strategies, such as using seismic wave velocity data to monitor stress state and fracture changes. Incorporating realistic fracture behavior into numerical models can enhance stability predictions, contributing to safer and more efficient mining operations.

To further advance the application of rock dynamics principles in mine stability analysis, future research in the US should focus on: Scaling up laboratory observations to field-scale behavior of fractured rock masses. Developing and validating advanced numerical models that incorporate the observed coupled hydro-mechanical and east-dynamic behaviors. Investigating the long-term evolution of fractured rock properties under repeated dynamic loading conditions typical of mining environments. Exploring innovative monitoring techniques that can leverage the observed relationships between stress state, fluid flow, and elastic wave properties to provide real-time assessment of min stability.

By continuing to advance our understanding of rock dynamics in the context of mine stability, the US mining industry can develop more efficient, safer, and environmentally sustainable mining practices.

8.2.2 Civil Engineering Applications

The principles of rock dynamics play a crucial role in various civil engineering applications in the United States, particularly in the design and construction of underground structures, foundations, and slope stability analysis. Recent research has provided new insights into the behavior of rocks under dynamic loading conditions, leading to improved design methodologies and risk assessment techniques.

Tunnel and Underground Structure Design

The design of tunnels and underground structures in the US increasingly relies on an advanced understanding of rock dynamics to ensure safety, stability, and long-term performance. Recent research has shed light on the complex behavior of fractured rock masses under dynamic loading, which is particularly relevant for underground structures in seismically active regions or those subject to blast-induced vibrations.

A study by Shokouhi et al. [13] investigated the relationship between elastic wave velocity changes and permeability changes in fractured rock samples subjected to dynamic stressing. This research has important implications for understanding the behavior of rock masses surrounding underground structures during seismic events or other dynamic loading scenarios. Key findings from this study that are applicable to tunnel and underground structure design include:

The research demonstrated that rapid stress cycling can significantly affect the dynamic compressive strength of rocks. For Westerly Granite, the study found that: Relative velocity change (Δ_c/c_0) scaled with oscillation amplitude for both normal stress (σ_n) and pore pressure (P_p) oscillations. Δ_c/c_0 generally increased with oscillation frequency, especially above 10 Hz. The amplitude of velocity oscillations (d_c/c_0) scaled linearly with oscillation amplitude (Fig. 8.13).

The study highlights the complexity of rock mass dynamic responses around underground structures, especially under high-frequency loading conditions. It was found that relative permeability changes generally scaled with oscillation amplitude, with small amplitude oscillations decreasing permeability and large amplitude oscillations increasing it. Pore pressure oscillations were about 2.5 times more effective

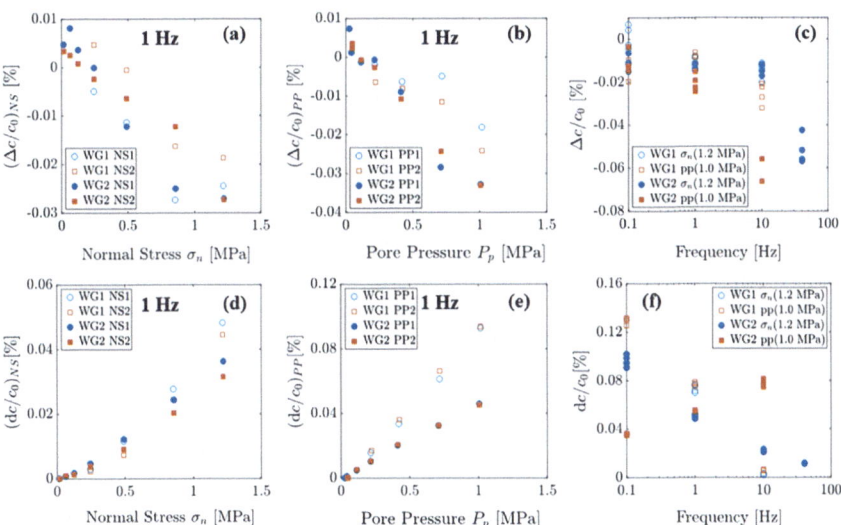

Fig. 8.13 Dependencies of relative velocity change and average velocity change amplitude on the amplitude and frequency of normal stress and pore water pressure oscillations [13]

than normal stress oscillations in enhancing permeability. These findings emphasize the importance of considering dynamic loading events in the design of underground structures within water-bearing rock masses, as such events can alter permeability and groundwater flow patterns. Additionally, the observed correlation between elastic and hydraulic property variations suggests that changes in rock elasticity can indicate potential shifts in water ingress and stability risks for underground structures [13].

Based on these findings regarding the dynamic response of rock masses, several recommendations can significantly enhance the design of tunnels and underground structures in the US. First and foremost, incorporating advanced dynamic analysis techniques is essential, as these approaches can effectively account for the nonlinear elastic dynamic response of fractured rock masses observed under dynamic loading conditions. Utilizing sophisticated numerical modeling tools capable of capturing the frequency-dependent behavior of rocks to improve structural resilience and reliability.

Additionally, engineers should also account for the potential evolution of rock mass permeability during dynamic loading events, particularly in water-sensitive environments where maintaining a dry interior is crucial. Design strategies might include implementing flexible waterproofing systems or adaptive drainage solutions to effectively address potential changes in groundwater flow patterns.

Furthermore, seismic monitoring systems may serve as valuable tools for assessing the hydraulic properties of the rock mass surrounding underground structures. The observed correlation between changes in elastic wave velocity changes and alterations in permeability suggests that such monitoring systems could provide early warnings of stability concerns or changes in groundwater conditions, thereby enabling proactive intervention.

Moreover, the design of support systems for underground structures should reflect the nonlinear behavior of fractured rock masses under dynamic loads. This may involve creating more robust support elements capable of withstanding significant deformations or designing adaptive support systems that can adjust to evolving rock mass conditions over time.

Finally, risk assessment frameworks for underground structures should integrate the potential impact of dynamic loading events on both the mechanical and hydraulic properties of the surrounding rock mass. Scenario-based assessments that evaluate the cumulative effects of multiple dynamic events over a structure's lifespan can provide a more comprehensive understanding of risks and guide the implementation of effective mitigation measures.

Foundation Design in Rock

The design of foundations in rock for large structures such as skyscrapers, bridges, and dams in the US must consider potential dynamic loading conditions, including those generated by earthquakes and man-made vibrations. Recent research in rock dynamics has yielded new insights that can improve foundation design practices.

Mc Beck et al. [14] conducted dynamic X-ray microtomography experiments coupled with machine learning techniques to study fracture development in rocks

8.2 Engineering Applications in the US

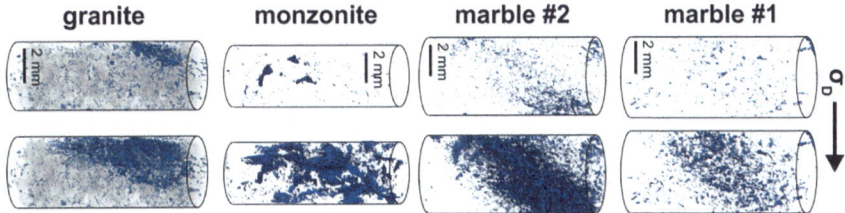

Fig. 8.14 Fracture development within in situ X-ray tomography triaxial compression experiments [14]

under compression. While not focused on foundation design, their findings are relevant for assessing rock mass behavior under foundation loads, particularly in dynamic conditions. Key predictors of fracture growth included fracture size, orientation, and network clustering. The study revealed that growing fractures tend to be smaller, shorter, thinner, more obliquely oriented (30°–60° from σ1), and more clustered compared to closing fractures. Notably, the distance to the nearest fracture emerged as a robust predictor for fracture propagation, simplifying the analysis of fracture systems in rock masses under dynamic loading scenarios (Fig. 8.14).

To improve foundation design in rock within the US, several recommendations arise from recent findings regarding fracture networks and their implications for structural stability. First, geotechnical site investigations should prioritize a detailed characterization of fracture networks, focusing on their size, orientation, and clustering. Advanced imaging techniques, such as borehole televiewers, can be employed to gather precise data on these features, enhancing the understanding of the subsurface conditions.

Furthermore, existing rock mass classification systems currently used in foundation design, such as the Rock Mass Rating (RMR) or Q-system, could be refined to integrate more detailed analyses of fracture network characteristics. This refinement would facilitate more accurate assessments of rock mass behavior under varying loading conditions.

In addition, the utilization of advanced numerical modeling techniques is crucial for foundation design, enabling the incorporation of realistic representations of fracture networks and their evolution under static and dynamic loads. Approaches like discrete element methods or hybrid continuum dis-continuum modeling can simulate the complex interactions within fractured rock masses, providing more accurate predictions of foundation performance.

In seismically active regions, dynamic analyses should account for the potential preferential growth of fractures along specific orientations and the impact of fracture clustering on the overall rock mass response. These considerations are critical for ensuring the resilience of foundations under seismic loading.

Moreover, risk assessment frameworks for foundation design should incorporate the potential for fracture propagation and coalescence during dynamic events. Developing probability-based approaches to evaluate the likelihood of such processes, utilizing predictors identified through detailed site analyses and modeling efforts, can enhance risk management strategies.

For critical structures, implementing monitoring systems capable of detecting changes in fracture network characteristics over time is recommended. These systems would provide valuable data for assessing long-term foundation performance and stability, enabling proactive maintenance and mitigation measures to address emerging risks.

Slope Stability Analysis

Rock dynamics principles are crucial for assessing the stability of both natural and engineered slopes in the US, particularly in regions prone to seismic activity or subject to blast-induced vibrations from nearby mining or construction activities. Recent research has provided new insights into the dynamic behavior of rock slopes, leading to improved analysis and design methodologies.

For instance, a study by Smith and Griffith [15] investigated the evolution of pulverized fault zone rocks under dynamic tensile loading during successive earthquakes. While this research focused on fault zone processes, its findings have relevant implications for understanding the behavior of rock slopes under repeated dynamic loading events.

The research revealed that with each successive loading event, fracture density increases while fragment size decreases with each successive loading event. This relationship follows a power law:

$$d = d_1 N^{-l}, F = F_1 N^m \qquad (8.19)$$

where d is fragment size, F is fracture density, N is the number of loading events, l and m are power law exponents.

This finding suggests that slopes subjected to repeated dynamic loading events, such as those in seismically active areas, may experience progressive damage accumulation over time [15] (Fig. 8.15).

The study also observed that pulverization can occur at strain rates as low as 10^{-3} s^{-1} under tensile loading, which is much lower than previously thought for compressive loading. This implies that rock slopes may be more susceptible to damage from dynamic loading events than previously assumed [15].

The research found that fragment size is primarily controlled by strain energy at low strain rates ($<10^{-3}$ s^{-1}). This suggests that energy-based approaches may be more appropriate for assessing slope stability under dynamic loading conditions [15].

To enhance slope stability analysis and design practices in the US, several key considerations should be addressed. In seismically active regions, the potential for cumulative damage accumulation over multiple loading events must be integrated into slope stability assessments. This integration necessitates the development of damage models capable of accounting for the observed power-law relationship between fracture density and the number of loading events. Furthermore, strain rate effects should be carefully evaluated, as rock pulverization can occur at lower strain rates under tensile loading. Therefore, incorporating sophisticated constitutive

8.2 Engineering Applications in the US

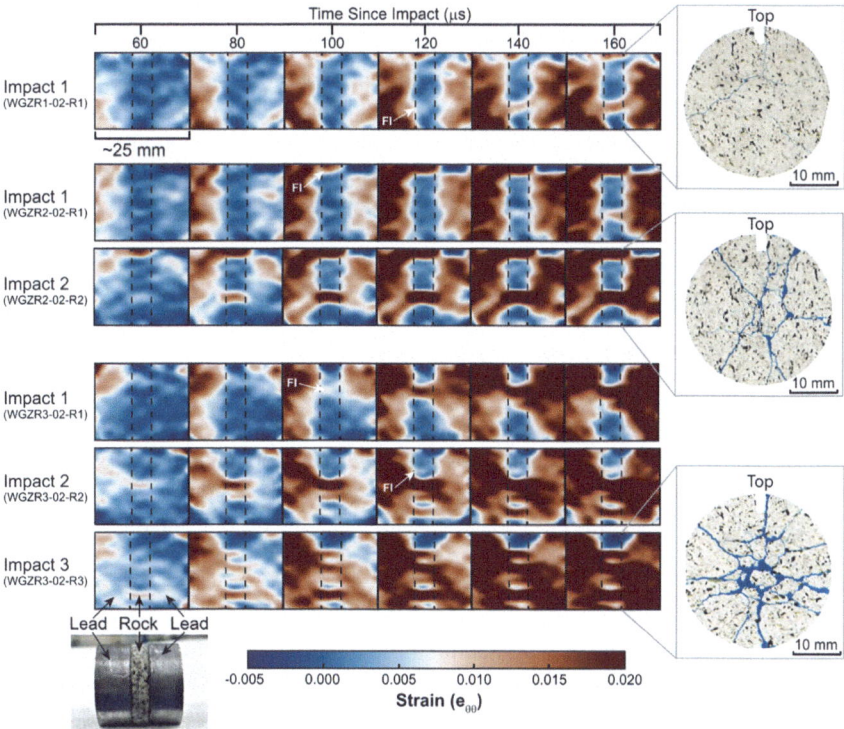

Fig. 8.15 Records of surface fracture initiation with successive tensile load in [15]

models that capture strain rate-dependent behavior across diverse loading conditions is essential for accurate analysis.

Energy-based approaches provide an effective framework for assessing rock mass damage, particularly given the finding that fragment size at low strain rates is primarily governed by strain energy. Implementing energy-based failure criteria in slope stability models can improve predictions of damage potential under dynamic loading conditions. Additionally, microstructural changes in the rock mass over time should be considered in stability analyses. This consideration may involve incorporating detailed representations of fracture networks into numerical models or developing methods to account for evolving rock mass properties.

Moreover, risk assessment frameworks for slope stability must address the potential for progressive damage accumulation due to repeated dynamic loading. Employing probabilistic approaches that evaluate the likelihood and consequences of slope failure under various loading scenarios and cumulative damage states are critical for effective risk management. To support this, monitoring systems should be implemented for critical slopes to detect changes in fracture density and rock mass properties over time. A combination of geophysical techniques, remote sensing, and in-situ instrumentation can provide valuable data for long-term stability assessments.

Finally, slope stabilization measures such as rock bolts, anchors, and shotcrete should be designed to accommodate progressive damage accumulation and evolving rock mass properties. This may involve the development of more robust or adaptive stabilization systems capable of responding to changes in rock conditions over time. By incorporating these advanced principles of rock dynamics into slope stability analysis and design, engineers can achieve more reliable and resilient slope designs, particularly in regions prone to seismic activity or frequent anthropogenic dynamic loading.

8.2.3 *Earthquake Engineering Applications*

The field of earthquake engineering in the US has greatly benefited from advancements in rock dynamics research, leading to improved understanding of seismic wave propagation, fault rupture mechanics, and ground motion prediction. These insights play a critical role in seismic hazard assessment, structural design, and risk mitigation strategies.

Seismic Wave Propagation and Ground Motion Prediction

Understanding seismic wave propagation through complex geological structures is crucial for accurate ground motion prediction and seismic hazard assessment. Recent research in rock dynamics has provided new insights into the factors that influence wave propagation and attenuation in fractured and heterogeneous rock masses.

For example, a study by Sleep [16] investigated the mechanics of rock damage and nonlinear attenuation during strong seismic shaking in brittle rocks. This research has important implications for understanding ground motion characteristics in regions with complex geological structures, such as those found in many parts of the western United States. Key findings from this study that are relevant to seismic wave propagation and ground motion prediction include:

The study presents a model of the shallow subsurface as fractured regolith that undergoes repeated failure during strong earthquakes. In this model, the rock self-organizes to exhibit a range of prestressed, which is represented by a scalar fractal distribution. This organization allows favorably oriented fractures to fail at low dynamic stresses, attenuating strong seismic waves [16]. The research derived a key equation for nonlinear attenuation $Q^{(-1)}$:

$$\Delta E_T = \Delta E_R + \Delta E_U \left[\tau_{norm} \left(\frac{1}{\tau_Y - \tau_D} - \frac{1}{\tau_Y} \right) \right] \left(\frac{\tau_D^2}{2G} \right) \qquad (8.20)$$

Where τ_{norm1} is a normalizing stress constant, τ_Y is the yield stress, τ_D is the dynamic stress, and G is the shear modulus.

This equation shows how attenuation increases rapidly with dynamic stress, which has implications for predicting ground motions in near-fault regions where strong shaking is expected. Furthermore, the study discusses how energy is partitioned

8.2 Engineering Applications in the US

during failure highlighting the creation of residual stress, the dilation of cracks against confining pressure, and the sliding of cracks during dynamic friction. Notably, these energy sinks are found to be comparable in magnitude, which is important for understanding the overall energy budget of seismic events [16] (Fig. 8.16).

The research incorporates rate and state friction laws to model the macroscopic behavior of fractured rock. A key equation presented is:

$$\tau = P\left[\mu_0 + a \ln(\varepsilon'/\varepsilon'_{ref}) + b \ln(\psi/\psi_{norm})\right] \quad (8.21)$$

where τ is shear traction, P is normal traction, μ_0 is the base friction coefficient, a and b are constants, ε' is strain rate, and ψ is the state variable.

This formulation provides a more realistic representation of fault behavior during seismic events, which is crucial for accurate ground motion prediction.

To enhance seismic wave propagation modeling and ground motion prediction practices in the US, several advanced considerations should be integrated into current methodologies. First, ground motion prediction models need to incorporate nonlinear site response effects, especially in near-fault regions where strong shaking is prevalent. This requires the development and implementation of sophisticated numerical models capable of capturing the observed nonlinear attenuation behavior.

Additionally, the fractal distribution of prestressed in the shallow subsurface, as identified in the study, should be taken into account during site response analyses. This necessitates new approaches to accurately characterize and model the stress state of near-surface rock masses. Moreover, seismic hazard assessments must incorporate the various energy dissipation mechanisms highlighted in the findings. Refining

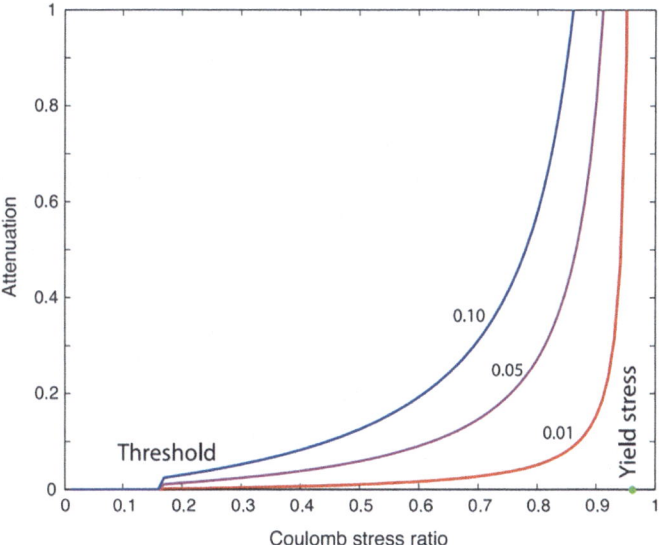

Fig. 8.16 Curves are shown for various values of t_{min} normalized to Coulomb stress ratio [16]

existing energy-based ground motion prediction equations to include these additional energy sinks will lead to more accurate assessments of the energy available for seismic wave propagation.

Furthermore, fault rupture models used in ground motion simulations should be updated to incorporate rate and state friction laws, thereby providing a more accurate representation of fault behavior during seismic events. This improvement may require updates to dynamic rupture simulation codes and thorough validation against observed ground motion data.

In near-fault regions, the findings on nonlinear attenuation and energy partitioning are particularly significant. Ground motion prediction practices in these areas should be revised to account for these effects, enabling more precise estimates of extreme ground motions. By integrating these principles into seismic wave modeling and ground motion prediction, engineers and researchers can achieve more reliable assessments of seismic hazards and improve the resilience of structures in seismically active areas.

The study also presents several useful scaling relationships, such as:

$$P_{lith} = \frac{\rho g c_S}{\omega}, \tau_S = \frac{\mu_S \rho g c_S}{\omega}, U_0 \approx \frac{\mu_S g}{\omega} \quad (8.22)$$

where ρ is density, P_{lith} is lithostatic pressure, c_S is S-wave velocity, ω is angular frequency, τ_S is failure shear stress, μ_S is coefficient of friction, and U_0 is particle velocity.

Incorporating these relationships into ground motion prediction models to improve their physical basis and predictive capabilities across different magnitude and distance ranges. Additionally, the complex nonlinear behavior observed in the study suggests that ground motion prediction models should place greater emphasis on uncertainty quantification, particularly for extreme events where empirical data is limited.

Fault Rupture Mechanics and Earthquake Source Characterization

Understanding fault rupture mechanics is crucial for accurately characterizing earthquake sources and predicting ground motions. Recent research in rock dynamics has provided new insights into the processes governing fault rupture initiation, propagation, and arrest.

Conrad et al. [17] studied the influence of frictional melt on fault behavior during large earthquakes, particularly on the San Andreas Fault, using novel experimental techniques. The study revealed four distinct phases of fault behavior: Phase A, where no melting occurs and low-amplitude stick–slip events dominate; Phase B, marked by the onset of melting and increased frequency of stick–slip events; Phase C, with significant melting and large stress drops; and Phase D, where stable sliding and consistent melt production occur. The researchers identified key mechanisms affecting fault strength, including flash weakening, melt lubrication, viscous braking, and fault welding. These processes control the dynamics of fault rupture and slip arrest. Additionally, the study provided insights into the energy budget of seismic events by analyzing energy flux and frictional power density.

8.3 Engineering Applications in Europe

The European continent has been at the forefront of rock dynamics research and applications, with numerous projects and studies contributing to our understanding of this field. This section explores key engineering applications of rock dynamics in Europe, emphasizing noteworthy case studies, technological advancements, and research findings.

8.3.1 Rock Dynamics in Geothermal Development

One of the most prominent areas of rock dynamics application in Europe is found in the geothermal energy development. The Rittershoffen geothermal field in Alsace, France, provides an excellent example of how rock dynamics principles are applied in the development and management of enhanced geothermal systems (EGS) [18]. The study of induced seismicity during the development of this geothermal field has provided valuable insights into the behavior of deep fractured reservoirs under various stimulation techniques (Fig. 8.17).

The Rittershoffen geothermal project involved drilling wells GRT-1 (2580 m) and GRT-2 (3200 m), and the implementation of thermal, chemical, and hydraulic stimulation on GRT-1. A seismic monitoring network with 43 stations recorded 1348 induced seismic events. Thermal stimulation of GRT-1 (4135 m^3 cold fluid injection) resulted in 146 seismic events, with significant activity occurring 39 h after injection began, indicating uncritically stressed zones and highlighting the importance of rock cohesion. Hydraulic stimulation (3180 m^3 fluid injection) caused 824 seismic events, with a delay in seismicity onset and evidence of the Kaiser effect, revealing stress memory in the rock. The seismic distribution aligned with the Rittershoffen fault, suggesting reactivation of this pre-existing fault structure. The study demonstrates the application of rock dynamics principles in mitigating induced seismicity risks in geothermal energy projects and optimizing stimulation techniques [18] (Fig. 8.18).

In addition to geothermal applications, rock dynamics also plays a critical role in underground construction and tunneling. A notable case study in this context is the piora adit in Switzerland, which showcases the use of seismic imaging techniques in challenging geological environments [19]. This study compared two novel seismic sources–a pneumatic impact source and a magneto strictive vibrator source – specifically for near-surface seismic imaging in the crystalline rocks of the piora adit (Fig. 8.19).

In seismic imaging for underground construction, the choice of seismic source significantly impacts data quality. The pneumatic impact source, generating signals

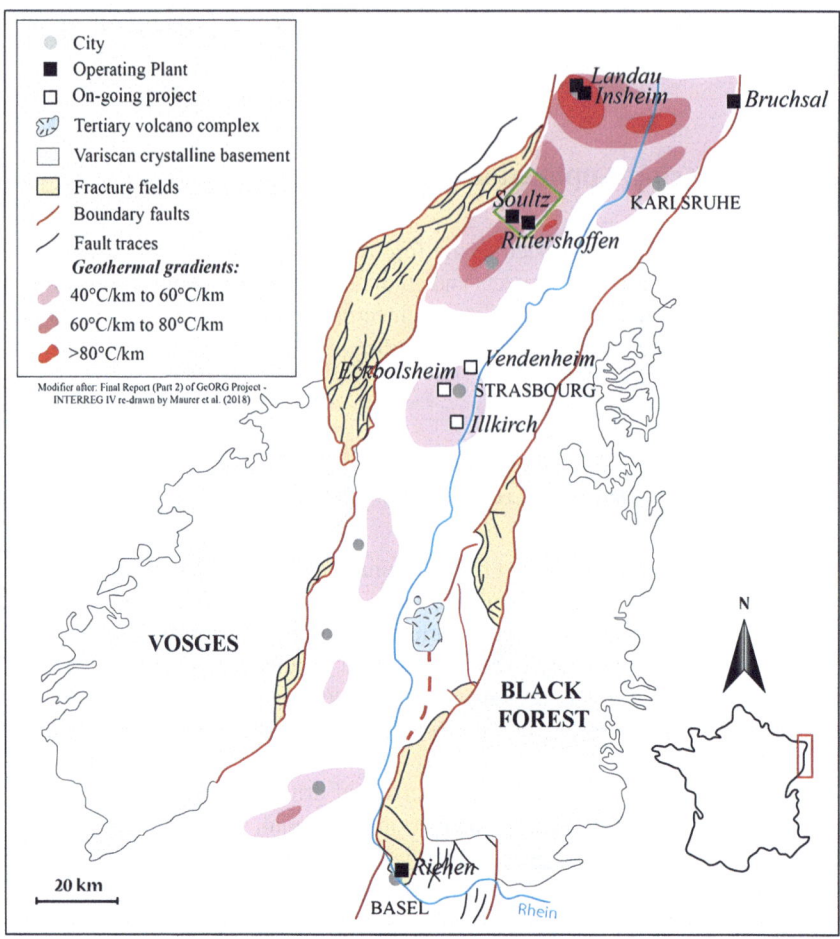

Fig. 8.17 Simplified geological map of the Upper Rhine Graben indicating the geothermal gradients and status of current deep geothermal fields modified from the final report of the GeORG Project INTERREG IV 2013 [18]

in the 100–2000 Hz range, is effective for hard rock with exploration up to 250 m, primarily producing surface waves along tunnel walls. In contrast, the magnetostrictive vibrator source, with a frequency range of 100–3000 Hz, provides better resolution for shallow imaging and explores up to 100 m. In the Piora adit, the vibrator source yielded clearer images of geological structures up to 80 m from the tunnel, demonstrating the value of source selection and processing techniques for underground rock characterization [19].

The study on thermo-hydro-mechanical (THM) processes in fractured geothermal reservoirs highlighted the significance of numerical modeling in geothermal energy extraction. By employing a discrete fracture network (DFN) model and conducting

8.3 Engineering Applications in Europe 471

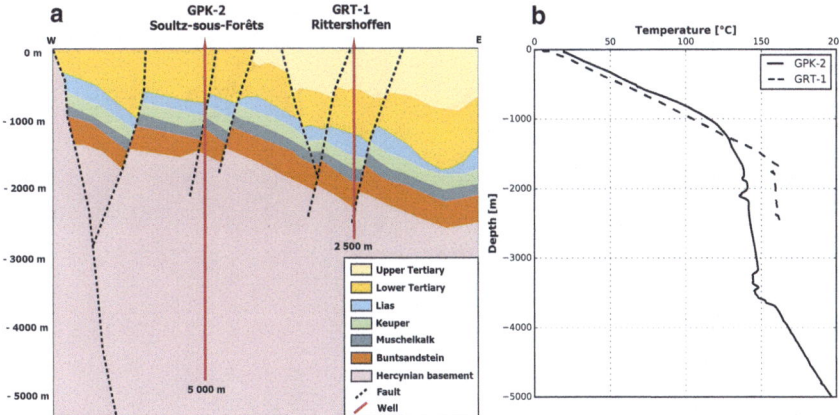

Fig. 8.18 a East–West simplified vertical section of the Rhine Graben crossing the Soultz-sous-Forêts GPK-2 and the Rittershoffen GRT-1 wells; **b** temperature logs, at equilibrated thermal conditions, for the Soultz-sous-Forêts GPK-2 well (from Genter et al. 2010) and for the Rittershoffen GRT-1 well (vertical scale is in MD) [18]

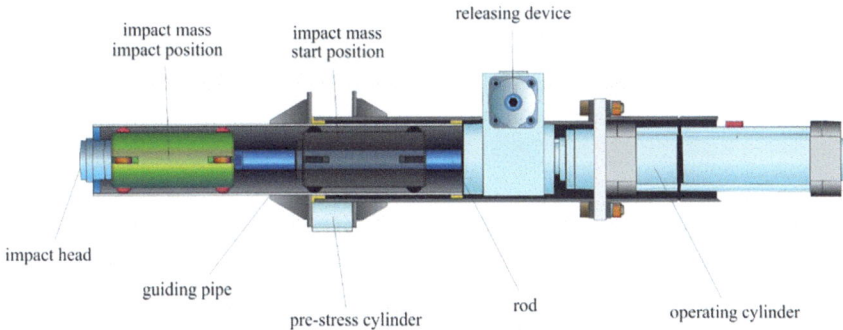

Fig. 8.19 Mechanical design of the pneumatic impact source. Total length of this impact system is 1.3 m, total system weight excluding auxiliary systems is nearly 44 kg [19]

sensitivity analyses, the research identified three key parameters: fracture aperture, rock matrix permeability, and wellbore radius. Fracture aperture and rock matrix permeability were found to reduce thermal breakthrough time, while wellbore radius positively affected mass flux and energy recovery. Additionally, the study emphasized that thermo-elastic effects play a more critical role than porous-elastic effects in determining reservoir stress distribution, offering valuable insights for optimizing geothermal energy production in fractured reservoirs, particularly in European projects [20] (Fig. 8.20).

The findings from the research on THM processes have important implications for the design and optimization of enhanced geothermal systems (EGS) in Europe. By concentrating on the identified key parameters, engineers can better predict and

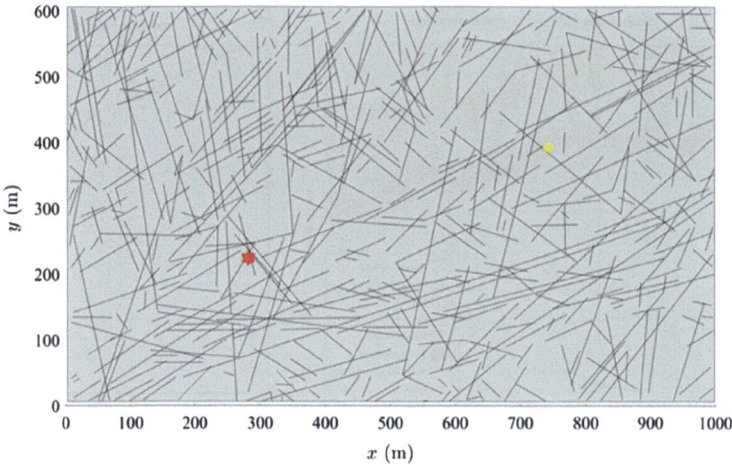

Fig. 8.20 Geometry of the model. Here, yellow circle is the injection well (x, y ¼ 740, 385) whereas red star is production well (x, y ¼ 280,215) [20]

control the thermal and hydraulic performance of geothermal reservoirs, while also managing the risk of induced seismicity. Moreover, the study provides a framework for optimizing EGS designs based on the most influential parameters, which can be applied to various geological settings throughout Europe [20].

8.3.2 Rock Dynamics Principles in Natural Hazard Studies

The application of rock dynamics principles is prominently illustrated in the study of natural hazards, particularly landslides. A notable case is the Vence landslide in southeastern France, which has been the subject of a long-term geophysical monitoring study, employing electrical resistivity tomography (ERT) to understand the hydrological dynamics of the unstable slope [21]. This 9.5-year multiparametric survey combined daily ERT measurements with rainfall records and groundwater level monitoring to characterize fluid circulation within the landslide mass.

The analysis identified three main clusters of resistivity data corresponding to distinct hydrological behaviors within the landslide. These clusters were linked to different geological units: one representing a consistently saturated landslide body composed of sandy-clay material, while the others were associated with underlying fractured limestone with water-filled faults. The long-term ERT monitoring revealed both rapid responses to rainfall events in shallow piezometers and deeper hydrological changes over time [21] (Fig. 8.21).

The application of geophysical monitoring techniques to landslide dynamics underscores the significance of long-term data collection in understanding complex slope hydrology. This approach allowed for the identification of distinct hydrological

8.3 Engineering Applications in Europe

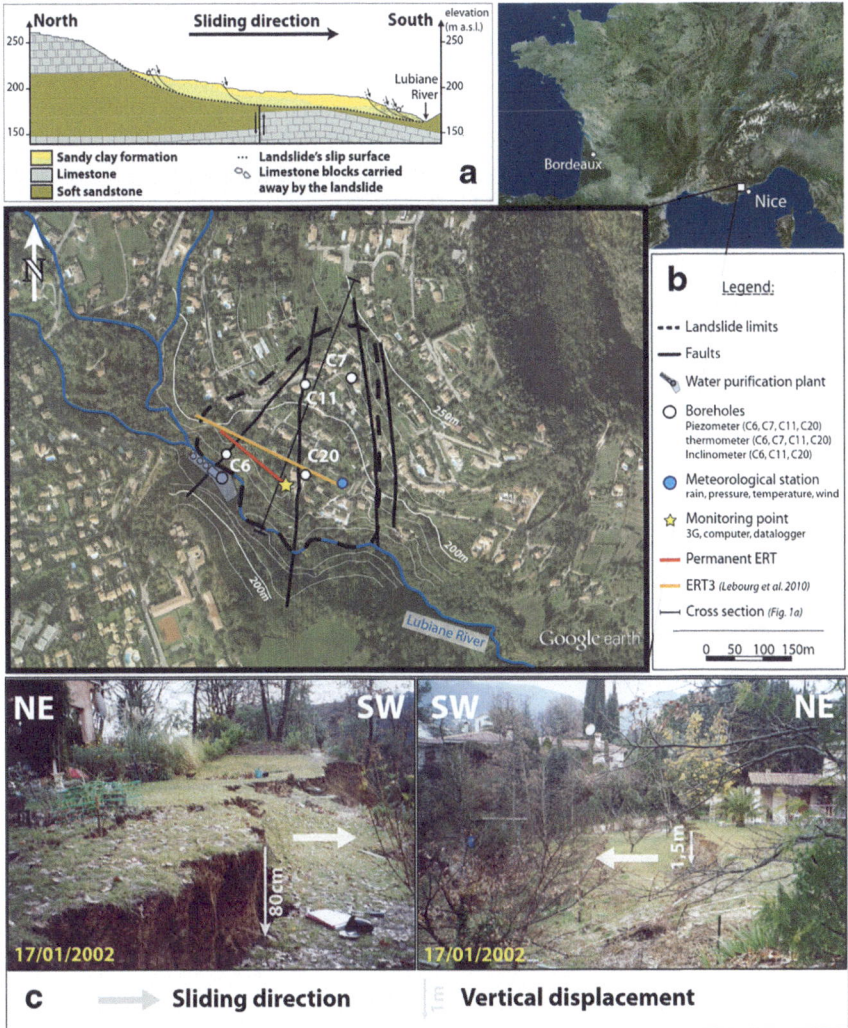

Fig. 8.21 Presentation of the Vence landslide. **a** simplified geological cross section of the landslide along sliding direction (see b for the location). **b** layout of the field observations and the instrumentation installed on site. **c** photographs of the head scarp after winter 2000 event [21]

units without relying on prior geological assumptions, providing crucial qualitative information on the spatial and temporal variability of slope hydrological behavior. Furthermore, such information is vital for optimizing the placement of additional instrumentation and improving landslide risk assessment [21].

The study on the Rochechouart impact structure highlights the versatility and effectiveness of electrical resistivity methods for characterizing impact-generated rocks. By combining various techniques at different scales, the researchers provided important insights into the complex geology of impact structures, including variations in melt content, porosity, and deformation within the basement rocks. This kind of multi-scale approach could be useful in other high-stress geological environments, such as fractured geothermal reservoirs, where understanding rock dynamics and fracture networks is crucial for optimizing energy extraction. The results from the study could also offer valuable methods for investigating complex geological settings with similar deformation processes [22] (Fig. 8.22).

The application of rock dynamics principles is also evident in the study of anthropogenic activities that can induce seismicity. An exemplary case is the Doseo basin in Central Africa, while not in Europe, offers valuable insights applicable to European oil and gas exploration. Research into post-rifting uplift and inversion within this intra-plate basin has identified two major exhumation events affecting a substantial portion of the African plate and southern Tethys realm [16].

The first major event, occurring from the late Eocene to Miocene event (~45–20 Ma), led to the removal of ~1050–1450 m of geological section. This exhumation resulted in the creation of angular unconformities, folds, and fault-related anticlines. Notably, this period coincided with the peak maturation of source rocks and the

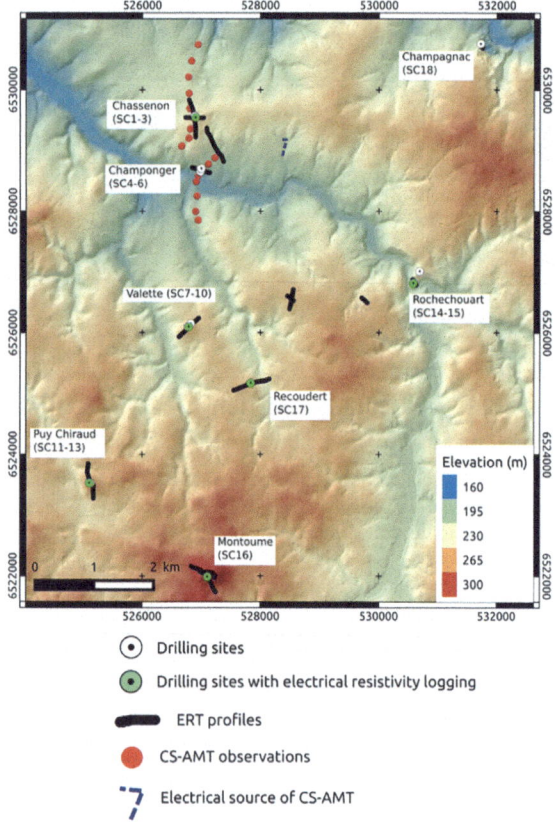

Fig. 8.22 Location of the electrical measurements performed via downhole logging, electrical resistivity tomography (ERT) soundings, and controlled-source audio-magneto telluric (CS-AMT) data. Background is shaded and colored digital elevation model grid with 5 m pixel size [22]

8.3 Engineering Applications in Europe

establishment of compressional traps, which are crucial components of the Cenozoic petroleum system [23] (Fig. 8.23).

The study concludes that the uplift within intra-plate basins is primarily driven by far-field lateral compression from active plate boundaries, rather than by mantle-sourced forces. This uplift is temporally associated with the convergence between Africa and Eurasia and the final Africa-Europe collision. Understanding these tectonic forces shaping basin evolution is crucial, as it has significant implications

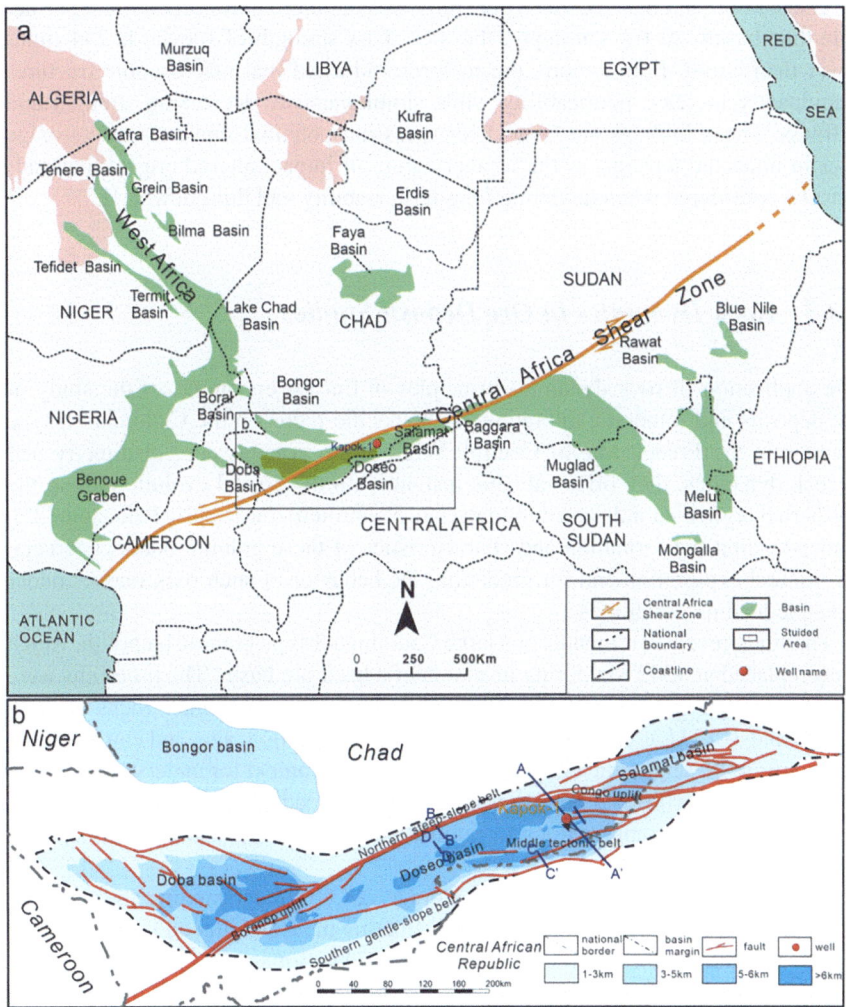

Fig. 8.23 **a** simplified tectonic map of the Western and Central African Rift System (WCARS); **b** tectonic framework and sedimentary thickness of the Doseo basin, and the location of Kapok-1 well [23]

for petroleum exploration in Europe, particularly in areas affected by similar far-field tectonic forces [23].

In addition to tectonic influences, rock dynamics plays a vital role in assessing the long-term stability and safety for nuclear waste disposal sites. Although not specifically focused on Europe, a study examining the petrophysical properties and mechanical behavior of impact melt-bearing breccia or suevite, from the Ries impact crater in Germany offers valuable insights relevant to nuclear waste storage in crystalline rocks [24].

The study found that suevite is generally weaker, more porous, and more permeable than basalt. At 0.5 km depth, the rock mass strength of suevite is 2–4 times lower than basalt. Furthermore, the research indicated that macroscopic fractures significantly increase permeability while simultaneously decreasing strength and stiffness. These findings are critical for assessing potential nuclear waste storage sites in impacted terrains, as the weaker nature of impact-altered crustal materials must be considered when modeling long-term stability and fluid flow [24].

8.3.3 Rock Dynamics in Ore Deposit Studies

The application of rock dynamics principles in Europe encompasses the study of ore deposits and mineral exploration, with specific focus on the Cadomian S-type granites in the French Massif Central. While these granites are not directly tied to rock dynamics, they offer valuable insights into the crustal evolution along the northern margin of Gondwana during the late Neoproterozoic to early Paleozoic [25]. Understanding the formation and characteristics of these granitic bodies is crucial for mineral exploration and for predicting the behavior of such rock masses under dynamic loading conditions.

The study revealed remnants of a large Cadomian S-type granitic batholith, which was emplaced at ~542 Ma during inversion of a back-arc basin. The protoliths were formed through the partial melting of Ediacaran metasedimentary rocks at 750–900 °C and 3–15 kbar. This information regarding the origin, age, and emplacement conditions of these granitic bodies provides valuable context for understanding their mechanical properties and behavior under dynamic loading. Such insights are critical for engineering applications in regions with similar geological histories [25] (Fig. 8.24).

In the field of carbon capture and storage (CCS), understanding the dynamic behavior of reservoir rocks is critical for evaluating the feasibility and safety of CO_2 injection. Although the specific study under consideration is not centered in Europe, it analyzes the formation of quartz and cristobalite ballen in impact melt rocks from the Ries impact structure in Germany, offering valuable insights into silica behavior under extreme pressure and temperature conditions [26]. This knowledge is pertinent for predicting the long-term behavior of silica-rich reservoir rocks during the pressure and temperature changes associated with CO_2 injection.

8.3 Engineering Applications in Europe 477

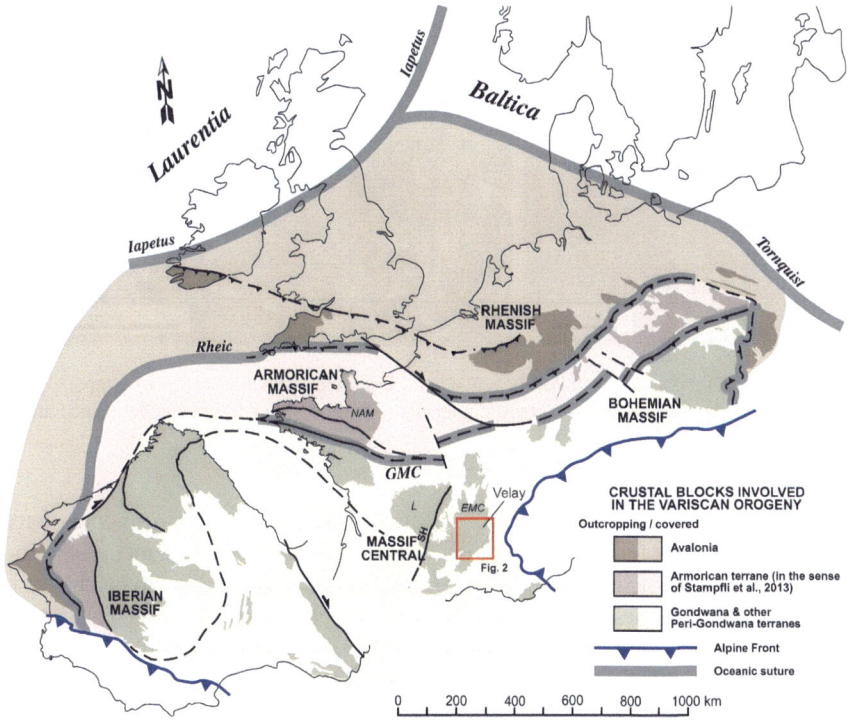

Fig. 8.24 Sketch map of the crustal blocks involved in the Variscan orogeny, adapted from Ballèvre et al. (2014). Yellow star highlights the location of the study area. Oceanic sutures: GMC, Galicia-Massif-Central Ocean. Regional subdivisions: NAM, North Armorican Massif; L, Limousin (Western Massif Central); EMC, Eastern Massif Central. Shear zones: SH, Sillon Houiller (coal line) [25]

The study proposes a formation mechanism for ballen structures analogous to perlitic structures observed in volcanic rocks. This mechanism involves shock transformation of fluid inclusion-rich quartz to hydrous amorphous silica, followed by rapid decompression and cooling causing dehydration and strain-induced globular fracture patterns. The crystallization that subsequently occurs within these globular domains results in the distinctive ballen texture [26]. Consequently, this detailed understanding of silica behavior under extreme conditions can inform models of reservoir rock response to CO_2 injection and long-term storage (Fig. 8.25).

The Pannonian-Carpathian-Alpine Seismic Experiment (PACASE) significantly advanced seismic monitoring techniques for investigating the crustal and lithospheric structure in Central Europe, deploying 214 broadband seismic stations across key geological regions. Over three years, the PACASE network amassed 7 TB of data, enabling studies including body and surface wave tomography, receiver function analysis, and regional earthquake detection. Initial results improved Moho depth mapping and analyzed the lithosphere-asthenosphere boundary in the Pannonian

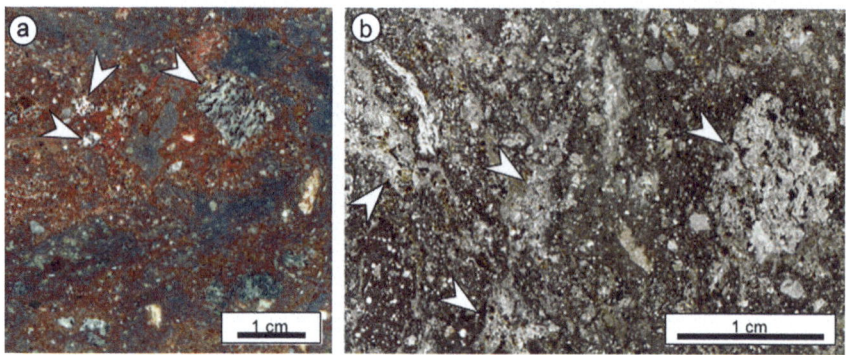

Fig. 8.25 Impact melt rock from Pol sin gen. **a** hand specimen and **b** thin section of sample CT917 showing a red matrix and clasts of granitic gneisses (arrows). (Color figure can be viewed at wil eyonlinelibrary.com) [26]

Basin, facilitating enhanced detection of small earthquakes. These findings have critical implications for seismic hazard assessment and geodynamic modeling. The project also highlighted the value of international collaboration, involving 13 institutions from 7 countries, and the development of technical methods such as remote seismic station monitoring, contributing to further initiatives like Adria Array [27].

Moreover, the PACASE project underscores the importance of international collaboration in large-scale geophysical studies. With participation from 13 institutions from 7 countries, it exemplifies the potential for coordinated research efforts in rock dynamics across Europe. The technical insights gained from the project, including advancements in remote monitoring and control of seismic stations, have been leveraged in subsequent initiatives such as Adria Array, further enhancing Europe's capabilities in seismic monitoring and rock dynamics research [27] (Figs. 8.26 and 8.27).

The study of ancient atmospheric noble gases in post-impact hydrothermal minerals from the Rochechouart impact structure illustrates how extreme impact conditions preserve past atmospheric compositions, offering valuable insights into paleoclimatology [28]. Specifically, quartz and agate minerals were identified as reliable proxies for paleo-atmospheric noble gas compositions, with one agate sample showing a $40Ar/36Ar$ ratio of 292.7 ± 3.6, lower than the modern value of 298.6. This finding highlights the potential of impact structures to serve as archives for studying Earth's atmospheric evolution [28] (Fig. 8.28).

From an engineering perspective, the research highlights the potential for impact structures to serve as natural laboratories for studying rock behavior under extreme conditions. The preservation of ancient atmospheric signatures within these rocks demonstrates their ability to retain information over geological timescales, which has implications for the long-term stability of engineered structures in impact-affected terrains.

In addition to this, the study of breccia formation by particle fluidization in fault zones offers valuable insights into the dynamics of fluid flow in fractured rock masses,

8.3 Engineering Applications in Europe   479

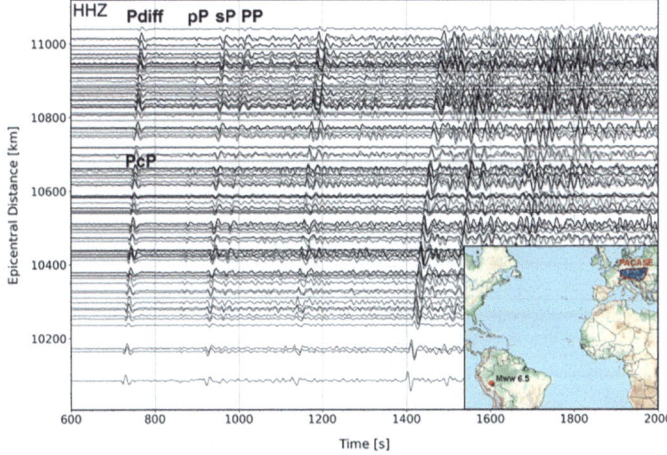

Fig. 8.26 Examples of tele seismic waveforms following the 2022–06–08 00:55:46 (UTC) M ww = 6.5 Peru earthquake. The waveforms are filtered with a bandpass filter from 0.01 to 0.08 Hz. The time axis is set with respect to the origin time of the earthquake [27]

which is relevant to various engineering applications including geothermal energy extraction, hydrocarbon production, and underground storage [29]. The research examined distinctive high-dilation, cement-rich breccias found in the Rusey Fault (Cornwall, UK) and Roamane Fault (Porgera gold deposit, Papua New Guinea).

The study proposes that these breccias form by fluidization of fault damage products in a high fluid flux regime. The estimated fluid velocities during fluidization were in the range of 0.1 m/s to 1 m/s, corresponding to fluid fluxes of 10 to 100 L/s per meter strike length of fault through dilatant fault apertures up to several tens of centimeters wide [29] (Fig. 8.29).

This understanding of fluid flow dynamics in fault zones has important implications for engineering projects in fractured rock masses. For geothermal energy development, it provides insights into potential fluid pathways and flow rates during reservoir stimulation. In hydrocarbon production, it can inform models of fluid migration and trapping mechanisms. For underground storage projects, including carbon capture and storage, it highlights potential risks of rapid fluid migration along reactivated fault zones.

The study of ilmenite phase transformations in suevite from the Ries impact structure in Germany provides insights into mineral behavior under extreme pressure and temperature conditions [30]. While primarily of geological interest, this research has potential implications for materials science and engineering applications involving high-pressure mineral processing.

The study found that ilmenite grains in the suevite samples underwent a series of phase transformations during the impact event, including transformation to the high-pressure polymorph liuite (>16 GPa), then to wangdaoite upon decompression, and finally back to ilmenite. This process resulted in a characteristic foam-like

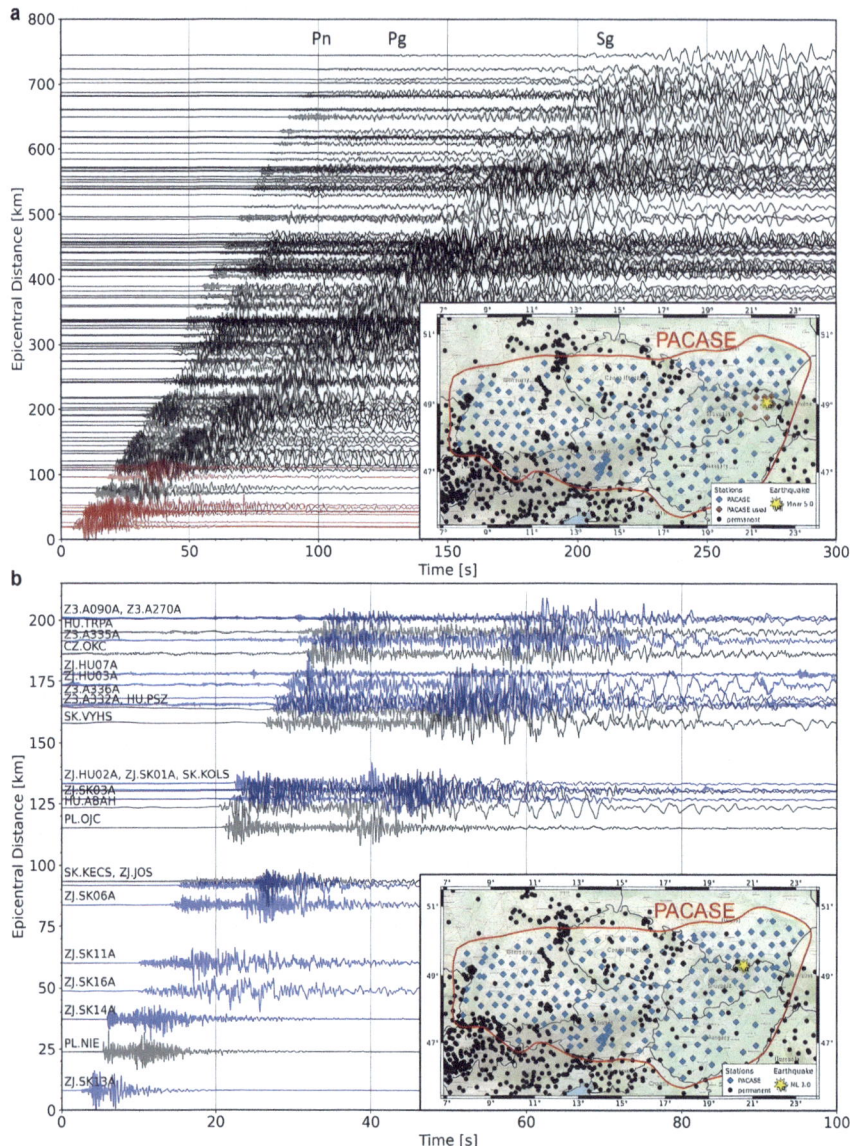

Fig. 8.27 **a** Example of raw waveforms following the 2023–10–09 18:23:09 (UTC) Mww = 5.0 easternmost Slovakia earthquake. **b** raw waveforms of selected PACASE (blue) and permanent (black) stations with a distance smaller than 200 km to the 2019–07–13 12:07:46 (UTC) Ml = 3.0 earthquake east of the High Tatras in Slovakia [27]

8.3 Engineering Applications in Europe

Fig. 8.28 Macrophotographs of the samples analyzed in this study. Quartz crystals appear in white. ROC20-1: quartz crystals with visible growth pattern are observed at the top. ROC20-2: the pink-colored crystals at the center correspond to carbonate. ROC-AGATA4 is an agate sample showing concentric zoning of silica. Scale bars correspond to 1 cm [28]

microstructure of small ilmenite grains with specific crystallographic orientation relationships [30].

From an engineering perspective, this research demonstrates the potential for using mineral microstructures as indicators of past pressure–temperature conditions in rocks. This could be valuable in assessing the extent of damage in rock masses subjected to dynamic loading, such as in underground explosions or impacts. Additionally, the understanding of these phase transformations could inform the development of new materials or processing techniques for industrial applications requiring high-pressure mineral transformations.

In conclusion, the application of rock dynamics principles in European engineering projects spans a wide range of fields, from geothermal energy development and underground construction to natural hazard assessment and the study of impact structures. The diverse geological settings across Europe provide unique opportunities for advancing our understanding of rock behavior under dynamic loading conditions.

Key themes emerging from this review include: The importance of comprehensive monitoring and data collection in understanding and managing induced seismicity in geothermal projects. The value of long-term geophysical monitoring in characterizing the hydrological dynamics of unstable slopes for landslide risk assessment. The application of advanced geophysical techniques, such as electrical resistivity methods, in characterizing complex geological structures like impact craters. The role of numerical modeling in optimizing geothermal reservoir development and

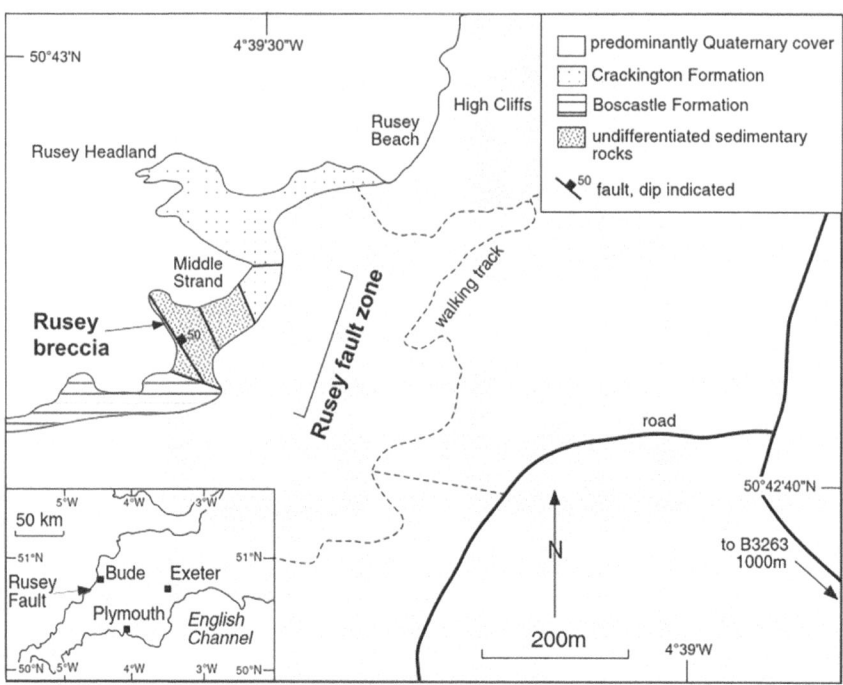

Fig. 8.29 Map showing the location and geological setting of the Rusey breccia in North Cornwall. Inset show location of the Rusey locality in SW England [29]

predicting long-term reservoir behavior. The potential for impact structures to serve as natural laboratories for studying rock behavior under extreme conditions. The importance of understanding fluid flow dynamics in fault zones for various engineering applications in fractured rock masses. The value of international collaboration in large-scale geophysical studies for advancing our understanding of crustal and lithospheric structure.

As Europe continues to lead in many areas of rock dynamics research and application, these insights will undoubtedly contribute to the development of more efficient, safer, and sustainable engineering practices in fields ranging from renewable energy production to underground construction and natural hazard mitigation. The ongoing integration of diverse research methods, from field-scale experiments to advanced numerical modeling, promises to further enhance our ability to predict and manage rock behavior under dynamic loading conditions in a variety of geological settings across the continent.

8.4 Engineering Applications in Other Countries

Engineering applications of rock dynamics have seen significant advancements in various countries outside China, the United States, and Europe. These applications span a wide range of industries and geological contexts, demonstrating the global importance of understanding and managing dynamic rock behavior. This section will explore the diverse engineering applications of rock dynamics in countries such as Canada, Australia, South Africa, and others, highlighting innovative approaches, case studies, and technological developments.

Canada

Canada, as one of the global leaders in mineral resource development, has faced significant challenges in its deep mining operations. These challenges stem primarily from high-stress conditions and seismicity, which are prevalent in underground mines. Rock dynamics, a discipline that examines the behavior of rock and support systems under dynamic conditions, has become central to addressing these issues. This chapter explores the contributions of Canadian research and testing facilities in advancing rock dynamics applications, focusing on innovative ground support systems, dynamic testing techniques, and the implications of these developments for seismicity management in mining operations.

Rock Dynamics in Deep Mining Operations

The abundance of mineral resources in Canada has necessitated the adoption of advanced engineering techniques to manage the stresses associated with deep mining operations. Among these challenges, mining-induced seismicity, including rock bursts, poses significant risks to safety and productivity. This has led to the development of cutting-edge research facilities and methodologies aimed at understanding and mitigating the effects of seismic activity on underground mines [31].

At the CANMET Experimental Mine in Val-d'Or, Quebec, research focused on the performance of rock support systems under dynamic loading, simulating seismic stresses. Tests on end-anchored mechanical rock bolts subjected to blast-induced dynamic loading revealed both axial and bending strains, with peak dynamic stresses reaching 66 MPa in tension. Strain gauges and geophones measured bolt shank strain and particle velocities, providing key insights into the bolts' dynamic behavior. These findings emphasize the necessity of resilient support systems for seismically active mines, capable of withstanding both static and dynamic loading conditions [31] (Fig. 8.30).

Innovative Dynamic Testing Facilities

Building on the foundational work at CANMET, Canadian researchers have developed sophisticated testing facilities, such as the Noranda/CANMET Dynamic Testing Apparatus, to evaluate the performance of rock support systems under dynamic loading. This apparatus employs a drop-weight system, capable of simulating impact loading on rock bolts [33]. Initially utilizing a 1,000 kg weight, which was later

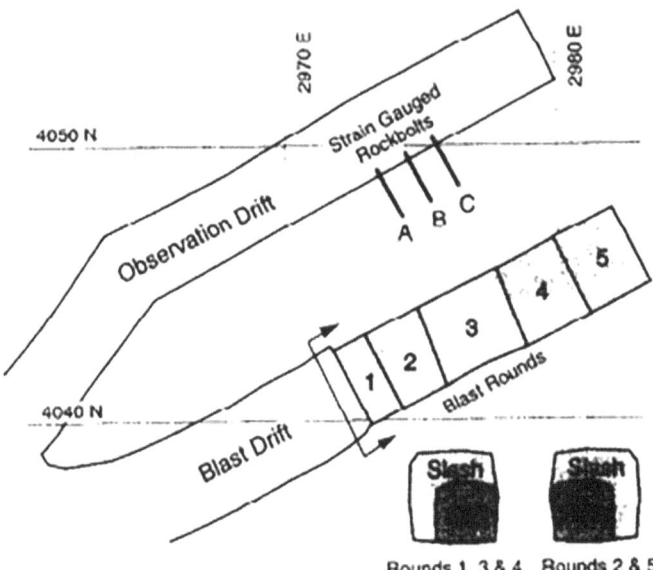

Fig. 8.30 Plan view of observation and blast drift showing location of monitored rock bolts and the five slash rounds (Case 1) [32]

increased to 3,000 kg, this system allowed researchers to test 1.7 m long bolts in simulated boreholes. These experiments have proven invaluable in understanding how different support technologies respond to high-energy, short-duration loads typical of seismic events.

A key advantage of the Noranda/CANMET apparatus lies in its instrumentation-rich setup, enabling precise measurements of load transfer, energy absorption, and deformation during dynamic loading. This level of detail allows researchers to study the complex interactions between rock bolts, surrounding rock masses, and applied loads. Such insights have driven the development of advanced ground support systems tailored to dynamic loading conditions.

Innovative Support Technologies

Research conducted at Canadian facilities has led to the development of innovative rock support technologies specifically designed to perform under dynamic loading conditions. One of the most significant advancements has been the creation of yielding rock bolts, which exhibit enhanced energy absorption capabilities. Unlike conventional rigid support elements, yielding bolts are designed to deform in a controlled mining under dynamic loads, dissipating energy and reducing the risk of sudden, catastrophic failure.

These yielding bolts have been shown to outperform traditional designs in dynamic environments, making them a preferred choice for high-stress, seismically active

8.4 Engineering Applications in Other Countries

mines. By allowing controlled deformation, these bolts not only absorb seismic energy but also maintain the structural stability of the surrounding rock mass.

Implications for Seismically Active Mining Operations

The contributions of Canadian research to rock dynamics extend beyond individual mines, shaping broader strategies for managing mining-induced seismicity. The ability to test full-scale rock support systems under dynamic conditions has provided engineers with the tools to design safer and more effective ground support strategies. By incorporating dynamic considerations into support design, mining operations can better mitigate the risks associated with rock bursts and seismic events.

Moreover, the focus on both axial and bending responses in rock bolt design has revealed that dynamic loading conditions often differ significantly from static scenarios. This understanding emphasizes the need for support systems that are not only strong but also flexible and energy-absorbing. Such systems can provide robust protection against the complex stresses encountered in deep underground mines.

Canada's contributions to the field of rock dynamics have been pivotal in advancing the understanding and application of dynamic ground support systems in mining operations. Research conducted at facilities like the CANMET Experimental Mine and the Noranda/CANMET Dynamic Testing Apparatus has provided critical insights into the behavior of rock support systems under seismic loading. These findings have informed the development of innovative technologies, such as yielding rock bolts, which are specifically designed to perform under dynamic conditions.

As mining operations continue to expand into deeper, more seismically active regions, the importance of rock dynamics research will only grow. The advancements made by Canadian researchers not only enhance mine safety and productivity but also serve as a model for addressing similar challenges worldwide. By integrating dynamic considerations into ground support design, the field of rock dynamics ensures the continued success of mining operations in even the most challenging environments.

Australia

Australia, renowned for its extensive mineral resources and varied geotechnical landscapes, has embraced rock dynamics as a cornerstone of its engineering and mining sectors. From addressing challenges in deep, high-stress underground mining environments to ensuring the stability of critical infrastructure under seismic and environmental loading, rock dynamics research has had a profound influence on safety, productivity, and sustainability [34]. This chapter explores the state-of-the-art research and applications of rock dynamics in Australia, focusing on advances in the mining industry, innovations in civil and coastal engineering, and their implications for design practices.

Rock Dynamics in Mining: Insights from WASM

The mining industry in Australia, particularly in deep, high-stress environments, has been a major driver of advancements in rock dynamics research. The Western Australian School of Mines (WASM) has established itself as a global leader in this field through the development of innovative testing facilities and methodologies for evaluating ground support systems under dynamic loading conditions [32].

1. The WASM Dynamic Test Facility

At the heart of WASM's contributions is the WASM Dynamic Test Facility, a cutting-edge installation designed to simulate rock burst conditions and evaluate the performance of integrated rock mass and reinforcement systems. The facility uses a drop-weight mechanism to release a reinforced simulated rock mass onto impact buffers, creating realistic dynamic loading conditions for testing ground support systems. Its state-of-the-art instrumentation-comprising load cells, motion sensors, and high-speed video cameras-allows researchers to precisely calculate the energy absorbed by various components of a support system [32].

The facility's ability to test both reinforcement elements, such as rock bolts, and surface support components, such as mesh panels and shotcrete, enables a comprehensive assessment of system performance. This holistic approach to testing reflects the complex realities of underground excavations, where the interaction between reinforcement and surface support plays a critical role in stability (Fig. 8.31).

2. Key Research Insights

Research at the Western Australian School of Mines (WASM) has provided valuable insights that have significantly advanced the field of mining engineering, particularly

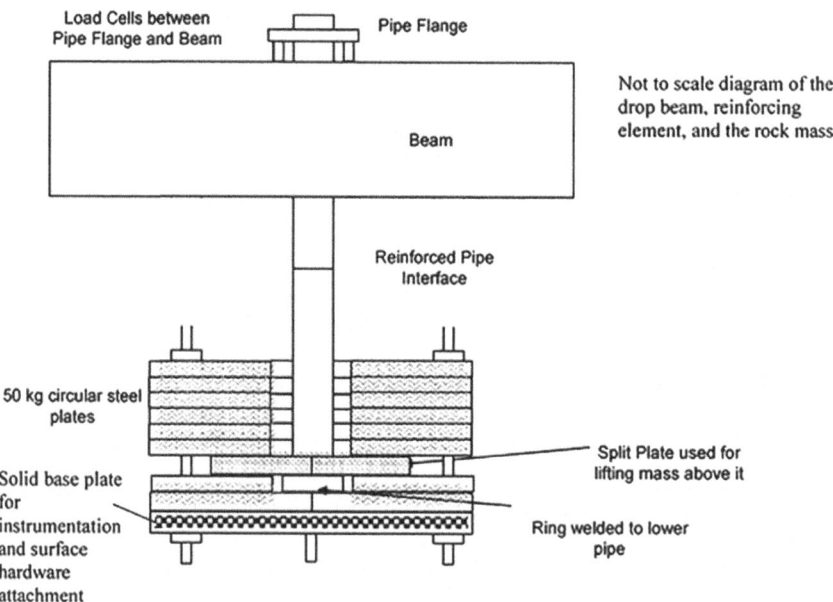

Fig. 8.31 WASM test facility (schematic) [33]

8.4 Engineering Applications in Other Countries

in the design and optimization of ground support systems for dynamic and high-seismicity environments. These findings have direct applications in enhancing the safety and resilience of underground mining operations.

One of the critical discoveries relates to the energy absorption capacity of support systems. WASM studies have demonstrated that energy absorption capacity increases with input energy, highlighting the necessity of designing systems that can accommodate a wide range of dynamic loading scenarios. This has shifted the focus of engineers from traditional static strength-based designs to prioritizing dynamic resilience, ensuring that support systems perform effectively under sudden, high-energy conditions such as seismic events or rock bursts.

Another significant area of research has been the dynamic behavior of shotcrete and mesh support systems, which play a crucial role in stabilizing fractured rock and preventing unraveling between reinforcement elements. Dynamic testing has revealed that these surface support systems exhibit remarkable energy absorption capabilities, making them essential for mitigating seismic risks. Their ability to contain fractured rock and sustain loads during dynamic events enhances the overall stability and safety of underground operations.

WASM's research has also been instrumental in the development of yield-based support design methodologies, which focus on accommodating the deformation and energy release associated with seismic activity. Unlike traditional static strength-based approaches, yield-based designs provide a more robust framework for managing dynamic stresses, enabling ground support systems to adapt to the high-deformation demands of seismically active mining environments.

Together, these contributions from WASM have informed the development of safer and more effective support systems, offering critical solutions for the challenges of underground mining in dynamic and high-stress conditions.

3. Practical Implications for the Mining Industry

The findings from WASM research have had profound impacts on the design and implementation of ground support systems in Australian mines. Comparative testing of support technologies has enabled mining companies to select systems best suited to their specific geotechnical conditions, enhancing both safety and efficiency. Additionally, advancements in dynamic testing methodologies have led to the development of innovative support systems, including energy-absorbing yielding bolts, which provide better protection against the extreme conditions often encountered in deep mines.

Rock Dynamics in Civil and Environmental Engineering

Beyond the mining sector, rock dynamics principles have found wide-ranging applications in civil and environmental engineering in Australia. These applications address critical challenges such as rock slope stability under seismic loading, the impact of rare but significant earthquakes, and the dynamic behavior of coastal rock masses under wave and tsunami impacts.

1. Rock Slope Stability and Earthquake Risks

Although Australia experiences relatively low seismic activity compared to regions such as Japan or California, rare but significant earthquakes can pose substantial risks to critical infrastructure. In response, Australian researchers have developed advanced numerical models to simulate the dynamic response of rock slopes under earthquake loading. These models incorporate detailed geological features, including joint sets and faults, while accounting for the nonlinear behavior of rock masses under cyclic loading. This approach allows for accurate predictions of slope stability during seismic events, providing valuable insights for infrastructure design and safety.

Key applications of this research include enhancing slope stability in open-pit mines, where deep excavations are subjected to a combination of mining-induced vibrations and potential seismic events. Additionally, the models have been used to ensure the safety of highway cuttings in mountainous terrain, minimizing risks to roadways that are vital for transportation and commerce. Furthermore, they play a critical role in safeguarding dam abutments, protecting essential water storage infrastructure from seismic-induced instability that could threaten water security and downstream communities.

The results of these studies have informed the development of seismic design criteria for rock slopes across various contexts, ensuring that infrastructure is designed to withstand rare but significant earthquake events. This research has strengthened the resilience of Australia's critical infrastructure, contributing to safer mining operations, transportation networks, and water management systems in the face of seismic risks.

2. Coastal Engineering and Rock Dynamics

Australia's extensive coastline, which supports much of its population and infrastructure, faces unique challenges from wave-induced forces and the potential impacts of tsunamis. To address these issues, researchers have applied principles of rock dynamics to investigate the stability of coastal cliffs and to design protective measures against erosion and rockfalls caused by wave loading.

A key advancement in this field has been the development of coupled hydromechanical models. These models simulate the propagation of wave-induced pressures through fractured rock masses and assess their potential to trigger instability. By integrating concepts from rock mechanics, hydraulics, and structural dynamics, these models offer a comprehensive approach to understanding and mitigating the complex interactions between ocean forces and coastal rock formations.

The research has led to significant practical outcomes, including improved stability assessment methods for coastal cliffs exposed to wave and tsunami loading. Additionally, the development of design guidelines for protective infrastructure, such as retaining walls and engineered barriers, has enhanced the ability to mitigate risks of rockfalls and erosion in vulnerable areas. These measures are particularly critical for protecting vital infrastructure, including ports, roadways, and residential developments, situated in high-risk coastal regions.

8.4 Engineering Applications in Other Countries

By addressing the dynamic interactions between ocean forces and geological structures, this research has provided essential tools and strategies for ensuring the stability and safety of Australia's coastal environments, safeguarding both human populations and critical infrastructure against the threats posed by natural hazards.

Australia's contributions to rock dynamics applications are notable for their depth, breadth, and innovative approach to solving practical challenges. The pioneering research conducted at institutions like WASM has significantly advanced our understanding of ground support systems under dynamic loading conditions, with direct contributions for mining safety and productivity. Similarly, the application of rock dynamics principles in civil and coastal engineering has enhanced the resilience of critical infrastructure against seismic and environmental risks.

By integrating advanced testing methodologies, numerical modeling, and innovative design approaches, Australian researchers have pushed the boundaries of rock dynamics. Their work not only addresses immediate engineering challenges but also establishes frameworks for future advancements in the field. As the demands on mining and infrastructure development continue to grow, the principles and applications of rock dynamics will remain central to ensuring safety, sustainability, and success in challenging environments.

South Africa

South Africa, with its extensive history of deep gold mining, has developed a robust research and application framework for rock dynamics. The unique challenges of ultra-deep mining, characterized by high stresses and seismic risks, have driven the development of innovative testing methodologies, support technologies, and design strategies. Beyond mining, rock dynamics principles have found applications in civil engineering projects, including dam design and tunnel support systems. This chapter explores South Africa's contributions to rock dynamics, highlighting their impact on safety, productivity, and design practices in challenging geological environments.

Rock Dynamics in Deep Gold Mining

Deep gold mining has been a cornerstone of South Africa's economy for decades, necessitating significant advancements in the management of seismic risks. The country has been at the forefront of rock dynamics research, developing both testing facilities and practical solutions to mitigate the hazards associated with rock bursts and mining-induced seismicity.

1. The SIMRAC Dynamic Stope Test Facility

One of South Africa's most notable contributions to rock dynamics research is the SIMRAC Dynamic Stope Test Facility, designed to replicate the conditions of a longwall stope-a common mining configuration in South African gold mines. This facility utilizes a 10,000 kg drop mass, capable of delivering up to 300 kJ of energy at velocities up to 7.7 m/s, allowing for the testing of large-scale support systems over an area of 5.5×6.5 m [35].

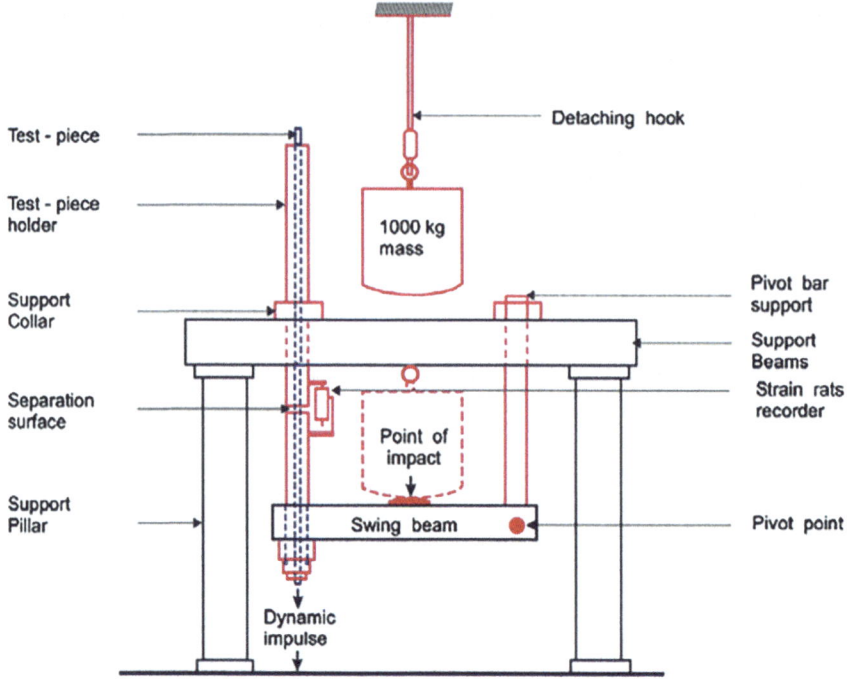

Fig. 8.32 General view of the dynamic testing facility for rock bolts [33]

The facility's design emphasizes repeatability, enabling systematic evaluation of different support configurations and loading scenarios. While its instrumentation was initially limited to basic measurements such as forces and video recordings, its robust structure provided critical insights into the performance of integrated support strategies under dynamic loading conditions (Fig. 8.32).

2. Key Research Insights

Research conducted at the SIMRAC (Safety in Mines Research Advisory Committee) facility in South Africa has made significant contributions to improving ground support systems in mining, particularly for managing the challenges posed by seismic activity in deep and ultra-deep mines. These findings have informed the development of innovative approaches to ensure safer and more efficient mining operations.

One of the major insights from SIMRAC studies is the importance of integrated support strategies. Researchers found that the interaction between different support elements, such as rock bolts, mesh, and shotcrete, is critical to the overall performance of the support system. This has led to the development of integrated support designs that combine localized reinforcement with areal coverage, effectively addressing both small-scale instabilities, such as loose rock fragments, and large-scale instabilities, like rock mass failures. These designs provide a cohesive system that enhances the stability of underground excavations.

8.4 Engineering Applications in Other Countries

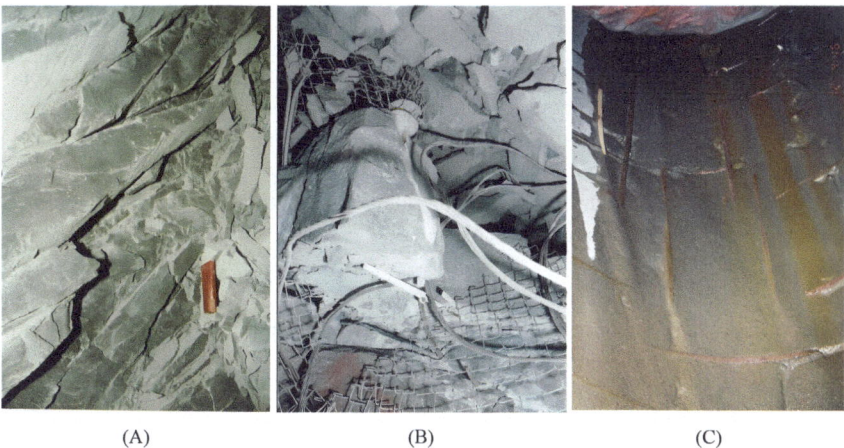

Fig. 8.33 Examples of **a** stress-induced fractures (photo taken by Dr. Ewan Sellers), **b** rock burst (photo taken by Dr. Ewan Sellers), and **c** using abrasion-resistant shotcrete for stabilising a vertical ore pass (7 years in operation) [36]

Recognizing the limitations of conventional strength-based approaches in managing rock burst risks, South African researchers pioneered energy-based design methodologies. This approach focuses on quantifying and controlling the energy release associated with mining-induced seismicity. A critical metric introduced through this research is the Energy Release Rate (ERR), which measures the potential for seismic activity by assessing changes in energy within the rock mass caused by mining operations [3]. By optimizing mining layouts and sequences to minimize ERR, engineers can reduce the concentration of high-stress zones and control energy release, mitigating seismic hazards and enhancing operational safety (Fig. 8.33).

Another groundbreaking innovation has been the development of yielding support systems, specifically designed to absorb significant amounts of kinetic energy while maintaining their load-bearing capacity. Examples of these systems include yielding rock bolts with sacrificial elements that deform plastically under high loads and advanced containment systems using high-strength mesh and energy-absorbing lacing. These technologies have proven indispensable in managing the high-energy loading conditions encountered in ultra-deep mines, enabling the safe extraction of ore bodies that were previously considered too hazardous due to their seismic potential.

Collectively, the research at SIMRAC has transformed ground support design in mining, particularly for high-stress, seismically active environments. By integrating support elements, adopting energy-based design principles, and utilizing yielding support technologies, these advancements have significantly improved the safety and resilience of mining operations, allowing for the extraction of deeper and more challenging ore bodies.

3. Global Influence of South African Research

South Africa's advancements in rock dynamics have had far-reaching impacts beyond its borders. The concepts of energy-based design and yielding support systems, developed for ultra-deep gold mining, have been adopted in other mining contexts worldwide, particularly in seismically active or high-stress environments. These innovations have set new standards for rock engineering practices, enhancing safety in challenging underground operations globally.

Rock Dynamics in Civil Engineering

While mining has been the primary focus of rock dynamics research in South Africa, the principles and methodologies developed in this field have found important applications in civil engineering. From dam design to tunnel support systems, rock dynamics research has contributed to the development of infrastructure in challenging geological environments.

1. Seismic Performance of Large Dams

South Africa's semi-arid climate and growing water demands have necessitated the construction of large dams, often in geologically complex areas. Researchers have focused on the dynamic rock-structure interaction in dam engineering, particularly the seismic performance of concrete arch dams founded on jointed rock masses.

2. Key Areas of Research

South African researchers have made significant advancements in key areas of rock dynamics, including dynamic wave propagation in jointed rock, numerical modeling, and dynamic tunnel support design. Studies on wave propagation have highlighted the critical influence of geological discontinuities, such as faults and joint sets, on seismic wave transmission and the dynamic response of dam-foundation systems, emphasizing the importance of considering the discontinuous nature of rock masses in seismic performance evaluations. To address these complexities, advanced numerical modeling techniques have been developed to simulate the intricate geometry of structures like arch dams, the nonlinear behavior of concrete under dynamic loads, and the impact of discontinuities in rock foundations. These models have informed seismic design criteria, ensuring the safety and stability of large dams during rare but significant seismic events. Additionally, dynamic tunnel support design research has focused on addressing South Africa's challenging geological conditions, where fractured and weathered rock masses are common. Innovative support systems, such as yielding rock bolts and deformable shotcrete linings, have been developed to absorb dynamic loads from seismic activity and construction-induced vibrations, enabling the safe construction of resilient tunnels in geologically complex regions. These advancements collectively enhance the safety and reliability of critical infrastructure in South Africa, while contributing valuable knowledge to global rock dynamics practices (Fig. 8.34).

South Africa's contributions to rock dynamics research are among the most advanced in the world, driven by the challenges of ultra-deep gold mining and the

8.4 Engineering Applications in Other Countries

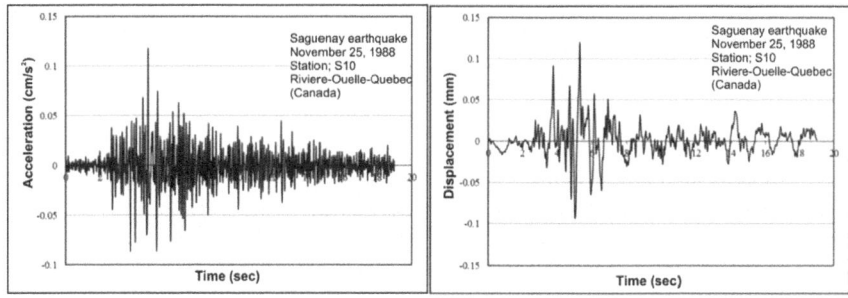

Fig. 8.34 Acceleration-time and displacement–time histories of saguenay earthquake record [37]

need to ensure the safety and stability of critical infrastructure. From the development of the SIMRAC Dynamic Stope Test Facility to the pioneering of energy-based design approaches and yielding support systems, South African researchers have pushed the boundaries of what is possible in rock engineering. These innovations have had profound impacts on the mining industry, enabling safer and more efficient extraction of resources in high-stress environments.

India

India's diverse geological environments and its rapid infrastructure development have led to the growing importance of rock dynamics in both research and engineering practice. From seismic stability of rock slopes in mountainous regions to the dynamic behavior of underground rock masses, Indian researchers have made significant advancements in applying rock dynamics principles to address the unique challenges posed by the country's geological and tectonic conditions. This chapter explores key applications of rock dynamics in India, with a focus on mining, civil engineering, and large-scale infrastructure projects.

Applications in Infrastructure Development

India's rapidly growing infrastructure sector, including hydroelectric power projects and transportation networks, has driven significant advancements in the application of rock dynamics. The dynamic performance of rock masses is critical in the design and safety of infrastructure in regions such as the Himalayan mountains and the Deccan Plateau.

1. Seismic Stability of Rock Slopes

In the tectonically active Himalayan region, seismic performance of large rock slopes is crucial for the safety of hydroelectric projects and transportation corridors. Indian researchers have developed comprehensive methodologies to assess the seismic stability of rock slopes, combining deterministic and probabilistic approaches. These methods take into account factors such as: Seismic ground motions, Rock mass discontinuities, and Groundwater conditions.

Advanced numerical modeling techniques, such as discrete element methods, have been employed to simulate the dynamic response of jointed rock masses. These models have revealed key failure mechanisms and provided insights into mitigation strategies, including: Engineered slope reinforcement, Seismic hazard zoning, and Early warning systems.

One notable application is the assessment of earthquake-induced landslide potential in the Himalayas. By combining field studies, laboratory testing, and numerical simulations, researchers have developed robust methodologies to predict and mitigate landslide risks in this high-seismicity, steep-topography region.

2. Underground Space Development

In the field of underground space development, Indian researchers have made significant contributions to addressing rock dynamics challenges associated with the construction and operation of large underground caverns. These efforts support critical infrastructure such as hydroelectric power plants, strategic storage facilities, and scientific laboratories, which often require excavations in complex geological settings.

A notable example is the development of underground pumped storage hydroelectric projects, which demand large cavern excavations that must remain stable under both static and seismic loading conditions. Key areas of research in this domain include optimizing the shapes and orientations of caverns to minimize stress concentrations and ensure stability, particularly in challenging geological environments. By refining cavern geometries, researchers have been able to reduce the likelihood of stress-induced instabilities, thereby improving the long-term performance of these structures (Fig. 8.35).

Another focus area has been the investigation of support system performance under dynamic loading conditions. Research into the behavior of rock bolts, shotcrete, and cable anchors has provided valuable insights into the interactions between support systems and surrounding rock masses during seismic or mining-induced disturbances. These studies have led to improved designs for support systems that enhance safety and structural integrity, even under adverse loading scenarios.

Additionally, Indian researchers have worked on refining design methodologies to account for the unique geological challenges present in India's diverse terrains. These methodologies integrate static and dynamic considerations to ensure cavern stability under a wide range of operational and environmental conditions. The findings have led to the adoption of advanced design practices, enabling the safe and efficient development of underground spaces for energy production, storage, and scientific research.

By addressing the complexities of underground excavation and support system performance, Indian researchers have contributed significantly to the development of robust solutions for large-scale underground projects. These advancements have paved the way for the safer and more efficient utilization of underground space in India's varied and often challenging geological environments.

8.4 Engineering Applications in Other Countries

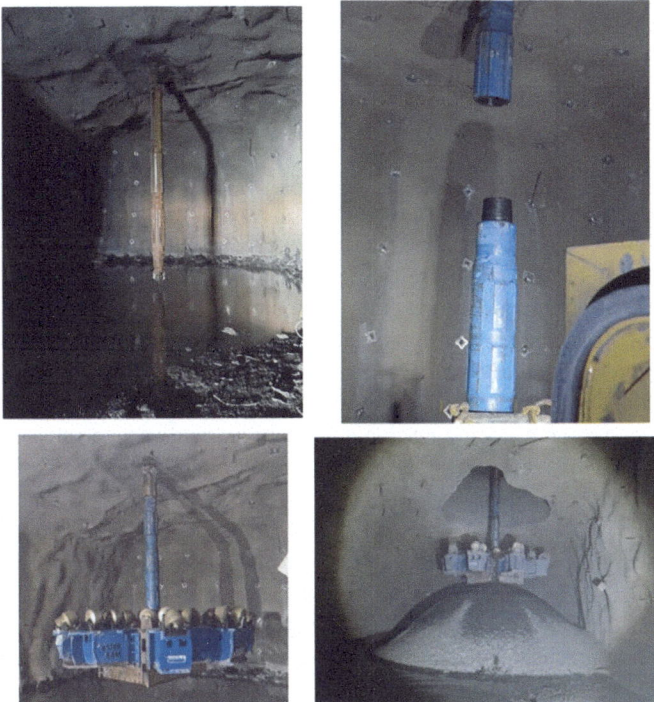

Fig. 8.35 The process of raise boring a 5-m diameter shaft with 742 m length (Photos courtesy of Phillip Viljoen, used with permission) [36]

Contributions to Rock Dynamics Research

India's contributions to rock dynamics research have significantly advanced the understanding of dynamic rock behavior, particularly in the context of geological variability and seismic hazards. Researchers have explored key areas such as the effects of earthquake records and specimen size on dynamic shear behavior, the influence of geological variability on the dynamic response of rock masses, and the development of site-specific seismic hazard models for infrastructure projects. In the Deccan Plateau, for example, studies have focused on the behavior of highly weathered rock formations under dynamic loading, addressing the unique challenges posed by this ancient geological region's weathered rocks and tectonic activity. These investigations provide a critical foundation for designing structures that can withstand complex dynamic stresses, enhancing the safety and resilience of infrastructure in seismically sensitive areas.

The application of rock dynamics principles in India has had a profound impact on the safety and efficiency of infrastructure development in diverse geological settings. From assessing the seismic stability of rock slopes in the Himalayas to optimizing underground caverns in the Deccan Plateau, Indian researchers have made significant strides in addressing the challenges posed by dynamic loading conditions. These advancements contribute to a deeper understanding of rock behavior and provide practical solutions for engineering design in seismically active and geologically complex regions.

Japan

Japan's location along the Pacific Ring of Fire makes it one of the most seismically active countries in the world. As such, rock dynamics research in Japan has been driven by the need to ensure the safety and resilience of critical infrastructure in the face of frequent earthquakes. From underground caverns and tunnels to nuclear power plants and bridges, Japanese engineers and researchers have developed advanced methodologies and technologies to understand and mitigate the effects of seismic loading on rock structures. This section explores key advancements in Japan's rock dynamics research, including dynamic stability analysis, seismic monitoring systems, and earthquake-resistant design for underground and surface structures.

Advanced Monitoring Systems for Underground Structures

To complement numerical modeling, Japanese researchers have developed advanced monitoring systems to study the real-time behavior of underground structures during seismic events. These systems integrate a variety of sensing technologies, including accelerometers to measure ground motion and structural vibrations, strain gauges to assess stress and strain distribution in support systems and rock masses, and fiber optic sensors to monitor deformation and temperature changes across large areas with high accuracy. The real-time data provided by these systems serve multiple critical purposes: validating design assumptions, assessing the performance of support systems under seismic loading, and informing decisions regarding maintenance, retrofitting, and emergency response. A notable application of this technology is the implementation of comprehensive monitoring systems in urban subway tunnels and critical underground facilities located in seismically active areas. These systems have proven invaluable for rapid damage assessment and post-earthquake recovery efforts, ensuring the continued functionality of vital transportation and utility networks.

Dynamic Soil-Structure Interaction in Rock Engineering

While primarily associated with soil mechanics, the principles of dynamic soil-structure interaction have been adapted and applied to rock engineering, particularly for structures founded on or embedded in rock masses. Japanese researchers have developed sophisticated numerical models to simulate the complex interactions between: Seismic waves, Foundation rocks, and Structural systems.

Advanced numerical models in Japan have significantly enhanced the understanding of the dynamic response of critical infrastructure by incorporating nonlinear material behavior, wave scattering effects, and geological discontinuities. Key contributions from this research include the development of seismic isolation systems, which employ innovative foundation designs to decouple structures from high-frequency ground motions while maintaining stability under static and low-frequency loads. These systems have been instrumental in improving the seismic resilience of

8.4 Engineering Applications in Other Countries

critical infrastructure, including nuclear power plants, long-span bridges, and high-rise buildings. For example, advanced foundation systems have been specifically designed to ensure the stability and resilience of nuclear power plants located in seismically active regions. The findings from these studies have informed seismic design criteria and construction practices, ensuring the safety, functionality, and longevity of critical infrastructure in Japan's earthquake-prone environment.

Earthquake-Resistant Design for Underground Structures

With limited land resources and a high population density, Japan has increasingly relied on underground space for urban development. However, the seismic vulnerability of underground structures presents unique challenges, particularly in soft rock or soil-like ground conditions. Researchers in Japan have conducted extensive studies on the seismic behavior of tunnels and subsurface structures to address these challenges.

1. Seismic Behavior of Tunnels

The seismic performance of tunnels is influenced by a range of factors, including surrounding ground conditions, tunnel shape and orientation, and the characteristics of the input ground motion. Understanding these factors is critical for designing tunnels that can withstand seismic events and maintain functionality in earthquake-prone regions.

Studies utilizing large-scale shaking table tests and numerical simulations have provided detailed insights into these influences. One of the key findings is that tunnels constructed in soft rock or soil-like conditions are significantly more susceptible to deformation during earthquakes compared to those in hard rock. The low stiffness and high deformability of these materials amplify seismic effects, increasing the likelihood of instability and damage.

The shape of the tunnel also plays a crucial role in its seismic performance. Circular tunnels have been shown to perform better under seismic loading due to their ability to evenly distribute stresses around their perimeter, reducing localized stress concentrations. In contrast, non-circular tunnels, such as rectangular or elliptical shapes, are more prone to stress buildup and deformation, particularly at corners or along flat sections.

These findings emphasize the importance of integrating ground conditions and tunnel geometry into seismic design practices. By optimizing tunnel shape and orientation and accounting for the characteristics of the surrounding ground, engineers can enhance the resilience of underground structures to seismic events, ensuring both safety and operational continuity.

2. Innovative Design Approaches

Building on insights into the seismic performance of tunnels, Japanese engineers have developed innovative design and construction approaches to enhance the seismic resilience of underground structures. These strategies focus on mitigating the effects of seismic forces, ensuring structural stability, and maintaining functionality during and after earthquakes.

One key innovation is the use of flexible tunnel linings, which are specifically designed to absorb and dissipate seismic energy. By reducing stress concentrations within the tunnel lining, these systems minimize the likelihood of cracking and structural failure, especially under dynamic loading conditions. This approach allows tunnels to adapt to seismic movements without compromising their integrity.

Another critical advancement is the implementation of seismic isolation systems, which are used to decouple tunnels from surrounding ground motions. These systems are particularly valuable in urban areas with high seismic risk, where minimizing ground motion transfer is essential to protect infrastructure. By isolating the tunnel structure from seismic forces, these systems significantly reduce the stresses imposed on the tunnel during earthquakes.

In addition to these design innovations, Japanese engineers have adopted advanced construction techniques to further improve tunnel stability in seismically active regions. For instance, the use of reinforced shotcrete linings and composite support systems has proven effective in strengthening tunnel perimeters, particularly in weak or soft geological conditions. These measures provide additional reinforcement, enabling tunnels to better withstand seismic forces and maintain stability during extreme events.

These combined strategies—innovative designs such as flexible linings and seismic isolation systems, along with advanced construction practices—represent a significant advancement in ensuring the safety and resilience of underground structures in earthquake-prone regions. By addressing both the dynamic forces of earthquakes and the geological challenges of tunnel construction, these approaches set new standards for the seismic performance of critical infrastructure.

3. Monitoring and Rapid Assessment

Japanese engineers have implemented advanced monitoring systems in critical underground facilities, such as urban subway networks and underground power plants. These systems provide real-time data during seismic events, enabling rapid assessment of structural performance and facilitating timely repairs or retrofits. This proactive approach ensures the continued safety and functionality of underground infrastructure in earthquake-prone regions.

Contributions to Global Rock Dynamics Research

Japan's advancements in rock dynamics have significantly influenced the global research community, setting new benchmarks in the analysis, design, and monitoring of rock structures under dynamic conditions. These contributions have not only advanced scientific understanding but also enhanced the safety and resilience of critical infrastructure in earthquake-prone regions worldwide.

A key area of impact has been the development of advanced numerical modeling techniques for the dynamic analysis of rock structures. Japanese researchers have pioneered the use of sophisticated simulation methods to understand the complex behavior of rock masses under seismic and dynamic loading. These models incorporate nonlinear material properties, fracture propagation, and stress redistribution, enabling more accurate predictions of structural performance during earthquakes.

8.4 Engineering Applications in Other Countries

Japan has also been at the forefront of applying seismic isolation systems to rock engineering. These systems, originally developed for above-ground structures, have been adapted to underground environments, where they decouple tunnels and other structures from surrounding ground motions. This innovation has greatly improved the seismic resilience of underground infrastructure, particularly in urban areas with high seismic risk.

In the field of monitoring and data analysis, Japanese researchers have introduced innovative technologies for real-time monitoring of underground structures. These systems use advanced sensors, such as fiber optic strain sensors and acoustic emission devices, combined with real-time data processing algorithms, to detect and analyze changes in stress, deformation, and potential failure points during seismic events. These technologies have set a global standard for proactive risk management in underground engineering.

Beyond these technical innovations, Japan's research has played a crucial role in shaping international seismic design standards, ensuring the safety and resilience of infrastructure in earthquake-prone regions worldwide. By sharing knowledge and collaborating with researchers globally, Japanese advancements have informed design practices for tunnels, dams, and other critical structures, enabling better preparation for seismic hazards.

Japan's leadership in rock dynamics research continues to inspire and drive progress in the field, fostering innovations that enhance the safety and reliability of infrastructure both domestically and internationally.

Brazil

Brazil's diverse geological environment, extensive mining activities, and reliance on hydroelectric power have driven significant advancements in the application of rock dynamics principles. From managing blast-induced vibrations in mining operations to addressing the challenges posed by dynamic loading in hydroelectric power plants and offshore oil fields, Brazilian researchers and engineers have developed innovative methodologies and technologies to ensure safety and efficiency in critical infrastructure projects. This section explores the unique contributions of Brazilian rock dynamics research in the fields of mining, hydroelectric power, and petroleum engineering.

Applications in the Mining Industry

The mining sector is a cornerstone of Brazil's economy, and rock dynamics plays a crucial role in ensuring the safety and stability of both open-pit and underground mining operations. One of the primary challenges is managing the dynamic effects of blast-induced vibrations on rock slopes and underground excavations.

1. Blast-Induced Vibrations and Slope Stability

In Brazil, the response of rock slopes to blast-induced vibrations has been a significant focus of research, particularly in mining and infrastructure development. Studies have examined the propagation of blast-induced waves through jointed rock masses and their potential to trigger slope instabilities. Key findings highlight the critical role of joint orientations and discontinuities in amplifying or attenuating vibration waves and have identified threshold vibration levels that can cause slope failure in different rock types. Based on these findings, Brazilian engineers have implemented improved blasting practices to mitigate the risk of blast-induced slope failures. These include tailoring blasting patterns to reduce ground vibration intensity, employing controlled blasting techniques near critical slopes, and integrating real-time monitoring systems to measure ground motion during blasting. Complementing these practices, advanced numerical models have been developed to simulate the dynamic response of rock slopes to both blasting and seismic loading. These models provide critical insights into failure mechanisms, enabling more precise risk assessments and the design of effective, targeted mitigation measures to enhance the safety and stability of rock slopes (Fig. 8.36).

2. Monitoring and Mitigation Strategies

To further enhance mine safety, Brazilian researchers have developed and implemented comprehensive monitoring systems to predict and prevent slope failures. These systems incorporate seismographs to record vibration levels during blasting or seismic events, ground movement sensors to detect early signs of slope deformations, and remote sensing technologies for large-scale monitoring of slope stability.

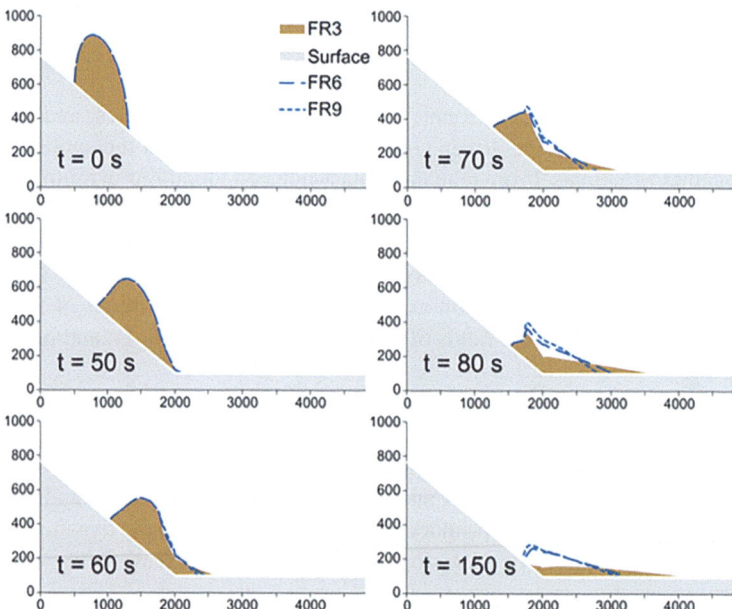

Fig. 8.36 Flow height evolution for all three experiments and six selected points in time, along the longitudinal profile illustrated in Fig. 8.12. Note that the aspect ratio is out of scale. For better visualization, the flow height is twofold exaggerated; offset: x-axis by -100 m, z-axis by 100 m [38]

By integrating these advanced tools, the ability to anticipate and mitigate potential slope failures has significantly improved, contributing to safer and more efficient mining operations across the country. These monitoring systems play a vital role in ensuring the stability of slopes in challenging geological conditions, thereby protecting both workers and infrastructure.

Hydroelectric Power Projects

Hydroelectric power accounts for a significant portion of Brazil's energy production, and rock dynamics research has played a critical role in the construction and operation of large dams and underground powerhouses. The dynamic loading induced by water level fluctuations, equipment vibrations, and seismic activity presents unique challenges for these structures.

1. Dynamic Stability of Rock Foundations and Abutments

Brazilian engineers have placed significant emphasis on assessing the dynamic stability of rock foundations and abutments for hydroelectric dams, a critical aspect of ensuring the safety and performance of these structures. Research in this area has focused on identifying zones of weakness in the rock mass, evaluating the impact of dynamic loading on jointed rock systems, and developing optimized rock support systems to enhance long-term stability. A notable application of this research is the use of advanced grouting techniques to improve the dynamic performance of rock foundations. By carefully characterizing joint systems and tailoring grouting programs to address specific zones of weakness, engineers have successfully minimized the risks of seepage and dynamic failures in large hydroelectric projects. These advancements have contributed to the resilience and reliability of Brazil's hydroelectric infrastructure, which plays a vital role in the country's energy supply.

2. Underground Powerhouses

The excavation of underground powerhouses for hydroelectric projects in Brazil poses unique challenges due to the complex dynamic loads generated by operational vibrations and seismic activity. Brazilian researchers have focused on optimizing the orientation and shape of caverns to minimize stress concentrations and evaluating the performance of support systems, such as rock bolts and shotcrete, under dynamic loading conditions. These studies have provided critical insights that have been applied to the design and construction of safer and more resilient underground powerhouses, ensuring the long-term stability and reliability of Brazil's hydroelectric infrastructure, which is essential to the country's energy security.

Petroleum Engineering Applications

The discovery of vast pre-salt oil reserves in ultra-deep waters off the Brazilian coast has presented a new frontier for rock dynamics research. These reserves, primarily composed of carbonate rocks, pose significant challenges due to their complex mechanical properties and the extreme conditions of deep offshore environments.

1. Dynamic Properties of Carbonate Reservoirs

Brazilian researchers have developed advanced laboratory testing methodologies to accurately measure the dynamic elastic properties and strength parameters of carbonate rocks under representative geological conditions. Key findings reveal significant velocity dispersion and attenuation of seismic waves in carbonate rocks, which complicate the interpretation of seismic data and well logs, as well as unique microstructures and pore geometries that significantly influence the dynamic behavior of carbonate reservoirs. To address these challenges, researchers have developed improved rock physics models tailored to the distinctive properties of carbonate rocks. These models enhance the accuracy of reservoir behavior predictions, improving the reliability of seismic interpretations and providing critical insights for more effective reservoir management strategies.

2. Wellbore Stability in Pre-Salt Reservoirs

Wellbore stability is a critical challenge in the development of Brazil's pre-salt oil reserves due to the complex stress regimes, weak rock formations, and dynamic loads encountered during drilling operations. Brazilian researchers have made significant advancements by developing numerical models to simulate the dynamic response of wellbores to various loading conditions, including drill string vibrations, pressure pulses from mud circulation, and seismic activity. These models also account for the anisotropic and time-dependent behavior of shale formations, which are prevalent in the overburden sections of pre-salt wells. The findings from these studies have informed the development of improved drilling practices and wellbore stability management strategies, enabling safer and more efficient extraction of oil from these geologically complex and economically important reserves.

Contributions to Global Rock Dynamics Research

Brazil's advancements in rock dynamics research have made significant contributions to the global understanding of dynamic rock behavior, addressing challenges in industries such as mining, energy, and offshore exploration. These innovations have not only improved practices within Brazil but have also informed international approaches to dynamic rock engineering.

A key area of impact has been the development of improved methodologies for managing blast-induced vibrations in mining operations. Brazilian researchers have introduced advanced vibration monitoring and predictive modeling techniques that enable better control of blasting impacts on surrounding rock masses and infrastructure. These methodologies have enhanced the safety and efficiency of mining activities while minimizing environmental and structural risks.

Another notable contribution is in the assessment of dynamic stability for rock foundations in large-scale hydroelectric projects. By applying state-of-the-art dynamic analysis techniques, Brazilian researchers have developed tools to evaluate the response of rock foundations under seismic and operational loads. These advancements have been critical in ensuring the long-term stability and resilience of hydroelectric dams in geologically diverse and seismically active regions.

Brazil has also provided new insights into the dynamic properties of carbonate rocks, a significant contribution to the field of offshore oil and gas exploration. Research into the behavior of carbonate formations under dynamic loading conditions has improved the understanding of reservoir performance, aiding in the design of more reliable extraction strategies for offshore fields, which are critical to global energy supply chains.

Through extensive collaboration with international research institutions and active participation in global engineering initiatives, Brazilian researchers have played a pivotal role in advancing the field of rock dynamics. By sharing knowledge and contributing to international projects, Brazil has positioned itself as a key player in addressing complex geo mechanical challenges worldwide. These contributions continue to influence best practices, ensuring safer and more efficient operations across diverse industries reliant on dynamic rock engineering solutions.

Chile

Chile, situated along the Pacific Ring of Fire, is one of the most seismically active countries in the world. Its extensive mining industry, particularly in copper production, operates in a uniquely challenging environment defined by frequent tectonic earthquakes and mining-induced seismicity. Rock dynamics research in Chile has been instrumental in ensuring the safety, efficiency, and sustainability of underground mining operations and critical infrastructure. This chapter examines the key contributions of Chilean researchers and engineers in the fields of seismic hazard assessment, real-time monitoring systems, dynamic support systems, and caving mining methods.

Seismic Hazard Assessment for Mining and Infrastructure

Given the dual threat of tectonic and mining-induced seismicity, Chilean researchers have developed advanced seismic hazard assessment methodologies tailored to the mining and infrastructure sectors.

1. Mining-Specific Seismic Hazard Assessment

Seismic hazard assessments for Chilean mines integrate regional tectonic stresses with localized stress changes caused by mining activities, providing a comprehensive understanding of seismic risks. These assessments involve the development of seismic source models that account for geological structures, stress redistribution from mining, and fault interactions, combined with the application of probabilistic seismic hazard analysis (PSHA) techniques to predict site-specific ground motion levels for various return periods. The insights gained from these assessments have been instrumental in optimizing mining sequences, excavation layouts, and support systems, effectively minimizing seismic risks. Notably, the ability to predict zones of elevated seismic hazard has enabled the implementation of more targeted mitigation strategies, significantly enhancing both safety and operational efficiency in Chilean mines (Fig. 8.37).

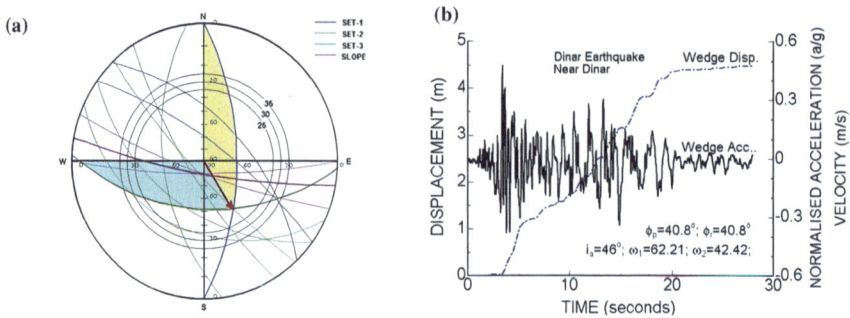

Fig. 8.37 Kinematic and dynamic analyses of the wedge sliding failure nearby Dinar. (**a**) A stereo projection of discontinuities, (**b**) computed responses of wedge block [39]

2. Hazard Assessment for Critical Infrastructure

Beyond mining, Chilean engineers have successfully applied rock dynamics principles to the design and assessment of critical infrastructure, including tailings dams, water supply systems, and transportation networks. Tailings dams, in particular, are highly vulnerable to seismic-induced failures, which could result in catastrophic environmental and economic consequences. To address these risks, Chilean researchers have pioneered performance-based design approaches for tailings dams. These methodologies incorporate both structural resilience under dynamic loading and the consequences of failure, including environmental, social, and economic impacts. By integrating these factors into the seismic design process, Chilean engineers have significantly enhanced the safety and reliability of critical infrastructure in one of the world's most seismically active regions, ensuring long-term resilience and sustainability.

Real-Time Seismic Monitoring and Early Warning Systems

Chilean researchers have been at the forefront of developing real-time seismic monitoring systems for underground mining operations. These systems are designed to detect and respond to both tectonic and mining-induced seismic events.

1. Seismic Monitoring Networks

Seismic monitoring networks in Chilean mines leverage distributed arrays of advanced sensors to enhance safety and operational efficiency. These networks include seismic sensors for detecting and locating seismic events, advanced signal processing systems for real-time seismic data analysis, and machine learning algorithms for classifying events and assessing potential hazards. These systems provide real-time hazard assessments and enable rapid responses to seismic events. Key benefits include improved safety through the timely evacuation of hazardous areas, a deeper understanding of spatiotemporal patterns of mining-induced seismicity, and data-driven optimization of mining sequences and support strategies, significantly reducing seismic risks while maintaining operational efficiency.

8.4 Engineering Applications in Other Countries 505

2. Early Warning Systems

Early warning systems have been deployed in several major Chilean mines to enhance safety by alerting workers and operators to potential seismic hazards. These systems utilize real-time data analysis to identify precursors to large seismic events, enabling the timely implementation of risk mitigation measures. They also provide actionable warnings that allow for the evacuation of personnel and the adjustment of operations to minimize damage. Furthermore, these systems facilitate post-event analysis, helping to refine hazard models and response strategies, ultimately improving the overall resilience of mining operations in seismically active regions.

Rock Dynamics in Caving Mining Methods

Caving methods, including block and panel caving, are widely used in Chile's copper mining industry. These methods involve the controlled collapse of rock masses to extract ore, creating unique challenges related to dynamic loading.

1. Seismic Effects on Cave Propagation

Chilean researchers have made significant contributions to the understanding of seismic activity's influence on caving operations, particularly in block and panel caving methods widely used in the country's mining sector. These studies have provided critical insights into the interplay between seismic forces, fragmentation processes, and material flow dynamics, leading to advancements in both safety and efficiency.

One key finding is the impact of seismic loading on cave propagation, where seismic events can accelerate or alter the direction of cave growth. This understanding has allowed engineers to predict and manage cave propagation more effectively, ensuring stability while optimizing ore extraction. Additionally, researchers have studied the role of dynamic stresses in fragmentation processes, revealing that seismic activity often enhances rock breakdown by promoting crack propagation. However, under certain stress conditions, seismic forces can temporarily stabilize fractured zones, presenting challenges for efficient material flow.

Chilean research has also focused on material flow dynamics under combined static and dynamic loading conditions, shedding light on how seismic events influence the movement of ore and waste within the cave. These insights have been instrumental in improving draw control strategies, ensuring more uniform ore recovery while reducing dilution and minimizing waste.

These findings have transformed the design and management of caving operations in Chile's seismically active regions, leading to safer, more efficient mining practices. By addressing the challenges of mining-induced seismicity, this research has enhanced operational safety and contributed to the long-term sustainability of the country's mining industry, a critical component of Chile's economy. Chile's advancements in understanding and managing dynamic rock behavior continue to influence global practices, particularly in mining operations located in highly seismic environments.

2. Dynamic Support Systems for Caving Mines

To address the challenges posed by the high-stress conditions encountered in deep caving operations, Chilean researchers have developed advanced support systems capable of withstanding the dynamic loading associated with seismic activity. These innovations have been critical in ensuring the safety and efficiency of deep mining operations in Chile's seismically active regions.

Key advancements include the development of energy-absorbing surface support systems, which provide flexibility under seismic loading conditions. These systems are designed to deform and dissipate energy during seismic events, reducing the likelihood of failure while maintaining structural stability. By accommodating sudden and extreme stress changes, these supports enhance the safety of underground workings.

Another innovation is the use of yielding rock bolts, which dissipate seismic energy through controlled deformation. These rock bolts are engineered to stretch plastically under high loads, effectively absorbing energy while preventing the collapse of surrounding rock masses. This capability is particularly important in deep operations where stress levels and seismicity are significant.

Chilean researchers have also developed hybrid support systems that combine stiff and yielding elements to achieve enhanced performance. These systems integrate the immediate stabilization provided by rigid components with the energy-dissipation capabilities of yielding elements, offering a balanced and robust approach to managing dynamic stresses.

The performance of these support systems has been extensively tested through a combination of laboratory experiments, numerical modeling, and in-situ monitoring in active mines. This comprehensive approach has ensured that the designs are both effective and practical for real-world applications.

The implementation of these advanced support strategies has enabled the safe and efficient extraction of deep ore bodies, even in highly seismically active regions. By addressing the unique challenges of deep caving operations, Chilean research has not only enhanced mining safety but also contributed to the sustainability and economic viability of mining in challenging geological environments. These innovations are now influencing support design practices in mining operations worldwide.

Contributions to Global Rock Dynamics Research

Chile's extensive expertise in rock dynamics has significantly advanced the global understanding of seismic hazards and dynamic loading conditions in mining and geotechnical engineering. Its unique geological setting, characterized by both tectonic and mining-induced seismicity, has driven the development of cutting-edge technologies and methodologies with worldwide applications.

One of Chile's key contributions is the integration of tectonic and mining-induced seismicity in hazard assessments. Researchers have developed sophisticated models that combine geological, geo mechanical, and seismic data to evaluate risks more comprehensively. This approach allows for the accurate prediction and management of seismic hazards in mining environments, enhancing both operational safety and productivity.

8.4 Engineering Applications in Other Countries

Another breakthrough is the development of real-time seismic monitoring and early warning systems. These systems use advanced sensor networks and data analytics to detect and analyze seismic activity as it occurs. Early warning capabilities enable timely evacuation and operational adjustments, minimizing risks to workers and infrastructure. Chile's expertise in this area has been adopted in mining regions around the world, including Australia, Canada, and South Africa.

In addition, Chilean researchers have pioneered innovative support systems designed for dynamic loading conditions, including energy-absorbing surface supports, yielding rock bolts, and hybrid systems. These technologies have proven critical for ensuring the stability of underground excavations in high-stress, seismically active environments. Their implementation has enabled the safe extraction of deep ore bodies, even in challenging geological conditions.

These advancements have demonstrated the global relevance of Chilean research and engineering practices. By addressing some of the most complex challenges in mining and rock mechanics, Chile has contributed valuable solutions that are now being utilized in other seismically active mining regions. This invention underscores Chile's position as a global leader in the field of rock dynamics, with ongoing contributions to the safety and sustainability of mining operations worldwide.

Russia

Russia's vast mineral and energy resources, combined with its challenging geological and environmental conditions, have driven significant advancements in rock dynamics research. The country's mining, tunneling, and petroleum industries have benefited from innovative solutions to address rock burst phenomena, dynamic stability in deep mines, hydraulic fracturing operations, and the unique challenges of Arctic and sub-Arctic environments. This chapter highlights the key contributions of Russian researchers to the field of rock dynamics, focusing on theoretical advancements, innovative support systems, and practical engineering applications.

Rock Dynamics in Mining and Tunneling

Russia's deep mining operations and extensive tunneling projects often encounter highly stressed rock masses, making the prediction and mitigation of dynamic rock failure a critical research area.

1. Rock Burst Phenomena and Energy-Based Criteria

Rock bursts—violent failures of highly stressed rock—pose significant safety challenges in deep mining operations. Russian researchers have developed energy-based criteria to assess the potential for rock bursts, focusing on the stress state of the rock mass, the rock's capacity to store elastic strain energy, and the mechanisms by which energy is released during failure. By quantifying the energy balance around excavations, this approach allows engineers to better predict the likelihood and severity of dynamic failure events. These criteria have been successfully applied in Russia's deep mining operations, enabling more accurate risk assessments and the implementation of effective mitigation strategies, significantly enhancing safety in high-stress mining environments.

2. Yielding Support Systems for Tunneling

To address dynamic loading conditions in tunneling, Russian engineers have pioneered the use of yielding steel arches as support elements. These arches are designed to deform plastically under high loads, effectively absorbing energy and maintaining stability in highly stressed or squeezing ground conditions. Key benefits of these systems include improved performance under intense dynamic loading, enhanced adaptability to ground movements, and a reduced risk of catastrophic failure in unstable rock masses. The successful implementation of these technologies has enabled the safe construction of tunnels in challenging geological environments, including water-bearing and highly fractured rock masses, ensuring stability and long-term functionality in difficult conditions.

Numerical Modeling of Dynamic Rock Behavior

Russian researchers have made significant advancements in numerical modeling techniques to simulate the dynamic behavior of fractured rock masses. A major focus has been the development of coupled hydromechanical models that account for the interactions between pore fluid pressure and rock stress states, fracture propagation and coalescence under dynamic loading, and the influence of fluid-rock interactions on rock mass stability. These models are particularly valuable for assessing the stability of underground structures in water-bearing rock masses and understanding the behavior of hydraulic fracturing operations in unconventional oil and gas reservoirs. By providing deeper insights into these complex processes, these modeling capabilities have significantly improved the design and management of mining and petroleum engineering projects in Russia, enabling safer and more efficient operations in challenging environments.

Applications in the Petroleum Industry

Russia's extensive oil and gas reserves, including those in Arctic and sub-Arctic regions, present unique challenges for rock dynamics research and engineering applications.

1. Hydraulic Fracturing in Unconventional Reservoirs

Russian researchers have developed advanced numerical models to characterize dynamic fracture propagation in heterogeneous and anisotropic rock formations, such as tight gas sandstones and shale reservoirs. These models incorporate the dynamic properties of rock masses, including rate-dependent strength and deformation, the interactions between natural and hydraulically induced fracture networks, and the effects of fluid pressure and stress redistribution during hydraulic stimulation. Key insights from these studies include strategies for optimizing hydraulic fracturing treatments to maximize reservoir contact, improved predictions of stimulated reservoir volume, and enhanced recovery rates in unconventional hydrocarbon resources. These advancements have significantly contributed to the development of Russia's unconventional oil and gas reserves, particularly in the vast pre-Caspian and Siberian basins, ensuring more efficient and sustainable resource extraction.

2. Wellbore Stability in Arctic Environments

The development of oil and gas resources in Arctic and sub-Arctic regions presents significant challenges due to the presence of permafrost and seasonally frozen ground. Russian researchers have investigated the dynamic behavior of frozen rock masses and their interactions with wellbores under thermal cycling, seismic activity, and mechanical loading during drilling and production. Their studies have revealed the viscoelastic behavior of frozen rocks under dynamic loading, influenced by ice within pores and fractures, as well as changes in stress wave transmission and wellbore stability caused by freeze–thaw cycles. These findings have led to recommendations for casing design and drilling strategies optimized for frozen ground conditions. This research has informed the development of specialized techniques for Arctic drilling, ensuring the safety, efficiency, and long-term reliability of operations in these extreme environments.

Innovations in Support Systems

Russian researchers have made significant advancements in the design of dynamic support systems for mining and tunneling in high-stress and seismically active environments. A key innovation is the development of hybrid support systems that effectively combine rigid elements for immediate stabilization with yielding components designed to absorb energy under dynamic loading. These systems have been successfully implemented in Russia's deep mining operations and challenging tunneling projects, ensuring the safe extraction of mineral resources and the construction of critical infrastructure in unstable ground conditions. This approach has enhanced both safety and efficiency in geotechnically complex environments.

Contributions to Global Rock Dynamics Research.

Russia's advancements in rock dynamics have significantly influenced the global research community, offering innovative solutions to some of the most complex challenges in mining and petroleum engineering. Drawing on the country's extensive experience with deep mining operations and resource extraction in challenging geological environments, Russian researchers have made several groundbreaking contributions.

A major theoretical advancement has been the development of energy-based models for assessing rock bursts. These models provide a framework for quantifying and managing the energy released during mining-induced seismic events, offering a predictive tool for evaluating the likelihood and severity of rock bursts. This energy-based approach has replaced traditional strength-focused methods, enabling more accurate risk assessments and improved safety protocols in deep and high-stress mining operations.

On the practical side, Russian engineers have introduced innovative yielding support systems tailored for dynamic loading conditions. These systems, including energy-absorbing rock bolts and hybrid supports, are designed to deform plastically under high loads, effectively dissipating seismic energy while maintaining structural stability. Such innovations have been instrumental in enhancing the safety and efficiency of mining operations in seismically active and high-stress environments.

In addition, Russia has pioneered the development of advanced numerical models for coupled hydromechanical processes, which simulate the interaction between fluid pressures and rock stresses. These models are particularly valuable for understanding and mitigating risks in resource extraction, including hydraulic fracturing in shale reservoirs and stability assessments in water-saturated rock masses. By accounting for both mechanical and hydraulic factors, these tools provide critical insights for optimizing operations and ensuring the safety of underground structures.

These contributions have been widely adopted in mining and petroleum engineering projects worldwide, highlighting the global relevance of Russian research in addressing complex geotechnical challenges. By combining theoretical advancements with practical innovations, Russia has solidified its position as a leader in rock dynamics, contributing to the development of safer and more sustainable resource extraction practices across the globe.

8.5 Conclusion

The global application of rock dynamics principles in engineering has significantly advanced our ability to manage and mitigate the risks associated with dynamic rock behavior. From the deep mining operations in China and South Africa to the seismic resilience of infrastructure in Japan and Chile, the research and innovations highlighted in this document demonstrate the critical role of rock dynamics in ensuring safety, efficiency, and sustainability across various industries. The development of advanced support systems, monitoring technologies, and numerical modeling techniques has not only improved our understanding of rock behavior under dynamic loading but also provided practical solutions for engineering challenges in mining, civil engineering, and energy production.

As we continue to push the boundaries of engineering in increasingly complex geological environments, the principles and applications of rock dynamics will remain at the forefront of our efforts. The ongoing integration of multidisciplinary approaches, from geophysics to materials science, will further enhance our predictive capabilities and engineering practices. The lessons learned and technologies developed in one region can be adapted and applied globally, fostering a collaborative approach to addressing the universal challenges of dynamic rock engineering. This collective progress will be essential in meeting the growing demands for resource extraction, infrastructure development, and environmental protection in the face of seismic and geological hazards.

References

1. He MC, Wang Q. Rock dynamics in deep mining. Int J Min Sci Technol. 2023;33:1065–82.
2. Wang GF, Gong SY, Dou LM, Wang H, Cai W, Cao AY. Rock burst characteristics in syncline regions and micro-seismic precursors based on energy density clouds. Tunn Undergr Space Technol. 2018;81:83–93.
3. Liu JW, Liu CY, Yao QL, Yao QL Si GY. The position of hydraulic fracturing to initiate vertical fractures in hard hanging roof for stress relief. Int J Rock Mech Min Sci 2020; 132:104328.
4. Ma K, Zhang JH, Zhou Z, Xu NW. Comprehensive analysis of the surrounding rock mass stability in the underground caverns of Jinping I hydropower station in Southwest China. Tunn Undergr Space Technol. 2020;104: 103525.
5. Meng ZP, Shi XC, Li GQ. Deformation, failure and permeability of coal-bearing strata during longwall mining. Eng Geol. 2016;208:69–80.
6. Liu XZ, Tang CA, Li LC, Lu PF, Sun R. Micro-seismic monitoring and stability analysis of the right bank slope at Dagangshan hydropower station after the initial impound-men. Int J Rock Mech Min Sci. 2018;108:128–41.
7. Tu HL, Zhou H, Qiao CS, Qiao CS, Gao Y. Excavation and kinematic analysis of a shallow large-span tunnel in an up-soft/low-hard rock stratum. Tunn Undergr Space Technol. 2020;97: 103245.
8. Lu CP, Liu Y, Liu GJ, Zhao TB. Stress evolution caused by hard roof fracturing and associated multi-parameter precursors. Tunn Undergr Space Technol. 2019;84:295–305.
9. Wang SF, Li XB, Wang SY. Separation and fracturing in overlying strata disturbed by longwall mining in a mineral deposit seam. Eng Geol. 2017;226:257–66.
10. Diddle B, Agioutantis Z, Maldonado E, Romero BJD, Parra VM. Prediction of dynamic and final vertical and horizontal movements due to longwall mining. Rock Mech Rock Eng 2024 (online)
11. Braunagel MJ, Griffith WA. The effect of dynamic stress cycling on the compressive strength of rocks. Geophys Res Lett. 2019;46(12):6479–86.
12. Wood CE, Manogharan P, Rathbun A, Rivière J, Elsworth D, Marone C, Shokouhi P. Relating hydro-mechanical and east-dynamic properties of dynamically stressed tensile-fractured rock in relation to applied normal stress, fracture aperture, and contact area. J Geophys Res 2024; 129(8):e2023JB027676.
13. Shokouhi P, Jin J, Wood C, Rivière J, Madara B, Elsworth D, Marone C. Dynamic stressing of naturally fractured rocks: on the relation between transient changes in permeability and elastic wave velocity. Geophys Res Lett 2020; 47(1):e2019GL083557.
14. Mc Beck J, Kandula N, Aiken JM, Cordonnier B, Renard F. Isolating the factors that govern fracture development in rocks throughout dynamic in situ X-ray tomography experiments. Geophys Res Lett. 2019;46(20):11127–35.
15. Smith ZD, Griffith WA. Evolution of pulverized fault zone rocks by dynamic tensile loading during successive earthquakes. Geophys Res Lett 2022; 49(19):e2022GL099971.
16. Sleep NH. Strong seismic shaking of randomly prestressed brittle rocks, rock damage, and nonlinear attenuation. Geochem. Geophys. Geosyst. 2010;11(10).
17. Conrad EM, Tisato N, Carpenter BM, Toro GD. Influence of frictional melt on the seismic cycle: Insights from experiments on rock analog material. J. Geophys. Res. Solid Earth 2023; 128(1):e2022JB025695.
18. Vincent M, Emmanuel G, Marc G, Rike K, Romain P, Nicolas C. Seismicity induced during the development of the Rittershoffen geothermal field, France. Geotherm Energy Sci-Soc-Technol. 2020;8(1–3):132–239.
19. Richter H, Hock S, Mikulla S, Krüger K, Lüth S, Polom U, Dickmann T, Giese R. Comparison of pneumatic impact and magneto strictive vibrator sources for near surface seismic imaging in geotechnical environments. J Appl Geophys. 2018;159:173–85.
20. Mahmoodpour S, Singh M, Turan A, Bär K, Sass I. Simulations and global sensitivity analysis of the thermo-hydraulic-mechanical processes in a fractured geothermal reservoir. Energy 2022; 247:123511.

21. Edouard P, Thomas L, Maurin V, Clara L, Emmanuel T, Mickael H. Multiyear time-lapse ERT to study short-and long-term landslide hydrological dynamics. Landslides. 2017;14:1333–43.
22. Quesnel Y, Sailhac P, Lofi J, Lambert P, Rochette P, Uehara M, Camerlynck C. Multiscale geoelectrical properties of the Rochechouart impact structure, France. Geochem Geophys Geosyst 2021; 22(9):e2021GC010036.
23. Song YF, Dou LR, Cheng DS, Zhang XS, Song S, Wang H, He ZY. Post-rifting uplift and inversion of an intra-plate basin: the Doseo basin, Central Africa. Mar Petrol Geol. 2024;164: 106751.
24. Zuo T, Li XL, Wang JG, Hu QW, Tao ZH, Hu T. Insights into natural tuff as a building material: effects of natural joints on fracture fractal characteristics and energy evolution of rocks under impact load. Eng Fail Anal 2024; 108584.
25. Simon C, Oscar L, Marc P, Michaël, Cyril C, Jean-François M, Pierre B, Adrien V, Linda M. Cadomian S-type granites as basement rocks of the variscan belt (Massif Central, France): implications for the crustal evolution of the north Gondwana margin. Lithos 2017; 286:16–34.
26. Trepmann C, Dellefant F, Kaliwoda M, Hess KU, Schmahl W, Hölzl S. Quartz and cristobalite ballen in impact melt rocks from the Ries impact structure, Germany, formed by dehydration of shock-generated amorphous phases. Meteorit Planet Sci. 2020;55(11):2360–74.
27. Schlömer A, Hetényi G, Plomerová J, Miroslav B, Götz B, Kristian C, Wojciech C, Lucia F, Wolfgang F. The Pannonian-Carpathian-Alpine seismic experiment (PACASE): network description and implementation. Acta Geod Geophys 2024; 1–22.
28. Avice G, Kendrick M, Richard A, Ferrière L. Ancient atmospheric noble gases preserved in post-impact hydrothermal minerals of the 200 Ma-old Rochechouart impact structure, France. Earth Planet Sci Lett. 2023;620: 118351.
29. Cox Stephen F, Munroe SM. Breccia formation by particle fluidization in fault zones: Implications for transitory, rupture-controlled fluid flow regimes in hydrothermal systems. Am J Sci. 2016;316(3):241–78.
30. Fabian D, Claudia A, Trepmann WW, Schmahl SA, Gilder, Iuliia VS, Melanie K. Ilmenite phase transformations in suevite from the Ries impact structure (Germany) record evolution in pressure, temperature, and oxygen fugacity conditions. Am Mineral 2024; 109(6):1005–23.
31. Selyutina NS, Petrov YV. Fracture of saturated concrete and rocks under dynamic loading. Eng Fract Mech 2020; 225:106265.
32. Tannant DD, Brummer RK, Yi X. Rock bolt behaviour under dynamic loading: field tests and modeling. Int J Rock Mech Min Sci Geomech Abs 1995; 32(6):537–50.
33. Hadjigeorgiou J, Potvin Y. A critical assessment of dynamic rock reinforcement and support testing facilities. Rock Mech Rock Eng 2011; 44:565–78.
34. Horst W. Deep mining: a rock engineering challenge. Rock Mech Rock Eng. 2019;52:1417–96.
35. Michel P, Ted A, Ken J. Rock bolts testing under dynamic conditions under dynamic conditions at CANMET-MMSL. In: The 6th International symposium on ground support in mining and civil engineering construction. 2007.
36. Salmi EF, Tan P, Sellers EJ, Stacey TR. A review on the geotechnical design and optimisation of ultra-long ore passes for deep mass mining. Environ Earth Sci 2024; 83:301.
37. Kayabali K, Habibzadeh F, Selçuk L. A comparative evaluation of shear behaviour of rock joints under static and dynamic loading. Arab J Geosci. 2022;15:1615.
38. Pudasaini SP, Martin M, Qiwen L. Dynamic simulation of rock-avalanche fragmentation. J Geophys Res Earth Surf 2024; 129:e2024JF007689.
39. Ömer A, Halil K. An experimental and theoretical approach on the modeling of sliding response of rock wedges under dynamic loading. Rock Mech Rock Eng. 2010;43:821–30.

References

Open Access This chapter is licensed under the terms of the Creative Commons Attribution-NonCommercial-NoDerivatives 4.0 International License (http://creativecommons.org/licenses/by-nc-nd/4.0/), which permits any noncommercial use, sharing, distribution and reproduction in any medium or format, as long as you give appropriate credit to the original author(s) and the source, provide a link to the Creative Commons license and indicate if you modified the licensed material. You do not have permission under this license to share adapted material derived from this chapter or parts of it.

The images or other third party material in this chapter are included in the chapter's Creative Commons license, unless indicated otherwise in a credit line to the material. If material is not included in the chapter's Creative Commons license and your intended use is not permitted by statutory regulation or exceeds the permitted use, you will need to obtain permission directly from the copyright holder.

Chapter 9
Main Achievements of Rock Dynamics

The field of rock dynamics has witnessed remarkable advancements in recent decades, encompassing theoretical developments, experimental techniques, numerical simulations, and practical applications. Theoretical models have become more sophisticated, incorporating strain rate effects, damage mechanics, and complex interactions between thermal, hydraulic, and mechanical processes. Experimental methods have evolved to provide more accurate and comprehensive data, from high-strain-rate testing to in-situ monitoring and non-destructive testing. Numerical simulation techniques have also seen significant improvements, with the development of advanced continuum, discrete, and hybrid methods that can handle complex rock dynamics problems more effectively. These advancements have collectively enhanced our ability to predict and analyze rock behavior under dynamic loading conditions, leading to significant improvements in engineering applications such as rockburst prediction, blast design, and earthquake engineering.

The progress made in rock dynamics has been instrumental in addressing complex engineering challenges in rock mechanics, particularly in deep underground environments and under extreme loading conditions. As we continue to push the boundaries of rock engineering, the ongoing development of rock dynamics will remain crucial for ensuring the safety, efficiency, and sustainability of future projects in challenging geological settings. The integration of these advancements will not only improve our understanding of rock behavior but also enable more effective and innovative solutions to the engineering problems we face in the field of rock mechanics.

9.1 Engineering Applications of Rock Dynamics

The engineering applications of rock dynamics have significantly transformed various fields of geotechnical and geological engineering. From mining and tunneling to petroleum and civil engineering, the integration of rock dynamics principles has led to innovative solutions that enhance safety, efficiency, and resilience. In mining, advancements such as microseismic monitoring and optimized blasting techniques have improved deep mining safety. In tunneling, adaptive excavation strategies and integrated support systems have enhanced the stability of deep underground structures. In the petroleum and gas industry, dynamic loading effects in wellbore stability modeling and hydraulic fracturing design have optimized extraction processes. In civil engineering, the incorporation of dynamic rock properties in foundation design and slope stability analysis has led to more resilient infrastructure. For nuclear waste disposal, rock dynamics ensures the long-term stability of repositories during seismic events. In geohazard mitigation, the field has advanced our understanding of seismic wave propagation and landslide risks, improving hazard assessments and mitigation strategies.

Rock dynamics has proven to be a pivotal field in addressing complex geotechnical challenges across multiple engineering disciplines. As technology continues to advance, particularly with the integration of artificial intelligence and machine learning, the future of rock dynamics holds promise for even more sophisticated approaches to predict and mitigate geotechnical risks. These developments will not only enhance our understanding of rock behavior under dynamic conditions but also lead to more effective and sustainable engineering practices in challenging geological settings.

9.2 Status Quo and Challenges of Rock Dynamics

The field of rock dynamics has achieved remarkable progress, establishing a robust theoretical framework, advancing experimental techniques, and enhancing numerical modeling capabilities. These advancements have significantly improved our understanding of rock behavior under dynamic loading conditions and have been successfully applied in various engineering fields such as mining, tunneling, petroleum engineering, earthquake engineering, and nuclear waste disposal. The development of comprehensive monitoring systems and risk management strategies has further enhanced our ability to identify and mitigate potential hazards in rock engineering projects. However, several challenges remain, including bridging the gap between laboratory-scale experiments and field-scale behavior, characterizing rate-dependent behavior, understanding coupled processes, accounting for heterogeneity and anisotropy, and addressing extreme loading conditions. These challenges highlight the need for continued research and innovation to improve the accuracy of predictive models and engineering designs.

The ongoing development of rock dynamics is crucial for addressing the complex and demanding requirements of modern geotechnical projects. Overcoming the current challenges will require interdisciplinary collaboration, investment in advanced experimental and computational tools, and a commitment to translating research findings into practical engineering solutions. As we continue to push the boundaries of rock engineering in increasingly complex environments, the field of rock dynamics will play a pivotal role in ensuring the safety, efficiency, and sustainability of future projects. By addressing these challenges, we can enhance our ability to predict and manage rock behavior, leading to more resilient and sustainable geotechnical infrastructure.

9.3 Future Trend of Rock Dynamics

The future of rock dynamics is poised for significant advancements, driven by technological innovations and interdisciplinary collaboration. Advanced monitoring and sensing technologies, such as Distributed Fiber Optic Sensing (DFOS) and microseismic monitoring, will provide unprecedented insights into rock behavior under dynamic loading conditions. These technologies, along with the development of smart dust and nanosensors, will enable more accurate and real-time data collection. Numerical modeling and simulation techniques will also see substantial improvements, with multi-scale and multi-physics modeling, machine learning, and high-performance computing playing crucial roles. Novel experimental techniques, including in-situ testing at great depths and real-time imaging, will enhance our understanding of rock behavior in complex environments. Sustainable and resilient design approaches, such as energy-absorbing support systems and biomimetic design principles, will become more prevalent, ensuring the long-term stability and safety of underground structures. The integration of rock dynamics into emerging fields like deep geothermal energy extraction, carbon capture and storage, and planetary mining will open new research directions and technological developments. Data-driven approaches and digital twins will revolutionize the way we analyze and manage rock dynamics phenomena, while interdisciplinary integration will provide a more holistic understanding of dynamic rock behavior.

In conclusion, the future of rock dynamics is bright and full of opportunities. As we continue to push the boundaries of underground engineering and resource extraction, the advancements in monitoring, modeling, and experimental techniques will be crucial in ensuring the safety and efficiency of these operations. The expansion into new fields and the integration of interdisciplinary approaches will not only enhance our technical capabilities but also align our efforts with broader societal goals of sustainability and resilience. By embracing these future trends, the field of rock dynamics will play a pivotal role in addressing the complex challenges of the twenty-first century, contributing to the advancement of science and the betterment of society.

Open Access This chapter is licensed under the terms of the Creative Commons Attribution-NonCommercial-NoDerivatives 4.0 International License (http://creativecommons.org/licenses/by-nc-nd/4.0/), which permits any noncommercial use, sharing, distribution and reproduction in any medium or format, as long as you give appropriate credit to the original author(s) and the source, provide a link to the Creative Commons license and indicate if you modified the licensed material. You do not have permission under this license to share adapted material derived from this chapter or parts of it.

The images or other third party material in this chapter are included in the chapter's Creative Commons license, unless indicated otherwise in a credit line to the material. If material is not included in the chapter's Creative Commons license and your intended use is not permitted by statutory regulation or exceeds the permitted use, you will need to obtain permission directly from the copyright holder.

The manufacturer's authorised representative in the EU is Springer Nature Customer Service Centre GmbH, Europaplatz 3, 69115 Heidelberg, Germany. If you have any concerns regarding our products, please contact ProductSafety@springernature.com

Printed and bound by CPI Group (UK) Ltd, Croydon, CR0 4YY

26/03/2026

02078991-0003